Advances in Stimuli-Responsive Nanosystems for Cancer Therapy

Padma V. Devarajan • Sathish Dyawanapelly
Vijayabhaskarreddy Junnuthula
Editors

Advances in Stimuli-Responsive Nanosystems for Cancer Therapy

Springer

Editors
Padma V. Devarajan
Department of Pharmaceutical Sciences
and Technology
Institute of Chemical Technology
Mumbai, Maharashtra, India

Sathish Dyawanapelly
Department of Pharmaceutical Sciences
and Technology
Institute of Chemical Technology
Mumbai, Maharashtra, India

Vijayabhaskarreddy Junnuthula
Faculty of Pharmacy
University of Helsinki
Helsinki, Finland

ISBN 978-981-95-2743-4 ISBN 978-981-95-2744-1 (eBook)
https://doi.org/10.1007/978-981-95-2744-1

This Springer imprint is published by the registered company Springer Nature Singapore Pte Ltd.
The registered company address is: 152 Beach Road, #21-01/04 Gateway East, Singapore 189721, Singapore

Foreword

The subject "Advances in Stimuli-Responsive Nanosystems for Cancer Therapy" is of high relevance in a world where the cancer burden is continuously escalating with a growth in incidence and mortality rates. Significant strides are being made in cancer therapy through new drug development, monoclonal antibody-based therapies, CAR-T cell therapies, and personalized medicine to enhance efficacy on the one hand and diminish toxicity on the other.

An approach that is practical and very promising is the design of Stimuli-Responsive Nanosystems for Cancer Therapy, to simultaneously address the issues of augmented efficacy with decreased toxicity. Such systems can transform cancer therapy through precise delivery of anticancer drugs at the desired site, with minimal exposure to healthy cells and tissues.

This book comprehensively addresses the plethora of Stimuli-Responsive Nanosystems that are under investigation, in clinical evaluation, and have been commercialized to date. This text will serve as a great resource to researchers and students working in this important field of cancer nanotherapy. It is a most relevant and useful text in a world where research in optimal cancer therapeutics relies extensively on drug delivery strategies.

I compliment the editorial team for inviting experts and putting together a treatise that is relevant, timely, and will serve as a great learning resource.

Best wishes to the team.

Board of Trustees Distinguished Professor of Pharmaceutics Diane J. Burgess
Pfizer Distinguished Chair of Pharmaceutical Technology,
University of Connecticut, Department of Pharmaceutical Sciences,
Storrs, CT, USA

Preface

Cancer a global scourge continues unabated across the globe. A silver lining however is the plethora of emerging therapies with great promise, which has led to improved treatments with enhanced survival rates. Rapid advances in cancer treatment, especially in targeted and immune-based therapies, as well as personalized medicine approaches promise enhanced therapeutic efficacy with reduced side effects—heralding a new paradigm in managing cancer.

Among the promising innovations, *stimuli-responsive nanosystems* offer the potential to revolutionize the treatment of cancer through targeted nanotherapies. Research in such systems is advancing rapidly with success stories emerging in the clinic. This book, *Advances in Stimuli-Responsive Nanosystems for Cancer Therapy*, is a comprehensive compendium of the various stimuli-responsive nanosystems being explored for the treatment of cancer. It is designed to provide an overview of the latest developments, challenges, and future directions in this rapidly evolving field.

The book is a systematic compilation of various aspects related to stimuli-responsive systems and covers a wide spectrum from fundamental concepts to advanced applications, as illustrated below:

- *Fundamental Principles*: The book commences with an overview of the mechanisms underlying stimuli-responsive delivery and elucidates both endogenous and exogenous stimuli that can be harnessed for precise cancer therapy.
- *Diverse Nanocarrier Platforms*: Subsequent chapters delve into the design and application of various nanosystems—including micelles, polymeric nanoparticles, polymersomes, dendrimers, liposomes, lipid nanoparticles, mesoporous silica nanoparticles, nanogels, hydrogels, and metal-organic frameworks—each tailored for specific therapeutic needs.
- *Emerging Frontiers*: The text also explores cutting-edge topics such as extracellular vesicles/exosomes, dual and multi-stimuli responsive systems, and preclinical studies that bridge the gap between laboratory innovation and clinical translation.

Each chapter is authored by distinguished scientists who have made significant contributions to their respective areas.

The book brings together leading experts and pioneering researchers to put together a compilation that could serve as a reference manual for researchers, clinicians, students, and industry professionals alike.

As editors, we express deep gratitude to the authors for their dedication and scholarly rigor, and to the entire scientific community who work relentlessly to advance the frontiers of cancer nanomedicine in their search for that ideal solution for cancer. We hope this book will inspire new ideas, foster collaborations, and catalyze breakthroughs in more effective and safer cancer therapies.

Mumbai, Maharashtra, India — Padma V. Devarajan
Sathish Dyawanapelly
Helsinki, Finland — Vijayabhaskarreddy Junnuthula

Contents

About the Editors

Padma V. Devarajan PhD (Tech), currently serves as Advisor and Chair Professor at the SIES School of Pharmaceutical Sciences in Mumbai, India. Previously, she served at the Institute of Chemical Technology (ICT), Mumbai, in various roles including Professor of Pharmacy, Head of the Department of Pharmaceutical Sciences and Technology, Dean of Research and Innovation, and Member of the Board of Governors. She is a distinguished academician and researcher with over 30 years of professional experience. Her research encompasses nanocarriers for targeted drug delivery in cancer and infectious diseases, the scale-up of nano drug delivery systems (DDS), bioenhancement strategies, and the development of innovative DDS for proteins and nucleic acids. Her work is supported by grants from both the Government of India and the pharmaceutical industry at national and international levels. She has edited and authored two books on targeted drug delivery, with this publication focusing on advanced aspects of this field. She holds several granted patents and has successfully licensed and commercialized patented technologies. She has contributed to the Controlled Release Society Inc., USA, as a board member, member of the Board of Scientific Advisors, and Chair of the Young Scientist Mentor Protégé Sub-committee. Her achievements have been recognized with numerous awards for her contributions as an academician and researcher. She represents India in the ISO working group on health, safety, and environmental aspects of nanotechnologies. Additionally, she has been appointed Chair of the National Innovation and Startup Policy Revision Committee by the Ministry of Education, Government of India.

Sathish Dyawanapelly is an Assistant Professor in Pharmaceutics at the Department of Pharmaceutical Sciences and Technology, Institute of Chemical Technology (ICT/UDCT), Mumbai, India. He holds a distinguished PhD (Tech) degree in Pharmaceutics (ICT-Mumbai) and master's degree in Pharmaceutics (NIPER). He has more than 7 years of teaching experience, and has been actively involved in research in the design and development of Drug Delivery Systems, Nanomedicines, and Nutraceuticals. He is widely recognized for his excellence and contribution in both academics and research. His scholarly output includes over 35 peer-reviewed publications and 3 book chapters. He developed and transferred three different technologies to industries. He is an editorial member for *Scientific Reports*

(Nature), *PLOS ONE*, *Frontiers in Drug Delivery*, *Discover Pharmaceutical Sciences* (Springer), and *Current Indian Science* (Bentham Science).

Vijayabhaskarreddy Junnuthula is a grant-funded researcher at the University of Helsinki, Finland. His research prominently revolves around the development of nanomedicine, spanning various domains within Pharmaceutical Sciences, most notably Pharmaceutics, Pharmacokinetics, and Local Drug Delivery for diverse therapeutic applications. He has been supported by the Finnish Cultural Foundation and was a recipient of the Marie Curie Fellowship from the European Commission under the Horizon 2020 program. He was awarded the Solvey chair prize at the French Society of Nanomedicine conference in Bordeaux, France. He has published in several high-impact journals as an author. He is a peer reviewer for several pharmaceutical science journals such as the *International Journal of Pharmaceutics*, *Journal of Drug Delivery Science and Technology*, *BMC Pharmacology*, and *Signal Transduction and Targeted Therapy*.

Contributors

Sanchit Arora Department of Pharmaceutics, Delhi Institute of Pharmaceutical Sciences and Research (DIPSAR), Delhi Pharmaceutical Sciences and Research University (DPSRU), New Delhi, India

Esha Attar Department of Pharmaceutical Sciences and Technology, Institute of Chemical Technology, Mumbai, Maharashtra, India

Ghanshamdas Bairagi Department of Pharmaceutical Sciences and Technology, Institute of Chemical Technology, Mumbai, Maharashtra, India

Derajram Benival Department of Pharmaceutics, National Institute of Pharmaceutical Education and Research - Ahmedabad (NIPER-A), An Institute of National Importance, Department of Pharmaceuticals, Ministry of Chemicals and Fertilizers, Government of India, Gandhinagar, Gujarat, India

Sunil Bharti University Institute of Pharmaceutical Sciences and UGC Centre of Advanced Studies, Panjab University, Chandigarh, India

Vanashree Chaudhari Department of Pharmaceutical Sciences and Technology, Institute of Chemical Technology, Mumbai, Maharashtra, India

Naveen Chella National Institute of Pharmaceutical Education and Research (NIPER), Guwahati, Assam, India

Manish Kumar Chourasia Division of Pharmaceutics and Pharmacokinetics, CSIR - Central Drug Research Institute, Lucknow, Uttar Pradesh, India

Academy of Scientific and Innovative Research (AcSIR), Ghaziabad, Uttar Pradesh, India

Sanat Kumar Dash Department of Pharmacy, Birla Institute of Technology and Science Pilani (BITS Pilani), Pilani Campus, Pilani, Rajasthan, India

Padma V. Devarajan Department of Pharmaceutical Sciences and Technology, Institute of Chemical Technology, Mumbai, Maharashtra, India

SIES School of Pharmaceutical Sciences, Navi Mumbai, India

Sathish Dyawanapelly Department of Pharmaceutical Sciences and Technology, Institute of Chemical Technology, Mumbai, Maharashtra, India

Rasha Elkanayati National Institute of Pharmaceutical Education and Research (NIPER), Guwahati, Assam, India

Department of Pharmaceutics and Drug Delivery, School of Pharmacy, The University of Mississippi, Mississippi, USA

Andrew Ezegbe Department of Pharmaceutical Technology and Industrial Pharmacy, University of Nigeria, Nsukka, Enugu State, Nigeria

Human and Natural Science Center, ABC Federal University, Santo Andre, Sao Paulo, Brazil

Ananya Goswami Department of Pharmaceutics, National Institute of Pharmaceutical Education and Research - Ahmedabad (NIPER-A), An Institute of National Importance, Department of Pharmaceuticals, Ministry of Chemicals and Fertilizers, Government of India, Gandhinagar, Gujarat, India

Dalapathi Gugulothu Department of Pharmaceutics, Delhi Institute of Pharmaceutical Sciences and Research (DIPSAR), Delhi Pharmaceutical Sciences and Research University (DPSRU), New Delhi, Delhi, India

Krishna Jadhav Institute of Nano Science and Technology (INST), Mohali, Punjab, India

Srishty Jaiswal Division of Pharmaceutics and Pharmacokinetics, CSIR - Central Drug Research Institute, Lucknow, Uttar Pradesh, India

Academy of Scientific and Innovative Research (AcSIR), Ghaziabad, Uttar Pradesh, India

Agrim Jhilta Institute of Nano Science and Technology (INST), Mohali, Punjab, India

Anil B. Jindal Department of Pharmacy, Birla Institute of Technology and Science Pilani (BITS Pilani), Pilani Campus, Pilani, Rajasthan, India

Vijayabhaskarreddy Junnuthula Drug Research Program, Faculty of Pharmacy, University of Helsinki, Helsinki, Finland

Shanmukh Kanchumurthy Department of Pharmacology and Toxicology, National Institute of Pharmaceutical Education and Research (NIPER), Hyderabad, Telangana, India

Indrajeet Karnik Department of Pharmaceutics and Drug Delivery, School of Pharmacy, The University of Mississippi, Mississippi, USA

Dhirendra S. Katti Department of Biological Sciences and Bioengineering, Indian Institute of Technology, Kanpur, Uttar Pradesh, India

The Mehta Family Centre for Engineering in Medicine, Indian Institute of Technology, Kanpur, Uttar Pradesh, India

School of Interdisciplinary Life Sciences, Indian Institute of Technology, Ponda, Goa, India

Pemal Preet Kaur University Institute of Pharmaceutical Sciences and UGC Centre of Advanced Studies, Panjab University, Chandigarh, India

Dharmendra Kumar Khatri Department of Pharmacology and Toxicology, National Institute of Pharmaceutical Education and Research (NIPER), Hyderabad, Telangana, India

Molecular and Cellular Neuroscience Lab, Department of Pharmacology and Toxicology, NIMS Institute of Pharmacy, NIMS University Rajasthan, Jaipur, Rajasthan, India

Preeya Negi Department of Pharmaceutical Sciences and Technology, Institute of Chemical Technology, Mumbai, Maharashtra, India

Swarnima Negi Institute of Nano Science and Technology (INST), Mohali, Punjab, India

Amruta Padakanti National Institute of Pharmaceutical Education and Research (NIPER), Guwahati, Assam, India

Vasavi Pasupuleti Department of Pharmacology and Toxicology, National Institute of Pharmaceutical Education and Research (NIPER), Hyderabad, Telangana, India

Ritu Patel Department of Pharmaceutics, National Institute of Pharmaceutical Education and Research - Ahmedabad (NIPER-A), An Institute of National Importance, Department of Pharmaceuticals, Ministry of Chemicals and Fertilizers, Government of India, Gandhinagar, Gujarat, India

Vandana B. Patravale Department of Pharmaceutical Sciences and Technology, Institute of Chemical Technology, Mumbai, Maharashtra, India

Rafquat Rana Division of Pharmaceutics and Pharmacokinetics, CSIR - Central Drug Research Institute, Lucknow, Uttar Pradesh, India

Jawaharlal Nehru University (JNU), New Delhi, Delhi, India

Dhwani Rana Department of Pharmaceutics, National Institute of Pharmaceutical Education and Research - Ahmedabad (NIPER-A), An Institute of National Importance, Department of Pharmaceuticals, Ministry of Chemicals and Fertilizers, Government of India, Gandhinagar, Gujarat, India

Shagun Rangra Department of Pharmaceutics, National Institute of Pharmaceutical Education and Research - Ahmedabad (NIPER-A), An Institute of National Importance, Department of Pharmaceuticals, Ministry of Chemicals and Fertilizers, Government of India, Gandhinagar, Gujarat, India

Michael Repka Department of Pharmaceutics and Drug Delivery, School of Pharmacy, The University of Mississippi, Mississippi, USA

Vaibhav Sadana Department of Pharmaceutics, Delhi Institute of Pharmaceutical Sciences and Research (DIPSAR), Delhi Pharmaceutical Sciences and Research University (DPSRU), New Delhi, India

Amita Sarwal University Institute of Pharmaceutical Sciences and UGC Centre of Advanced Studies, Panjab University, Chandigarh, India

Abhishek Saxena Department of Pharmaceutics, Delhi Institute of Pharmaceutical Sciences and Research (DIPSAR), Delhi Pharmaceutical Sciences and Research University (DPSRU), New Delhi, India

Gaurav Shanbhag National Institute of Pharmaceutical Education and Research (NIPER), Guwahati, Assam, India

Shivani Sharma University Institute of Pharmaceutical Sciences and UGC Centre of Advanced Studies, Panjab University, Chandigarh, India

Rahul Shukla Department of Pharmaceutics, National Institute of Pharmaceutical Education and Research (NIPER) Raebareli, Lucknow, Uttar Pradesh, India

Shipra Shukla Department of Pharmaceutics, Indo-Soviet Friendship College of Pharmacy, Moga, Punjab, India

Raghuraj Singh Institute of Nano Science and Technology (INST), Mohali, Punjab, India

Department of Pharmaceutics, Teerthanker Mahaveer College of Pharmacy (TMCOP), Teerthanker Mahaveer University (TMU), Moradabad, Uttar Pradesh, India

Dadi A. Srinivasarao Department of Pharmaceutics, National Institute of Pharmaceutical Education and Research (NIPER), Hyderabad, Telangana, India

Ashish Dilip Sutar Department of Pharmaceutics, National Institute of Pharmaceutical Education and Research (NIPER) Raebareli, Lucknow, Uttar Pradesh, India

Sai Veerabhadra Nikhil Teki Department of Biological Sciences and Bioengineering, Indian Institute of Technology, Kanpur, Uttar Pradesh, India

The Mehta Family Centre for Engineering in Medicine, Indian Institute of Technology, Kanpur, Uttar Pradesh, India

Department of Chemistry, Indian Institute of Technology, Kanpur, Uttar Pradesh, India

Namarta Thakur University Institute of Pharmaceutical Sciences and UGC Centre of Advanced Studies, Panjab University, Chandigarh, India

Pooja Todke Department of Pharmaceutical Sciences and Technology, Institute of Chemical Technology, Mumbai, Maharashtra, India

Shourya Tripathi Division of Pharmaceutics and Pharmacokinetics, CSIR - Central Drug Research Institute, Lucknow, Uttar Pradesh, India

Jawaharlal Nehru University (JNU), New Delhi, Delhi, India

Chukwuebuka Emmanuel Umeyor Department of Pharmaceutics and Pharmaceutical Technology, Faculty of Pharmaceutical Sciences, Nnamdi Azikiwe University, Awka, Anambra State, Nigeria

Prateek Uttreja Department of Pharmaceutics and Drug Delivery, School of Pharmacy, The University of Mississippi, Mississippi, USA

Sateesh Kumar Vemula Department of Pharmaceutics and Drug Delivery, School of Pharmacy, The University of Mississippi, Mississippi, USA

Rahul Kumar Verma Institute of Nano Science and Technology (INST), Mohali, Punjab, India

Department of Pharmaceutics, Teerthanker Mahaveer College of Pharmacy (TMCOP), Teerthanker Mahaveer University (TMU), Moradabad, Uttar Pradesh, India

Nithin Vidiyala Drug Product Design and Development, Cerevel Therapeutics, Cambridge, MA, USA

Pooja Yadav Division of Pharmaceutics and Pharmacokinetics, CSIR - Central Drug Research Institute, Lucknow, Uttar Pradesh, India

Department of Pharmaceutics, Teerthanker Mahaveer College of Pharmacy (TMCOP), Teerthanker Mahaveer University (TMU), Moradabad, Uttar Pradesh, India

Zahra Sifat Zaidi Department of Biological Sciences and Bioengineering, Indian Institute of Technology, Kanpur, Uttar Pradesh, India

The Mehta Family Centre for Engineering in Medicine, Indian Institute of Technology, Kanpur, Uttar Pradesh, India

Overview of Stimuli-Responsive Nanosystems: Endogenous and Exogenous Triggers for Stimuli-Responsive Delivery for Cancer Therapy

1

Pooja Todke, Esha Attar, Vanashree Chaudhari, Vijayabhaskarreddy Junnuthula, Sathish Dyawanapelly, and Padma V. Devarajan

Abstract

Effective cancer therapy is daunted with challenges such as limited drug concentration at the target site, non-specific drug distribution, rapid clearance, and drug resistance. A smart approach to overcome most of these challenges is through targeting drugs to tumors. The tumor microenvironment (TME) offers unique features like acidosis, hypoxia, elevated reactive oxygen species (ROS), and glutathione (GSH), which can be harnessed for site-specific targeted drug delivery. Stimuli-responsive nanosystems exploit these endogenous conditions (pH, enzymes, redox imbalance) or rely on exogenous triggers (ultrasound, temperature, magnetic fields) to achieve precise, controlled drug release. Endogenous stimuli-responsive systems leverage TME-specific conditions for selective drug activation, while external triggers enable spatiotemporal control via physical energy sources. This chapter discusses recent advances in smart nanocarriers, detailing their mechanism of action and the therapeutic benefits they offer. Stimuli-responsive nanosystems enhance tumor targeting, reduce off-target effects, and improve treatment efficacy. The chapter provides an overview of the design and application of these advanced delivery platforms, highlighting their potential to overcome limitations in conventional cancer therapy, and serves as a

P. Todke · E. Attar · V. Chaudhari · S. Dyawanapelly (✉) · P. V. Devarajan (✉)
Department of Pharmaceutical Sciences and Technology, Institute of Chemical Technology, Mumbai, Maharashtra, India
e-mail: sa.dyawanapelly@ictmumbai.edu.in

V. Junnuthula
Drug Research Program, Faculty of Pharmacy, University of Helsinki, Helsinki, Finland

P. V. Devarajan et al. (eds.), *Advances in Stimuli-Responsive Nanosystems for Cancer Therapy*, https://doi.org/10.1007/978-981-95-2744-1_1

prelude to the book, which is a detailed coverage of stimuli-responsive nanosystems.

Keywords

Nanosystems · Endogenous stimuli · Exogenous stimuli · Delivery · Cancer Therapy

Abbreviations

DDS	Drug delivery system
EChT	Electrochemical therapy
EDT	Electrodynamic therapy
EPR	Enhanced permeability and retention
ETT	Electrothermal therapy
GSH	Glutathione
ICD	Immunogenic cell death
IFP	Interstitial fluid pressure
LCST	Lower critical solution temperature
MCT	Monocarboxylate transporter
MRI	Magnetic resonance imaging
NIR	Near-infrared
PNIPAM	Poly(N-isopropylacrylamide)
ROS	Reactive oxygen species
TME	Tumor microenvironment
UCST	Upper critical solution temperature
VEGF	Vascular endothelial growth factor

1 Introduction

Cancer continues unabated as a leading cause of death across the globe, with a continuous rise in incidence and mortality rates. Decades of research have revealed that cancer is not a monolithic disease but a complex system where malignant cells interact dynamically with diverse stromal components in the tumor microenvironment (TME). This intricate interplay creates formidable challenges for conventional single-modality treatments, which often prove inadequate against both early-stage and advanced tumors due to profound inter- and intratumoral heterogeneity. While combined approaches integrating surgery, radiotherapy, chemotherapy, and immunotherapy represent the current clinical standard, patient-specific variability and inherent resistance mechanisms frequently lead to suboptimal outcomes accompanied by severe treatment-related toxicities. These limitations emphasize the urgent need for optimized, personalized therapeutic strategies that account for each tumor's unique biological context.

Central to this challenge is the evolving understanding of the TME's multifaceted role in cancer progression. The TME comprises not only cancer cells but also immune cells, fibroblasts, vascular networks, and extracellular matrix components that collectively create a permissive niche for tumor growth and dissemination. This microenvironment exhibits distinct physicochemical properties, including acidosis (pH 6.5–7.0), hypoxia, elevated temperature, and oxidative stress that drive malignant progression while simultaneously erecting barriers to effective treatment. Acidosis, resulting from aerobic glycolysis (Warburg effect), carbonic anhydrase IX activity, and proton pump overexpression, promotes invasion and immune evasion. Similarly, metabolic hyperactivity generates localized hyperthermia (0.5–2 °C above adjacent tissue), while chaotic vasculature creates hypoxic zones that fuel treatment resistance.

Remarkably, these pathological features can be transformed into therapeutic opportunities through innovative drug delivery strategies. The classic enhanced permeability and retention (EPR) effect exploits tumor vasculature's leakiness for passive nanoparticle accumulation, though its clinical translation has been hampered by heterogeneity. Modern nanomedicine now prioritizes active targeting using stimuli-responsive systems that leverage both endogenous triggers (pH gradients, overexpressed enzymes, redox imbalances) and exogenous stimuli (light, magnetic fields, and ultrasound). For instance, pH-sensitive linkers selectively release chemotherapeutics in acidic tumor regions, while thermosensitive materials activate only upon reaching tumor-specific temperatures. Enzyme-cleavable nanocarriers and glutathione-sensitive particles further exemplify how TME biology can be harnessed for precision therapy.

These advances highlight a paradigm shift from generic cytotoxic approaches to sophisticated, microenvironment-aware strategies. Endogenous stimulus-responsive systems exploit inherent pathological conditions without external intervention, offering intrinsic specificity, while exogenous approaches using physical energy sources provide spatiotemporal control unaffected by biological variability. Particularly promising are combinatorial systems integrating multiple triggers, such as thermo-responsive materials that also respond to tumor-specific enzymes or tumor pH, which may overcome the limitations of single-stimulus approaches. As research continues to unravel the complexity of the TME, smart drug delivery platforms that are dynamically adapted to both tumor biology and individual patient characteristics can maximize efficacy while minimizing collateral damage to healthy tissues [1].

2 Features of the Tumor Microenvironment

A complex and dynamic bionetwork found inside solid tumors, the tumor microenvironment (TME) is made up of stromal cells, immune cells, blood vessels, extracellular matrix (ECM), and signaling chemicals that interact together to promote the spread of cancer. Tumor aggressiveness, treatment resistance, and immunosuppression are all influenced by the unique physicochemical abnormalities that the TME displays in contrast to healthy tissues. These abnormalities include increased interstitial fluid pressure (IFP), hypoxia, decreased pH, and higher reactive oxygen species (ROS). Among these features, the acidic pH of the TME in the range of 6.5–7.0, which is lower than the pH in normal tissues (pH 7.4), is a hallmark of solid tumors, primarily driven by the Warburg effect, where cancer cells preferentially undergo aerobic glycolysis, producing excessive lactic acid even in the presence of oxygen. This metabolic shift is further exacerbated by hypoxia due to the tumor's chaotic vasculature, which upregulates glycolysis via hypoxia-inducible factor-1α (HIF-1α). Additionally, tumor cells overexpress carbonic anhydrase IX (CA-IX), which converts CO_2 and H_2O into protons and bicarbonate, further acidifying the extracellular space while maintaining a neutral intracellular pH through proton pumps like V-ATPases and Na^+/H^+ exchangers. This acidic milieu promotes cancer progression by enhancing extracellular matrix degradation, suppressing immune cell activity, and reducing the efficacy of weakly basic chemotherapeutics. Leveraging this pH gradient, researchers have developed pH-responsive drug delivery systems (DDS) such as dynamic intelligent nanoassemblies, which utilize acid-labile linkers (e.g., hydrazone, acetal) or pH-sensitive polymers to achieve tumor-specific drug release. These systems exploit not only the mildly acidic extracellular TME but also the more acidic subcellular compartments like the endosomes (pH 5.5–6.0) and lysosomes (pH 4.0–5.0) for precise intracellular drug delivery, thereby improving tumor accumulation and penetration while minimizing off-target effects [2].

Similarly, the elevated temperature in tumors (0.5–2 °C higher than normal tissues) arises from hypermetabolic activity due to rapid proliferation and inefficient aerobic glycolysis, generating excess heat. Poor vascularization impedes heat dissipation, while hypoxia-induced anaerobic metabolism and immune cell-mediated inflammation further contribute to localized hyperthermia. This thermal differential has been exploited for thermographic diagnostics and hyperthermia-based therapies, where controlled heating enhances radiation and chemotherapy efficacy. Moreover, thermo-responsive nanocarriers, such as those incorporating poly(N-isopropylacrylamide) (PNIPAM), release drugs upon mild heating, offering spatially controlled treatment [3].

A key mechanism enabling passive tumor targeting is the EPR effect, where macromolecules and nanoparticles (50–200 nm) exhibit preferential accumulation due to leaky tumor vasculature (endothelial gaps up to 2 μm) and impaired lymphatic drainage. The EPR effect is amplified by inflammatory mediators like VEGF, bradykinin, and nitric oxide, which increase vascular permeability. However, heterogeneity in tumor perfusion, high IFP, and dense ECM limit uniform drug distribution, prompting strategies such as vascular normalization (e.g., anti-angiogenic therapy) and nanocarrier optimization (size, surface charge) to enhance penetration [4].

Stimuli-responsive drug delivery systems capitalize on endogenous TME triggers (pH, ROS, enzymes, hypoxia) and exogenous stimuli (light, magnetic fields, and ultrasound) to achieve spatiotemporal control over drug release. For instance, redox-sensitive nanocarriers disassemble in the high glutathione (GSH) environment of tumors, while enzyme-responsive systems release payloads upon cleavage by tumor-specific proteases like matrix metalloproteinases. Combining multiple stimuli, for instance, pH + temperature + ROS, further refines targeting precision. Despite challenges like TME heterogeneity and clinical scalability, these advances represent a paradigm shift in oncology, moving toward personalized nanomedicine that exploits the unique biology of the TMEs for safer, more effective cancer therapy [5].

3 Endogenous (Internal) Stimuli-Responsive Nanosystems

Strategies for targeting tumor-specific functional moieties rely on differences between normal and tumor cells. Selective tumor targeting and drug release based on the intracellular environment conditions, including pH, ionic microenvironment, and enzymes at the tumor site, can be achieved through prolonged blood circulation, controlled release of therapeutic payload, and improved internalization via endocytosis. Tumor microenvironment, including hypoxia, acidic pH, and presence of glutathione, adenosine triphosphate (ATP), is favorable for stimuli-responsive delivery systems for high efficacy [6].

3.1 pH-Responsive Nanosystems

In healthy tissues, the extracellular pH is around 7.4, and the intracellular pH is around 7.2. Inverted pH gradient in cancer cells is highly associated with activated glycolysis, which produces lactate and protons (H^+), creating an acidic extracellular (pH 6.4–7.1) and alkaline intracellular pH (7.1–7.8). This inverted pH gradient depends on increased activity/expression of various acid efflux proteins and plasma membrane transporters that regulate pH homeostasis, including monocarboxylate transporters (MCTs), Na^+–H^+ exchangers (NHEs), and carbonic anhydrases (CAs). Glutaminolysis further contributes to lactate and H^+ production. To survive acid stress, tumors upregulate H^+ transporters and reduce CO_2 via suppressed oxidative metabolism, aided by carbonic anhydrases. The acidic tumor microenvironment (TME) not only supports initial tumor cell proliferation but also promotes aggressiveness, metastasis, drug resistance, and immune evasion [7].

The pH-responsive targeted drug delivery systems developed to deliver a variety of therapeutics employ pH-sensitive polymers. These polymers exhibit pH-dependent properties and release the cargo at target sites in response to pH. Cancer-targeted pH-responsive nanosystems are successfully explored using natural biopolymers, and synthetic polymers. Based on the mechanism of protonation/ionization, pH-sensitive polymers are divided into polyacids and polybases, which are determined by their charge at physiological pH. Protonation of acidic groups reduces the acidity considerably. Acidic pH-responsive systems designed with acid-labile bonds between drug and polymer, or within the polymer represent a promising strategy to achieve tumor targeting. Acid-labile bond are amides, imine, hydrazone, orthoetsers and phenyl vinyl ether. Further, with increase in pH anionic polymers swells. On the other hand cationic polymers exhibit ionization with a decrease in pH. Cationic pH-responsive polyacids are poly(vinyl-phosphonic acid), poly(methacrylic acid), poly(vinyl-sulfonic acid), poly(glutamic acid), poly(aspartic acid). Polybases includes, poly(4-vinyl pyridine), poly[2-dimethylaminoethyl methacrylate], poly[(2-diethylamino) ethyl acrylamide], poly(acryloylmorpholine), poly(N-vinylimidazole). The release of therapeutic payload through common pH-response modes is physical dissociation, chemical bond cleavage-resulting nanoparticle dissociation, chemical bond cleavage-promoting payload release, and nanoparticle swelling. In the case of physical dissociation mode, it refers to the response behavior of nanoparticles to a specific pH environment, which leads to hydrophobic core protonation or ionization, further causing the physical dissociation of nanoparticles. Chemical bond cleavage dissociation mode refers to the mode in which the chemical bonds of hydrophobic and hydrophilic fragments in the nanoparticles cleave and lead to the dissolution of the nanoparticles [8].

To date, many pH-sensitive nanocarriers have been designed based on triggering low extracellular/endosomal pH. The drugs could be non-covalently or covalently conjugated with pH-sensitive carriers, including liposomes, polymer–drug conjugates, polymeric nanoparticles, micelles, and polymersomes [9].

3.2 Enzyme-Responsive Nanosystems

Mammalian physiological and metabolic processes depend on enzymes. Enzymes play an important role in catalyzing vital substrates in a cellular environment. Dysregulation of enzyme expression could also be a hallmark of cancer. Compared with normal tissues, enzymes are upregulated in tumor cells, thus resulting extreme secretion in the TME, such as cathepsin, matrix metalloproteinases, hyaluronidases, glucosidase, phospholipase, and esterase. Enzyme-responsive nanosystems can be triggered by target enzymes with high specificity and drug release in the tumor microenvironment, limiting off-site target toxicity. Protease enzyme that cleave the peptide bonds of proteins which observed high in cancerous tissues compared to health cells. The enzyme cathepsin B, azoreductase, and endopeptidase enzymes are overexpressed in diseased conditions that helps triggering in enzyme-sensitive nanocarriers for caner [10].

The key enzymes, including hydrolases such as oxidoreductases, lipases, and proteases, are relied on for the development of stimuli-responsive nanomaterials for drug delivery and diagnostics.

- *Lipase:* Phospholipases are enzymes responsible for cleaving fatty acids in mammalian cells. Three groups of PLA2 isozymes are identified: group I, group II, and cytosolic PLA2. Elevated levels of group II PLA2 are found in advanced stages of cancers, particularly in digestive organ cancers. Additionally, increased concentrations of PLA2 are observed in pleural effusion before cancer progression in gastric cancer patients. Secreted phospholipase A2 (sPLA2) is overexpressed in prostate, breast, and pancreatic cancers [11].

- *Protease:* Cancer is often linked to abnormal activity in certain proteases, including cathepsins, matrix metalloproteinases, and urokinase-type plasminogen activators. Protease-sensitive systems can provide the enhanced selectivity needed for targeting cancer cells while minimizing adverse effects.
- *Glycosidase:* Glycosidases play key roles by breaking down glycosidic bonds in polysaccharides. This led to the development of important enzyme-responsive materials from biodegradable carbohydrate polymers. An example is β-Mannanase, found in the gastrointestinal tract and secreted by colon microflora.
- *Oxidoreductase:* Two-electron cytosolic reductases and NAD(P)H:quinone oxidoreductase1 (NQO1) are overexpressed in numerous cancer cells compared to healthy cells. These enzymes will play a key role in targeting of hypoxic tumor tissues [12].

3.3 Redox-Responsive Nanosystems

Redox homeostasis is a dynamic balance between oxidation (ROS production) and reduction (antioxidant defense). Under normal cell conditions, mitochondria and NADPH oxidases generate reactive oxygen species (ROS), including free radicals like superoxide ($O_2{\bullet}^-$) and hydroxyl (OH•), as well as non-radicals like hydrogen peroxide (H_2O_2) [13]. These molecules act as signaling intermediates, regulating processes such as cell survival, proliferation, and gene expression. However, when ROS production overwhelms cellular antioxidant defenses (e.g., superoxide dismutase, catalase), oxidative stress occurs, leading to damage to lipids, proteins, and DNA. This disruption of redox balance contributes to various diseases, including cancer. ROS levels are 100-fold higher in tumor cells than in normal cells because of oxidative stress, which is characterized by an imbalance between ROS generation and antioxidant defense. Thus, the elevated level of ROS in TME is a unique feature that can be used in the rational design of oxidative-responsive cancer-targeted drug delivery. Oxidative responsive groups are thioketal, phenylnorate ester, thioether, selenoether, aminoacrylate, and peroxalate esters. Thioethers are another often utilized reductive group that could be reduced to ketones [14, 15].

3.4 Glutathione-Responsive Nanosystems

The tumor microenvironment usually displays an elevated level of glutathione (GSH) than normal cells to survive in an oxidative stress environment and maintain aberrant metabolism, rapid proliferation, and mitochondrial dysfunction. Elevated GSH acts as a critical antioxidant, neutralizing ROS to prevent oxidative damage to DNA, proteins, and lipids, which could otherwise trigger cell death. Additionally, GSH supports tumor-promoting signaling pathways (e.g., NF-κB, HIF-1α) and provides reducing power for nucleotide synthesis, essential for cancer cell division. GSH is a tripeptide that is composed of glutamate (Glu), cysteine (Cys), and glycine (Gly). The concentrations of intracellular glutathione (GSH) range from 0.5 to

10 mM in normal cell types, about 100–1000 times that in the extracellular environment. The concentration of GSH in cancer cells ranges from 5 to 20 millimolar (mM) or even higher in some cases, and the intracellular GSH concentration (1–10 mM) is about 1000 times higher than that in the extracellular compartment (2–20 μM). Due to its high intracellular level, GSH plays a vital role in tumor cell proliferation, metastasis, and drug resistance [16].

Owing to the unique features of GSH in cancer cells have been developed for stimuli-responsive cancer therapy has been developed by employing of intracellular GSH as an effective target. For instance, widely used localized GSH-activated strategy by employing disulfide bonds (-SS-) and diselenide bonds (-SeSe-) react with GSH via redox for effective targeting of drug to tumor cells. GSH deprivation is another effective strategy for overcoming drug resistance and improving treatment outcomes. In addition to redox reaction, competitive coordination is used to trigger drug release in the GSH-responsive drug delivery system. In the tumor microenvironment, the stronger competitive coordination between GSH and certain components of the drug delivery system can trigger disassembly of the nanosystems and release the loaded cargos [17].

4 Exogenous (External) Stimuli-Responsive Nanosystems

Smart nanocarriers activated by external stimuli (ultrasound, electrical, temperature, magnetic field, light) enable precise, tumor-targeted drug delivery while minimizing systemic toxicity. These systems respond to penetrating physical energies, triggering drug release in a controlled manner, only at tumor sites. Such smart technologies overcome biological barriers, enhance therapeutic efficacy, and enable combination strategies representing the next frontier in precision oncology [6, 7].

4.1 Ultrasound-Responsive Nanosystems

Ultrasound-responsive nanosystems represent an advanced approach in cancer therapy, leveraging the unique properties of ultrasound (US), a non-ionizing mechanical wave with deep tissue penetration and minimal side effects to enable precise, localized drug delivery. These systems utilize nanocarriers such as micelles, mesoporous silica nanoparticles, or perfluorocarbon droplets to encapsulate therapeutics, ensuring stability in circulation while exploiting the EPR effect to accumulate at tumor sites. Upon US exposure (frequencies >20 kHz), controlled drug release is triggered through mechanical and thermal effects, including cavitation (microbubble-induced shear forces enhancing vascular permeability), acoustic radiation forces (non-cavitation mechanisms like streaming and pressure improving drug transport), and mild hyperthermia (temperature rise from absorbed US energy, quantified by the thermal index) [18]. This spatiotemporal control allows selective tumor targeting while sparing healthy tissues, addressing key limitations of conventional chemotherapy. The non-invasive nature, absence of ionizing radiation, and tunability (via frequency, duty cycle, and exposure time adjustments) make US-responsive systems clinically attractive [19]. However, challenges persist, such as the need for contrast agents to improve imaging clarity due to vascular compression effects and the risk of overheating at high frequencies. Advanced nanocarriers, including microbubbles and phase-shift emulsions, further enhance efficacy by enabling direct cytosolic drug delivery through transient membrane pores, bypassing endosomal degradation. By integrating these mechanisms, US-responsive nanosystems offer a promising paradigm for on-demand, localized cancer therapy with reduced systemic toxicity. Ongoing research focuses on optimizing acoustic parameters and nanomaterial designs to maximize therapeutic outcomes, positioning this technology as a transformative tool in precision oncology [20].

4.2 Electro-Responsive Nanosystems

Electro-responsive nanotherapeutics represent a radical approach in cancer treatment, employing electric fields to enhance drug delivery, induce selective tumor cell death, and stimulate anti-tumor immunity. These systems utilize conductive or dielectric nanomaterials (e.g., gold nanoparticles, graphene oxide, or polypyrrole-based polymers) that respond to external electrical stimuli, enabling precise spatiotemporal control over therapy. Upon exposure to an electric field, these nanocarriers can trigger on-demand drug release, generate localized heat, or catalyze cytotoxic reactions, minimizing off-target effects while maximizing therapeutic efficacy. A key advantage lies in their multimodal mechanisms of action. In electrothermal therapy (ETT), nanoparticles convert electrical energy into heat, ablating tumors through hyperthermia-induced necrosis or apoptosis. Electrochemical therapy (EChT) exploits redox reactions to produce reactive oxygen species (ROS) or alter pH, selectively killing cancer cells. Meanwhile, electrodynamic therapy (EDT) enhances ROS generation via nanocatalytic processes, promoting ferroptosis or immunogenic cell death (ICD), which activates systemic anti-tumor immunity. Beyond direct tumor ablation, electro-responsive systems improve drug delivery by overcoming biological barriers. For instance, electroporation—brief electrical pulses temporarily permeabilize cell membranes, enhancing nanoparticle uptake. Similarly, stimuli-responsive polymers release chemotherapeutics only at the tumor site upon electrical activation, reducing systemic toxicity [21].

4.3 Temperature-Responsive Nanosystems

Thermoresponsive drug delivery systems leverage the slight temperature elevation in tumor cells (40–42 °C) compared to normal cells (37 °C) to enable targeted cancer therapy. The increased metabolic activity of rapidly dividing tumor cells generates higher energy, creating a detectable thermal gradient that can be exploited using thermoresponsive polymers. These polymers, such as poly(N-isopropylacrylamide) (PNIPAAm), exhibit reversible physicochemical changes at specific temperatures, allowing controlled drug release at tumor sites. Polymers with a lower critical solution temperature (LCST) become hydrophobic above their LCST, aggregating and releasing encapsulated drugs, while upper critical solution temperature (UCST) polymers dissolve at higher temperatures, facilitating drug delivery. To enhance efficacy, mild hyperthermia can be induced to further elevate tumor temperature, improving polymer responsiveness and promoting sustained drug release while also directly killing cancer cells. Various delivery systems, including liposomes, micelles, nanoparticles, and hydrogels, incorporate these polymers for temperature-triggered drug release, increasing bioavailability and reducing off-target toxicity. Advancements like 4D bioprinted thermoresponsive polymers and smart drug delivery systems are refining this approach, overcoming limitations of traditional chemotherapy such as multidrug resistance and cytotoxicity. By capitalizing on the inherent thermal differences between tumor and healthy cells, thermoresponsive polymers offer a promising, targeted strategy for cancer treatment with improved specificity and therapeutic outcomes [22].

4.4 Magnetic-Responsive Nanosystems

Magnetic-responsive drug delivery systems represent a transformative approach in oncology, leveraging the unique properties of magnetic nanoparticles (MNPs) to enhance tumor targeting, controlled drug release, and multimodal therapy. Superparamagnetic iron oxide nanoparticles (SPIONs), particularly magnetite (Fe_3O_4) and maghemite (γ-Fe_2O_3), are widely used due to their biocompatibility, biodegradability, and responsiveness to external magnetic fields. These MNPs enable two key therapeutic strategies: (1) magnetic guidance, where an external field directs drug-loaded carriers to tumor sites, improving accumulation via passive (EPR effect) or active (ligand-mediated) targeting; and (2) magnetic hyperthermia, where alternating magnetic fields (AMF) induce localized heat, triggering drug release or directly ablating cancer cells [23]. The integration of MNPs with nanocarriers such as hydrogels, liposomes, or dendrimers further enhances functionality, combining deep-tumor penetration, prolonged circulation, and stimuli-responsive release. For instance, SPIONs embedded in thermosensitive liposomes can release chemotherapeutics upon AMF exposure, while MNP-doped hydrogels enable sustained drug delivery with spatial precision. Preclinical studies demonstrate efficacy in breast, liver, and colon cancers, with hyperthermia at temperatures >47 °C achieving complete tumor regression in some models. Additionally, MNPs serve as MRI contrast agents, facilitating real-time monitoring of drug delivery. However, challenges remain, including optimizing AMF parameters, minimizing off-target MNP accumulation, and addressing scalability and cost barriers. While most systems are in preclinical stages, their potential to synergize chemotherapy with hyperthermia and reduce systemic toxicity underscores their promise for clinical translation, pending further research into long-term safety and efficacy [8].

4.5 Light-Responsive Nanosystems

Light-responsive nanosystems represent an emerging class of phototherapeutic platforms that can be activated upon exposure to specific wavelengths of light. These systems have attracted significant attention for their potential applications in targeted drug delivery, photodynamic therapy, and photothermal therapy. The design of light-responsive drug delivery systems is usually built on the mechanisms that include photolysis, photoisomerization, photo-crosslinking/de-crosslinking, and photoreduction, which are mainly responsive to the irradiation of light in either the UV or Visible range. However, the low tissue penetration depth and potential phototoxicity resulting from the high-energy nature of these photons frequently hinder the clinical translation of light-responsive systems [24].

To overcome these shortcomings, the recent development has centered on the nanocarrier that responds to low-energy photon irradiation, especially within the near-infrared (NIR) region. NIR light offers deeper tissue penetration and reduced photodamage, making it more suitable for in vivo applications. Drug release NIR triggered is typically accomplished using mechanisms like two-photon absorption or photon upconversion, allowing for precise and localized delivery of the drug within biological tissues [25].

5 Conclusion and Future Perspectives

The development of stimuli-responsive nanosystems represents a paradigm shift in cancer therapy, offering precise, tumor-targeted drug delivery while minimizing systemic toxicity. By exploiting endogenous triggers (pH, enzymes, redox imbalance, GSH) and exogenous stimuli (ultrasound, electric fields, temperature, magnetic fields), these smart nanocarriers enable spatiotemporal control over drug release, overcoming biological barriers and heterogeneity in the tumor microenvironment. Stimuli-responsive nanosystems herald great promise for effective cancer chemotherapy.

Acknowledgments The authors would like to acknowledge the Institute of Chemical Technology, Mumbai, and the University of Helsinki for the infrastructural facilities (iThenticate software for similarity index and other software tool for editing the manuscript) provided for the completion of the work.

Author Contribution Pooja Todke, Esha Attar, and Vanashree Chaudhari: writing—original draft, writing—review and editing; Vijayabhaskarreddy Junnuthula: data collection, writing—review and editing; Sathish Dyawanapelly and Padma V. Devarajan: conceptualization, supervision, writing—review and editing.

Declarations of Generative AI and AI-Assisted Technologies The authors declare that all sections of this chapter were written using human intellect. However, we have used a few AI-based tools for grammatical and English language corrections.

References

1. Dyawanapelly S, Junnuthula V, Singh A. The holy grail of polymer therapeutics for cancer therapy: an overview on the pharmacokinetics and bio distribution. Curr Drug Metab. 2015;16(7):522–37.
2. Anderson NM, Simon MC. The tumor microenvironment, vol. 30. Current Biology. Cell Press; 2020. p. R921–5.
3. Wang M, Zhao J, Zhang L, Wei F, Lian Y, Wu Y, et al. Role of tumor microenvironment in tumorigenesis. J Cancer. 2017;8:761–73.
4. Wu T, Dai Y. Tumor microenvironment and therapeutic response, vol. 387. Cancer Letters. Elsevier Ireland Ltd; 2017. p. 61–8.
5. Arneth B. Tumor microenvironment, vol. 56. Medicina (Lithuania): MDPI AG; 2020.
6. Raza A, Rasheed T, Nabeel F, Hayat U, Bilal M, Iqbal HMN. Endogenous and exogenous stimuli-responsive drug delivery systems for programmed site-specific release, vol. 24. MDPI AG: Molecules; 2019.
7. Wu JM, Liu Y, Han HY, Song ZY. Recent advances in endogenous and exogenous stimuli-responsive nanoplatforms for bacterial infection treatment. Biomed Eng Commun. 2023;2(1):2.
8. Mai BT, Fernandes S, Balakrishnan PB, Pellegrino T. Nanosystems based on magnetic nanoparticles and thermo- or pH-responsive polymers: an update and future perspectives. Acc Chem Res. 2018;51(5):999–1013.
9. Mu Y, Gong L, Peng T, Yao J, Lin Z. Advances in pH-responsive drug delivery systems. OpenNano. 2021;1:5.
10. Irshad S, Siddiqui B, ur Rehman A, Farooq RK, Ahmed N. Recent trends and development in targeted delivery of therapeutics through enzyme responsive intelligent nanoplatform. Int J Polymeric Material Polymeric Biomaterials. 2022;71(6):403–13.
11. Mu J, Lin J, Huang P, Chen X. Development of endogenous enzyme-responsive nanomaterials for theranostics, vol. 47. Chem Soc Rev; 2018. p. 5554–73.
12. Fathi M, Safary A, Barar J. Therapeutic impacts of enzyme-responsive smart nanobiosystems. Bioimpacts. 2020;10(1):1–4.
13. Han L, Zhang XY, Wang YL, Li X, Yang XH, Huang M, et al. Redox-responsive theranostic nanoplatforms based on inorganic nanomaterials. J Control Release. 2017;259:40–52.
14. Raza A, Hayat U, Rasheed T, Bilal M, Iqbal HMN. Redox-responsive nano-carriers as tumor-targeted drug delivery systems. Eur J Med Chem Elsevier Masson srl. 2018;157:705–15.
15. Guo X, Cheng Y, Zhao X, Luo Y, Chen J, Yuan WE. Advances in redox-responsive drug delivery systems of tumor microenvironment, vol. 16. J Nanobiotechnol BioMed Central Ltd; 2018.
16. Szliszka E, Czuba ZP, Domino M, Mazur B, Zydowicz G, Krol W. Ethanolic Extract of Propolis (EEP) enhances the apoptosis- inducing potential of TRAIL in cancer cells. Molecules. 2009;14(2):738–54.
17. Liu P, Hao L, Liu M, Hu S. Glutathione-responsive and -exhausting metal nanomedicines for robust synergistic cancer therapy, vol. 11. Frontiers in Bioengineering and Biotechnology. Frontiers Media S.A.; 2023.
18. Zardad AZ, Choonara YE, du Toit LC, Kumar P, Mabrouk M, Demarco Kondiah PP, et al. A review of thermo- and ultrasound-responsive polymeric systems for delivery of chemotherapeutic agents, vol. 8. MDPI AG: Polymers; 2016.
19. Tang JJ, Zhang X, Cheng L, Liu Y, Chen Y, Jiang Z, et al. Multiple stimuli-responsive nanosystem for potent, ROS-amplifying, chemo-sonodynamic antitumor therapy. Bioact Mater. 2022;15:355–71.
20. Fateh ST, Moradi L, Kohan E, Hamblin MR, Dezfuli AS. Comprehensive review on ultrasound-responsive theranostic nanomaterials: mechanisms, structures and medical applications. Beilstein J Nanotechnol. 2021;12:808–62.
21. Kang H, Hu S, Hyeon Cho M, Ho Hong S, Choi Y, Soo Choi H. Theranostic nanosystems for targeted cancer therapy, vol. 23; 2018.

22. Jha A, Rama A, Ladani B, Verma N, Kannan S, Naha A. Temperature and pH-responsive nanogels as intelligent drug delivery systems: A comprehensive review. J Appl Pharm Sci. 2021;11(12):001–16.
23. Yang HY, Li Y, Lee DS. Multifunctional and stimuli-responsive magnetic nanoparticle-based delivery systems for biomedical applications, vol. 1. Advanced Therapeutics. Blackwell Publishing Ltd; 2018.
24. Martínez-Carmona M, Lozano D, Baeza A, Colilla M, Vallet-Regí M. A novel visible light responsive nanosystem for cancer treatment. Nanoscale. 2017;9(41):15967–73.
25. Zeng Fiona Phua S, Xue C, Qi Lim W, Yang G, Chen H, Zhang Y, et al. Light-responsive prodrug-based supramolecular nanosystems for site-specific combination therapy of cancer.

[illegible]

2 Stimuli-Responsive Micelles for Cancer Therapy

Shourya Tripathi, Srishty Jaiswal, Pooja Yadav,
Rafquat Rana, and Manish Kumar Chourasia

Abstract

Cancer, being a major global health concern, accounts for nearly ten million deaths every year. Traditional treatment strategies, including chemotherapy, surgery and radiotherapy, remain the primary therapeutic measures but are associated with severe adverse effects, low selectivity and incomplete tumour eradication. Nanotechnology-based methods have potential to emerge as safer and more effective therapies for cancer. Nanoparticles offer targeted delivery of chemotherapeutic agents along with improved pharmacokinetics and reduced toxicity. Among the various types of nanocarriers, micelles have gained prominence due to their amphiphilic structure and their ability to encapsulate hydrophobic drugs with an enhanced solubility and circulation time. Their small size further facilitates tumour accumulation via the enhanced permeability and retention effect. In addition to this, stimuli-responsive micelles further allow for precise, site-specific drug release triggered by tumour-specific cues such as pH, redox conditions, enzymes or temperature. Despite these advantages, challenges such as premature drug leakage persist, necessitating improved micelle design for greater in vivo stability. Functionalization with targeting ligands and incorporation of environmentally sensitive components can further enhance their thera-

S. Tripathi · R. Rana
Division of Pharmaceutics and Pharmacokinetics, CSIR - Central Drug Research Institute, Lucknow, Uttar Pradesh, India

Jawaharlal Nehru University (JNU), New Delhi, Delhi, India

S. Jaiswal · P. Yadav · M. K. Chourasia (✉)
Division of Pharmaceutics and Pharmacokinetics, CSIR - Central Drug Research Institute, Lucknow, Uttar Pradesh, India

Academy of Scientific and Innovative Research (AcSIR), Ghaziabad, Uttar Pradesh, India
e-mail: manish_chourasia@cdri.res.in

P. V. Devarajan et al. (eds.), *Advances in Stimuli-Responsive Nanosystems for Cancer Therapy*, https://doi.org/10.1007/978-981-95-2744-1_2

peutic potential. Overall, micelle-based nanocarriers represent a promising strategy for safer and more effective cancer treatment.

Keywords
Tumour targeting · Stimuli-responsive drug delivery system · Responsive micelles · Cancer therapy · Smart drug delivery system · Nanomedicine

Abbreviations

CMC	Critical micelle concentration
DLS	Dynamic light scattering
DOX	Doxorubicin
EPR	Enhanced permeability and retention
LCST	Lower critical solution temperature
NIR	Near-infrared light
PEG	Polyethylene glycol
PEI	Polyethylenimine
PMLA	Poly(β-L-malic acid)
PTX	Paclitaxel
ROS	Reactive oxygen species
TEM	Transmission electron microscopy

1 Introduction

Cancer continues to affect millions of lives globally, killing almost ten million people annually. The treatment modality still remains the decade-old concoction of chemotherapy, radiation and surgery. Despite the advancements in medical science, cancer biology and clinical oncology, the people diagnosed with cancer live in uncertainty and suffer detrimental side effects in the name of traditional therapy approaches, with damage to healthy tissues and incomplete eradication of cancer cells. The survival of patients still remains low as the available methods are unsatisfactory. It is the need of the hour to devise new treatment strategies to change this horrid picture [1]. The potential of nanotechnology for targeted and more effective treatment and diagnosis of cancer has been widely explored. With the use of nanotechnology, not only can the therapies be guided selectively to the neoplasm, but also drug resistance can be overcome. In addition to this, nanotechnology aids the usage of innovative immunotherapies and augment the therapeutic efficacy of radiation-based treatments. These factors together can contribute to better management of cancer with decreased risk and increased life expectancy to the patient [2].

Nanoparticulate drug delivery systems are playing an important role in cancer research. They help overcome drawbacks of poor drug bioavailability and short circulation times. They help in affording controlled and targeted drug release. They are known to exhibit good pharmacokinetics, reduction of side effects and drug resistance [3]. Numerous nanocarriers have been developed of which most commonly explored are liposomes, micelles, dendrimers, inorganic nanoparticles such as silica nanoparticles, carbon nanotubes, gold nanoparticles, quantum dots and magnetic nanoparticles [4]. These nanosystems have helped overcome the non-selective nature of existing treatment approaches. This is because such specialized drug delivery systems can differentiate between healthy and cancerous tissue, a property that lacks in traditional chemotherapy, leading to significant mortality and morbidity. A specifically targeted drug delivery system releases the chemotherapeutic agent only at the desired site in a controlled manner. This tumour targeting has emerged as a promising approach to enable access to tumours and at the same time prevent penetration into healthy tissues [5].

Recent progress in materials chemistry and drug delivery has led to the design of stimuli-responsive devices for the precise delivery of drugs based on spatiotemporal control and dose. Stimuli-responsive drug delivery systems identify cancer cells by sensing the abnormal conditions present explicitly at the tumour site, as the spatial-temporal profile of up- and down-regulated hallmarks of cancer is relatively tumour specific [6]. Such devices require biocompatible materials that can undergo activation in response to specific biophysical triggers. This leads to their protonation, hydrolytic cleavage, or (supra)molecular conformational change, allowing the controlled release of medication in a targeted manner [7].

Micelles are amphiphilic colloidal structures with particle sizes ranging from 5 to 100 nm. These structures are composed of molecules with two characteristic regions: one being hydrophilic (water loving) and the other being hydrophobic (water repelling). The assembly of micelles takes place at a certain concentration and temperature. The concentration at which the formation of micelles begins is referred to as the critical micelle concentration (CMC), and the temperature above which micelle aggregation occurs is known as the critical micellar temperature. Micelles are known to exist in monomer form below this temperature. The number of monomer molecules that come together to form a micelle is known as the aggregation number. In the pharmaceutical industry, micelles serve as very good drug carriers. The lipophilic drugs can be incorporated in the hydrophobic core, whereas the hydrophilic surface interacts with the polar molecules. Polymeric micelles, which are amphiphilic block copolymers, provide a very high degree of value in therapeutics, owing to their high solubilization and better intestinal permeability, making them excellent carriers for drug delivery [8]. Polymeric micelles are formed when amphiphilic copolymers self-assemble in aqueous solutions and are characterized by a hydrophobic core formed from hydrophobic segments and a stabilizing hydrophilic outer shell derived from hydrophilic segments. Polymeric micelles are generally more stable than surfactant micelles, have a lower CMC, and dissociate slowly. This stability allows higher accumulation of the drug in targeted sites. The size range of polymeric micelles, being comparable to that of the virus and the lipoproteins, is of particular importance when the enhanced permeability and retention (EPR) effect is of concern in body distribution. By combining intrinsic environmentally sensitive groups into these copolymers, the micelles become responsive to numerous environmental stimuli like pH, redox conditions, temperature and light at the tumour environment. Modification of these micelles with active targeting groups allows for targeted and triggered release of the drugs. The hydrophilic segments of the micelles afford water solubility and improve biocompatibility which leads to extended circulation in the bloodstream. The superior biocompatibility also contributes to reduced accumulation of residual micelles inside the body. Micelles also afford their functionalization with various targeting ligands owing to the various rich surface functional groups, allowing targeted release of therapeutic moiety.

Despite the great success of micelles as drug-delivery system for anticancer drugs, some issues persist regarding drug release in the circulation and at the target site. The major issue is the premature release of drug into the circulation with insufficient release at the desired target site. Ideally, the drugs should stay completely encapsulated inside the circulating micelles and be released only after reaching the tumour site. Premature release can be prohibited by improving micelle stability or enhancing drug interactions inside the micelle core, such as hydrogen bonding or specific bonding strength with cleavable covalent bonds [9]. Designing stimuli-responsive micelles is an effective strategy to achieve successful and adequate release of a drug to the target tumour site (Fig. 2.1). Based on the progressively better understanding of the biology of the tumour, it is known that the unique characteristics at tumours include low pH in solid tumours, even lower pH in the tumour cellular environment, the expression of specific enzymes, the presence of high levels of glutathione in the cytoplasm, and temperatures that are higher than those of surrounding normal tissues. These unique features are being utilized to target

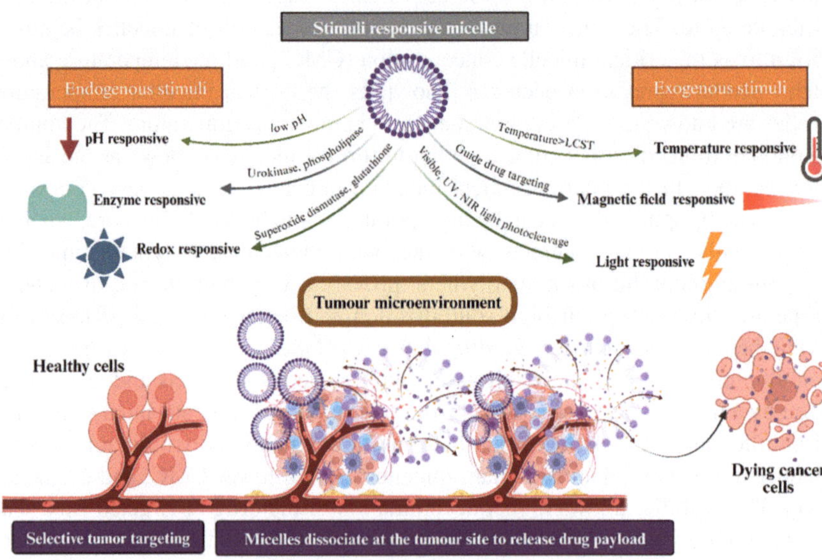

Fig. 2.1 Schematic depicting different types of endogenous and exogenous stimuli-responsive micelles and their selective tumour targeting to kill cancer cells

internal stimuli, and at the same time, external stimuli, such as magnetic fields, ultrasound, and light, can also be used to break down stimuli-responsive micelles, which will allow for spatially and temporally controlled drug delivery and release [10].

2 Micelles for Cancer Therapy

2.1 pH-Responsive Micelles

pH-responsive micelles have recently emerged as important carriers of anticancer drugs [11]. The physiological blood and healthy tissues pH is known to be 7.4; however, the microenvironment of tumours is acidic with pH values ranging from 6.5 to 7.2. In addition to this, cellular compartments such as endosomes and lysosomes are even more acidic with pH values of 6.5 to 5.5 and 4.5 to 5, respectively. This difference in pH has been widely exploited to achieve specific activation of micelles at desired organelle and this leads to tumour-targeted drug delivery. This pH sensitivity arises due to degradation of ionizable groups and cleavage of pH-sensitive linkages present in the polymeric micelles, leading to degradation of polymeric micelles and subsequent release of the drug payload at the tumour site [12]. pH-sensitive polymeric micelles have been prepared by Wu and team, to release paclitaxel (PTX) in the acidic environment of endosomes. The polymeric micelles were prepared using three grafted copolymers, including 1,2-distearoyl-sn-glycero-3-phosphoethanolamine-polyethylene glycol-2000 (DSPE-PEG2000), antinucleosome antibody (mAb 2C5)-modified DSPE-PEG3400 (DSPE-PEG3400-2C5), and poly(ethylene glycol)-coupled poly(L-histidine) (PHIS-PEG2000). As the fabricated PHIS-PEG copolymer is insoluble in water, the mixed micelles were prepared using dialysis method. The average diameter of the fabricated polymeric micelles was found to be around 112 nm. The drug release was found to be pH dependent with much faster release being obtained at pH 5 and 6, as compared to physiological pH of 7.4, where the micellar structures remained stable. The imidazole groups present in the PHIS polymer was found to be responsible for this pH-sensitive behaviour. When the pH dropped to around 5, PHIS of the micelles completely protonated, dissociated from the micellar core as a result of phase separation, which ultimately led to drug release. This behaviour is indicative that such polymeric micelles can respond to small pH variations and cause targeted drug release within the tumour cells [13].

In another study employing the polymer PHIS for its amphoteric and pH-sensitive nature, Hu and co-workers utilized polysulfadimethoxine and PHIS to fabricate polymeric micellar system. In vivo testing in mouse model xenografted with MCF-7 cell line, showed deep penetration of the micellar system into the tumour tissue, supporting their pH-responsive behaviour. This occurred through protonation deprotonation of the fabricated system, wherein, a shielded micelle, which was obtained by coating with a block copolymer consisting of α-methoxy ω-hydroxy poly(ethylene glycol) (mPEG) and a pH-sensitive anionic block in polysulfadimethoxin. The sulfonamide group being anionic at a neutral pH loses the charge to separate from micelle in slightly acidic environment, exposing the positive charge. The shielded micelle that is designed for circulation in the bloodstream, reduces the odds of nonspecific interaction. However, upon exposure to the slightly acidic pH, the micelle transforms into a deshielded micelle and becomes positively charged thus facilitating entry into the tumour cell. As it reaches the cell, the PHIS present in micellar core protonates in the acidic endosomal environment, disassembles and finally leads to endosomal membrane destabilization, subsequently causing more efficient drug delivery into the cytosol of the cell [14].

Another approach combines the pH-targeted drug delivery with electrostatic interactions to achieve efficient cell internalization. It is known that positively charged nanocarriers can interact with the negatively charged cell membrane through electrostatic interaction, to achieve efficient internalization into the cells. However, this is not feasible inside the living body due to the nonspecific interactions and short circulation times when inside the body. Thus, for pH-sensitive drug delivery, charge reversal and charge conversion are being employed. In such systems, the pH-responsive micelles remain neutral or negatively charged at physiological pH, but when exposed to the lower pH of tumour microenvironment, the charge is converted to the positive one. In one similar study, poly(β-L-malic acid) (PMLA), a natural and biocompatible aliphatic polyester, that is known for being an excellent carrier for anti-cancer drugs has limited usage due to its strong negative charge that leads to its low cellular uptake efficiency. Thus, in this study, PMLA was first functionalized with polyethylenimine (PEI) to impart a positive charge and then conjugated with a pH-sensitive version of DOX, DOX-CA, forming micelles (PMLA-PEI-DOX). Then, TAT peptide was incorporated to improve cellular uptake, yielding PMLA-PEI-DOX-TAT. The micelles were further encapsulated with PEG-DMMA to camouflage the positive charge and TAT peptide, which led to a dual pH-responsive nanocomplex PMLA-PEI-DOX-TAT@PEG-DMMA. Therefore, at normal pH, the nanocomplex carried a negatively charged surface affording prolonged circulation time whereas in the acidic pH environment of the tumour, PEG was detached, exposing the positive charge and TAT peptide. Both in vitro and in vivo studies of the developed formulation, showed effective inhibition of tumour growth with minimal systemic toxicity [15].

Another popular strategy in pH-responsive micelles is the use of linkages that degrade in the lower pH environment of the tumour tissue. In a study performed by Yang and group, TAT peptide was delivered using pH-sensitive polymeric micelles. They developed a conjugate of DOX and TAT (Dox-TAT) and loaded it into micelles made of luteinizing hormone-releasing hormone-modified poly(ethylene glycol)-poly(L-histidine)-DOX (LHRH-PEG-PHIS-Dox). DOX was linked to the polymer backbone to enhance micelle stability and increase drug loading efficiency. It was found that the micelles release Dox-TAT in response to the acidic pH of the tumour environment, allowing access even to the cytosol of multidrug-resistant cancer cells. It was also reported that the micelles themselves could be taken up by cells through receptor-mediated endocytosis, leading to high cytotoxicity. In both in vitro and in vivo experiments, LHRH-PEG-PHIS-Dox/Dox-TAT showed the most significant anti-tumour effects among the tested groups and had minimal impact on body weight compared to the control [16].

In another study exploiting the linker cleavage in micelles at acidic pH, Zhou and co-workers prepared a micellar drug delivery system from an N-(2-hydroxypropyl) methacrylamide (HPMA) copolymer. For this, DOX and β-sitosterol were coupled to the water-soluble HPMA polymer backbone through a pH-sensitive hydrazone linker. This linker was found to be stable at physiological pH but underwent hydrolysis readily at acidic pH. In vitro studies demonstrated that the micelles could disintegrate easily to release 80% of the encapsulated drug at pH 5, characteristic of the lyso/endosomal environment. Furthermore, in vivo studies supported the fact that these micelles possessed a significantly enhanced anticancer effect [17].

2.2 Thermo-Responsive Micelles

Thermo-responsive drug delivery systems offer a controlled and targeted delivery of therapeutic agents in response to temperature change. Polymers exhibiting thermo-responsive activity make up a significant class of stimuli-responsive drug delivery system. Thermo-responsive polymeric micelles exhibit significant structural changes when subjected to temperature variations. Amphiphilic block copolymers, which can self-assemble to form a core-shell structure in aqueous media are the building blocks of micelles. Drugs that are hydrophobic are usually found in the core, whereas the hydrophilic shell of the micelles helps in stabilization of micelles in the biological fluids [18]. Thermo-responsive micelles release the encapsulated drug when polymers undergo a phase transition at a specific temperature, known as lower critical solution temperature (LCST) or the cloud point. The micelles are stable below the cloud temperature, retaining the encapsulated drug within the micelles. Above the LCST, the polymers become water-insoluble and the formed micelles are shrunken. However, after reaching the temperature above the cloud point, the micelles get destabilized and the encapsulated drug is released at the targeted site. The capacity of thermo-responsive polymers to modify their LCST is a significant and helpful characteristic. The chemical composition of the monomers is generally changed in order to achieve this temperature. Through random copolymerization using optimum amount of monomers with varying degrees of hydrophilicity or hydrophobicity, the LCST of thermo-responsive polymers can be adjusted [19]. Generally, incorporating hydrophobic co-monomers results in a lower cloud point temperature, while incorporating hydrophilic co-monomers leads to a higher cloud point temperature. This temperature sensitivity makes it possible to deliver drugs in a regulated and targeted manner. This is especially helpful in situations where drug release might be triggered by localized temperature changes, including those in tumour tissues or inflammatory areas.

Depending on their chemical structures, N-substituted acrylamide polymers usually demonstrate phase transition behaviour at different temperatures. Poly(N-isopropyl acrylamide) polymer is widely used because of its LCST, which is extremely near to the average body temperature (32 °C) [20]. Pluronics, which are triblock copolymers made of a polypropylene oxide (PPO) centre block surrounded by polyethylene glycol (PEG) blocks (PEG-b-PPO-b-PEG), and poly(N-isopropylacrylamide) (pNIPAAm) are two well-known thermosensitive polymers. Using reversible addition-fragmentation chain-transfer (RAFT) polymerization, Ghasemi and co-workers created poly(N-isopropylacrylamide-b-lauryl acrylate) (PNIPAAm-b-PLA), a thermo-responsive amphiphilic block copolymer. This technique made it possible to accurately regulate the molecular weight, which is necessary for efficient drug delivery. The amphiphilic character of the copolymer enabled the development of micelles in aqueous solutions, loaded with docetaxel in its hydrophobic core through Van der Waals interactions and resulting in a high drug loading efficiency. Thermo-responsive activity was demonstrated by the PNIPAAm segment, making temperature-triggered drug release possible and suitable for targeted cancer therapy. Using MTT assay on MCF-7 breast cancer cells, the effectiveness of docetaxel-loaded micelles was confirmed, showing a notable reduction in cell proliferation. The results of this work demonstrate the potential of PNIPAAm-b-PLA copolymers as efficient anti-cancer drug carriers that provide regulated release and enhanced therapeutic results. To properly confirm the system's safety and effectiveness for more widespread clinical applications, more in vivo study is required [21].

Thermo-responsive micelles have also been used a gatekeeper to target other drug delivery carriers loaded with therapeutic agents at the desired site. Long and group prepared F127-PCL250 (FP250) micelle which was attached to mesoporous silica nanoparticles. This micelle-attached nanoparticle was loaded with chemotherapeutic agent DOX. The uncapping of FP250 micelle from mesoporous silica nanoparticle following shrinkage of FP250 micelle at the elevated temperature of 40 °C resulted in a rapid release of DOX from both mesoporous silica nanoparticle and micelle, which significantly enhanced the drug concentration. The drug release was inhibited below cloud point by the thick micelles of FP250 on mesoporous silica nanoparticles. The shrinking of FP250 allowed DOX to escape as the temperature reached above cloud point [22]. Thermo-responsive poly(ε-caprolactone)-based amphiphilic block copolymers were synthesized via living ring-opening polymerization of functionalized ε-caprolactone monomers by Soltantabar and team. These polymers showed thermo-responsive characteristics; by varying the length of their oligo (ethylene glycol) (OEG) chains, the LCST values could be adjusted. The developed polymer was used to formulate micellar drug delivery system loaded with DOX and quercetin. Temperature-dependent drug release from the micelles was seen in physiological drug release studies. The cloud point temperature of the produced PME3CL-b-PBnCL polymer was 41 °C, which is marginally above but close to the physiological temperature. The favourable release profile led to its selection for additional biological investigations. H9c2 rat heart cells and HepG2 human liver cancer cells were used in in vitro cell culture studies [23]. A thermo-responsive amphiphilic copolymer functionalized with a propargyl group and tri (ethylene) glycol was formulated by Bhadran and associates. The polymer obtained thermo-responsive nature by the addition of the tri (ethylene) glycol group. Cross-linked micelles were formed by the lipoic azide-functionalized PCL. DOX was encapsulated within the prepared micelle. Interestingly, the hydrophobic micelle core's cross-linking raised the polymer's cloud point value from 37.5 to 40.5 °C. Compared to uncross-linked micelles, release experiments carried out in physiological buffer revealed that cross-linked micelles lowered the premature leakage of DOX by half [24]. The anticancer drug zinc protoporphyrin (ZnPP) has problems with its poor solubility and undesirable side effects. Bagheri and co-workers created thermo-responsive polymeric micelles based on the PNIPAM–PEG–PNIPAM tri-block copolymer to target prostate cancer cells and improve delivery and efficacy of the drug ZnPP. The structural stability and decreased drug leakage from these nanocarriers was demonstrated. An LCST of 36.20 °C was found after the polymer characterization. Moreover, the particle size of developed material was found to be 148 nm with entrapment of more than 98%. A higher cell uptake was confirmed for the developed micelles, thus suggesting an effective drug delivery system for cancer therapy [25].

2.3 Magnetic Field-Responsive Micelles

In the area of cancer therapy, magnetic responsive micelles are gaining interest as an innovative approach that combines the complementary advantages of nanotechnology and magnetic responsiveness to improve therapeutic efficacy and selectivity. These micelles are usually formed of a hydrophobic core and a hydrophilic shell, which is made of biocompatible materials such as polyethylene glycol (PEG), which assures stability and extends the duration that the micelle remains in circulation in the bloodstream. Magnetic nanoparticles, typically iron oxide, are embedded in these micelles and can be targeted to tumour locations with the help of external magnetic fields [26]. By significantly boosting drug accumulation at the malignant site, magnetic targeting reduces systemic toxicity and enhances therapeutic outcomes. Magnetic responsive micelles have a complex mode of operation. Chemotherapeutic drugs are encapsulated in a hydrophobic core that prevents early breakdown and minimizes undesirable adverse effects. Once inside the tumour, the micelles can react to various stimuli like temperature variations, pH shifts, or enzymatic activity to release their payload, which guarantees a steady and regulated release of the therapeutic contents. Furthermore, when exposed to an alternating magnetic field, the magnetic nanoparticles within the micelles have the ability to cause localized hyperthermia. In addition to destroying cancer cells directly, this hyperthermia increases their susceptibility to chemotherapy and radiation, improving the overall therapeutic effect. Adding imaging and diagnostic capabilities to these micelles is yet another noteworthy benefit. When used in conjunction with magnetic resonance imaging, magnetic nanoparticles provide superior contrast, enabling theragnostic which is the simultaneous monitoring of therapy outcomes and diagnosis [27]. Real-time monitoring of drug distribution and tumour response is made possible by this dual feature, which provides vital data that allows treatment procedures to be dynamically changed. The capacity of magnetic responsive micelles to lessen adverse effects and boost specificity is one of its main advantages.

Healthy tissues are saved from excessive dosages of hazardous chemotherapeutic medicines by focusing drug delivery at the tumour site through magnetic targeting. This leads to fewer adverse effects and an enhanced quality of life for patients. Moreover, the controlled release mechanisms incorporated into these micelles guarantee that the drug is released within the most optimal therapeutic window, preserving its effectiveness for an extended period of time. The goal of this field's current research is to maximize the performance of these micelles by improving their chemical and physical characteristics. To ensure effective targeting and drug release, efforts are being made to fine-tune the size, surface features, and magnetic properties [28]. Ensuring the safety and biocompatibility of these micelles for extended use in the human body is a crucial field of study. To ease the transfer from laboratory research to clinical use, scientists are also investigating scalable and reasonably priced production technologies. Despite their promising potential, there are a number of obstacles to overcome, such as the requirement for thorough in vivo testing to confirm the security and effectiveness of these systems. There are also regulatory approval procedures to follow, which can be difficult and time-consuming. However, magnetic-responsive micelles appear to have a bright future in cancer treatment. Magnetic-responsive micelles have the potential to revolutionize cancer treatment by providing a highly targeted, efficient, and minimally invasive therapeutic option as research advances.

A rod-like magnetic micelle was formulated by Gao and group that provides an ideal nanoplatform for the accurate and highly effective delivery of therapeutic medicines to tumour locations, improving the treatment of cancer. Phenylboronic acid functionalization is used to make these micelles, which are created by a copolymer self-assembling. They are loaded with magnetic nanoparticles and the chemotherapeutic drug DOX. When an external magnetic field is provided, the presence of magnetic nanoparticles increases the micelles' concentration at the tumour location. These nanoparticles also enhance the contrast between the cell compartments and the polymer micelles, which makes it easier to assess the distribution of nanocarriers. The DOX-loaded nanoparticles are endocytosed by sialic acid-positive tumour cells, which are specifically recognized by the phenylboronic acid targeting ligands [29].

Huang and group created a potential platform for dual-targeted drug delivery that makes use of magnetic-guided and folate-mediated mechanisms to improve the accuracy and potency of delivering therapeutic drugs to tumour locations. These magnetic nanocarriers are made from biocompatible Pluronic F127 and poly(dl-lactic acid) copolymer, tumour-targeting ligand-folic acid using a simple chemical conjugation process, and superparamagnetic iron oxide nanoparticles (SPIONs). In vitro drug release and antiproliferative effects on tumour cells were assessed using DOX hydrochloride as a model anticancer agent, both in vivo and in vitro, with or without an external magnetic field of 0.1 T. These magnetic nano micelles show impressive cell-specific targeting in vitro, in accordance to the Alamar blue assay. This smart device demonstrated a robust response to magnetic field and permits folate receptor-mediated absorption into tumour cells. The application of magnetic field led the nano micelles to the tumour site in an in vivo tumour model employing rabbits with VX2 tumours, thereby exhibiting considerable therapeutic efficacy against tumours [30]. Likewise, Mahdieh et al. used the solvent evaporation method to create a magnetic targeted drug delivery system that encapsulates hydrophobic DOX and hydrophobic oleic acid-superparamagnetic nanoparticles within their hydrophobic core. The system is based on waterborne polyurethane nano micelles. The resulting drug-loaded magneto micelles had an average diameter of 158 nm and were spherical in shape. These magneto micelles demonstrated superparamagnetic characteristics and produced remarkable contrast effects in magnetic resonance imaging. The findings demonstrated a considerable improvement in targeting efficacy upon delivery via magnetic guidance [31]. Tang and colleagues created redox-responsive, star-shaped magnetic micelles with magnetic guidance and active targeting. These micelles are created by the self-assembly of four-arm disulfide-bonded poly(ethylene glycol)-poly(ε-caprolactone) copolymers. These micelles' hydrophobic cores contain magnetic iron oxide nanoparticles and the chemotherapeutic drug DOX at the same time. Phenylboronic acid ligands are chemically bonded to the ends of the hydrophilic PEG segments to facilitate active targeting [32]. Similarly, dual-responsive magnetic drug delivery nano micelles loaded with DOX were developed and characterized by Zhu and associates. The EPR effect is used in the design of these magnetic nanocarriers to passively target tumour tissues, while folic acid is used to aggressively target the acidic tumour environment. When the PEG layer sheds upon arrival, the drug-loaded nanoparticles are released by the nano micelles, which triggers the initial pH reaction and increases the uptake of DOX by tumour cells. Once inside the tumour cells, the magnetic particles produce a magnetothermal effect in reaction to an external magnetic field, which sets off the second response. The tumour cells' temperature is raised by this magnetic hyperthermia, which facilitates drug absorption even more. It is envisaged that the dual response mechanism will increase the anti-cancer efficacy of magnetothermal therapy and chemotherapy in a synergistic manner [33].

2.4 Light-Responsive Micelles

Light as an external stimulus in the form of UV, visible, infrared/near-infrared (NIR) light has become popular means to deliver drugs specifically at the target tumour site due to its precise spatiotemporal resolution, non-invasive nature and precise remote control [34]. Out of the different types of light as external stimuli, NIR light is particularly noteworthy as it can penetrate as deep as about 10 cm inside human tissues, without causing severe damage. On the other hand, UV light has limited penetration owing to its absorption by the skin, blood, and other tissues [35]. Another advantage of this system is that light-induced reactions can be fine-tuned by adjusting parameters in terms of wavelength, intensity, and exposure time, as per the desirability. These advantages have caused an upsurge in the development of several light-responsive nanodrug delivery systems for controlled drug delivery in cancer therapy. Among them, micelles are being widely developed as light-responsive vehicles by incorporating chromophores into their structures [36]. The light-responsive micelles work by releasing their payloads by dissociating upon illumination with light. This dissociation is either an irreversible photocleavage or a reversible structural change [37].

In a study, Saravanakumar et al., to increase the photodynamic effect, synthesized a biocompatible amphiphilic block copolymer, ^{1}O2-PEG-b-PCL, which was used to create micelles incorporating a singlet oxygen ($^1O_2$)-sensitive vinyldithioether cleavable linker at the junction of the core and shell of the micelle. On exposure to visible light, the system could undergo cleavage and release the loaded drug. Double bond in the vinyldithioether linker reacts with $^1O_2$ to form an unstable transitional dioxetane, which extemporaneously decomposes. This innovative approach allows for hitch-free, controlled release of the therapeutic using light for an efficient and specific process to treat cancer [38].

In another study, Chen and co-workers designed light-responsive micelles to deliver DOX to the tumour site with enhanced cellular uptake. Firstly, DOX-conjugated polymer, called Poly-Dox, was self-assembled into micelles (Poly-Dox-M) in water. These micelles were then coated with PEG to impart stability in blood circulation, help in shielding protein adsorption and reduce the cardiotoxicity of the drug, ultimately improving the tumour targeting and stability. The system, when exposed to UV light, undergoes lysis of the amide bond linking DOX with PEG, which causes the detachment of PEG and finally enhances DOX cellular uptake due to increased positive charge. In vitro investigation on K7M2wt osteosarcoma cells under UV light showed significant high uptake. In vivo studies in tumour-bearing nude mice showed that micelles, when irradiated with UV, achieved the efficient suppression of tumours with negligible damage to normal organs, which was confirmed by immunohistochemistry. The dual functionality of this system, targeted delivery and light-responsive drug release shows great promise for enhancing anticancer therapy, particularly for osteosarcoma, by maximizing drug efficacy while minimizing adverse effects [39].

Sun and co-workers reported an amphiphilic dendritic polymer, PEO-D3DNQ, for drug release triggered by both UV and NIR light. The polymer was prepared through click conjugation between DNQ-decorated poly(amidoamine) dendron D3 (generation 3) and azide-terminated PEO (PEO-N3) with molecular weights of 2 and 5 kDa. The formed micelles, characterized using TEM, DLS and UV-visible spectroscopy, showed sensitivity towards both UV and NIR light upon dispersion in a PBS solution. Upon UV_{365} nm irradiation, the micelle absorbed light and managed to reduce the intensity of DNQ at about 8 min, which indicated the completion of the Wolff rearrangement. In contrast, when irradiated with 808 nm NIR light, the rearrangement took 40–60 min. The micelles were largely destroyed after 30 min irradiation of 808 nm, which increased the drug-release rate almost eight times and improved the NIR triggered drug release capability [40].

2.5 Enzyme-Responsive Micelles

Nanocarriers are crucial in biomedical applications, especially for anticancer therapies [41]. Enzymes play a critical role in catalysing biological reactions, and abnormal enzyme activity is often associated with various diseases, especially in cancer cells. A key strategy for targeted drug delivery and enhanced therapeutic efficacy is the use of enzyme-responsive drug delivery systems. Leveraging enzyme-responsive micelles has shown promise in enhancing the effectiveness of cancer therapy and improving diagnostic accuracy. These systems leverage enzymatic degradation to control drug release, taking advantage of the unique biorecognition and catalytic properties of enzymes. Despite progress, challenges in substrate specificity, imaging accuracy, and safety remain to be addressed in case of enzyme-responsive micelles [42].

Cancer, often characterized by abnormal enzyme expression, provides opportunities to design nanocarriers with enzyme-labile linkages. These nanocarriers can alter their physical or chemical properties or release drugs in response to enzymatic activity, allowing for precise, on-demand drug delivery. Enzyme-responsive micelles are modified on the surface to make it active for the bio-catalytic action of enzymes. Several classes of enzymes, including oxidoreductases, hydrolases, and lipases, play critical roles in cancer and are key targets for the development of enzyme-responsive micelles in cancer therapy [43]. Solid tumours are known to express higher levels of enzymes such as peptidases, proteases, and lipases in comparison to normal tissues, as these enzymes play crucial roles in invasion, angiogenesis, tumour cell growth and metastasis [44].

Matrix metalloproteinases (MMPs), a class of extracellular matrix (ECM)-remodelling proteases, are closely linked to tumour invasion and metastasis. These zinc- and calcium-dependent endopeptidases are responsible for ECM degradation and remodelling and are overexpressed in all tumours compared to the normal tissues. Hence, MMPs are promising targets for enzyme-responsive, site-specific drug release, with MMP2 and MMP9 being the most extensively studied for targeted cancer therapy [45]. In a recent study conducted by Zhu and team, novel tumour-targeted micellar drug delivery platform was formulated using block copolymers that respond to MMP2 [46]. This system consisted of an MMP2-sensitive PEG2000-PTX conjugate, a transactivating transcriptional activator peptide-PEG1000-phosphoethanolamine, and PEG1000-phosphoethanolamine, serving as an MMP2-sensitive functional polymer, cell-penetrating enhancer, and nanocarrier building block, respectively. Both in vitro and in vivo studies demonstrated that this design outperformed non-sensitive counterparts like free PTX and conventional micelles with respect to cell cytotoxicity, tumour targeting, cell internalization and anti-tumour efficacy [47].

A multifunctional MMP2-sensitive polymeric micelles were designed by Jahveri and colleagues for the co-delivery of siRNA and hydrophobic drugs. This novel formulation, constructed from an MMP2-sensitive copolymer (PEG-pp-PEI-PE) via self-assembly, exhibited excellent stability, efficient siRNA condensation by

PEI, PTX solubilization in the lipid core, and tumour targeting through both the enhanced permeability and retention effect and MMP2 sensitivity [48].

Cabazitaxel, a second-generation taxane, shows great potential for treating metastatic castration-resistant prostate cancer but is limited by poor solubility and lack of targeted delivery. Barve et al. developed and reported nano-system which is biodegradable, enzyme-responsive, and targeted polymeric micelle for cabazitaxel delivery, composed of two amphiphilic block copolymers. One copolymer includes PEG, an enzyme-responsive peptide cleavable by MMP-2, and cholesterol, while the other includes a targeting ligand, PEG, and cholesterol. The resulting micelle exhibited low critical micelle concentration, high drug loading, and efficient entrapment. It significantly improved cellular uptake and tumour growth inhibition in prostate cancer models compared to free cabazitaxel. These results indicated that this targeted micelle system has high potential, enhances therapeutic efficacy of active ingredient and reduces systemic toxicity, offering a promising platform for advanced prostate cancer treatment and potentially for other chemotherapeutic agents [49].

The block copolymer CH_3O-PEG43-b-P(AA20-g-TPE35) was synthesized via ATRP, hydrolysis, and esterification for targeted delivery of DOX by Yan and the group. This system was later characterized by 1H NMR and FT-IR, where it was found that they were able to self-assemble into fluorescent micelles. These micelles when encapsulated with DOX, led to an increase in micelle size from 79.4 to 103.8 nm and were found compatible in in vitro studies. They exhibited enzyme-responsive degradation and pH-sensitive drug release. In vitro, DOX-loaded micelles efficiently delivered DOX to MCF-7 cancer cells, with significant intracellular accumulation which was confirmed by CLSM and flow cytometry. Biocompatibility assays showed that the copolymer was non-toxic, while the DOX-loaded micelles significantly reduce cancer cell viability, indicating strong anticancer potential [50].

From the reported observations, MMPs are found to be vital for ECM remodelling in tumours and offers a promising mechanism for targeted drug delivery. Recent advancements in MMP2-sensitive micellar systems show significant potential for improving the efficacy and specificity of cancer therapeutics.

2.6 Redox-Responsive Micelles for Cancer Therapy

Recently, various stimuli-responsive polymeric micelles, such as those responsive to temperature, pH, and redox conditions, have garnered significant attention. Among these stimuli, redox potential stands out due to the stark contrast in GSH concentration between the reducing intracellular environment (approximately 2–10 mM) and the mildly oxidizing extracellular environment (approximately 2–20 μM) [51]. Redox-sensitive polymers typically feature disulfide linkages or diselenide groups, which further ensure sufficient stability in the circulation and the extracellular space also undergo rapid cleavage in the reducing intracellular environment [52].

Study done by Wang and associates entails about a micellar system which is fast ATP-depleting and gets activated by intracellular redox to co-deliver PTX and siRNA targeting polo-like kinase 1 (PLK1) for multidrug-resistant cancer therapy. These redox-responsive amphiphilic polymer self-assembles into nano micelles that release PTX and siRNA in glutathione-rich environments, depleting ATP to prevent PTX efflux. This dual modulation enhances PTX efficacy, which was found to be very effective and it significantly inhibited tumour growth in multidrug-resistant cancer models in vivo [53]. Redox-activatable ATP-depleting micelles for simultaneous delivery of PTX and PLK1 siRNA, providing dual modulation to resensitize MDR cancer cells to PTX, effectively inhibiting tumour growth through intracellular ATP depletion and siRNA inhibition has been reported.

In another study a novel conjugate, chitosan-cystamine-retinoic acid (CS-ss-RA), which forms redox-sensitive micelles in water with a low critical micelle concentration was reported for cancer therapy. These micelles were found spherical, with ideal particle size, negative charge, and were found to be efficient to load and encapsulate the cancer drug PTX. Safety tests showed that these micelles were safe for intravenous use, causing minimal haemolysis and cytotoxicity. In studies with liver cancer cells (HepG2) and 3D tumour models, PTX-loaded CS-ss-RA micelles were more effective at killing cancer cells than free PTX. These newly formulated micelles improved the uptake of PTX into cells through clathrin-mediated endocytosis and macropinocytosis. These positive outcomes indicated that CS-ss-RA micelles could be a promising tool for delivering hydrophobic cancer drugs, offering high drug loading capacity, low toxicity, and enhanced anti-tumour activity in vitro [54].

The dual redox-responsive micelles, which were sensitive to both reactive oxygen species (ROS) and glutathione, to encapsulate the anticancer drug camptothecin were formulated by Chiang and group These micelles safeguarded camptothecin in the bloodstream and released it selectively in cancer cells with elevated ROS or GSH levels, effectively preventing toxicity in healthy cells and organs. Using mPEG-b-P(Des-alt-Cys) amphiphilic diblock copolymers, the micelles increased the solubility and activity of campotothecin which was further confirmed by its potent anti-tumour activity in cancer cells both in vitro and in vivo. Overall, the results highlighted the potential outcome of dual redox-responsive micelles as a promising nanomedicine platform for enhancing cancer therapy efficacy while minimizing adverse effects [55].

An innovative redox-responsive drug delivery system utilizing an amphiphilic PEG-based ether-anhydride copolymer (mPEG-ss-CPP-SA), which incorporates disulfide bonds between the mPEG and CPP/SA segments was reported. This system was found capable to self-assemble into micelles in an aqueous solution, and its reduction-triggered disassembly was thoroughly examined using GSH. Curcumin, which is a lipophilic polyphenol compound, was loaded into the micelles via physical mixing. When compared to redox-insensitive micelles, it was systematically evaluated for the intracellular drug release profile, cell apoptosis, cell cycle effects, anti-tumour efficacy both in vitro and in vivo, and histological assessment of the redox-responsive mPEG-ss-CPP-SA micelles in BALB/c mice bearing 4 T1 tumours. Cytotoxicity analysis revealed that these curcumin-loaded redox-responsive micelles were more effective at inhibiting tumour cell growth due to the rapid release of the drug in the tumour microenvironment, compared to redox-insensitive micelles. In vivo studies demonstrated that these redox-responsive micelles offered greater safety and a more effective therapeutic outcome, particularly in inducing tumour necrosis, than their redox-insensitive counterparts. This study, therefore, presents a scalable strategy for designing and engineering advanced nanomedicines for cancer therapy [56].

A recent study by Yang et al. reported an impressive outcome which involved formulating DOX (DOX)/HA-ss-PSA nano micelles. The prepared nano micelles exhibited uniform particle size distribution overall and an outstanding drug release rate of DOX of up to 80% within 48 h in tumour microenvironment. DOX/HA-ss-PSA nanoparticles demonstrated significant cytotoxic effects. Further, in in vitro studies, these HA-ss-PSA nano micelles showed marked tumour cell targeting and inhibition of cell migration, which was further confirmed by cell uptake and scratch assays. The HA-ss-PSA nano micelles effectively inhibited the growth of HepG2 cells and tumours in vivo, indicating to a new approach, that PSA could serve as a reliable backbone for a drug delivery system. In vivo studies revealed high tumour inhibition rates, effective targeting of liver and tumour tissues, and significantly reduced toxicity to other tissues, which further adds to the outcome of this wonderful study. This advancement provided a promising and insightful strategy for liver cancer treatment [57].

Immunotherapy with immune checkpoint blockade for glioblastoma shows promise but faces significant challenges due to the blood-tumour barrier and an immunosuppressive tumour microenvironment. A recent study developed redox-sensitive micelles co-encapsulating anti-programmed death-ligand 1 antibodies (aPD-L1) and PTX to enhance immune checkpoint blockade efficacy through PTX-induced immunogenic cell death. These micelles efficiently crossed the blood-tumour barrier and were retained in the tumour microenvironment, preserving the bioactivity of aPD-L1. The combined delivery of aPD-L1 and PTX improved immune responses, suppressed primary and recurrent glioblastoma, and induced long-lasting immunological memory in mouse models. This co-encapsulation approach offered a promising strategy for chemo-immunotherapy, potentially overcoming the limitations of antibody delivery and immunosuppressive conditions in glioblastoma. The positive outcomes of this study suggest that this nano-delivery system could enhance the effectiveness of immune checkpoint blockade therapies and may be applicable to other immunosuppressive cancer [58].

2.7 Dual/Multi-responsive Micelles

Stimuli-sensitive micelles that respond to various biological stimuli, such as upregulated enzymes, altered redox potential, acidic pH, along with outside artificial stimuli like light, magnetic fields, temperature, and ultrasound, are considered "smart" nanocarriers/multi-responsive micelles [44]. These multi-responsive micelles are used for delivering anticancer drugs and imaging agents for various therapeutic and diagnostic applications [59]. Incorporating a variety of endogenous and external stimuli can enhance the design of multi-responsive nanomedicines. Theoretically, incorporating multiple stimuli into these nano-micelles enables more precise and specific controlled drug release [60]. As multi-responsive micelles can achieve higher specificity in delivering therapeutic drugs compared to mono-responsive nanomedicines, leading to more effective and safer cancer treatments [61]. Multi-stimuli responsive systems represent a promising advancement, offering controllable, smart delivery systems with precise drug release and detection capabilities. These advanced multi-responsive offer new strategies for enhancing current nanoparticulate drug formulations, addressing aspects such as preparation and stability, intra-tumoural retention and penetration, and overall efficacy. However, designing these complex systems poses greater challenges compared to single-stimulus sensitive systems [62].

In a recent study by Maghsoudian and the group, a multi-responsive micellar system, which is both pH and redox responsive, has been reported. In this particular study, researchers have developed nano micelles for the responsive delivery of DOX, consisting of a zwitterionic polymer of methyl methacryloyl phosphoryl choline and polycaprolactone with a gold nanoparticle core sensitive to pH and GSH. These nano micelles have a lower CMC than conventional ones, ensuring high efficacy and precision in targeting drugs to the intended site. Designed to combine chemotherapy and sonodynamic therapy, the nano micelles incorporate the TMT peptide, significantly enhancing cellular uptake in highly metastatic cells through receptor-mediated endocytosis. A critical challenge with injectable micellar carriers is their dismantling into free surfactants in vivo, leading to dilution below the CMC. To address this, researchers developed copolymer micelles with a superhydrophilic zwitterionic block and a superhydrophobic block featuring a disulfide bond. These copolymer micelles, with 2-methacryloyloxyethyl phosphorylcholine (MPC) and polycaprolactone (PCL) as the shell and gold(Au) nanoparticles as the core, were designed to deliver DOX to tumour cells while minimizing the side effects of the free cytotoxic agent. The addition of gold nanoparticles to the micelle core greatly improves their stability in the bloodstream. These nano micelles demonstrated excellent biocompatibility and responsiveness to stimuli, releasing drugs in response to GSH and pH. Notably, the PCL-SS-MPC nano micelles exhibited high DOX release at 10 mM GSH and pH 5. Further results of qualitative testing of these nano micelles confirmed that the AuS-PM-TMTM-DOX micelles successfully induced apoptosis and improved cellular uptake in MCF7 and MDA-MB-231 breast cancer cell lines. The cytotoxic effects of AuS-PM-DOX combined with ultrasound were approximately 38% higher than AuS-PM-DOX alone at an IC50 concentration of 0.68 nM, primarily due to enhanced apoptosis. The disulfide linkages in AuSNPs augmented ROS production upon ultrasound stimulation, due to reduced GSH levels. Compared to other nanosensitizers like titanium dioxide, AuS-PM exposed to ultrasound radiation (1.0 W/cm, 2 min) substantially increased cavitation effects and produced 3–5 times higher ROS levels [63].

The study details the core-cross-linked poly(thiolether ester) micelles, which were designed for targeted drug delivery to cancer cells. These micelles were sensitive to ROS, glutathione, and pH for intracellular drug delivery. The poly(thiolether ester) was synthesized using ethanedithiol and glycidylmethacrylate through thiol-ene/thiol-epoxy polymerization, which resulted in an abundant hydroxyl group in side chains. They were able to cross-link into nano micelles in specific conditions, leading to enhanced drug loading efficiency. Later on, these nano micelles exhibited accelerated drug release in environments typical of cancer cells, such as high levels of ROS, acidity, and reducing conditions. Cellular assays confirmed the efficacy of the developed multi-responsive micelles in targeting and releasing the drugs, making this innovative approach a promising candidate for smart drug delivery in cancer therapy [64].

A research study entails about injectable multi-responsive micelle hybrid hydrogel to tackle both skin cancer and bacterial infections. This pH-responsive hydrogel continuously releases encapsulated glucose oxidase, promoting intratumoural glucose consumption, hydrogen peroxide production, and an acidic environment. Additionally, the hydrogel facilitates the rapid release of catalytic iron ions, which catalyze hydrogen peroxide decomposition into toxic radicals, inducing cell death. When this system was combined with spatiotemporal photothermal therapy, it enhanced therapeutic efficacy synergistically. In vitro and in vivo studies demonstrated remarkable efficacy in skin cancer treatment (98.8% and 97.6% suppression, respectively) and bacterial eradication (≥99%). This pioneering approach integrates bioenzyme, pH and photothermal enhancement for simultaneous cancer therapy and anti-infection, offering promising prospects for clinical therapeutics [65].

The research work introduces a new approach to creating a multi-responsive micelle, which is a hybrid-brush polymer obtained through the aminolysis of poly(succinimide) with various amino-containing compounds. This new polymer demonstrates responsiveness to changes in pH, temperature, or both, and self-assembles into micelles in aqueous solutions, making it a potential drug delivery platform. When this nanosystem is loaded with DOX, these multi-responsive micelles remain stable under normal physiological conditions and start releasing the drug in acidic or reducing environments, mimicking conditions found in cancer tissues. In vitro studies show accelerated DOX release at lower pH and higher reductant concentrations due to the combined effects of protonation and disulfide cleavage. In vitro cell line studies revealed low toxicity for blank micelles and effective anticancer activity for DOX-loaded micelles against HeLa cells, similar to free DOX. This study marks the first reported development of such a multi-responsive micellar drug delivery system, suggesting its promise for delivering hydrophobic anti-tumour drugs [66]. Some of the studies describing various stimuli-responsive micellar systems for cancer therapy are given in Table 2.1.

Table 2.1 Composition of different responsive micelles and their therapeutic outcomes

Composition of responsive micelles	Drug	Outcomes	References
Methoxy-poly(ethylene glycol)-block-poly(α-carboxyl-ε-caprolactone) (mPEG45-b-PCCL15@ Fe3O4)	PTX	Biodegradable anionic copolymer with superparamagnetic behaviour Reduced burst release pH-responsive release pattern	[67]
Poly-Dox (light-responsive DOX-conjugated polymer tethered to PEG)	DOX	Amide bond cleavage upon UV radiation, enhanced cellular uptake of Dox; optimal illumination duration for maximal effect around 5 min	[39]
PBA-PEG-Azo-PCL/ mPEG-ss-CPT	CPT	Dual targeting polymer micelles with redox and enzyme-responsive linkers; rapid intracellular release; 80% CPT release within 48 h	[68]
Hyaluronic acid derivative	PTX	At high GSH levels, micelles released the active	[69]
Polyethylene glycol segment and, photo-responsive hydrophobic donor–acceptor Stenhouse adduct	DOX	NIR light-responsive, micelles disassemble after NIR exposure, 83.7% drug release within 30 min after 5 min of NIR irradiation, production of NO inhibits P-gp	[70]
RGD-PEG-ss-PCL	DOC/ ICG	RGD acts as a ligand; efficiency of anticancer drugs was high in in vivo study	[71]
PEO113-b-PCL35-b-PEO113 and PAA13-b-PCL35-b-PAA13	CAPE	Cargo release due of disulfide-group destruction	[72]
HA-ss-FA	DOX	20 mM GSH resulted in fast DOX release, the ligand was CD44 Prominent in vivo tumour targetability observed	[73]
Methoxy poly(ethylene oxide)-b-poly(aspartate-hydrazide)	DOX	Hydrolysis of hydrazone bond and delayed drug release	[74]
Gal-PEEP-a-PCL-ss-PDMAEMA	DOX	Redox-triggered drug release observed	[75]
$MnFe_2O_4$, PCL-b-p(N-co-D)	DOX	High DOX (9.1%) and MnFe2O4 (38.7%) contents; superparamagnetism; magnetothermal-dependent drug release; enhanced DOX endocytosis and cytotoxicity; 84% tumour growth inhibition in vivo	[76]
HA-ss-BF	DOX	Micelles were observed to disassemble in redox condition	[77]
Dimethylmaleic anhydride, pluronic P85, triphenylphosphonium	PTX	At high-level GSH, fast PTX release was found	[78]

(continued)

Table 2.1 (continued)

Composition of responsive micelles	Drug	Outcomes	References
Folic acid, dextran	PTX	Folic acid as a ligand; high uptake and efficient anti-tumour effect for MDR was reported	[79]
HA-ss-TOS	PTX	Used for breast cancer and melanoma treatment	[80]
Phenylboronic acid, Fe_3O_4	DOX	Enhanced tumour site accumulation with external magnetic field; improved contrast for distribution evaluation; selective recognition by sialic acid-positive tumour cells; 83% tumour growth inhibition in H22 hepatocarcinoma model	[81]

Abbreviations: *Poly-Dox* Doxorubicin-conjugated polymer, *CPT* Camptothecin, PBA-*PEG-Azo-PCL/mPEG-ss-CPT* phenylboronic acid poly(ethylene glycol)-4,4′-(diazene-1,2-diyl)benzoyl-poly(ε-caprolactone)/monomethylpoly(ethylene-glycol)-ss-camptothecin, *NIR* Near infrared light, *RGD-PEG-ss-PCL* RGD peptide (arginine-glycine-aspartic acid), polyethylene glycol (PEG), polycaprolactone (PCL) micelle, *CAPE* caffeic acid phenyl ester, *HA-ss-FA* hyaluronic acid conjugation with folid acid, *Gal-PEEP-a-PCL-ss-PDMAEMA* poly(ethylethylene phosphate)-a-poly(ε-caprolactone)-ss-poly[2-(dimethylamino)ethyl methacrylate], *PCL-b-p(N-co-D)* polycaprolactone-b-poly(N-isopropylacrylamide-co-N,N-dimethylacrylamide), *$MnFe_2O_4$* manganese ferrite, *HA-ss-BF* hyaluronic acid-ibuprofen prodrug, *PTX* paclitaxel, *HA-ss-TOS* vitamin E succinate conjugated hyaluronic acid

3 Concluding Remarks and Perspectives

Cancer remains to haunt the mankind with existing treatment modalities producing heavy side effects and providing low survival rates. Innovative treatment strategies have been devised in the past decades and research in this area is only growing. Nanotechnology presents a promising approach in the form of nanoparticles that can be tailored to produce spatiotemporal and targeted drug delivery so as to eradicate the tumour without systemic toxicity to healthy tissues. Amongst the various nanoparticles, micelles have evolved as a promising drug delivery system that can incorporate both hydrophilic and hydrophobic drugs. In addition to this, stimuli-responsive micelles in particular provide targeted drug delivery specifically to the tumour site. Various types of stimuli-responsive micelles have emerged including pH responsive, thermo responsive, magnetically responsive, redox responsive, light responsive and so on. To enhance therapeutic efficacy, the drug release can be triggered by external stimuli such as light or sound, or by internal microenvironmental factors like pH or enzymatic activity, depending on the drug's pharmacodynamic properties. Such systems utilize either an endogenous stimuli that is characteristic hallmark of tumour tissue or exogenous stimuli that can guide the release of drugs at the target tumour site. Despite these strides that such systems have made and shown promising potential at the preclinical stage, the picture at the clinical scale is disappointing. Nevertheless, promising preclinical studies on the responsive micelles have led to several formulations now being tested in clinical trials for various cancer treatments, showing encouraging outcomes. Poly(ethyleneglycol)-block-poly(aspartic acid) loaded with DOX (NK911) or with PTX (NK105) and poly(ethylene glycol)-block-poly(glutamic acid) entrapped with cisplatin (NC-6004), SN-38 (NK012) or oxaliplatin (NC-4016) are the major examples of formulations that are under clinical trials. This proves that stimuli-responsive micelles can potentially contribute to the way cancer will be managed in the near future. Furthermore, a comprehensive understanding about the nano formulation, their dose, ways for its scale up and its clinical fate in vivo will help bridge the gap.

References

1. Brody H, Hodson R, Beckett J, Haggart A. Cancer treatment. 2024;629(May):2024.
2. Irvine DJ, Dane EL. Enhancing cancer immunotherapy with nanomedicine. Nat Rev Immunol. 2020;20:321–34.
3. Yao Y, Zhou Y, Liu L, Xu Y, Chen Q, Wang Y, et al. Nanoparticle-based drug delivery in cancer therapy and its role in overcoming drug resistance. Front Mol Biosci. 2020;7
4. Niculescu AG, Grumezescu AM. Novel tumor-targeting nanoparticles for cancer treatment— a review. Int J Mol Sci. 2022;23
5. Xu Y, Wu H, Huang J, Qian W, Martinson DE, Ji B, et al. Probing and enhancing ligand-mediated active targeting of tumors using sub-5 nm ultrafine iron oxide nanoparticles. Theranostics. 2020;10(6):2479–94.
6. Li Z, Song N, Yang YW. Stimuli-responsive drug-delivery systems based on supramolecular nanovalves. Matter. 2019;1:345–68.
7. Mura S, Nicolas J, Couvreur P. Stimuli-responsive nanocarriers for drug delivery. Nat Mater. 2013;12:991–1003.
8. Bose A, Burman DR, Sikdar B, Patra P. Nanomicelles: types, properties and applications in drug delivery. IET Nanobiotechnol. 2021;15:19–27.
9. Miller T, Breyer S, Van Colen G, Mier W, Haberkorn U, Geissler S, et al. Premature drug release of polymeric micelles and its effects on tumor targeting. Int J Pharm. 2013;445(1–2):117–24.
10. Nakayama M, Akimoto J, Okano T. Polymeric micelles with stimuli-triggering systems for advanced cancer drug targeting. J Drug Target. 2014;22:584–99.
11. Zheng Y, Oz Y, Gu Y, Ahamad N, Shariati K, Chevalier J, et al. Rational design of polymeric micelles for targeted therapeutic delivery. Nano Today. 2024;55:102147.
12. Kanamala M, Wilson WR, Yang M, Palmer BD, Wu Z. Mechanisms and biomaterials in pH-responsive tumour targeted drug delivery: a review, vol. 85. Biomaterials; 2016. p. 152–67.
13. Wu H, Zhu L, Torchilin VP. PH-sensitive poly(histidine)-PEG/DSPE-PEG co-polymer micelles for cytosolic drug delivery. Biomaterials. 2013;34(4):1213–22.
14. Hu J, Miura S, Na K, Bae YH. PH-responsive and charge shielded cationic micelle of poly(l-histidine)- block-short branched PEI for acidic cancer treatment. J Control Release. 2013;172(1):69–76.
15. Zhou Q, Hou Y, Zhang L, Wang J, Qiao Y, Guo S, et al. Dual-pH sensitive charge-reversal nanocomplex for tumor-targeted drug delivery with enhanced anticancer activity. Theranostics. 2017;7(7):1806–19.
16. Yang T, Li F, Zhang H, Fan L, Qiao Y, Tan G, et al. Multifunctional pH-sensitive micelles for tumor-specific uptake and cellular delivery. Polym Chem. 2015;6(8):1373–82.
17. Zhou Z, Li L, Yang Y, Xu X, Huang Y. Tumor targeting by pH-sensitive, biodegradable, cross-linked N-(2-hydroxypropyl) methacrylamide copolymer micelles. Biomaterials. 2014;35(24):6622–35.
18. Bian F, Xiang M, Yu W, Liu M. Preparation and self-assembly behavior of thermosensitive polymeric micelles comprising poly(styrene-b-N,N-diethylacrylamide). J Appl Polym Sci. 2008;110(2):900–7.

19. Yang J, Zhang D, Jiang S, Yang J, Nie J. Synthesis of Y-shaped poly(solketal acrylate)-containing block copolymers and study on the thermoresponsive behavior for micellar aggregates. J Colloid Interface Sci. 2010;352(2):405–14.
20. Rijcken CJ, Snel CJ, Schiffelers RM, van Nostrum CF, Hennink WE. Hydrolysable core-crosslinked thermosensitive polymeric micelles: synthesis, characterisation and in vivo studies. Biomaterials. 2007;28(36):5581–93.
21. Ghasemi S, Ahmadi L, Farjadian F. Thermo-responsive PNIPAAm-b-PLA amphiphilic block copolymer micelle as nanoplatform for docetaxel drug release. J Mater Sci. 2022;57(36):17433–47.
22. Long H, Tian W, Jiang S, Zhao J, Zhou J, He Q, et al. A dual drug delivery platform based on thermo-responsive polymeric micelle capped mesoporous silica nanoparticles for cancer therapy. Microporous Mesoporous Mater. 2022;338:111943.
23. Soltantabar P, Calubaquib EL, Mostafavi E, Biewer MC, Stefan MC. Enhancement of loading efficiency by coloading of doxorubicin and quercetin in thermoresponsive polymeric micelles. Biomacromolecules. 2020;21(4):1427–36.
24. Bhadran A, Polara H, Calubaquib EL, Wang H, Babanyinah GK, Shah T, et al. Reversible cross-linked thermoresponsive polycaprolactone micelles for enhanced stability and controlled release. Biomacromolecules. 2023;24(12):5823–35.
25. Bagheri-Meyabad M, Motasadizadeh H, Norouzi P, Fatahi Y, Asadi H, Varshochian R, et al. Self-assembled NIPAM–PEG–NIPAM polymeric nanomicelles for the delivery of zinc protoporphyrin: a potential stimuli-triggered cancer treatment approach. J Mater Sci. 2024;59(7):3049–65.
26. Yang X, Chen Y, Yuan R, Chen G, Blanco E, Gao J, et al. Folate-encoded and Fe3O4-loaded polymeric micelles for dual targeting of cancer cells. Polymer. 2008;49(16):3477–85.
27. Jaganathan H, Gieseck RL, Ivanisevic A. Characterizing proton relaxation times for metallic and magnetic layer-by-layer-coated, DNA-templated nanoparticle chains. Nanotechnology. 2010;21(24):245103.
28. Xu Z, Hou Y, Sun S. Magnetic core/shell Fe3O4/Au and Fe3O 4/Au/Ag nanoparticles with tunable plasmonic properties. J Am Chem Soc. 2007;129(28):8698–9.
29. Gao Y, Zhao J, Zhang X, Wei X, Xiong X, Guo X, et al. A rod bacterium-like magnetic polymer micelle for strongly enhancing selective accumulation and internalization of nanocarriers. J Mater Chem B. 2017;5(25):4943–54.
30. Huang C, Tang Z, Zhou Y, Zhou X, Jin Y, Li D, et al. Magnetic micelles as a potential platform for dual targeted drug delivery in cancer therapy. Int J Pharm. 2012;429(1–2):113–22.
31. Ahmadi Kamalabadi M, Neshastehriz A, Ghaznavi H, Amini SM. Folate functionalized gold-coated magnetic nanoparticles effect in combined electroporation and radiation treatment of HPV-positive oropharyngeal cancer. Med Oncol. 2022;39(12)
32. Tang Z, Zhang L, Wang Y, Li D, Zhong Z, Zhou S. Redox-responsive star-shaped magnetic micelles with active-targeted and magnetic-guided functions for cancer therapy. Acta Biomater. 2016;42:232–46.
33. Zhu J, Yang Y, Wang J, Hong W, Li Y, Wang Z, et al. Dual responsive magnetic drug delivery nanomicelles with tumor targeting for enhanced cancer chemo/magnetothermal synergistic therapy. Int J Nanomedicine. 2023;18:7647–60.
34. Shanmugam V, Selvakumar S, Yeh CS. Near-infrared light-responsive nanomaterials in cancer therapeutics. Chem Soc Rev. 2014;43:6254–87.
35. Jochum FD, Theato P. Temperature- and light-responsive smart polymer materials. Chem Soc Rev. 2013;42(17):7468–83.
36. Zhao Y. Rational design of light-controllable polymer micelles. Chem Record. 2007;7:286–94.
37. Cui J, Del Campo A. Photo-responsive polymers: properties, synthesis and applications. In: Smart polymers and their applications; 2014. p. 93–133.
38. Saravanakumar G, Lee J, Kim J, Kim WJ. Visible light-induced singlet oxygen-mediated intracellular disassembly of polymeric micelles co-loaded with a photosensitizer and an anticancer drug for enhanced photodynamic therapy. Chem Commun. 2015;51(49):9995–8.
39. Chen J, Qian C, Ren P, Yu H, Kong X, Huang C, et al. Light-responsive micelles loaded with doxorubicin for osteosarcoma suppression. Front Pharmacol. 2021;12

40. Sun L, Zhu B, Su Y, Dong CM. Light-responsive linear-dendritic amphiphiles and their nanomedicines for NIR-triggered drug release. Polym Chem. 2014;5(5):1605–13.
41. Kapalatiya H, Madav Y, Tambe VS, Wairkar S. Enzyme-responsive smart nanocarriers for targeted chemotherapy: an overview. Drug Deliv Transl Res. 2022;12:1293–305.
42. Shahriari M, Zahiri M, Abnous K, Taghdisi SM, Ramezani M, Alibolandi M. Enzyme responsive drug delivery systems in cancer treatment. J Control Release. 2019;308:172–89.
43. Cheng W, Gu L, Ren W, Liu Y. Stimuli-responsive polymers for anti-cancer drug delivery. Mater Sci Eng C. 2015;45:600–8.
44. Zhou Q, Zhang L, Yang TH, Wu H. Stimuli-responsive polymeric micelles for drug delivery and cancer therapy. Int J Nanomedicine. 2018;13:2921–42.
45. Kessenbrock K, Plaks V, Werb Z. Matrix metalloproteinases: regulators of the tumor microenvironment. Cell. 2010;141:52–67.
46. Egeblad M, Werb Z. New functions for the matrix metalloproteinases in cancer progression. Nat Rev Cancer. 2002;2:161–74.
47. Park JB, Lee CS, Jang JH, Ghim J, Kim YJ, You S, et al. Phospholipase signalling networks in cancer. Nat Rev Cancer. 2012;12:782–92.
48. Jhaveri AM, Torchilin VP. Multifunctional polymeric micelles for delivery of drugs and siRNA. APR Front Pharmacol. 2014;5
49. Barve A, Jain A, Liu H, Zhao Z, Cheng K. Enzyme-responsive polymeric micelles of cabazitaxel for prostate cancer targeted therapy. Acta Biomater. 2020;113:501–11.
50. Yan K, Zhang S, Zhang K, Miao Y, Qiu Y, Zhang P, et al. Enzyme-responsive polymeric micelles with fluorescence fabricated through aggregation-induced copolymer self-assembly for anticancer drug delivery. Polym Chem. 2020;11(48):7704–13.
51. Moghaddam FD, Zare EN, Hassanpour M, Bertani FR, Serajian A, Ziaei SF, et al. Chitosan-based nanosystems for cancer diagnosis and therapy: stimuli-responsive, immune response, and clinical studies. Carbohydr Polym. 2024;330:121839.
52. Maboudi AH, Lotfipour MH, Rasouli M, Azhdari MH, MacLoughlin R, Bekeschus S, et al. Micelle-based nanoparticles with stimuli-responsive properties for drug delivery. Nanotechnol Rev. 2024;13
53. Wang H, Li Y, Zhang M, Wu D, Shen Y, Tang G, et al. Redox-activatable ATP-depleting micelles with dual modulation characteristics for multidrug-resistant cancer therapy. Adv Healthc Mater. 2017;6(8)
54. Luo T, Han J, Zhao F, Pan X, Tian B, Ding X, et al. Redox-sensitive micelles based on retinoic acid modified chitosan conjugate for intracellular drug delivery and smart drug release in cancer therapy. Carbohydr Polym. 2019;215:8–19.
55. Chiang YT, Yen YW, Lo CL. Reactive oxygen species and glutathione dual redox-responsive micelles for selective cytotoxicity of cancer. Biomaterials. 2015;61:150–61.
56. Wang J, Yang G, Guo X, Tang Z, Zhong Z, Zhou S. Redox-responsive polyanhydride micelles for cancer therapy. Biomaterials. 2014;35(9):3080–90.
57. Yang Y, Shi W, Zhang Z, Gong F, Feng X, Guo C, et al. Redox-responsive nanomicelles with intracellular targeting and programmable drug release for targeted tumor therapy. Curr Drug Deliv. 2023;21(2):295–307.
58. Zhang Z, Xu X, Du J, Chen X, Xue Y, Zhang J, et al. Redox-responsive polymer micelles co-encapsulating immune checkpoint inhibitors and chemotherapeutic agents for glioblastoma therapy. Nat Commun. 2024;15(1)
59. Liu G, Lovell JF, Zhang L, Zhang Y. Stimulus-responsive nanomedicines for disease diagnosis and treatment. Int J Mol Sci. 2020;21:1–44.
60. Negut I, Bita B. Polymeric micellar systems—a special emphasis on “smart” drug delivery. Pharmaceutics. 2023;15
61. Zhang Y, Li J, Pu K. Recent advances in dual- and multi-responsive nanomedicines for precision cancer therapy. Biomaterials. 2022;291:121906.
62. Ding C, Tong L, Feng J, Fu J. Recent advances in stimuli-responsive release function drug delivery systems for tumor treatment. Molecules. 2016;21

63. Maghsoudian S, Motasadizadeh H, Farhadnejad H, Fatahi Y, Fathian Nasab MH, Mahdieh A, et al. Targeted pH- and redox-responsive AuS/micelles with low CMC for highly efficient sonodynamic therapy of metastatic breast cancer. Biomater Adv. 2024;158:158.
64. Hu Y, Deng M, Yang H, Chen L, Xiao C, Zhuang X, et al. Multi-responsive core-crosslinked poly (thiolether ester) micelles for smart drug delivery. Polymer. 2017;110:235–41.
65. Zheng H, Wang S, Zhou L, He X, Cheng Z, Cheng F, et al. Injectable multi-responsive micelle/nanocomposite hybrid hydrogel for bioenzyme and photothermal augmented chemodynamic therapy of skin cancer and bacterial infection. Chem Eng J. 2021;404:126439.
66. Zhang G, Yi H, Bao C. Stimuli-responsive poly(Aspartamide) derivatives and their applications as drug carriers. Int J Mol Sci. 2021;22
67. Liu Y, Li Y, Rao Z, Xu J, Zhao S, Zhao C, et al. Physiochemical properties and paclitaxel release behaviors of dual-stimuli responsive copolymer-magnetite superparamagnetic nanocomposites. Nanotechnology. 2019;30(10)
68. Zhang L, Wang Y, Zhang X, Wei X, Xiong X, Zhou S. Enzyme and redox dual-triggered intracellular release from actively targeted polymeric micelles. ACS Appl Mater Interfaces. 2017;9(4):3388–99.
69. Yang Y, Yun K, Li Y, Zhang L, Zhao W, Zhu Z, et al. Self-assembled multifunctional polymeric micelles for tumor-specific bioimaging and synergistic chemo-phototherapy of cancer. Int J Pharm. 2021;602:120651.
70. Zhang Y, Zhang X, Chen W, He Y, Liu Y, Ju H. Self-assembled micelle responsive to quick NIR light irradiation for fast drug release and highly efficient cancer therapy. J Control Release. 2021;336:469–79.
71. Ren L, Nie J, Wei J, Li Y, Yin J, Yang X, et al. RGD-targeted redox responsive nano micelle: co-loading docetaxel and indocyanine green to treat the tumor. Drug Deliv. 2021;28(1):2024–32.
72. Kamenova K, Grancharov G, Kortenova V, Petrov PD. Redox-responsive crosslinked mixed micelles for controllable release of caffeic acid phenethyl ester. Pharmaceutics. 2022;14(3)
73. Yang Y, Zhao Y, Lan J, Kang Y, Zhang T, Ding Y, et al. Reduction-sensitive CD44 receptor-targeted hyaluronic acid derivative micelles for doxorubicin delivery. Int J Nanomedicine. 2018;13:4361–78.
74. Gebrie HT, Addisu KD, Darge HF, Birhan YS, Thankachan D, Tsai HC, et al. pH/redox-responsive core cross-linked based prodrug micelle for enhancing micellar stability and controlling delivery of chemo drugs: an effective combination drug delivery platform for cancer therapy. Biomater Adv. 2022;139
75. Zhang Y, He J, Cao D, Zhang M, Ni P. Galactosylated reduction and pH dual-responsive triblock terpolymer Gal-PEEP-a-PCL-ss-PDMAEMA: a multifunctional carrier for the targeted and simultaneous delivery of doxorubicin and DNA. Polym Chem. 2014;5(17):5124–38.
76. Wang Y, Wang R, Chen L, Chen L, Zheng Y, Xin Y, et al. Enhanced tumor penetration for efficient chemotherapy by a magnetothermally sensitive micelle combined with magnetic targeting and magnetic hyperthermia. Front Pharmacol. 2022;13
77. Chai Z, Teng C, Yang L, Ren L, Yuan Z, Xu S, et al. Doxorubicin delivered by redox-responsive Hyaluronic Acid–Ibuprofen prodrug micelles for treatment of metastatic breast cancer. Carbohydr Polym. 2020;245:116527.
78. Wang H, Shi W, Zeng D, Huang Q, Xie J, Wen H, et al. pH-activated, mitochondria-targeted, and redox-responsive delivery of paclitaxel nanomicelles to overcome drug resistance and suppress metastasis in lung cancer. J Nanobiotechnol. 2021;19(1)
79. Chen D, Ge S, Zuo L, Wang S, Liu M, Li S. Adjudin-loaded redox-sensitive paclitaxel-prodrug micelles for overcoming multidrug resistance with efficient targeted Colon cancer therapy. Drug Deliv. 2020;27(1):1094–105.
80. Du Y, Wang S, Zhang T, He D, Tu J, Shen Y. Enhanced cytotoxicity of a redox-sensitive hyaluronic acid-based nanomedicine toward different oncocytes via various internalization mechanisms. Drug Deliv. 2020;27(1):128–36.
81. Chen G, Zheng Q, Dai J, Liu J, Yin J, Xu X, et al. Reduction-sensitive mixed micelles based on mPEG-SS-PzLL /TPGS to enhance anticancer efficiency of doxorubicin. React Funct Polym. 2022;174:105242.

Stimuli-Responsive Polymeric Nanoparticles for Cancer Therapy

3

Dadi A. Srinivasarao and Nithin Vidiyala

Abstract

According to the World Health Organization (WHO), cancer is a leading cause of death worldwide. Despite the availability of various chemotherapeutic agents, delivery of such drugs to the tumor site in a spatio-temporally controlled manner is still an unmet need. Stimuli-responsive polymeric nanoparticulate delivery systems showed improved therapeutic efficacy as compared to conventional systems, due to their unique properties such as cancer tissue-specific targeting and/or stimuli-responsive drug release within the tumor microenvironment. These nanoparticles can respond to various endogenous and/or exogenous stimuli, thereby releasing entrapped drugs at the tumor site. This chapter discusses the characteristics of tumor microenvironment, such as alterations in oxygen concentration (hypoxia), pH, redox state, and oxidative stress of the cancer cells, that provide insights for stimuli-responsive targeted drug delivery. Further, this chapter mentions about widely explored exogenous stimuli, such as magnetic fields, ultrasound, light, and temperature, that are employed for targeted therapy. In addition, targeted drug delivery strategies that explored mono- or multi-stimuli responsive polymeric nanoparticulate drug delivery systems have been mentioned. In a nutshell, this chapter provides a deep understanding of cancer-targeted stimulus-responsive polymer nanoparticles that solve the challenges encountered in conventional chemotherapy.

D. A. Srinivasarao (✉)
Department of Pharmaceutics, National Institute of Pharmaceutical Education and Research (NIPER), Hyderabad, Telangana, India
e-mail: srinivasarao.ananda@niperhyd.ac.in

N. Vidiyala
Drug Product Design and Development, Cerevel Therapeutics, Cambridge, MA, USA

P. V. Devarajan et al. (eds.), *Advances in Stimuli-Responsive Nanosystems for Cancer Therapy*, https://doi.org/10.1007/978-981-95-2744-1_3

Keywords

Cancer · Nanoparticles · Stimuli-responsiveness · Tumor microenvironment · Controlled release drug delivery · Targeting

Abbreviations

AAc	Acrylic acid
AC	Alternating current
ATRAM	Acidity-triggered rational membrane
BA	Betulinic acid
BDD	1,4-butanediol diacrylate
BSA	Bovine serum albumin
CAR	Chimeric antigen receptor
CRAIT	Click reaction-assisted immune cell targeting
DCs	Dendritic cells
DiR	1,1′-dioctadecyl-3,3,3′,3′-tetramethylindotricarbocyanine iodide
DOX	Doxorubicin
DTT	Dithiothreitol
ECM	Extracellular matrix
EPR	Enhanced permeation and retention
FUS	Focused ultrasound
HA	Hyaluronic acid
HIFU	High intensity focused ultrasound
HWH	Hot-water hyperthermia
IFN-γ	Interferon gamma
IL	Interleukin
LCST	Lower critical solution temperature
LFAMF	Low-frequency alternating magnetic field
MFH	Magnetic fluid hyperthermia
MFs	Multifunctional ferrofluids
MMPs	Matrix metalloproteinases
MRI	Magnetic resonance imaging
NIR	Near-infrared
NK cells	Natural killer cells
PATK	Poly(amino thioketal)
PBTE	Poly(β-thioester)
PEG	Polyethylene glycol
PGED-PPS	Poly(glycidyl methacrylate)-polyacrylamide sulfide
PLA	Polylactic acid
PLGA	Polylactic-co-glycolic acid
PMVE	Poly(methyl vinyl ether)
PNIPAAm	Poly(N-isopropyl acrylamide)
PNVCL	Poly(vinyl caprolactam)

POxs	Poly(2-oxazoline)s
PPADT	Poly(1,4-phenylacetone dimethylene thioacetone)
PTKU	Polythioacetone ethyl carbamate
PTX	Paclitaxel
PU	Polyurethane
PVAX	Poly(vanillanol-co-oxalate)
ROS	Reactive oxygen species
TAMs	Tumor-associated macrophages
TCO	Trans-cyclooctene
TIMPs	Tissue inhibitors of metalloproteases
TIR	Tumor inhibition rate
TME	Tumor microenvironment
TPGS	D-α-tocopheryl polyethylene glycol 1000 succinate
TPP	Triphenylphosphine
UCST	Upper critical solution temperature
USFDA	United States Food and Drug Administration
UV	Ultraviolet

1 Introduction

Cancer is a dreadful disease characterized by uncontrolled cell proliferation and a consequential increase in the tissue mass. According to the Global Cancer Observatory, cancer is the leading cause of death worldwide, accounting for approximately ten million deaths in the year 2020 [1, 2]. The most commonly occurring cancers include breast, lung, colon, rectum, and prostate cancers [2]. The cancerous cells differ from healthy cells in biochemical pathways, thereby creating a different physiological microenvironment in the vicinity of the tumor. Such a microenvironment is known as tumor microenvironment (TME). The TME is a complex and heterogeneous milieu characterized by varying physical, chemical, and biological characteristics as compared to the healthy tissue microenvironment [3–5]. Reports reveal that uncontrolled cell proliferation and enhanced extracellular matrix (ECM) deposition in tumors imparts stiffness to the tumor tissue. As a result, blood and lymphatic vessels get compressed, ultimately causing hypoxia and impairment in lymphatic drainage [6]. Further, temperature, pH, partial pressure of oxygen, concentration of reactive oxygen species (ROS), and redox state of the cancer cells differ from the healthy cells that promote cancer progression, metastasis, and immune suppression. Metastasis involves the dislodgement of cancer cells from the initial site of cancer progression and settling at a distant site that favors cancer cell adhesion and survival. This process is governed by various chemoattractants, chemokines, and cancer cell receptors [7]. The resultant inhabited cells promote tissue remodeling at the metastatic site by secreting cytokines, growth factors, and angiogenic molecules, ultimately resulting in tumor cell colonization and proliferation [8]. Such metastatic tumors pose numerous challenges for the elimination of cancerous cells from the body. Conventional therapeutic interventions for the treatment of cancer majorly involve (i) surgical methods, (ii) radiation therapy, (iii) chemotherapy, and (iv) immunotherapy. Surgical methods are majorly based on the resection of the tumor tissue via surgical means. Radiation therapy involves the delivery of intense radiation at the site of the tumor to cause cell death. These two therapies showed beneficial results in eradicating benign tumors [9, 10]. Apart from these, other therapies, namely immunotherapy and chemotherapy, are also used for the eradication of tumors [11, 12].

Immunotherapy involves harnessing the body's immune system to elicit immune responses against the cancer cells, which results in the elimination of pathogenic tumor cells. These immune responses are mediated by a variety of immune cells, such as T lymphocytes, B lymphocytes, natural killer (NK) cells, tumor-associated macrophages (TAMs), myeloid-derived suppressor cells, dendritic cells (DCs), and tumor-associated neutrophils [13]. Literature reveals that T lymphocytes are known to play a pivotal role in tumor regression by mediating cytotoxic effects toward tumor cells. In this context, T cells can be classified into two types based on the type of cell surface proteins present on them. These include CD8+ (killer T cells) and CD4+ (helper T cells) [14]. CD8+ T cells are able to kill tumor cells and, therefore, help to suppress tumor progression. These cells sometimes act in coordination with CD4+ T helper 1 (TH1) cells and cause cytotoxic effects via the mediation of interleukin-2 (IL-2) and interferon-gamma (IFN-γ) [15]. Whereas CD4+ T cells (TH2 cells) are known to support B lymphocyte responses and promote tumor growth via the mediation of IL-4, IL-5, and IL-13 production [15]. T-cell-based cancer therapeutics showed encouraging results in pre-clinical studies for the treatment of various cancers. These therapies elicited beneficial immune responses by activating T-cells, thereby promoting (i) modification of TME to elicit strong immune responses against tumor cells, (ii) reduction of metastasis, and (iii) prevention of recurrence/relapse. A successful T-cell-based therapy approved by the United States Food and Drug Administration (USFDA) is chimeric antigen receptor (CAR) T-cell therapy. CAR T cell therapy is an immunotherapy that utilizes genetically altered immune cells (T cells) that can destroy cancer cells efficiently.

Chemotherapy involves the administration of cytotoxic chemotherapeutic agents, such as paclitaxel, doxorubicin, cisplatin, docetaxel, carboplatin, etoposide, gemcitabine, vinorelbine, camptothecin, podophyllotoxin, and cabazitaxel, to cause cancer cell death. A major drawback of chemotherapy is the lack of drug specificity toward tumor cells, thereby causing toxic effects, not only toward the tumor cells but also healthy cells. Due to this, healthy tissues having constantly proliferating cells (such as hair follicles, bone marrow, and gastrointestinal tract) are adversely affected, which results in severe tissue toxicity [16–18]. Therefore, there is a need for the development of suitable drug delivery strategies that enable more selectivity toward tumor cells while sparing healthy cells/tissues.

The evolution of polymer chemistry has revolutionized the field of drug delivery by offering controlled drug release at the site of interest. Biocompatible, biodegradable, and stimuli-responsive polymers showed improved therapeutic benefits by enabling drug release in a spatio-temporally controlled manner. Stimuli-responsive polymers consist of repeated monomeric units that undergo accelerated degradation when they come in contact with either physical, chemical, biological, or pathological stimuli. Such degradation helps to release entrapped therapeutic agents at the specific part of the body, thereby providing therapeutic drug concentration at the site of interest while avoiding drug release at off-target sites. As a consequence, undesirable tissue toxicity can be minimized. Therefore, these delivery systems are known as smart, intelligent, or on-demand drug delivery systems [19–22]. Such delivery systems outperformed during the treatment of benign and malignant cancers by offering improved specificity and selective toxicity [23]. Since uncontrolled cell proliferation during cancer can lead to malignancy, it is essential to design controlled-release drug delivery systems that respond to specific pathological stimuli or even external stimuli so as to release the chemotherapeutic agents at the site of interest.

In addition to the composition of the delivery system, particle size also plays an important role in the delivery of chemotherapeutic agents at the tumor site. Reports demonstrated that nanoparticles, due to their small size and high surface area-to-volume ratio, allow better interaction and subsequent internalization by mammalian cells. This can potentially enhance drug penetration across the cellular and extracellular barriers of TME [24]. Therefore, nanoparticulate drug delivery systems having a size <100 nm showed enhanced permeability and passive accumulation at the tumor site through tumor's leaky blood vasculature. Further, the accumulated nanoparticles also showed improved tumor retention due to impaired lymphatic drainage. This effect is termed the enhanced permeation and retention effect (EPR effect) [25]. Various nanoparticles, such as micelles, dendrimers, cyclodextrin-based drug conjugates, and polymeric nanoparticles, have been explored thus far for the delivery of anticancer agents so as to achieve the EPR effect. Since the EPR effect allows tumor accumulation, nanoparticles composed of stimuli-responsive polymers may improve drug delivery, more precisely at the tumor site, thereby enabling enhanced therapeutic effects. The following section elaborates on various stimuli characteristic of TME (internal stimuli) and other external stimuli that are explored for the delivery of anticancer drugs at the tumor site.

2 Characteristics of Tumor Microenvironment and Stimuli-Responsive Controlled Release Drug Delivery Strategies

The tumor microenvironment is characterized by hypoxia, lower pH gradient, altered redox environment, excessive production of ROS, high interstitial pressure, altered tissue temperature, stiffened ECM, high protease activity, higher expression of drug efflux transporters, overexpression of specific receptors, alterations in the cytokine or growth factor secretion, variations in immune cell infiltration, etc. [26]. Researchers explored these pathological conditions to control drug delivery at the tumor site. This section elaborates on various endogenous (internal) and exogenous (external) stimuli and nanoparticulate drug delivery systems that respond to respective stimuli so as to offer controlled drug delivery at the tumor site.

2.1 Endogenous Stimuli-Responsive Drug Delivery Strategies

2.1.1 Hypoxia and Hypoxia-Responsive Polymeric Nanoparticles

Hypoxia is characterized by a reduction in the tissue oxygen concentration up to <2%, corresponding to its partial pressure of ~10 mmHg [27] as compared to 3.0–7.4% oxygen concentration (corresponding to its partial pressure between 40 and 60 mmHg) in the healthy tissues. Hypoxia is a characteristic factor of TME that was well explored by the researchers. Since tumor cells show high proliferation and increased tissue mass, they experience reduced blood supply, and as a consequence, a decrease in oxygen concentration can be observed. The consequential hypoxia is known to promote enhanced cell proliferation, angiogenesis, invasion, or metastasis. In this regard, researchers explored hypoxia-responsive polymeric materials to control drug delivery at the tumor site. In a study, Mamnoon et al. developed doxorubicin-loaded hypoxia-responsive polylactic acid (PLA_{8500})—diazobenzene—polyethylene glycol (PEG_{2000}) based polymeric nanoparticles (polymersomes) for targeted drug delivery applications. In this study, the authors conjugated estradiol moiety on the surface of nanoparticles so as to achieve better targetability toward estrogen receptor-positive breast cancer cells. The developed polymerosomes, when exposed to hypoxic conditions (2% oxygen), showed ~90% release of doxorubicin in a sustained manner for 12 h. Whereas doxorubicin release at normoxic conditions (21% oxygen) was observed to be ~30% at the end of 12 h, indicating hypoxia responsiveness of the polymerosomes. Such responsiveness can be attributed to the reduction of diazobenzene moiety when exposed to hypoxic conditions. Further, estradiol-conjugated polymerosomes showed improved cellular uptake in MCF7 cells, as these cells overexpress the estrogen receptors. As a consequence, the polymerosomes showed improved cell death against MCF7 cells when incubated in 2D cell culture plates and 3D spheroids under hypoxic conditions. Taken together, these results demonstrated that hypoxia-responsive and receptor-targeted polymerosomes showed improved anticancer activity [28].

2.1.2 pH of TME and pH-Responsive Polymeric Nanoparticles

The pH of the TME differs from that of a healthy tissue microenvironment due to various factors, such as hypoxia and anaerobic metabolism (via glycolysis), that cause accumulation of lactate and decrease tumor pH as low as 5.6–6.8) [29]. In addition to these, the presence of $Na^+/HCO3^-$ cotransporters, carbonic anhydrase, Na^+/H^+ exchangers, ATPases, and monocarboxylate transporters also shifts the tumor pH toward acidic scale [30]. The resultant pH causes alterations in cell fate processes and ultimately leads to uncontrolled cell proliferation, metastasis, and tumor progression. These pathological events can be controlled if the delivery system is able to release entrapped therapeutic agents at the tumor site. Therefore, pH-responsive polymers have been explored by the researchers to control drug release at the site of the tumor. pH-responsive polymers typically contain ionizable weak bases or weak acids that undergo protonation or deprotonation and become soluble in aqueous biological fluids. The commonly used functional groups include amine, aconitic acid, Schiff base, etc. Poly(amidoamine), a weak base, is routinely used for pH-responsive drug delivery applications. Since amine functional groups readily get protonated at the acidic pH of TME, they can transform their charge and acquire a positive charge. Whereas, aconitic acid and Schiff base are susceptible to hydrolysis at lower pH that results in the generation of amine groups, which subsequently get protonated and transform their charge. Several reports revealed that such aconitic acid-, amine-, and Schiff-base-based polymers showed controlled drug release at TME [26, 31–34]. Recent developments in the design of pH-responsive nanoparticles include the exploration of pH-responsive peptide molecules that can respond to the acidic pH of TME. In a study, Palanikumar et al. developed acidity-triggered rational membrane (ATRAM) peptide-functionalized polylactic-co-glycolic acid (PLGA)—bovine serum albumin (BSA) core-shell nanoparticles (of ~130 nm size) for targeted drug delivery applications. In this study, the authors used mitochondria-targeting doxorubicin-triphenylphosphine (DOX-TPP) as a therapeutic agent. DOX-TPP-loaded BSA-grafted PLGA nanoparticles were fabricated and functionalized (on the nanoparticle surface) with pH-responsive ATRAM peptide. In vitro cellular uptake studies demonstrated improved internalization (>five-fold) of PLGA-BSA-ATRAM nanoparticles at acidic pH (pH 6.5) as compared to physiological pH (pH 7.4) in MCF-7 cells. Further, the resultant drug-loaded nanoparticles showed improved toxicity toward different human cancer cell lines such as HeLa, MCF-7, and pancreatic carcinoma MIA PaCa-2. Moreover, reduced uptake of PLGA-BSA-ATRAM nanoparticles by immune cells (human monocytic leukemia THP-1 cells) was reported. This study indicates that the fabricated nanoparticles may evade immune clearance, and as a consequence, tumor accumulation can be improved. Further in vivo efficacy studies demonstrated that the developed PLGA-BSA-ATRAM nanoparticles extended in vivo circulation time of the therapeutic agent (DOX-TPP) and effectively reduced tumor mass and volume in 4 T1 cells implanted breast cancer mouse model. In addition, the treatment of mice with the developed nanoparticles not only improved survival time but also reduced nanoparticle toxicity toward healthy tissues. Taken together, these studies demonstrated the potential of pH-responsive polymeric nanoparticles for targeted drug delivery applications in breast cancer therapy [35].

2.1.3 Redox State of the Cancer Cells and Redox-Responsive Polymeric Nanoparticles

Various metabolic pathways cause the generation of excessive ROS in tumor cells. As a consequence, upregulation of various antioxidant enzyme systems takes place, which helps to counteract the effects of ROS. Therefore, tumor cells exhibit a reductive environment intracellularly. Literature reveals that the upregulation of thioredoxins, peroxiredoxins [36, 37], and glutathione [38] takes place in cancer that results in the generation of reductive stress [39, 40]. Among these, glutathione plays a predominant role in making the intracellular microenvironment in a reductive state. Glutathione is a tripeptide that contains a thiol (-SH) group. Glutathione levels inside the normal healthy cells may range from 1 to 10 milliMolar, whereas extracellular glutathione levels are a few orders of magnitude less than these values (2–20 microMolar) [41]. Further, this tripeptide levels get elevated in cancer cells, which provide important insights for scientists to develop redox-responsive delivery systems. The important chemical groups that act in response to the redox environment are succinimide-thioether linkages, disulfide bonds, diselenide bonds, tetrasulfide bonds, etc. [42]. In a study, Gautam et al. developed paclitaxel (PTX)-loaded redox-responsive micelles using betulinic acid–disulfide–D-α-tocopheryl poly(ethylene glycol) succinate (BA-Cys-T). The resultant micellar delivery system was conjugated with hyaluronic acid (HA) so as to improve its targeting toward CD44 receptors that are overexpressed on cancer cells. Further, the authors demonstrated the synergistic effect of a combinatorial therapy using PTX and betulinic acid (BA) against cancer cells (MDA-MB-231 and 4 T1 cells). The developed micellar system showed redox responsiveness (due to the presence of the disulfide group) and enhanced cellular uptake (due to the presence of HA). This micelle-based formulation, when administered to tumor-bearing mice, showed improved systemic half-life (body retention) and enhanced tumor accumulation, as compared to free drug administration. Further, in vivo efficacy studies that were conducted in the 4 T1 cell-induced cancer xenograft model of female BALB/c mice revealed improved therapeutic efficacy of micellar formulation as evidenced by reduced tumor weight, tumor volume, and improved survival of animals. Taken together, the developed redox-responsive drug delivery system showed improved therapeutic efficacy as compared to free drug administration [43].

2.1.4 Oxidative Stress in Cancer Cells and ROS-Responsive Polymeric Nanoparticles

Reactive oxygen species (ROS) are relatively reactive moieties that contain oxygen. These moieties are single-electron reduction products of oxygen, which include hydroxyl radical, hydrogen peroxide (H_2O_2), peroxynitrite, and hypochlorite. These moieties are generated during cell metabolism as by-products via the mediation of various enzymes such as NADPH oxidase, nitric oxide synthase, cyclooxygenase, lipoxygenase, xanthine oxidase, and cytochrome P450. In tumors, ROS production is exacerbated due to disturbances in metabolic pathways, oncogenic mutations, and hypoxia [44]. In this context, ROS-responsive materials can potentially enhance the release of entrapped therapeutic agents, specifically at the tumor site, thereby offering improved therapeutic effect. The linker groups responsible for ROS-induced degradation of polymer backbone include thiokesterone, polyproline, thioether, selenium (Se), amino acrylate, peroxalate, tellurium, borate, etc. Based on these linkers various ROS responsive polymers, such as poly(1, 4-phenylacetone dimethylene thioacetone) (PPADT), polythioacetone ethyl carbamate (PTKU), poly(glycidyl methacrylate)-polyacrylamide sulfide (PGED-PPS), poly(amino thioketal) (PATK), poly(ethylene glycol)-b-polyurethane-Se-Se-b-poly(ethylene glycol) (PEG-PU-Se-Se-PEG), PEG-polyurethane-Se-PEG triblock copolymers, PEG-PU-Se-PEG, poly(vanillanol-co-oxalate) (PVAX), and phenyl boric acid based polymers, have been developed [45–48]. These polymers showed promising results for site-specific drug delivery. In a study, Tu et al. developed PTX-loaded nanoparticles (shown in Fig. 3.1) using a ROS-responsive polymer, D-α-tocopheryl polyethylene glycol 1000 succinate (TPGS)-poly(β-thioester) [TPGS-PBTE] for

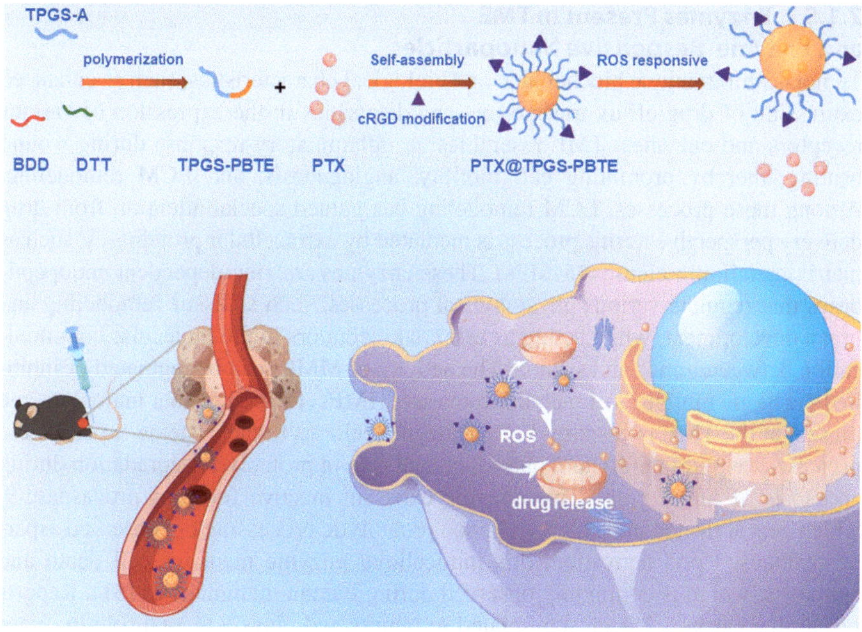

Fig. 3.1 Schematic representation of paclitaxel-loaded ROS-responsive nanoparticles for cancer therapy. D-α-tocopheryl polyethylene glycol 1000 succinate (TPGS), dithiothreitol (DTT), 1, 4-butanediol diacrylate (BDD), poly(β-thioester) [PBTE], paclitaxel (PTX), reactive oxygen species (ROS). (Reproduced with permission from Ref. [49], Creative Commons CC BY license)

head and neck cancer treatment. The resultant TPGS-PBTE nanoparticles showed optimal ROS-responsiveness as evidenced by a substantial amount of drug (PTX) release (~73% in 12 h) when exposed to 10 mM H2O2, as compared to lower drug release (~4.79% in 12 h) under nonoxidative conditions. As a consequence, the developed nanoparticles showed improved cytotoxicity and apoptotic effect toward squamous cell carcinoma (SCC-7) cells. Further, the authors fabricated a fluorescent dye—1,1′-dioctadecyl-3,3,3′,3′-tetramethylindotricarbocyanine iodide (DiR) loaded nanoparticles and functionalized the DiR nanoparticle surface using arginine-glycine-aspartate (Arg-Gly-Asp) peptide sequences (cRGD). The resultant nanoparticles were then intravenously administered to SCC-7-bearing mice to study their biodistribution. In vivo biodistribution data revealed enhanced accumulation of developed nanoparticles in tumors as evidenced by higher area under the curve ($AUC_{0\text{-}t}$) values (676.27 ± 55.98 mg/L.h for RGD-DIR@TPGS-PBTE nanoparticles, 530.48 ± 64.55 mg/L.h for DIR@TPGS-PBTE nanoparticles, and 22.34 ± 5.58 mg/L.h for free DIR). Further, in vivo efficacy studies revealed that cRGD functionalized TPGS-PBTE nanoparticles showed a higher tumor inhibition rate (TIR) with minimal toxicity. Taken together, the developed ROS-responsive nanoparticles showed improved tumor targeting and higher therapeutic efficacy [49].

2.1.5 Enzymes Present in TME and Enzyme-Responsive Nanoparticles

Tumors exhibit unique biochemical and biological characteristics, such as enhanced expression of drug efflux transporters and disparities in the expression of various receptors and enzymes. TME resembles an inflammatory response during wound healing, thereby promoting cell motility, angiogenesis, and ECM remodeling. Among these processes, ECM remodeling has gained special attention from drug delivery perspective as this process is mediated by extracellular proteinases, such as matrix metalloproteinases (MMPs). These enzymes are zinc-dependent endopeptidases that regulate various physiological processes, such as tissue remodeling and organ development, while acting as essential mediators in the molecular communication between tumor and stroma. The activity of MMPs can be regulated or inhibited via tissue inhibitors of metalloproteases (TIMPs) [50]. The other major enzyme present at the TME is caspase-3. This enzyme belongs to the cysteine-aspartic acid protease (caspase) family and plays a pivotal role in proteolytic degradation during apoptosis of tumor cells. This enzyme exists in inactive form, as procaspase-3, which gets activated into caspase-3 upon proteolytic processing of conserved aspartic residues. Upon activation, this intracellular enzyme mediates cell death and apoptosis, which is commonly observed during human malignancies [51]. Reports reveal that caspase-3 is closely related to tumors and plays a pivotal role in tumor development by promoting metastasis, invasion, and immune suppression [52]. Therefore, researchers explored these enzymes to control drug delivery at the tumor site so as to enhance drug accumulation in the tumors. In a study, Wan et al. developed prodrugs, such as D-α-tocopherol polyethylene glycol 3350 succinate (TPGS3350)-Gly-Pro-Leu-Gly-Val-Arg (GPLGVR)-DOX [TPD] and FA-Asp-Glu-Val-Asp (DEVD)-DOX [FPD], that respond to MMP-9 (extracellular enzyme) and caspase-3 (intracellular enzyme), respectively. Subsequently, the authors fabricated DOX-loaded TPD and FPD micelles using the dialysis method and evaluated in vitro and in vivo efficacy of developed micelles. In vitro studies revealed that the fabricated enzyme-responsive micelles showed improved cellular internalization and cytotoxicity in 4 T1 cells. In vivo studies demonstrated that DOX-loaded TPD and FPD showed improved tumor inhibition rate as evidenced by a reduction in tumor weight and volume in enzyme-responsive micelle-treated animals as compared to control animals. Taken together, these results demonstrated that enzyme-responsive nanocarriers show improved therapeutic effect in cancer [44].

In addition to the above stimuli, the immune microenvironment at the tumor site has also been explored by researchers to achieve localized drug delivery. Since the extent of immune cell infiltration gets elevated during tumor progression, drug delivery vehicles that show affinity toward immune cells or their receptors show improved tumor accumulation. Further, infiltrated immune cells stimulate inflammatory responses via the mediation of various cytokines [15, 53–55]. Such physiological responses have been explored to improve the accumulation of nanoparticles in tumors. In a study, Lee et al. showed a click reaction-assisted immune cell targeting (CRAIT) strategy that improved the delivery of DOX-loaded nanoparticles at the deep avascular tumor sites. In this study, the authors performed 1,2,4,5-tetrazine

(Tz) and trans-cyclooctene (TCO) mediated cycloaddition reaction for targeting nanoparticles to CD11b + cells of TME and systemic circulation [56]. In this study, the authors initially synthesized amine-terminated mesoporous silica nanoparticles and subsequently functionalized them with 1,2,4,5-tetrazine (MSNs-Tz). Following this, the authors labeled anti-CD11b antibodies with trans-cyclooctene (TCO) and confirmed a click reaction between fluorescently labeled MSNs-Tz and TCO in vitro (bone marrow-derived cells) and in vivo. Further, intravital imaging studies were performed after sequential injections of anti-CD11b-TCO and MSNs-Tz in 4 T1 tumor-bearing mice. The imaging data showed brightly stained CD11b + cells at a 3-h time point in the tumor blood vessels, which could be due to the conjugation of anti-CD11b-TCO and MSNs-Tz nanoparticles. Whereas the fluorescence intensity was substantially decreased and became undetectable after 24 h of injection, which indicates the absence of nanoparticles in blood vessels. Further, imaging of tumor tissue revealed accumulation of MSNs-Tz in the interstitium and other cells of TME, which suggests that the CRAIT strategy offered improved tumor accumulation and retention of MSNs-Tz. These results suggested that both the EPR effect and CD11b + cell tagging caused the accumulation of MSNs-Tz in tumors. Further, in vivo bio-distribution studies in 4 T1-tumor-bearing mice revealed improved tumor accumulation of near-infrared (NIR) dye-loaded nanoparticles. Taken together, this study demonstrated enhanced tumor accumulation of nanoparticles at the tumor site [56].

In addition to the aforementioned approaches, researchers explored various combinatorial approaches wherein multiple endogenous stimuli are used to further enhance drug delivery at the tumor site. Such combinatorial approaches that explored dual endogenous stimuli for tumor-targeted drug delivery have been mentioned in Table 3.1.

2.2 External (Exogenous/Extrinsic) Stimuli-Responsive Controlled Release Drug Delivery Strategies

Extrinsic stimuli-responsive drug delivery strategies are based on the application of external stimuli such as magnetic, ultrasound, heat (hyperthermia), light, radiation, temperature, or electrical stimuli. These extrinsic stimuli-responsive drug delivery strategies offered substantial advancements in cancer therapy, as the extent and duration of extrinsic stimuli can be precisely controlled. Such precision control enables the desired dose of drug delivery at the site of interest in a spatially and temporally controlled manner, as and when required (on-demand), which, in turn, helps to eradicate tumors effectively. The following section describes various exogenous stimuli-responsive drug delivery strategies along with a few case studies.

2.2.1 Magnetic Stimuli-Responsive Drug Delivery

Magnetism is a physical attribute that occurs through a magnetic field and influences moving electric charges, electric current, and magnetic materials. Upon application of a magnetic field, ferromagnetic materials get pulled into the field, thereby

Table 3.1 Endogenous dual stimuli-responsive nanoparticle for drug delivery

S. No.	Endogenous stimuli	Animal model	Reference(s)
1	Hypoxia + pH	MCF-7 xenograft tumor-bearing BALB/c nude mice	[57]
2	Hypoxia + redox	PANC1 tumor-bearing mice	[58]
3	Hypoxia + ROS	4 T1 xenograft tumor-bearing BABL/c mice	[59]
4	Hypoxia + enzyme	HeLa xenograft tumor-bearing mice	[60]
5	pH + redox	4 T1 xenograft tumor-bearing BABL/c mice	[61]
6	pH + ROS	4 T1 xenograft tumor-bearing BABL/c mice	[62]
7	pH + enzyme	Nude mice bearing tumors of MCF-7/ADR cells	[63]
8	Redox + ROS	HCT116 tumor-bearing nude mice	[64]
9	Redox + enzyme	MC38 tumor-bearing mice	[65]
10	ROS + enzyme	Nude mice bearing tumors of A549 cells	[66]

causing their retention within the magnetic field [67]. Therefore, the accumulation of magnetic materials at the tumor site can be achieved by applying an external magnetic field at the tumor site. Magnetic nanoparticles have gained special attention in various biomedical applications, as these can be localized at a particular site in the body by the application of an external magnetic field. As a consequence, spatial targeting toward the tumors can be achieved, which helps to eradicate tumor cells while minimizing the adverse effects of chemotherapeutic agents on healthy tissues. Hence, magnetic nanoparticles composed of iron, cobalt, manganese, and nickel [68, 69] have been explored by researchers to achieve passive, active, and magnetically controlled targeting. Advancements in the field include the usage of multifunctional ferrofluids (MFs) that behave as 'liquid magnet' when the magnetic field is applied. Further, these ferrofluids are amenable to surface modification with various functional groups (both biological and non-biological origin), which enables enhanced specificity and targetability for the drug delivery system. Therefore, magnetic nanoparticles can be used for the conjugation of therapeutic, diagnostic, theranostic, and imaging molecules for the diagnosis, imaging, or treatment of cancer [70]. External magnetic field (alternative magnetic field) also helps to attain hyperthermia, which in turn causes cell death and irreversible damage to tumor tissues. The most commonly used magnetically responsive nanoparticles are iron oxide nanoparticles and hybrid ferrite nanoparticles. In a study, Belyanina et al. demonstrated improved targetability and therapeutic efficacy of DNA aptamer-functionalized gold-coated magnetic nanoparticles (~50 nm in size). In this study, the authors developed nanoparticles composed of 8–12 nm sized magnetic core and 30–40 nm gold shell. The resultant nanoparticles were surface functionalized with cancer cell (Ehrlich carcinoma) specific AS-14 aptamer (5′-TCCTCTGACTGTAACC ACGAAGGTGTCGGCCTTAGTAAGGCTACAGCCAAGGGAACGTAGC ATAGGTAGTCCAGAAGCC-3′). In vitro studies demonstrated that the aptamer-functionalized nanoparticles, when exposed to the low-frequency alternating magnetic field (LFAMF), caused enhanced carcinoma cell death via apoptotic phosphatidylserine translocation, followed by necrosis. Whereas control treatment experiments, wherein cells were treated with non-functionalized nanoparticles without a magnetic field, did not cause cell death. Further, in vivo experiments that were performed in solid Ehrlich carcinoma tumor-transplanted mice demonstrated a decrease in tumor size. These results demonstrated that the developed aptamer-functionalized gold-coated magnetic nanoparticles showed improved therapeutic efficacy in cancer [71]. In yet another study, Rodríguez-Luccioni et al. studied comparative cell viability in human epithelial colorectal adenocarcinoma (Caco-2) and human breast cancer (MCF-7) cells after exposing them to magnetic fluid hyperthermia (MFH) and hot-water hyperthermia (HWH). Cytotoxicity studies demonstrated that cell viability was not affected in the absence of an oscillating magnetic field, whereas MFH caused a substantial reduction in cell viability when exposed to a magnetic field in both cell lines. Further, the extent of cell death was higher in MFH than in HWH. These results demonstrated that the MFH effectively induced apoptosis in cancer cells [72]. In addition to iron-based nanoparticles, hybrid ferrite-based nanoparticles, such as maghemite (γ-Fe_2O_3), magnetite (Fe_3O_4), and hematite (α-Fe_2O_3), have also been explored by researchers for inducing magnetic hyperthermia and for magnetic resonance imaging (MRI) applications in cancer theranostics [73].

2.2.2 Ultrasound-Responsive Drug Delivery

Ultrasound is a mechanical sound wave (with >20 kHz frequency) that mediates both thermal and non-thermal effects while showing tissue penetration due to its non-radiative and low tissue-attenuation properties. The extent of ultrasound energy can be altered to achieve the required level of intensity, which in turn can be focused at the site of interest in the body. High-intensity focused ultrasound (HIFU) has been explored by researchers for multiple objectives, such as tumor imaging, tumor eradication, and modulating drug release. The drug release is mediated majorly by (i) acoustic cavitation, (ii) heat generation, or (iii) acoustic radiation forces. Therefore, ultrasound has been explored widely for the treatment of cancer and has become a choice for non-invasive therapy during the ablation of tumor mass. Recently, focused ultrasound (FUS) is being explored for tumor ablation via thermal effects. In a study, various cancer cells, such as head and neck cancer (FaDu) cells, prostate cancer (PC-3) cells, and glioblastoma (T98G) cells, were studied to exhibit a sensitization effect toward short FUS with cavitation (FUS-Cav) or without cavitation in the presence of single-dose X-ray radiation therapy or water bath hyperthermia. Results of this study demonstrated that FUS-Cav substantially increased the sensitivity of cancer cells to X-ray therapy or water bath hyperthermia, as evidenced by a reduction in long-term clonogenic survival, cellular invasion, metabolic activity, and induction of sonoporation. Taken together, this study provided valuable insights for the exploration of FUS for the eradication of solid tumors [74].

2.2.3 Electro-Responsive Drug Delivery Systems

The electric field is a physical attribute that originates from electric charges and electric currents. The electric field surrounds electrically charged matter and elicits attractive forces, thereby transducing electrical energy into mechanical form. Few polymers, such as ferrocene, polypyrrole, and carbon nanotubes, show the inherent ability of conductivity upon electric stimulation. Such nanomaterials have been explored as controlled-release drug delivery systems for various therapeutic applications. In a study, Ge et al. developed daunorubicin-loaded polypyrrole nanoparticles that can respond to an electric field, thereby causing the release of entrapped therapeutic agent daunorubicin. The authors speculated that the release of the drug occurs due to electrochemical reduction/oxidation or electric-field-driven movement of charged molecules that cause conformational changes in the polymeric matrix and promote repulsion of the drug molecules from the nanoparticles. The drug release can be controlled by the extent of change in the overall charge of the polymeric molecules, which in turn depends on the extent of the applied electric charge [75]. Further progress in the field includes the development of electrodynamic therapies using electro-responsive nanoparticles. In a study, Gu et al. developed platinum-coated mesoporous silica nanoparticles to achieve a synergistic therapeutic effect during electrodynamic chemotherapy. In this study, the authors prepared pH-responsive DOX-loaded nanoparticles using silica and chitosan polymer. The resultant nanoparticle's surface was decorated using platinum via in situ chemical reductive growth. Such a platinum coating is expected to promote the decomposition of water molecules so as to generate cytotoxic hydroxyl radicals (·OH) when subjected to a square-wave alternating current (AC) field, thereby enabling effective elimination of large-sized tumors. The resultant radicals (reactive oxygen species) augmented cellular apoptosis. Further, the extent of free radical generation and consequential cell death was enhanced under electronic current. In vivo efficacy studies that were conducted in the 4 T1 BALB/c mouse tumor model revealed that combination therapy using nanoparticles and an electric field (0.01 Hz square-wave AC electric field at a current of 5 mA for 10 min) substantially minimized tumor growth by causing tumor cell death. Taken together, these results demonstrated the synergistic activity of electro-chemodynamic therapy for cancer treatment [76].

2.2.4 Light-Responsive Drug Delivery Systems

Light is electromagnetic radiation having a specific wavelength. Based on the wavelength, light can be categorized into ultraviolet (UV, having 200–400 nm wavelength), visible (400–750 nm wavelength), or near-infrared (NIR, 750–2000 nm wavelength) light. Among these, UV light shows deleterious effects on the tissues, whereas visible light is safer. However, visible light shows poor penetration. Further, NIR light exhibits improved tissue penetration with lower phototoxicity. Therefore, visible light and NIR light possess the ideal characteristics for biological applications. Literature reveals that photo-activatable polymeric nanoparticle-based drug delivery systems showed improved therapeutics as compared to conventional systems in the treatment of cancer. In a study, Zhang et al. developed photoactivatable polymeric nanoparticles backboned with a prodrug for the delivery of gene in the treatment of platinum-resistant ovarian cancer. In this study, the authors used blue light (visible light) with a wavelength of 430 nm for minimizing deactivation of gene via endo/lysosomal uptake. This light subsequently causes platinum [Pt(IV)] prodrug dissociation to release Pt (II) while promoting the unpacking of siRNA (c-fos) so as to elicit synergistic therapeutic efficacy toward platinum-resistant ovarian cancer. The developed light-responsive delivery system showed improved therapeutic effect in vitro and in vivo [77]. In yet another study, Fan et al. developed acid and NIR light-responsive polydopamine-based nanoparticles for dual drug chemo and photothermal therapy. In this study, polydopamine is used to achieve improved photothermal efficiency under NIR light irradiation. The results demonstrated that an increase in the concentration of polydopamine nanoparticles from 25 to 200 μg/ml caused a rise in temperature from 12.8 to 32.7 °C thus showing encouraging results for its use in photothermal therapy. Subsequently, the authors prepared dual drug (DOX and epigallocatechin-3-gallate) loaded folate decorated nanoparticles and studied their photothermal and stimuli-responsive release profile. The results demonstrated that the prepared nanoparticles showed a temperature change of ~28 °C under NIR laser light (808 nm) irradiation while showing higher drug release (compared to non-irradiated particles) due to heat generation, thereby causing cleavage/destruction of intermolecular forces and hydrogen bonding between drugs and polymeric nanoparticles. In vivo biodistribution studies revealed enhanced accumulation and intra-tumoral penetration of folate-decorated nanoparticles in the 4 T1 breast cancer mice model after intravenous injection. In vivo efficacy studies revealed tumor-specific distribution of dual drug-loaded nanoparticles and consequential antitumor efficacy as evidenced by delayed cancer progression and reduced tumor volume in dual drug-treated animals upon light irradiation, as compared to other groups such as mono/dual drug therapy in the absence of light irradiation. Taken together, these results demonstrated that co-delivery of DOX and epigallocatechin-3-gallate using polydopamine nanoparticles enhanced the anticancer effect upon laser irradiation [78].

2.2.5 Temperature-Responsive Drug Delivery Systems

Temperature is a key factor that governs the solubility of various polymers, thereby influencing their drug-release properties. Based on their solubility at a specific temperature, polymers can be categorized into two types: [1] Polymers that exhibit lower critical solution temperature (LCST), wherein the polymer is insoluble in the solvent when ambient temperature exceeds LCST thereby becoming turbid; [2]

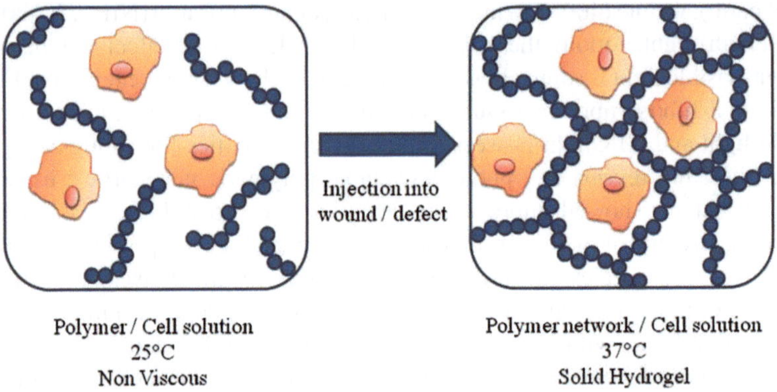

Fig. 3.2 Formation of a scaffold in response to variations in temperature. (Reproduced with permission from Ref. [79] Creative Commons CC BY license)

Polymers that exhibit upper critical solution temperature (UCST) wherein the polymer is insoluble in the solvent at the ambient temperature and becomes soluble when temperature exceeds UCST [79, 80]. Various thermo-responsive polymers have been employed in drug and gene delivery and for tissue engineering applications. Formation of a polymer network in response to variations in the temperature has been mentioned in Fig. 3.2.

Based on these properties, drug release can be augmented from the nanoparticle-based delivery systems using temperature as a trigger. The most widely explored polymer showing LCST is poly(N-isopropyl acrylamide) (PNIPAAm). This polymer is known to form hydrogen bonds with water below its LCST, thus exhibiting solubility. Whereas PNIPAAm expels water molecules above 32 °C, thereby undergoing dehydration and contraction of its structure [81]. Considering its properties, copolymerization of PNIPAAm with different materials has been explored to alter thermo-responsiveness and swelling behavior [82]. The other polymers that show thermo-responsiveness are pluronic F127 (poloxamer 407), which is an amphiphilic triblock copolymer consisting of poly(ethylene oxide)-poly(propylene oxide)-poly(ethylene oxide) blocks, poly(2-oxazoline)s (POxs), poly(vinyl caprolactam) (PNVCL), poly(methyl vinyl ether) (PMVE), etc. [83–86]. These polymers can be explored for the fabrication of thermo-sensitive nanoparticles to achieve localized drug delivery upon stimulation by thermal energy. In a study, Shin et al. developed dual (light and temperature) responsive poly(N-isopropylacrylamide)-pyrrole nanocomposites for the delivery of doxorubicin. In this study, the authors used acrylic acid (AAc) during the synthesis of PNIPAAm and subsequently prepared polypyrrole nanoparticles and decorated them in PNIPAM-AAc. Further, folic acid was conjugated to the carboxylic acid groups of PNIPAAm. The resultant nanocomposites were loaded with DOX and evaluated for stimuli responsiveness against NIR light and temperature. The drug release study data indicated higher drug release when the temperature is increased from 25 to 37 °C. The extent of drug release was further increased upon NIR light stimuli. The resultant nanoparticles showed

Table 3.2 Exogenous dual stimuli-responsive nanoparticle for drug delivery

S. No.	Exogenous stimuli	Animal model	Reference(s)
1	Magnetic + ultrasound	SKOV3 cells injected xenograft tumor-bearing nude mice	[88]
2	Magnetic + electro-responsive	–	[89]
3	Magnetic + light	MCF-7 xenograft tumor-bearing mice	[90]
4	Magnetic + temperature	–	[91]
5	Ultrasound + electro-responsive	C6 tumor-bearing BALB/c nude mice	[92]
6	Ultrasound + light	RIF-1 cells injected tumor-bearing C3H/HeN mice	[93]
7	Ultrasound + temperature	MCF-7 xenograft tumor-bearing mice	[94]
8	Electro-responsive + light	4 T1 tumor-bearing BALB/c mice	[95]
9	Electro-responsive + temperature	–	[96]
10	Light + temperature	4 T1 xenograft tumor-bearing BALB/c mice	[97]

enhanced internalization in MDA-MB-231 cells. Further, the developed nanoparticles showed higher cytotoxicity toward cancer cells when subjected to temperature and NIR stimuli. Taken together, the developed dual-responsive nanoparticulate drug delivery system showed improved therapeutic effect [87].

Drug delivery at the tumor site can be further improved using multiple stimuli-responsive nanoparticulate delivery systems. Therefore, researchers explored various approaches wherein multiple exogenous stimuli or a combination of endogenous and exogenous stimuli are used to further enhance drug delivery at the tumor site. Such combinatorial approaches that explored dual exogenous stimuli have been mentioned in Tables 3.2 and 3.3.

3 Concluding Remarks and Perspectives

Conventional chemotherapy is associated with severe adverse effects for the treatment of cancer. Such complications can be circumvented by exploring stimuli-responsive polymeric materials and the development of stimuli-responsive nanoparticulate drug delivery systems. Such delivery systems can respond to the pathological conditions of the TME, thereby controlling drug delivery at the site of interest in a spatiotemporally controlled manner. As discussed in this chapter, targeted drug delivery strategies that explored mono- or multi-stimuli responsive polymeric nanoparticulate drug delivery systems have improved the specificity and

Table 3.3 Combination of endogenous and exogenous stimuli responsive nanoparticle-mediated drug delivery strategies

S. No.	Endogenous stimuli	Exogenous stimuli	Animal model	Reference(s)
1	Hypoxia	Magnetic	4 T1 xenograft tumor-bearing BABL/c mice	[98]
2	Hypoxia	Ultrasound	4 T1 xenograft tumor-bearing BALB/c nude mice	[99]
3	Hypoxia	Light	4 T1 xenograft tumor-bearing mice	[100]
4	Hypoxia	Temperature	BALB/c nude mice bearing breast cancer stem cell tumors	[101]
5	pH	Magnetic	BALB/c nude mice bearing MCF-7/ADR cells transplanted tumors	[102]
6	pH	Ultrasound	MCF-7 xenograft tumor-bearing BALB/c nude mice	[103]
7	pH	Light	4 T1 xenograft tumor-bearing mice	[104]
8	pH	Temperature	CT26 tumor-bearing mice	[105]
9	pH	Temperature + magnetic field	–	[106]
10	pH	Electric field	4 T1 xenograft tumor bearing BALB/c mice	[76]
11	Redox	Magnetic	MDA-MB231 xenograft tumor-bearing BALB/c nude mice	[107]
12	Redox	Ultrasound	Sesame oil-diluted DMBA administered mammary tumor-bearing Sprague–Dawley rats	[108]

(continued)

Table 3.3 (continued)

S. No.	Endogenous stimuli	Exogenous stimuli	Animal model	Reference(s)
13	Redox	Electro-responsive	MCF-7 xenograft tumor-bearing nude mice	[109]
14	Redox	Light	A549/T tumor-bearing BALB/c nude mice	[110]
15	ROS	Magnetic	GBM6 xenograft tumor-bearing nude mice	[111]
16	ROS	Ultrasound	BxPC-3-Luci cells injected xenograft tumor-bearing nude mice	[112]
17	ROS	Light	A549/T tumor-bearing BALB/c nude mice	[110]
18	ROS	Temperature	–	[113]
19	Enzyme	Magnetic	–	[114]
20	Enzyme	Ultrasound	U87 xenograft tumor-bearing nude mice	[115]
21	Enzyme	Electro-responsive		
22	Enzyme	Light	A549 tumor-bearing nude mice	[116]
23	Enzyme	Temperature	A549 tumor-bearing nude mice	[117]
24	pH + redox	Temperature	–	[118]
25	pH + ROS	Light	MDA-MB231 xenograft tumor-bearing BALB/c nude mice	[119]
26	pH + redox + ROS	Ultrasound	Subcutaneous tumor model of HepG2 BALB/c nude mice	[120]
27	Enzyme	Light + magnetic stimuli	Tumor-bearing nude mice	[121]

consequential therapeutic efficacy of drugs. Therefore, the development of stimuli-responsive nanoparticle-based drug delivery systems that exhibit preferential tumor accumulation holds tremendous potential for cancer therapies.

References

1. Global Cancer Observatory: Cancer Today. Lyon: International Agency for Research on Cancer.
2. Sung H, Ferlay J, Siegel RL, Laversanne M, Soerjomataram I, Jemal A, et al. Global cancer statistics 2020: GLOBOCAN estimates of incidence and mortality worldwide for 36 cancers in 185 countries. CA Cancer J Clin. 2021;71(3):209–49.
3. Anderson NM, Simon MC. The tumor microenvironment. Curr Biol. 2020;30(16):R921–5.
4. de Visser KE, Joyce JA. The evolving tumor microenvironment: from cancer initiation to metastatic outgrowth. Cancer Cell. 2023;41(3):374–403.
5. Srinivasarao DA, Shah S, Famta P, Vambhurkar G, Jain N, Pindiprolu SKSS, et al. Unravelling the role of tumor microenvironment responsive nanobiomaterials in spatiotemporal controlled drug delivery for lung cancer therapy. Drug Deliv Transl Res. 2025;15(2):407–35.
6. Stylianopoulos T, Martin JD, Snuderl M, Mpekris F, Jain SR, Jain RK. Coevolution of solid stress and interstitial fluid pressure in tumors during progression: implications for vascular collapse. Cancer Res. 2013;73(13):3833–41.
7. Joyce JA, Pollard JW. Microenvironmental regulation of metastasis. Nat Rev Cancer. 2009;9(4):239–52.
8. Kaplan RN, Riba RD, Zacharoulis S, Bramley AH, Vincent L, Costa C, et al. VEGFR1-positive haematopoietic bone marrow progenitors initiate the pre-metastatic niche. Nature. 2005;438(7069):820–7.
9. Tohme S, Simmons RL, Tsung A. Surgery for cancer: a trigger for metastases. Cancer Res. 2017;77(7):1548–52.
10. Baskar R, Lee KA, Yeo R, Yeoh KW. Cancer and radiation therapy: current advances and future directions. Int J Med Sci. 2012;9(3):193–9.
11. Shibata H, Saito S, Uppaluri R. Immunotherapy for head and neck cancer: a paradigm shift from induction chemotherapy to neoadjuvant immunotherapy. Front Oncol. 2021;11
12. Bailly C, Thuru X, Quesnel B. Combined cytotoxic chemotherapy and immunotherapy of cancer: modern times. NAR Cancer. 2020;2(1)
13. Chen DS, Mellman I. Oncology meets immunology: the cancer-immunity cycle. Immunity. 2013;39(1):1–10.
14. Lo Presti E, Dieli F, Meraviglia S. Tumor-infiltrating Î³Î´ T lymphocytes: pathogenic role, clinical significance, and differential programing in the tumor microenvironment. Front Immunol. 2014;5
15. Fridman WH, Pagès F, Sautès-Fridman C, Galon J. The immune contexture in human tumours: impact on clinical outcome. Nat Rev Cancer. 2012;12(4):298–306.
16. Sakai H, Kokura S, Ishikawa T, Tsuchiya R, Okajima M, Matsuyama T, et al. Effects of anti-cancer agents on cell viability, proliferative activity and cytokine production of peripheral blood mononuclear cells. J Clin Biochem Nutr. 2013;52(1):64–71.
17. Taşkın-Tok T, Gowder SJT. Anticancer drug—friend or foe. In: Pharmacology and therapeutics. InTech; 2014.
18. Basak D, Arrighi S, Darwiche Y, Deb S. Comparison of anticancer drug toxicities: paradigm shift in adverse effect profile. Life. 2021;12(1):48.
19. Unsoy G, Gunduz U. Smart drug delivery systems in cancer therapy. Curr Drug Targets. 2018;19(3)
20. Alvarez-Lorenzo C, Concheiro A. Smart drug delivery systems: from fundamentals to the clinic. Chem Commun. 2014;50(58):7743–65.

21. Dyawanapelly S, Junnuthula VR, Singh A. The holy grail of polymer therapeutics for cancer therapy: an overview on the pharmacokinetics and bio distribution. Curr Drug Metab. 2015;16(7):522–37.
22. Singh AVAMR and SD. Conjugates, polymer–drug: targeted cancer treatment. In: Encyclopedia of biomedical polymers and polymeric biomaterial; 2015. p. 2089–112.
23. Mitchell MJ, Billingsley MM, Haley RM, Wechsler ME, Peppas NA, Langer R. Engineering precision nanoparticles for drug delivery. Nat Rev Drug Discov. 2021;20(2):101–24.
24. Blanco E, Shen H, Ferrari M. Principles of nanoparticle design for overcoming biological barriers to drug delivery. Nat Biotechnol. 2015;33(9):941–51.
25. Matsumura Y, Maeda H. A new concept for macromolecular therapeutics in cancer chemotherapy: mechanism of tumoritropic accumulation of proteins and the antitumor agent smancs. Cancer Res. 1986;46(12 Pt 1):6387–92.
26. Zhao R, Fu C, Wang Z, Pan M, Ma B, Yin Q, et al. A pH-responsive nanoparticle library with precise pH Tunability by co-polymerization with non-Ionizable monomers. Angew Chem Int Ed. 2022;61(19)
27. McKeown SR. Defining normoxia, physoxia and hypoxia in tumours—implications for treatment response. Br J Radiol. 2014;87(1035):20130676.
28. Mamnoon B, Feng L, Froberg J, Choi Y, Sathish V, Mallik S. Hypoxia-responsive, polymeric nanocarriers for targeted drug delivery to estrogen receptor-positive breast cancer cell spheroids. Mol Pharm. 2020;17(11):4312–22.
29. Zhang X, Lin Y, Gillies RJ. Tumor pH and its measurement. J Nucl Med. 2010;51(8):1167–70.
30. Neri D, Supuran CT. Interfering with pH regulation in tumours as a therapeutic strategy. Nat Rev Drug Discov. 2011;10(10):767–77.
31. Mighani H. Schiff Base polymers: synthesis and characterization. J Polym Res. 2020;27(6):168.
32. Verma R, Lamba NP, Dandia A, Srivastava A, Modi K, Chauhan MS, et al. Synthesis of N-Benzylideneaniline by Schiff base reaction using Kinnow peel powder as Green catalyst and comparative study of derivatives through ANOVA techniques. Sci Rep. 2022;12(1):9636.
33. Raczuk E, Dmochowska B, Samaszko-Fiertek J, Madaj J. Different Schiff bases—structure, importance and classification. Molecules. 2022;27(3):787.
34. Zhang C, Kermaniyan S, Smith SA, Gillies ER, Such GK. Acid-responsive poly(glyoxylate) self-immolative star polymers. Biomacromolecules. 2021;22(9):3892–900.
35. Palanikumar L, Al-Hosani S, Kalmouni M, Nguyen VP, Ali L, Pasricha R, et al. pH-responsive high stability polymeric nanoparticles for targeted delivery of anticancer therapeutics. Commun Biol. 2020;3(1):95.
36. Vartanian S, Lee J, Klijn C, Gnad F, Bagniewska M, Schaefer G, et al. ERBB3 and IGF1R signaling are required for Nrf2-dependent growth in KEAP1-mutant lung cancer. Cancer Res. 2019;79(19):4828–39.
37. Lee MK, Zhang X, Kim HJ, Hwang YS. Peroxiredoxin 5 is involved in cancer cell invasion and tumor growth of oral squamous cell carcinoma. Oral Dis. 2021;
38. Kasherman Y, Sturup S, Gibson D. Is glutathione the major cellular target of cisplatin? A study of the interactions of cisplatin with cancer cell extracts. J Med Chem. 2009;52(14):4319–28.
39. Scalcon V, Bindoli A, Rigobello MP. Significance of the mitochondrial thioredoxin reductase in cancer cells: an update on role, targets and inhibitors. Free Radic Biol Med. 2018;127:62–79.
40. Hampton MB, Vick KA, Skoko JJ, Neumann CA. Peroxiredoxin involvement in the initiation and progression of human cancer. Antioxid Redox Signal. 2018;28(7):591–608.
41. Schafer FQ, Buettner GR. Redox environment of the cell as viewed through the redox state of the glutathione disulfide/glutathione couple. Free Radic Biol Med. 2001;30(11):1191–212.
42. Abed HF, Abuwatfa WH, Husseini GA. Redox-responsive drug delivery systems: a chemical perspective. Nano. 2022;12(18):3183.

43. Gautam S, Marwaha D, Singh N, Rai N, Sharma M, Tiwari P, et al. Self-assembled redox-sensitive polymeric nanostructures facilitate the intracellular delivery of paclitaxel for improved breast cancer therapy. Mol Pharm. 2023;20(4):1914–32.
44. Wan D, Zhu Q, Zhang J, Chen X, Li F, Liu Y, et al. Intracellular and extracellular enzymatic responsive micelle for intelligent therapy of cancer. Nano Res. 2023;16(2):2851–8.
45. Zhai Z, Ouyang W, Yao Y, Zhang Y, Zhang H, Xu F, et al. Dexamethasone-loaded ROS-responsive poly(thioketal) nanoparticles suppress inflammation and oxidative stress of acute lung injury. Bioact Mater. 2022;14:430–42.
46. Ma N, Li Y, Xu H, Wang Z, Zhang X. Dual redox responsive assemblies formed from diselenide block copolymers. J Am Chem Soc. 2010;132(2):442–3.
47. Li F, Li T, Cao W, Wang L, Xu H. Near-infrared light stimuli-responsive synergistic therapy nanoplatforms based on the coordination of tellurium-containing block polymer and cisplatin for cancer treatment. Biomaterials. 2017;133:208–18.
48. Liu J, Jia B, Li Z, Li W. Reactive oxygen species-responsive polymer drug delivery systems. Front Bioeng Biotechnol. 2023;11
49. Tu Y, Zhang W, Fan G, Zou C, Zhang J, Wu N, et al. Paclitaxel-loaded ROS-responsive nanoparticles for head and neck cancer therapy. Drug Deliv. 2023;30(1)
50. Kessenbrock K, Plaks V, Werb Z. Matrix Metalloproteinases: regulators of the tumor microenvironment. Cell. 2010;141(1):52–67.
51. Huang KH, Fang WL, Li AFY, Liang PH, Wu CW, Shyr YM, et al. Caspase-3, a key apoptotic protein, as a prognostic marker in gastric cancer after curative surgery. Int J Surg. 2018;52:258–63.
52. Dou H, Yang YP, Qi LY, Zhu Y, Cheng LF, Yu WY, et al. Recent advances in caspase-3, breast cancer, and traditional Chinese medicine: a review. J Chemother. 2023:1–19.
53. Jain N, Srinivasarao DA, Famta P, Shah S, Vambhurkar G, Shahrukh S, et al. The portrayal of macrophages as tools and targets: a paradigm shift in cancer management. Life Sci. 2023;316:121399.
54. Binnewies M, Roberts EW, Kersten K, Chan V, Fearon DF, Merad M, et al. Understanding the tumor immune microenvironment (TIME) for effective therapy. Nat Med. 2018;24(5):541–50.
55. Dasanu CA, Sethi N, Ahmed N. Immune alterations and emerging immunotherapeutic approaches in lung cancer. Expert Opin Biol Ther. 2012;12(7):923–37.
56. Lee SH, Park OK, Kim J, Shin K, Pack CG, Kim K, et al. Deep tumor penetration of drug-loaded nanoparticles by click reaction-assisted immune cell targeting strategy. J Am Chem Soc. 2019;141(35):13829–40.
57. Zhang Z, Feng J, Zhang T, Gao A, Sun C. Application of tumor pH/hypoxia-responsive nanoparticles for combined photodynamic therapy and hypoxia-activated chemotherapy. Front Bioeng Biotechnol. 2023;11
58. Guo JS, Li JJ, Wang ZH, Liu Y, Yue YX, Li HB, et al. Dual hypoxia-responsive supramolecular complex for cancer target therapy. Nat Commun. 2023;14(1):5634.
59. Guo X, Han L, Chen W, He H, Zhang W, Huang C, et al. Hypoxia and singlet oxygen dual-responsive micelles for photodynamic and chemotherapy therapy featured with enhanced cellular uptake and triggered cargo delivery. Int J Nanomedicine. 2024;19:247–61.
60. Li D, Ren T, Ge Y, Wang X, Sun G, Zhang N, et al. A multi-functional hypoxia/esterase dual stimulus responsive and hyaluronic acid-based nanomicelle for targeting delivery of chloroethylnitrosouea. J Nanobiotechnol. 2023;21(1):291.
61. Wan L, Chen Z, Deng Y, Liao T, Kuang Y, Liu J, et al. A novel intratumoral pH/redox-dual-responsive nanoplatform for cancer MR imaging and therapy. J Colloid Interface Sci. 2020;573:263–77.
62. Li X, Ji Q, Yan C, Zhu Z, Yan Z, Chen P, et al. H2O2/pH dual-responsive biomimetic nanoenzyme drugs delivery system for enhanced tumor photodynamic therapy. Nanoscale Res Lett. 2022;17(1):103.
63. Ni R, Zhu J, Xu Z, Chen Y. A self-assembled pH/enzyme dual-responsive prodrug with PEG deshielding for multidrug-resistant tumor therapy. J Mater Chem B. 2020;8(6):1290–301.

64. Chiang YT, Yen YW, Lo CL. Reactive oxygen species and glutathione dual redox-responsive micelles for selective cytotoxicity of cancer. Biomaterials. 2015;61:150–61.
65. Shen P, Zhang X, Ding N, Zhou Y, Wu C, Xing C, et al. Glutathione and esterase dual-responsive smart Nano-drug delivery system capable of breaking the redox balance for enhanced tumor therapy. ACS Appl Mater Interfaces. 2023;15(17):20697–711.
66. Zhang M, Zhang X, Cai S, Mei H, He Y, Huang D, et al. Photo-induced specific intracellular release EGFR inhibitor from enzyme/ROS-dual sensitive nano-platforms for molecular targeted-photodynamic combinational therapy of non-small cell lung cancer. J Mater Chem B. 2020;8(35):7931–40.
67. Skomski R, Balasubramanian B, Sellmyer DJ. Magnetism of nanomaterials. In: Magnetic nanomaterials—fundamentals, synthesis and applications. Weinheim, Germany: Wiley-VCH Verlag GmbH & Co. KGaA; 2017. p. 29–80.
68. Xie J, Jon S. Magnetic nanoparticle-based theranostics. Theranostics. 2012;2(1):122–4.
69. Shokrollahi H. Structure, synthetic methods, magnetic properties and biomedical applications of ferrofluids. Mater Sci Eng C. 2013;33(5):2476–87.
70. Farzin A, Etesami SA, Quint J, Memic A, Tamayol A. Magnetic nanoparticles in cancer therapy and diagnosis. Adv Healthc Mater. 2020;9(9):e1901058.
71. Belyanina IV, Zamay TN, Zamay GS, Zamay SS, Kolovskaya OS, Ivanchenko TI, et al. *In vivo* cancer cells elimination guided by aptamer-functionalized gold-coated magnetic nanoparticles and controlled with low frequency alternating magnetic field. Theranostics. 2017;7(13):3326–37.
72. Torres-Lugo M, Rodriguez HL, Latorre-Esteves M, Mendez J, Soto O, Rodriguez AR, et al. Enhanced reduction in cell viability by hyperthermia induced by magnetic nanoparticles. Int J Nanomedicine. 2011;373
73. Can MM, Coşkun M, Fırat T. A comparative study of nanosized iron oxide particles; magnetite (Fe3O4), maghemite (γ-Fe2O3) and hematite (α-Fe2O3), using ferromagnetic resonance. J Alloys Compd. 2012 Nov;542:241–7.
74. Hu S, Zhang X, Unger M, Patties I, Melzer A, Landgraf L. Focused ultrasound-induced cavitation sensitizes cancer cells to radiation therapy and hyperthermia. Cells. 2020;9(12):2595.
75. Ge J, Neofytou E, Cahill TJ, Beygui RE, Zare RN. Drug release from electric-field-responsive nanoparticles. ACS Nano. 2012;6(1):227–33.
76. Gu T, Chen T, Cheng L, Li X, Han G, Liu Z. Mesoporous silica decorated with platinum nanoparticles for drug delivery and synergistic electrodynamic-chemotherapy. Nano Res. 2020;13(8):2209–15.
77. Zhang Q, Kuang G, He S, Lu H, Cheng Y, Zhou D, et al. Photoactivatable prodrug-backboned polymeric nanoparticles for efficient light-controlled gene delivery and synergistic treatment of platinum-resistant ovarian cancer. Nano Lett. 2020;20(5):3039–49.
78. Fan R, Chen C, Hou H, Chuan D, Mu M, Liu Z, et al. Tumor acidity and near-infrared light responsive dual drug delivery polydopamine-based nanoparticles for chemo-photothermal therapy. Adv Funct Mater. 2021;31(18)
79. Ward MA, Georgiou TK. Thermoresponsive polymers for biomedical applications. Polymers (Basel). 2011;3(3):1215–42.
80. Pasparakis G, Tsitsilianis C. LCST polymers: Thermoresponsive nanostructured assemblies towards bioapplications. Polymer (Guildf). 2020;211:123146.
81. Jeong B, Kim SW, Bae YH. Thermosensitive sol–gel reversible hydrogels. Adv Drug Deliv Rev. 2012;64:154–62.
82. Hertle Y, Hellweg T. Thermoresponsive copolymer microgels. J Mater Chem B. 2013;1(43):5874.
83. Akash MSH, Rehman K, Chen S. Pluronic F127-based Thermosensitive gels for delivery of therapeutic proteins and peptides. Polym Rev. 2014;54(4):573–97.
84. Hoogenboom R, Schlaad H. Thermoresponsive poly(2-oxazoline)s, polypeptoids, and polypeptides. Polym Chem. 2017;8(1):24–40.
85. Mohammed MN, Bin YK, Shariffuddin JHBH. Poly(N-vinyl caprolactam) thermoresponsive polymer in novel drug delivery systems: a review. Mater Express. 2018;8(1):21–34.

86. Verdonck B, Gohy JF, Khousakoun E, Jérôme R, Du Prez F. Association behavior of thermo-responsive block copolymers based on poly(vinyl ethers). Polymer (Guildf). 2005;46(23):9899–907.
87. Shin HH, Choi HW, Lim JH, Kim JW, Chung BG. Near-infrared light-triggered thermo-responsive poly(N-isopropylacrylamide)-pyrrole nanocomposites for chemo-photothermal cancer therapy. Nanoscale Res Lett. 2020;15(1):214.
88. Du M, Chen Y, Tu J, Liufu C, Yu J, Yuan Z, et al. Ultrasound responsive magnetic mesoporous silica nanoparticle-loaded microbubbles for efficient gene delivery. ACS Biomater Sci Eng. 2020;6(5):2904–12.
89. Guduru R, Liang P, Runowicz C, Nair M, Atluri V, Khizroev S. Magneto-electric nanoparticles to enable field-controlled high-specificity drug delivery to eradicate ovarian cancer cells. Sci Rep. 2013;3(1):2953.
90. Feng Q, Zhang Y, Zhang W, Hao Y, Wang Y, Zhang H, et al. Programmed near-infrared light-responsive drug delivery system for combined magnetic tumor-targeting magnetic resonance imaging and chemo-phototherapy. Acta Biomater. 2017;49:402–13.
91. Purushotham S, Ramanujan RV. Thermoresponsive magnetic composite nanomaterials for multimodal cancer therapy. Acta Biomater. 2010;6(2):502–10.
92. Ma K, Qi G, Wang B, Yu T, Zhang Y, Li H, et al. Ultrasound-activated au/ZnO-based Trojan nanogenerators for combined targeted electro-stimulation and enhanced catalytic therapy of tumor. Nano Energy. 2021;87:106208.
93. Nomikou N, Curtis K, McEwan C, O'Hagan BMG, Callan B, Callan JF, et al. A versatile, stimulus-responsive nanoparticle-based platform for use in both sonodynamic and photodynamic cancer therapy. Acta Biomater. 2017;49:414–21.
94. Zhang L, Zhang S, Chen H, Liang Y, Zhao B, Luo W, et al. An acoustic/thermo-responsive hybrid system for advanced doxorubicin delivery in tumor treatment. Biomater Sci. 2020;8(8):2202–11.
95. Gangrade A, Gawali B, Jadi PK, Naidu VGM, Mandal BB. Photo-electro active nanocomposite silk hydrogel for spatiotemporal controlled release of chemotherapeutics: an in vivo approach toward suppressing solid tumor growth. ACS Appl Mater Interfaces. 2020;12(25):27905–16.
96. Tsang YW, Chi KH, Huang CC, Chi MS, Chiang HC, Yang KL, et al. Modulated electro-hyperthermia-enhanced liposomal drug uptake by cancer cells. Int J Nanomedicine. 2019;14:1269–79.
97. Sun X, Liu D, Xu X, Shen Y, Huang Y, Zeng Z, et al. NIR-triggered thermo-responsive biodegradable hydrogel with combination of photothermal and thermodynamic therapy for hypoxic tumor. Asian J Pharm Sci. 2020;15(6):713–27.
98. Zhou H, Guo M, Li J, Qin F, Wang Y, Liu T, et al. Hypoxia-triggered self-assembly of Ultrasmall iron oxide nanoparticles to amplify the imaging signal of a tumor. J Am Chem Soc. 2021;143(4):1846–53.
99. Zhuang F, Ma Q, Dong C, Xiang H, Shen Y, Sun P, et al. Sequential ultrasound-triggered and hypoxia-sensitive nanoprodrug for cascade amplification of sonochemotherapy. ACS Nano. 2022;16(4):5439–53.
100. Zhu R, He H, Liu Y, Cao D, Yan J, Duan S, et al. Cancer-selective bioreductive chemotherapy mediated by dual hypoxia-responsive nanomedicine upon photodynamic therapy-induced hypoxia aggravation. Biomacromolecules. 2019;20(7):2649–56.
101. Li H, Yan W, Suo X, Peng H, Yang X, Li Z, et al. Nucleus-targeted nano delivery system eradicates cancer stem cells by combined thermotherapy and hypoxia-activated chemotherapy. Biomaterials. 2019;200:1–14.
102. Wang D, Li X, Li X, Kang A, Sun L, Sun M, et al. Magnetic and pH dual-responsive nanoparticles for synergistic drug-resistant breast cancer chemo/photodynamic therapy. Int J Nanomedicine. 2019;14:7665–79.
103. Luo W, Wen G, Yang L, Tang J, Wang J, Wang J, et al. Dual-targeted and pH-sensitive doxorubicin prodrug-microbubble complex with ultrasound for tumor treatment. Theranostics. 2017;7(2):452–65.

104. Yu C, Kong L, Tian J, Zhang Y, Jia X, Dang W, et al. Photoacoustic imaging-guided triple-responsive nanoparticles with tumor hypoxia relief for improving chemotherapy/photothermal/photodynamic synergistic therapy against breast cancer. Biomed Pharmacother. 2023;164:114928.
105. Qi J, Li W, Lu K, Jin F, Liu D, Xu X, et al. pH and thermal dual-sensitive nanoparticle-mediated synergistic antitumor effect of immunotherapy and microwave thermotherapy. Nano Lett. 2019;19(8):4949–59.
106. Hervault A, Dunn AE, Lim M, Boyer C, Mott D, Maenosono S, et al. Doxorubicin loaded dual pH- and thermo-responsive magnetic nanocarrier for combined magnetic hyperthermia and targeted controlled drug delivery applications. Nanoscale. 2016;8(24):12152–61.
107. Yang HY, Jang MS, Li Y, Lee JH, Lee DS. Multifunctional and redox-responsive self-assembled magnetic nanovectors for protein delivery and dual-modal imaging. ACS Appl Mater Interfaces. 2017;9(22):19184–92.
108. Dhamija P, Mehata AK, Tamang R, Bonlawar J, Vaishali MAK, et al. Redox-sensitive poly(lactic- *co* -glycolic acid) nanoparticles of Palbociclib: development, ultrasound/photoacoustic imaging, and smart breast cancer therapy. Mol Pharm. 2024;21(6):2713–26.
109. Chen Y, Chen M, Zhai T, Zhou H, Zhou Z, Liu X, et al. Glutathione-responsive chemodynamic therapy of manganese(III/IV) cluster nanoparticles enhanced by electrochemical stimulation via oxidative stress pathway. Bioconjug Chem. 2022;33(1):152–63.
110. Zhong D, Wu H, Wu Y, Li Y, Yang J, Gong Q, et al. Redox dual-responsive dendrimeric nanoparticles for mutually synergistic chemo-photodynamic therapy to overcome drug resistance. J Control Release. 2021;329:1210–21.
111. Stephen ZR, Kievit FM, Veiseh O, Chiarelli PA, Fang C, Wang K, et al. Redox-responsive magnetic nanoparticle for targeted convection-enhanced delivery of O^6 -benzylguanine to brain tumors. ACS Nano. 2014;8(10):10383–95.
112. Zhang C, Chen J, Song Y, Luo J, Jin P, Wang X, et al. Ultrasound-enhanced reactive oxygen species responsive charge-reversal polymeric nanocarriers for efficient pancreatic cancer gene delivery. ACS Appl Mater Interfaces. 2022;14(2):2587–96.
113. Kashyap S, Singh N, Surnar B, Jayakannan M. Enzyme and thermal dual responsive amphiphilic polymer core–shell nanoparticle for doxorubicin delivery to cancer cells. Biomacromolecules. 2016;17(1):384–98.
114. Nosrati H, Mojtahedi A, Danafar H, Kheiri MH. Enzymatic stimuli-responsive methotrexate-conjugated magnetic nanoparticles for target delivery to breast cancer cells and release study in lysosomal condition. J Biomed Mater Res A. 2018;106(6):1646–54.
115. Zhu P, Chen Y, Shi J. Nanoenzyme-augmented cancer sonodynamic therapy by catalytic tumor oxygenation. ACS Nano. 2018;12(4):3780–95.
116. Tam LKB, Chu JCH, He L, Yang C, Han KC, Cheung PCK, et al. Enzyme-responsive double-locked photodynamic molecular beacon for targeted photodynamic anticancer therapy. J Am Chem Soc. 2023;145(13):7361–75.
117. Gao G, Jiang Y, Guo Y, Jia H, Cheng X, Deng Y, et al. Enzyme-mediated tumor starvation and phototherapy enhance mild-temperature photothermal therapy. Adv Funct Mater. 2020;30(16)
118. Duan L, Wang Y, Zhang Y, Wang Z, Li Y, He P. pH/redox/thermo-stimulative nanogels with enhanced thermosensitivity via incorporation of cationic and anionic components for anticancer drug delivery. Int J Polym Mater Polym Biomater. 2018;67(5):288–96.
119. Chen M, Yang J, Zhou L, Hu X, Wang C, Chai K, et al. Dual-responsive and ROS-augmented nanoplatform for chemo/photodynamic/chemodynamic combination therapy of triple negative breast cancer. ACS Appl Mater Interfaces. 2022;14(1):57–68.
120. Tang J, Zhang X, Cheng L, Liu Y, Chen Y, Jiang Z, et al. Multiple stimuli-responsive nanosystem for potent, ROS-amplifying, chemo-sonodynamic antitumor therapy. Bioact Mater. 2022;15:355–71.
121. Yuan P, Yang T, Liu T, Yu X, Bai Y, Zhang Y, et al. Nanocomposite hydrogel with NIR/magnet/enzyme multiple responsiveness to accurately manipulate local drugs for on-demand tumor therapy. Biomaterials. 2020;262:120357.

Stimuli-Responsive Polymersomes for Cancer Therapy

4

Amruta Padakanti, Prateek Uttreja, Indrajeet Karnik, Rasha Elkanayati, Gaurav Shanbhag, Sateesh Kumar Vemula, Michael Repka, and Naveen Chella

Abstract

Cancer remains a complex and life-threatening disease, and mortality is projected to increase significantly by 2060. Current therapeutic modalities such as chemotherapy, radiation therapy, and surgical interventions are challenged by drug resistance and adverse effects, necessitating the development of innovative strategies. Nanotechnology-based drug delivery systems, particularly stimuli-responsive polymeric vesicles, present a promising avenue for advancing cancer treatment. These systems increase efficacy and safety by improving drug bioavailability, targeting, and payload protection. Polymersomes are vesicular structures that are hollow and surrounded by bilayer membranes formed by the self-assembly of amphiphilic block copolymers. Their tunable mechanical robustness, encapsulation capacity, and easy surface modification make them attractive candidates for next-generation drug delivery systems. In this chapter, we have explored the versatility of stimuli-responsive polymersomes in cancer therapy. It provides an overview of the structure and formation mechanisms (including solvent-free and solvent-displacement methods) and the advantages

A. Padakanti · G. Shanbhag · N. Chella (✉)
National Institute of Pharmaceutical Education and Research (NIPER), Guwahati, Assam, India
e-mail: naveen@niperguwahati.in

P. Uttreja · I. Karnik · S. K. Vemula · M. Repka
Department of Pharmaceutics and Drug Delivery, School of Pharmacy, The University of Mississippi, Mississippi, USA

R. Elkanayati
National Institute of Pharmaceutical Education and Research (NIPER), Guwahati, Assam, India

Department of Pharmaceutics and Drug Delivery, School of Pharmacy, The University of Mississippi, Mississippi, USA

P. V. Devarajan et al. (eds.), *Advances in Stimuli-Responsive Nanosystems for Cancer Therapy*, https://doi.org/10.1007/978-981-95-2744-1_4

and limitations of each approach for controlling size, morphology, and encapsulation efficiency. The key properties that distinguish polymersomes from classical liposomes are critically analyzed. The focus on stimuli-responsive drug release strategies allows the spatiotemporal control of drug delivery to tumor microenvironments. This chapter reviews preclinical studies showing the efficacy of polymersomes in cancer models, including enhanced tumor accumulation and therapeutic efficacy. Challenges hindering clinical translation, including biocompatibility, stability, scalability, and regulatory barriers, are also addressed. Future research directions, including multistimuli-responsive systems, combination therapies, and personalized nanomedicine approaches, are also explored. The potential of polymersomes to advance cancer treatment was critically assessed, emphasizing their capacity to enable precise drug delivery, reduce off-target effects, and address drug resistance. However, the evaluation also underscores the necessity for further research to fully establish their clinical applicability.

Keywords

Polymersomes · Stimuli-responsive · Cancer therapy · Preclinical studies · Personalized nanomedicine

Abbreviations

LCST	Lower critical solution temperature
PISA	Polymerization-induced self-assembly
RAFT	Reversible addition-fragmentation chain-transfer
SPECT	Single-photon emission computed tomography
SPION	Superparamagnetic iron oxide nanoparticles
TBP	Transferrin-binding peptide

1 Amruta Padakanti and Prateek Uttreja contributed equally with all other contributors.Introduction

Cancer is characterized by uncontrolled cellular proliferation, resulting in the development of tumors that damage adjacent healthy tissue [1–3]. Cancer is projected to remain the leading cause of mortality by 2060 with a 2.08-fold increase in deaths [4]. Primary treatment options include chemotherapy, radiation therapy, surgical procedures for advanced and solid tumors, or a combination of these [5]. However, invasive surgery can adversely affect the patients [6–10]. Additionally, a major obstacle in chemotherapy is a conventional cancer treatment approach, is the resistance of patients to multiple drugs [11–14]. This resistance reduces the efficacy of the therapeutic agents or causes the efflux of the drugs. Consequently, new approaches such as nanosystems are being studied to address the challenges associated with current treatment plans [15].

Nanotechnology is rapidly growing in the biopharmaceutical sector as a replacement for conventional therapeutics. Nanomaterial-based systems function as drug delivery vehicles to enhance efficacy, safety, and bioavailability by facilitating the transcytosis of drugs across tight barriers and delivering them to specific sites [16, 17]. Furthermore, nanoparticulate systems enable the delivery of biomolecules such as DNA, siRNA, and mRNA [18, 19]. These systems can also improve circulation, mitigate toxicity, and protect therapeutic agents from degradation [20]. Additionally, such systems can provide sustained or stimulus-based drug release [21]. The surfaces of nanoparticle-based carriers can be modified to modulate the release of encapsulated drug molecules or cargo [17, 22]. These advantages of nanosystems render them a promising option for cancer therapies. Current nanosystems for cancer treatment include liposomes, polymeric micelles, lipid-polymer hybrid nanoparticles, and dendrimers [23–26]. Inorganic therapeutics include iron nanoparticles, carbon nanotubes, and gold nanorods [27].

In addition to modified or prolonged drug release patterns, surface modification or altered chemistry of the materials can also enable nanocarriers to release cargo in response to external or internal stimuli, referred to as triggers. The "triggers" can be divided into two categories—"intrinsic stimuli" and "extrinsic stimuli." Intrinsic stimuli shoot from or within cancer cells and are characteristic of the tumor environment, such as pH, temperature, and enzymatic content [28–30].

Conversely, the stimuli can be triggered outside the tumor cells to unload the drug from the nanoparticles. These external stimuli involve applying local heat, ultrasound, magnetic fields, near-infrared, or UV-visible light to release drugs near the desired site [31]. The release of the payload to the affected site is triggered by internal or external stimuli that activate stimuli-sensitive nanoparticles, depending on how the biomolecules are included. Suppose that the drug is entrapped within the nanoparticle. In this case, the stimulus renders a structural change to the surface of the nanoparticles, which might be in the form of cleavage or degradation of the carrier. When the therapeutic moiety is attached as a conjugate, the connection between the carrier and the therapeutic entity is cleaved to release the latter [32–34]. Consequently, the combination of diverse stimuli and a wide range of biocompatible materials available for drug delivery facilitates the development of customizable stimuli-responsive vehicles, enabling enhanced spatiotemporal control of drug transport to tumors while minimizing potential effects on surrounding healthy cells [35].

1.1 Overview of Stimuli-Responsive Polymersomes in Cancer Treatment

Polymer-based drug delivery systems are widely used across biopharmaceutical landscape to treat and diagnose various disease conditions [36]. Because of their highly tunable, bioresorbable, and biocompatible characteristics, polymers are ideal candidates for drug delivery vehicles [37, 38]. Polymer nanotechnology has also been widely investigated for various biomedical applications including cancer diagnosis and therapy [39]. Polymersomes, also known as polymeric micelles, were first described by Discher et al. in 1999 [40]. Polymersomes are self-assembled hollow structures composed of block or graft copolymers. A robust polymeric bilayer forms an amphiphilic exterior wall enclosing an aqueous interior. Polymersomes demonstrate mechanochemical robustness and can encapsulate hydrophilic (inside the core) and hydrophobic (in the membrane) molecules for cancer therapy [41]. The unique membrane also offers highly functional surface modifications to not only facilitate the transport of vehicles across membranes, which are challenging for permeation, but also modulate the release of the entrapped entity [42]. Several studies have explored the multifunctional nature of polymers for cancer treatment. Polymersomes engineered with stimuli-responsive and biocompatible polymers offer a promising avenue for cancer treatment. Their compartmental structure allows for precise drug encapsulation, while responsive elements enable controlled release in response to specific stimuli such as pH and temperature changes [43]. Targeting moieties such as folic acid ensures the polymersomes selectively attack cancer cells, thereby minimizing adverse effects on healthy tissues [44, 45]. The biodegradability of polypeptide-based copolymers further enhances their biocompatibility [46, 47]. These characteristics coupled with the ability to modulate drug release through environmental factors make polymersomes adaptable and efficient delivery systems for anticancer drugs.

1.2 Mechanism of Polymersome Formation

Polymersomes are self-assembled hollow sacs formed from block copolymer building blocks. The self-assembly of these nanoparticles is possible owing to the amphiphilic properties of the block copolymers. A pre-synthesized hydrophilic polymer chain activator is polymerized with a hydrophobic monomer. The hydrophobic block expands in length, and the overall amphiphilicity of the block copolymer changes with the morphology of the polymersome, which ultimately results in a vesicle with a morphology by controlling the degree of polymerization of the hydrophobic block [48, 49].

There are two commonly accepted mechanisms through which polymersomes are formed. The mechanism I proposed by Lasic and Fromherz is a "bilayer-to-vesicle" model. This is a two-step process wherein micelles aggregate to form a bilayer sheet in the first step. The bilayers start bending due to edge energy, forming spherical vesicles to reduce the free energy [50–53].

Mechanism II explains a different pathway in which micelles self-assemble and grow due to solvent diffusion into the micellar core. Bilayer sheets were not formed via this mechanism. As the bending energy decreases, micelles rearrange into vesicles [40, 54].

However, these two mechanisms are not the only ones that describe the assembly of copolymer-based vesicles. As mentioned earlier, current innovations in block copolymerization techniques can induce numerous variations in the structure and function of the copolymer. Ianiro et al. revealed that self-assembly may begin with liquid–liquid phase separation, forming copolymer-rich droplets. These droplets then nucleate micelles at the liquid–liquid interface, eventually merging to form spherical vesicles [55].

1.3 Manufacturing Methods of Polymersomes

The formulation of polymersomes can be achieved using several techniques, which can be divided into two major groups based on the solvent used to solubilize the block copolymer: solvent-free techniques and solvent displacement methods, in which block copolymers are formed in aqueous and organic solutions, respectively. Once these solvent-polymer solutions are formed, they are further mixed with an aqueous solution, and the organic fractions are eventually removed to obtain polymersomes [56].

1.3.1 Solvent-Free Techniques

Polymerization-Induced Self-Assembly (PISA) Polymerization-induced self-assembly utilizes the differential solubility between the hydrophilic monomer activator and the polymer, which progressively decreases upon the addition of hydrophobic molecules. This phenomenon results in liquid–liquid phase separation and the self-assembly of the amphiphilic block to minimize interfacial tension.

Phase separation constitutes the critical step governing the size, structure of the polymersomes, and membrane thickness, which can be precisely controlled by modifying the hydrophilic and hydrophobic block ratios or the degree of polymerization. Furthermore, stimuli such as temperature and pH can induce the self-assembly of amphiphilic copolymers, enabling additional control over polymersome morphology [55, 57–60]. PISA is an efficacious and scalable methodology for fabricating self-assembled polymersomes, offering advantages such as high solid concentration, one-pot synthesis, tunable morphology, and biocompatibility. However, it encounters challenges related to elevated reaction temperatures, solvent restrictions, catalyst toxicity, and long-term stability. Further investigation into biodegradable monomers, green chemistry approaches, and industrial scalability will enhance the applicability of PISA in the biomedical and pharmaceutical fields [61].

Rehydration Rehydration is a straightforward and cost-effective technique utilized for the preparation of polymersomes. This method involves the dissolution of a copolymer in an organic solvent, followed by evaporation to produce a thin film on a glass or metal surface. Subsequent rehydration of this thin film with aqueous solutions results in the formation of polymersomes. The primary advantage of this technique lies in its simplicity and cost-effectiveness. However, the vesicles produced by this method exhibit a broad size distribution, which can be refined to a narrower distribution by passing them through a filter with defined pore sizes [48, 61, 62]. Furthermore, challenges such as multilamellar aggregate formation and inadequate rehydration of highly hydrophobic polymers can adversely affect uniformity and drug-loading efficiency [63].

Electroformation Electroformation is a controlled version of rehydration utilized to generate vesicles with a narrow size distribution. This technique involves the deposition of block copolymers onto electrodes and subsequent rehydration under alternating current to regulate water diffusion, thereby controlling vesicle size. Electroformation enables precise manipulation of the size distribution of polymersomes through the monitoring of vesicle bulging and separation. Polymersomes produced via this method exhibit high encapsulation efficiency and hollow core structures, rendering them suitable for controlled drug delivery applications. However, the feasibility for scale-up necessitates specialized equipment, presenting challenges for industrial applications. Furthermore, process optimization is critical, as parameters such as voltage, frequency, and hydration conditions significantly influence vesicle formation and stability [64].

1.3.2 Solvent Displacement Techniques

Direct Injection The copolymer is solubilized in a suitable organic solvent, and the resultant polymer solution is introduced into an aqueous medium. The difference in the solubility of hydrophobic copolymer block between organic and aqueous phases increases interfacial tension, resulting in self-assembly and formation of polymersomes. This is a simple, cost-effective, and widely used method that relies on the solubility differences between polymers in organic and aqueous solvents. However, it produces a broad size distribution, which requires further refinement through extrusion or solvent selection [65–67]. The direct injection technique can be further refined using an emulsion phase transfer method.

Emulsion Phase Transfer This procedure introduces a water-in-oil emulsion comprising a polymer with an organic solvent into an oil-in-water emulsion rich in amphiphiles. Thus, the resultant mixture is a water-oil-water (w/o/w) triphasic system. This technique allows for encapsulation of hydrophilic molecules in the aqueous core and hydrophobic molecules in the oil phase [68, 69]. The primary advantage lies in its ability to generate vesicles with a uniform and narrow size distribution, while simultaneously enabling the efficient encapsulation of both hydrophilic and hydrophobic drug molecules within the vesicular system. However, it suffers from limitations, such as potential instability due to oil retention in lipid bilayers, which may alter membrane properties. It also has lower yields than other fabrication technologies [69, 70].

Microfluidics Microfluidics is a well-regulated and advanced manufacturing technique that offers superior control over the size distribution and morphology of the polymersomes. The principle involved in this technique is the creation of double emulsion droplets (w/o/w) in a microfluidic device. Upon removal of organic solvent, these are converted to polymersomes. A stream of organic amphiphiles dissolved in the organic solvent is extruded into the aqueous phase to produce stable amphiphilic water-in-oil droplets. Subsequently, these droplets are introduced into an amphiphile-coated aqueous medium, forming double emulsion vesicles of the water-oil-water (w/o/w) phase [71–73]. This technology helps achieve precise control over the polymersome size, morphology, and composition with high reproducibility and efficient encapsulation of both hydrophilic and hydrophobic drugs. This is a continuous, high-throughput fabrication technique. However, the device fabrication and operational setup are complex, which necessitates specialized expertise and equipment. Additionally, scaling up microfluidic systems for large-scale industrial production is challenging, and fluidic channel clogging can affect consistency in manufacturing [72].

2 Design of Stimuli-Responsive Polymersomes

2.1 Structural Features and Composition

The phase behavior of amphiphilic copolymers serves as a framework for polymeric vesicles. Polymersomes offer numerous advantages over liposomes due to their unique properties. They exhibit enhanced stability at biological pH, improved functionalization, and adjustable mechanical properties. Additionally, polymersomes possess an appealing drug release profile, biodegradability, bioavailability, and ability to respond to stimuli for drug release.

A spherical hollow structure containing an aqueous core of polymersomes, enclosed in a polymer shell composed of diblock polymers [74]. Polymersomes can be synthesized using an AB-type diblock copolymer, where A is a hydrophobic copolymer and B is a hydrophilic copolymer [40, 75]. Triblock copolymers, in which the innermost structures are formed from an additional hydrophobic block, are another potential component of polymersomes in addition to the AB variety. With the help of amphiphilic block copolymers, polymersomes can be formed according to chain length and molecular weight [76]. Polymersomes can have membrane thickness ranging from 2 nm to 50 nm [77]. The stability, stiffness, and effectiveness of entrapping hydrophobic drugs are improved with thicker membranes than polymersomes with thinner membranes.

Polymersomes offer clear benefits compared to polymeric nanoparticles owing to their unique vesicular structural arrangement, which enables the encapsulation of drugs with both hydrophobic and hydrophilic characteristics. The greater stability and controlled drug release provided by the bilayer shape can be subsequently optimized for targeted therapeutic outcomes by varying the thickness and composition of the membrane. Additionally, polymersomes can be customized for multiple purposes, allowing them to transport different medications and diagnostic agents. They can also be modified to precisely target and deliver these substances. Their design closely resembles that of biological membranes, which can enhance biocompatibility and minimize immunogenicity. Polymersomes are an excellent choice for drug delivery in various medical applications owing to their versatility and strong structure [78, 79].

Two essential features of the tumor microenvironment and vasculature, increased retention and improved permeability, are responsible for the passive targeting effect called the enhanced permeation and retention (EPR) effect [80]. Polymersomes that fall within the size range significant for the EPR effect have a bilayer thickness of approximately 50 nm and an overall diameter of approximately 100 nm. Particles within a size range of approximately 10–200 nm can effectively use the EPR effect [81]. Particles within this range are preferable to exploit the tumor's leaky vasculature, avoid quick clearance of kidney (particles less than 5–6 nm), and gain advantage from the liver and spleen's efficiency capture (size larger than 200 nm) [82]. Polymersomes with a diameter of 100 nm can endure the dynamic conditions of the bloodstream, allowing them to reach tumor sites and navigate the tumor vasculature, and can have a diameter ranging from 100 to 780 nm or even more significant in specific cancers. Because lymphatic drainage is restricted, polymersomes can be retained within the tumor at this size. However, the direct influence of the bilayer thickness on the EPR phenomenon is unclear. Polymersome stability and stiffness are affected by their thickness, affecting how long they stay in circulation and how they interact with the tumor environment. Polymersomes with a higher bilayer thickness have a better chance of reaching tumor tissue and accumulating there because of their greater stability against the early release of encapsulated drugs and their shielding from destruction in the bloodstream [83].

2.2 Stimuli-Responsive Mechanisms

Conventional nanocarriers, without responsive linkers, gradually release chemotherapy drugs through a diffusion-controlled mechanism. The application of these nanocarriers may not provide tumor tissues with sufficient amounts of anticancer drugs to be effective. Drug molecules have a plasma level that, when exceeded, can be harmful and, when not reached, can be ineffective. Therefore, nanocarriers play a crucial role in reducing the overall drug levels in the body, while ensuring that the therapeutic range is maintained following a single dose [84]. As a result, many advancements have been made in the development of polymersomes that can respond to various stimuli, such as temperature, light, electrochemical signals, magnetic fields, pH, redox reactions, ionic conditions, inflammation, ultrasound, enzymes, and glucose. These advancements are aimed at enhancing the effectiveness of anticancer drugs. Temperature, pH, and redox-responsive nanocarriers have gained significant popularity in biomedical applications, particularly cancer therapy [85]. These carriers are designed to respond to specific stimuli, allowing targeted and effective treatment. This chapter explores the latest advances in polymersomes that respond to chemical and biological stimulus. These innovative polymersomes have significant potential in cancer therapy and show promise for future applications in this field.

2.2.1 pH-Sensitive Polymersomes

In the context of polymersomes that deliver cytotoxic drugs, the pH gradient is one of the most common environmental triggers. The glycolytic metabolism of cancer cells results in a lower pH in the surrounding environment than in healthy tissues. The acidic environments commonly encountered in tumor tissues or certain cellular segments, such as lysosomes, and endosomes often have pH values lower than the normal physiological pH of 7.4 [86]. Studies have demonstrated that pH levels below 4.5 exist within the tumor microenvironment [87]. Polymersomes are designed by incorporating pH-responsive substances to exploit an acidic condition, which can trigger drug release. These polymersomes are engineered with structures incorporating acid-sensitive linkers and bonds including hydrazine, acetals, ortho esters, and imines [88]. By harnessing these processes, pH-responsive substances can achieve site-specific drug release at tumor locations or within target cells, thereby lowering systemic adverse effects and increasing therapeutic effectiveness.

Drug release may be due to one of the following mechanisms:

(a) Protonation and Deprotonation

- Ionizable groups within the polymer accept or donate protons at different pH levels, thereby changing their charge state.
- The alteration affects polymer solubility, leading to swelling or disassembly of the carrier, thereby releasing encapsulated drug molecules [87].

(b) Bond Breakage

- Acid-sensitive bonds such as hydrazine, imines, and acetal become prone to hydrolysis under acidic conditions.
- Hydrolysis leads to material degradation and subsequent drug release [89]

(c) Conformational Changes

- Certain polymer transitions between expanded and collapsed forms in response to pH fluctuations.
- These conformational changes trigger drug release by altering the stability and permeability of the polymersome structure [90]

A recent study reported the synthesis of an amphiphilic triblock copolymer, methoxyl poly(ethylene glycol)-block-poly(L-lysine)-block-poly(2-(diisopropyl amino)ethyl methacrylate). This copolymer comprises a hydrophilic diblock with a hydrophobic block (mPEG-PLys) and (PDPA). The resulting polymersomes demonstrated efficient codelivery of doxorubicin (DOX), a polar chemotherapeutic agent, and siRNA targeting ADP-ribosylation factor 6 (siArf6) into tumor cells in response to changes in lysosomal pH. The combination of DOX and siArf6 within the PLD (PLD@DOX/siArf6) polymersome resulted in a synergistic cytotoxic impact on PC-3 prostate carcinoma cells [91].

Similarly, Zhu et al. fabricated polymersomes with pH-responsive effects by the co-assembly of an acid-cleavable polymer-drug conjugate (PEG-doxorubicin conjugate) and amphiphilic cholate-grafted poly(L-lysine) linked by an acid-labile benzoic imine chain. This linkage conferred pH-dependent membrane permeability, leading to rupture of the polymersome in response to a decrease in pH [92].

Polymers containing proton-responsive groups such as guanidine and tertiary amines can absorb protons in acidic environments and release them in basic media, resulting in varying solubility profiles in different solvents. Notable examples include poly(dimethylaminoethyl methacrylate) (PDMAEMA) and poly(diethylaminoethyl methacrylate) (PDEAEMA). Additionally, a drop in pH can induce degradation of some polymersomes. Polymersomes, wherein drugs are linked to polymers via maleic acid amide, hydrazine, ketal, imine, β-thiopropionate, acetal, or ortho ester linkages, may release in acid-based environment due to the cleavage of these linkages [93].

2.2.2 Temperature-Sensitive Polymersomes

Temperature is a commonly used stimulus for activating the specific responsiveness of polymersomes in drug delivery systems. Thermoresponsive polymers undergo a phase transition following fluctuations in temperature, especially in terms of solubility, swellability, and permeability that alters the release of the drug loaded into the nanocarrier. Certain polymers exhibit temperature-dependent solubility changes [94]. Some possess an upper critical solution temperature (UCST), meaning that they remain insoluble at lower temperatures but dissolve upon heating. On the other hand, a few polymers have a lower critical solution temperature (LCST), which is soluble at a lower temperature. Various applications in the field of biomedicine have exploited this particular property. Specifically, with respect to drug delivery, the release of a medication can be triggered by a natural increase in temperature that causes the thermosensitive polymer to contract, or by externally applied elevated temperatures. Because the local temperature in solid tumors is marginally higher than the average body temperature, this finding holds great promise for cancer therapy. Modifying the LCST of the thermosensitive polymer to fall somewhere within the two extremes makes it feasible for drug delivery systems to accumulate specifically in the tumor [95].

Various temperature-responsive block copolymers have been successfully synthesized, with poly (N-isopropyl acrylamide) (PNIPAM) being the most widely utilized block for creating temperature-sensitive copolymers. Specifically, PNIPAM can switch from hydrophilic to hydrophobic when the temperature is adjusted to the lower critical solution temperature (LCST) of 32 °C. Below the LCST, it shows a swollen behavior, and at a temperature exceeding the LCST, PNIPAM exists in dehydrated and collapsed conditions and releases the loaded drugs. However, the LCST values can be enhanced by mixing with hydrophilic polymers and adjusting the environment to the body temperature [96, 97].

Polymerization process using controlled reversible addition-fragmentation chain-transfer (RAFT), Liu et al. synthesized a temperature-responsive copolymer poly(N-vinylcaprolactam)n-poly(dimethylsiloxane)65-poly(N-vinylcaprolactam)n. By adjusting the polyvinyl caprolactam chain length, stable and monodisperse polymersomes that penetrate doxorubicin at body temperatures between 37 and 42 °C are created. The vesicle membrane developed pores at higher temperatures, which regulated the release of DOX from the vesicles. The developed polymersomes showed cytotoxicity against human alveolar adenocarcinoma cells. With their formidable drug-loading capacity (40%) and ability to undergo temperature-response morphological alterations without losing structural shape, these polymersomes show great potential as drug delivery systems for cancer treatment [98].

2.2.3 Redox-Sensitive Polymersomes

Redox potentials exhibit significant variations within the extracellular and intracellular environments, as well as between cancer cells and normal tissues. The concentration of glutathione (GSH), a well-recognized reductive agent, is lower (2–20 μM) in body fluids (such as blood) as the extracellular region and on the surface of the cell than in the nuclei and cytosol, where the amount of agents is much higher (10 mM), creating a highly favorable microenvironment for reduction. As a result, by exploiting the notable disparity in the redox potential between the external and internal compartments, responsive nanovescicles can be manipulated to alter the characteristics of the polymersome layers or provide these sections with innovative capabilities. This includes the controlled release of various biologically active substances from polymeric vesicles within cells [99, 100]. One common feature of reduction-responsive nanocarriers is the presence of GSH-breakable chains in the linker groups of cysteine derivatives such as succinimidyl 2-pyridyldithiol (SPDP), and disulfide-based dimethacrylate [101].

Wu et al. used an S-nitrosoglutathione functionalized poly(ethylene glycol)-block-poly(propylene sulfide) (PEG-PPS) block is a copolymer to prepare reactive oxygen species (ROS) and glutathione (GSH) sensitive polymersomes. In HepG2/ADR cell lines, DOX-loaded polymersomes showed substantial growth inhibition and ROS-triggered release, with increased intracellular accumulation. This important therapeutic endeavor raises the possibility that these stimuli-responsive polymersomes might be useful for treating malignancies with multiple treatment resistance [102].

Similarly, Jia et al. demonstrated sustained delivery of calcein in the intracellular compartment in response to the reduction potential of the polymersomes made up of cholesterol-based amphiphilic block copolymer PEG_{45}-SS-PACho. The block copolymer PEG-SS-polyacrylate/cholesterol (PAChol) was the basis for a strong reduction-responsive polymersome. The sensitivity to reduction was brought about by the disulfide bridge connecting the hydrophilic PEG block and the hydrophobic PAChol block [103].

2.2.4 Light-Sensitive Polymersomes

Photosensitive polymersomes, which can respond to ultraviolet, visible, and near-infrared light, have been utilized as external stimuli for drug delivery. This technique relies on light-dependent carriers that release their cargo in stimuli to light stimulation [104]. Rapid response to various stimuli is possible only after the introduction of acids, bases, or other reagents into the surrounding environment. Nonetheless, light requires no external chemicals because it is a distant stimulation.

To create light-responsive polymersomes, amphiphilic polymer systems have been modified by adding the photolabile chromophores such as azobenzene, stilbene, triphenylmethane, and 2-nitrophenyl alanine [105–107].

Light-sensitive polymersomes are produced by self-assembly of amphiphilic polymers functionalized with light-activatable substances (photothermal conversion agents, photosensitizers, and chromophores). Any part of the polymersome can include light-activatable components or be covalently linked to the amphiphilic copolymer. When exposed to light, the photosensitive moieties inside the amphiphilic block polymers undergo various processes that convert the radiation into a chemical response, including photoreactions, photosensitization-induced oxidation, heat generation, and photoconversion. Disintegration and breakdown of the polymersome structure facilitate the release of the contained drugs [108].

Wang et al. used light-sensitive polymersomes made from diblock copolymers of poly(ethylene glycol)-poly(D, L-lactic acid) to examine the effects of protoporphyrin IX (PpIX) on the viability of normal and tumor cells. A light source in the form of ionized cold atmospheric plasma (CAP) gas that radiates at ambient temperature was employed to initiate drug release. When irradiated, the healthy cells around the tumor suffered less damage when using this CAP technique than UV radiation. Owing to the significant difference in toxic PpIX degradation between cancer cells and normal cells, the release of PpIX from polymersomes specifically induces cytotoxicity in malignant cells. Notably, they exhibit a significant decrease in cell viability in tumor cells compared to that in normal cells [109].

In order to determine if polymersomes might release DOX and Nile Red (NR) in response to photo triggering, Hou and colleagues synthesized an amphiphilic block copolymer poly(N,N′-dimethylacrylamide)-b-poly(n-butyl acrylate) (PDMA-b-PNBA). Emulsification yielded after the block copolymer polymersomes was generated in bulk by the reversible addition-fragmentation chain transfer (RAFT) approach. After being exposed to 365 nm UV light for 15 min, the hydrophobic portion of the copolymer broke down, causing the polymersome to break down and the medication to escape. The rate of emission was governed by the exposure time [110].

2.2.5 Enzyme-Sensitive Polymersomes

Amphiphilic polymers undergo a relatively slow hydrolytic breakdown process, which eventually leads to the degradation of the polymersomes. As a result, under typical physiological conditions, the rate of drug release is limited. Polymersomes have been engineered to be more sensitive to lysosomal and microorganism-secreting enzymes to speed up the hydrolytic disintegration rate and release their payload [111]. Several disorders, such as cancer, thrombosis, inflammation, and infectious pathogens, are associated with enzymes overexpression. The presence of specific enzymes in inflammatory tissues causes these polymersomes to undergo physicochemical modification [112]. This method involves developing polymersomes by covalently bonded enzyme-responsive targets to the copolymer.

Yao et al. designed polymersomes with quinone trimethyl lock-capped slef-immolative side linkages stabilized photosensitizers, Nile blue, and coumarin loaded into the hydrophobic portion of the polymer. In normal cells, the double quenching process prevents premature activation and release of loaded drug molecules without NAD(P)H: quinone oxidoreductase isozyme 1 (NQO1) or endogenous enzymes. In cancer cells with elevated levels of NQO1, the enzyme disintegrates quinone trimethyl links, causing the release of photosensitizers along with coumarin and Nile blue in a controlled manner. The release mechanism triggers the transition of near-infrared (NIR) emission from inactive to active, effectively stimulating photosensitizers for photodynamic treatment. This groundbreaking method utilizes the unique overexpression of NQO1 in tumor cells to precisely activate PDT, reduce harm to nearby healthy cells, and improve the effectiveness of the therapy [113].

Enzyme responsiveness was further utilized to enhance the therapeutic effectiveness of water-soluble anticancer drugs by inhibiting premature drug release in systemic circulation and improving drug localization inside cancer cells. A previous study compared the performance of polymersomes prepared from an amphiphilic polymer (PEAMP) containing methacrylate chains responsive to the esterases. DoX-loaded polymersomes prepared by self-assembly demonstrated significant leakage of Dox.HCl at neutral and acidic pH, and the presence of esterases under in vitro conditions. This leakage was due to esterase-induced hydrolysis in the polymer structure. The improvement in the therapy was further confirmed by the improved uptake of DoX.HCl by HeLa cells, which have a rich esterase fraction, greater circulation half-life of polymersomes with reduced clearance in pharmacokinetic studies, and finally, effective tumor inhibition rate in tumor-bearing mice models [114].

2.2.6 Ultrasound and Magnetic Field-Sensitive Polymersomes

Ultrasound is a highly versatile medical tool that has been extensively developed for a wide range of applications in the field of biomedicine. It is a noninvasive, cost-effective, and safe method that provides real-time results. This approach has been proven to be a valuable tool for enhancing the effectiveness of chemotherapy. It can penetrate deep tissues, making it a promising option for treating cancers that have shown resistance to conventional treatments. In addition, its cost-effectiveness adds to its appeal as a professional treatment option. It can also be used for drug release, hyperthermia induction, and tumor imaging [115].

Polymersomes activated by pH changes and ultrasonications were generated by Chen and Du. A triblock copolymer of PEG-b-poly[2 (diethylamino)ethyl methacrylate-stat-tetrahydrofuranyloxy)ethylmethacrylate] (PEG b-p(DEA-stat-TMA)) was used to construct the polymersomes. An experiment was performed to investigate the effects of the solution pH and ultrasonic radiation duration on the assembly' size and shape. When exposed to 180 W ultrasonic radiation, the vesicles shrank and released doxorubicin more quickly. The polymersomes contracted due to hydrolysis of the PTMA and changes in their hydrophobic balance under the effect of ultrasound [116].

Several polymersomes have been engineered to release the cargo in stimulus to a magnetic stimulus. In most cases, magnetic nanoparticles are trapped within polymersomes, and their release is triggered by factors such as mobility, contrast, or magnetothermal induction. It is possible to package magnetic agents embedded in vesicular layer or hollow chambers. The release mechanism occurs when the polymersomes shrink due to stretching caused by an extrinsic magnetic field, causing the magnetic fragments to aggregate within the vesicle membrane [117]. A significant improvement in DOX release and cytotoxicity was reported under a magnetic field when DOX and superparamagnetic iron oxide nanoparticles (SPION) were loaded into the polymersomes of poly(trimethylene carbonate)-block-poly(L-glutamic acid) (PTMC-PGA) copolymers in cancer cells [118].

3 Preclinical and Clinical Advances

3.1 In Vitro Studies

In vitro assays are often used to assess the efficacy, toxicity, and mechanisms of action of potential drug candidates before proceeding to more complex and expensive studies. These findings support the hypothesis that polymersomes are effective anticancer drug delivery systems. In the case of stimuli-responsive polymersomes, primarily in vitro studies are conducted to evaluate the release of drugs loaded in the presence of triggering mechanisms under in vitro conditions, such as pH, temperature, the presence of enzymes, and the application of magnetic field. Furthermore, therapeutic effectiveness against cancer can be confirmed by cell line studies specific to cancer.

For instance, a dissolution study at various pH conditions (5.5, 6.5, and 7.4) was performed to understand the pH sensitivity of the Angiopep-2 conjugated polymersomes. The studies indicated that DOX release was dependent on pH stimuli, with significantly less release at pH 7.4 (25.89 ± 3.35% at 96 h), indicating the stability of polymersomes in plasma. The release rate increased with a decrease in pH and was found to be higher at pH 5.5 (83.11 ± 3.18%) with a burst release. Furthermore, an in vitro blood–brain barrier (BBB) model was created to evaluate to which the polymersomes penetrated the BBB. The targeted polymersomes had a greater transport ratio (17.74% ± 3.57) compared to the non-targeted polymersomes, which had a ratio of 3.24% ± 1.78. The antitumor activity of the polymersomes was assessed using Cell-Counting Kit-8 (CCK-8) and U87-MG cells. The outcomes showed that the combination of radiation and DOX substantially affected the inhibition of tumor cells [119].

A dialysis in vitro technique was implemented to analyze the release of curcumin (CUR) from the polymersomes in the release medium, while the effectiveness of NIR-responsive drug release was confirmed by irradiating the samples for 5 min with a laser before transferring them the dialysis bag and initiating the release study. The MTT assay was used to assess the formulations' cytotoxicity. Flow cytometry and fluorescence microscopy were used to analyze the cellular accumulation of the targeted (Apt-P-AuNR-CUR) and non-targeted polymersomes in cultured B16F0 and CHO cells. The release from the polymersomes was found to be higher after NIR radiation owing to the breaking of the hydrogen bonds and disruption of the polymersome self-assembled structure, resulting in leakage of CUR from the polymersome. In vitro, investigations demonstrated that Apt-P-AuNR-CUR exhibited superior cellular uptake and therapeutic efficacy compared with P-AuNR-CUR. This was attributed to the strong binding affinity of the Sgc-8c aptamer on the outermost layer of the polymersomes to the PT7 receptor, which is highly expressed on the outer membrane of melanoma carcinoma cells [120].

3.2 Efficacy in Animal Models

A significant portion of the existing knowledge regarding the evaluation of polymersomes depends solely on in vitro assays. Therefore, it is fundamental to expand and explore these delivery systems via approaches to solidify the reliability of polymersome-based anticancer therapy and to provide comprehensive insights into the pharmacokinetics, biodistribution, and toxicity of these nanoformulations. Furthermore, the following section discusses recent in vivo studies conducted on polymersomes to achieve the aforementioned goals.

For instance, the effectiveness and safety of newly developed polymersomes were assessed in vivo using male athymic nude BALB/c mice xenografted with malignant (A375) melanoma cells. Polymersomes were developed using poly (propylene sulfide)20-bl-poly (ethylene glycol)12 (PPS20-b-PEG12), which self-assembles into structures capable of dual photodynamic (PDT) and chemotherapeutic effects. They incorporated a hydrophobic photosensitizer, zinc phthalocyanine, into their shell while encapsulating doxorubicin hydrochloride (DOX) in their inside compartment. Biodistribution studies conducted via fluorescence imaging revealed marked accumulation of polymersomes at the tumor site, as evidenced by increased fluorescence intensity and significantly lower accumulation in other organs. Upon irradiation, the polymersomes demonstrated notable suppression of tumor progression, reducing it by approximately 70% over 14 days. This outcome was significantly superior to the results observed in control groups treated with saline, free DOX, and irradiated polymersomes devoid of DOX [121].

The shape, size, surface charge, and elasticity of nanocarriers considerably impact their in vivo biological performance. These physical attributes can enhance the therapeutic delivery efficiency. Extensive research has been conducted on the performance of polymersomes, focusing on factors such as their shape, size, surface charge, and elasticity. These insights offer valuable guidance for understanding the connection between polymersome elasticity and accumulation in the brain tumors. Zheng M. et al. developed polymersomes using poly(ethylene glycol)-block-poly(pentafluorophenyl methacrylate) block copolymers (PEG-b-P(PFPMA) cross-linked with diamines and decorated them with Angiopep-2. The elasticity of the polymersomes was assessed by measuring the Young's modulus using atomic force microscopy (AFM). In vivo experiments were conducted on nude mice implanted with orthotopic luciferase-tagged U87MG glioblastoma cells. These results demonstrated that soft polymersomes had a relatively improved circulation rate in the blood. However, the stiffest polymersomes exhibited the highest brain tumor accumulation. These results indicate that the combined impact of cellular uptake, tumor penetration, and BBB permeation outweighs the benefits of prolonged blood circulation in achieving optimal delivery to brain tumors [122].

The pH-stimulus polymersomes developed by incorporating poly([N-(2-hydroxypropyl)] methacrylamide) 35-b-poly[2-(diisopropylamino)ethyl methacrylate]75 (PHPMA35-b-PDPA75) were loaded with 10% doxorubicin and successfully released DOX in the acidic tumor microenvironment due to the pH switchable property of the PDPA. The antitumor activity and biodistribution were evaluated using female black C57BL/6 mice bearing EL4 lymphoma tumors injected subcutaneously (heterotopic model). The biodistribution assay performed using the fluorescent dye DBCO-Cyanine7 (Cy7) exhibited enhanced accumulation in the tumor region of the loaded polymersomes. In addition, the definite decrease in tumor volume was clearly visible for the DOX-loaded polymersomes with a 100% survival rate compared to the mice groups treated with saline or free DOX [123].

Targeted drug delivery is one of the major advantages of nanoformulations, as it can reduce side effects and enhance efficacy. The targeting ability was evaluated in vivo in animal models. Administration of functionalized polymersomes with a transferrin-binding peptide (TBP) to the animals (xenografted HCT-116 colorectal tumor model) resulted in a significant improvement in the cellular absorption of doxorubicin with inhibition of the cancer cell volume. Simultaneously, the off-target adverse effects were minimized [124].

3.3 Clinical Trials and Human Studies

Despite the advantages of polymersomes over lipid-based nanoparticles, including liposomes and solid lipid nanoparticles (LNPs), polymersomes have not yet been used in clinical trials. Lipid nanoparticles continue to dominate the clinical drug delivery landscape [66]. Searching PubMed for the term "liposome" yielded 991 results, with publications starting in 1991. However, a more specific search for "polymersome" and "cancer" yields only 377 results from 2004 to 2024, including 55 reviews and meta-analyses. None of the searches returned results for clinical trials involving polymersomes; however, a similar search for clinical trials involving liposomes yielded 1090 studies. Additionally, there are several liposome-based marketed products for cancer therapy, such as Doxil®, Myocet Liposomal®, and Vyxeos Liposomal® [125]. To confirm the results from PubMed, ClinicalTrials.gov was accessed in May 2024. The search was conducted using the keywords "polymersome" and "cancer," where only one study that was withdrawn in 2015 was reported. This clinical study involved a minimal-risk blood-drawing protocol to evaluate a new method for capturing Circulating Tumor Cells (CTCs) using three types of near-infrared emissive polymersomes (NIR-EPs). This study aimed to determine the effectiveness of NIR-EPs targeting mesenchymal markers, prostate-specific membrane antigen (PSMA), and epithelial cell adhesion molecule (EpCAM) in capturing CTCs. The study was planned to be further expanded, if successful, to validate its effectiveness in broader clinical use. However, this study was terminated for several reasons.

4 Challenges and Opportunities

Polymersomes represent a potential novel frontier in the field of drug delivery. Moving polymersomes from laboratory research to clinical applications requires addressing several key challenges. First and foremost is ensuring biocompatibility and safety through comprehensive long-term studies assessing degradation products and clearance from the body. These studies are essential for meeting regulatory standards. Additionally, efforts must be focused on overcoming obstacles related to stability, storage, and large-scale manufacturing to make polymersome-based treatments economically viable and accessible to a broader population [126]. However, their incorporation into biological systems poses several challenges that must be addressed to ensure that the therapy is effective and safe for therapeutic applications. Some of these challenges include:

4.1 Biocompatibility and Toxicity Concerns

Polymersomes, which are vesicles produced by the amphiphilic self-assembly of block copolymers, often require cross-linking of hydrophobic segments to improve their stability. Different cross-linkers have been used to fabricate various polymersomes, such as PBD-b-PEO or PEO-b-PAA-b-poly (N-isopropylacryl-amide) [127]. However, once the cross-linking procedure is completed, it may be difficult to remove harmful residues because of the potential for retention in the membrane [128].

4.2 Stability and Shelf Life

Polymersomes need to maintain their stability throughout storage. Multiple factors including aggregation and rupture impact both the efficacy and security of final products through early medication release. It is essential to maintain polymersome stability regarding shape, size, and surface charge over extended periods through environmental variations during both processing and storage. The interaction of polymersomes with cellular components may influence drug release at targeted therapeutic sites and affect their overall performance [126].

4.3 Scalability of Production

Developing manufacturing methods that can produce polymersomes with consistent quality, size, and functional characteristics at a reasonable cost remains a significant challenge. The complexity of polymersome formulation techniques, precise regulation of their vesicles, and lack of automation can make large-scale manufacturing processes challenging [129].

4.4 Regulatory and Ethical Considerations

Drug delivery systems need to comply with regulations to be in compliance. Drug delivery systems need to follow Good Manufacturing Practices (GMPs) and get regulatory clearance from authorities such as the Food and Drug Administration at the right time. High quality, safety, and efficacy standards in polymersome-based therapeutic systems can be maintained by strictly following the established standard protocols during their development and manufacturing processes. Unlike liposomes, polymersome systems have progressed at a slower rate than liposomes because they are less available in preclinical and clinical studies. Choosing the appropriate material for the packaging of the product remains an important consideration. Liposomes typically use phospholipids that are easily accessible, among which phosphatidylcholine is most commonly used due to its natural abundance.

Polymersomes are primarily constructed using synthetic materials. The lack of pre-clinical and clinical studies on polymersomes creates substantial challenges for obtaining clinical approval. Regulatory agencies require comprehensive assessments of toxicology profiles along with immunogenicity levels, biocompatibility, biodistribution, and biodegradation profiles of polymersomes prior to granting approval. Polymeric chain weight distribution variations at structural and molecular levels present extra obstacles when seeking regulatory approval [130].

5 Multi-Responsive Polymersomes for Cancer Therapy

Stimuli-responsive polymerosomes have been employed to deliver the drugs in response to stimulation. However, it is sometimes insufficient to address the increasingly complicated needs of cancer treatments. The multi-responsive polymersomes have dual responsiveness to combine a variety of stimuli in the single entity by including pH-temperature, light-temperature, pH-redox, or pH-light. Furthermore, these distinctive characteristics aim to provide accurate drug targeting inside the tumor environment, with programmable stimuli-triggered behavior, decreasing off-target effects, and lower toxicity [126]. Combining dual responsiveness provides a synergistic strategy by concurrently using a variety of triggers to address the heterogeneity and complexity of cancer cells. Aimed to minimize the possibility of resistance development and tumor growth, which is a significant benefit. Detailed information on recent studies that demonstrated the possibility of combining polymersomes with two stimuli-responsiveness block copolymers for targeted drug administration and enhanced outcomes for cancer therapy is presented in Table 4.1.

5.1 Hierarchy Level

Cancer is characterized by a high degree of complexity, involving significant genetic, molecular, and cellular heterogeneity that influences tumor development, advancement, and metastatic dissemination. It remains the second leading cause of

Table 4.1 Details of a few studies from the literature using multi-stimuli responsiveness against cancer

Sr. No.	Type of polymer	Stimuli responsible	Therapeutic agent	Key findings	Ref
1.	Poly (ethylene glycol)–poly(acrylic acid)–poly(2-(diethyl amino) ethyl methacrylate) (PEG–PAA–PDEA)	Reduction and pH	Cytochrome C and bovine serum albumin	Intra-cellular delivery of protein and cancer cell apoptosis	[131]
2.	Poly[(mPEG-SS-amino) (N, N-di isopropyl ethylene diamino) phosphazenes] (PPDPs)	Reduction and pH	Doxorubicin hydrochloride And doxorubicin	Improved stability, targeted drug delivery	[132]
3.	Diblock copolymers poly(ethylene glycol)-SS-poly(2-(diethyl amino)ethyl methacrylate) (PEG-SS-PDEA)	Reduction and pH	FITC-BSA & FITC-CC	Controlled drug delivery of DoX	[133]
4.	Poly (ethylene glycol)-mono-2,4,6-trimethoxybenzilidene-pentaerythritol carbonate (PEG-PTMBPEC)	pH and light	Doxorubicin	Sustained drug release & enhanced passive targeting, superior antitumor activity. Lower systemic toxicity compared to free DOX	[133]
5.	Poly(ethylene glycol)-b-poly(N,N-diethylamino ethyl methacrylate) (PEG-b-PDEAEM)	pH and light	Doxorubicin	Higher stability with pH and light-induced drug release	[134]
6.	Methoxyl poly(ethylene glycol)-*b*-poly(*N*-isopropyl acrylamide)-*b*-poly[2-(diethylamino)ethyl methacrylate-*co*-2-hydroxy-4-(methacryloyloxy) benzophenone] (mPEG-*b*-PNIPAM-*b*-P(DEAEMA-*co*-BMA))	pH and temperature	Doxorubicin & paclitaxel	Showed Efficient uptake by MCF-7 & HeLa cancer cells with non-cytotoxic towards L929 Normal cells	[135]
7.	Poly(N-vinyl caprolactam) (PVCL)-b-poly(dimethylsiloxane) (PDMS)-b-PVCL	pH and temperature	Doxorubicin	Triggered DOX release w.r.t pH and temperature providing controlled delivery. No significant toxicity was observed in subacute toxicity studies	[136]

(continued)

Table 4.1 (continued)

Sr. No.	Type of polymer	Stimuli responsible	Therapeutic agent	Key findings	Ref
8.	UCNP-PNSP@DOX	UV-redox	DOX	Targeted drug delivery	[137]
9.	PCL8000-PEG8000-PCL8000 triblock polymer	pH and light (NIR radiation)	ICG (Indocyanine green) & DOX	Improved tumor growth inhibition and reduced side effects	[138]

mortality globally, reflecting the significant challenge posed by its multifaceted nature [139]. Despite remarkable advancements gained through anticancer approaches, the basic phenomenon of conventional cancer management is still defined by genetic irregularities in cells, leading to the uncontrolled proliferation of tumors and subsequent use of standard therapies to treat it. There are approximately a hundred distinct forms of cancer, each of which can be found in a variety of organs and sub-tissues. Moreover, the same tumor may express different protein markers based on its specific site or stage. Currently, surgery, chemotherapy, radiation, and immunotherapy are the predominant standard therapies for cancer treatment, depending on the type and grade, either separately or in combination. Surgery is an appropriate choice for treating local solid tumors that are in readily accessible regions of the body and do not metastasize. Nevertheless, owing to the tendency of several malignancies to metastasize, it is imperative to provide intensive therapies, including radiation and chemotherapy. However, these therapies have notable drawbacks, such as poor targeting effectiveness, inadequate tumor penetration, multidrug resistance, inefficient systemic bio-distribution, and unregulated drug release [127, 139]. Immunotherapy has evolved as a successful and precise therapeutic technique for cancer treatment, offering an alternative approach to traditional medicines. It utilizes the immune system of an individual to combat cancer. Immunotherapeutic strategies include various methods such as cancer vaccines, monoclonal antibodies, gene editing, immune checkpoint inhibitors, and oncolytic viruses [140]. The inherent variability of cancer has witnessed a substantial change with the introduction of immunotherapies, which has led to the development of the concepts of precision and personalized medicine. This strategy prioritizes tailoring the possibilities for therapy based on the distinct requirements of each person.

5.1.1 Polymersomes in Personalized Nanomedicine

Over a decade, it has become more prominent that each patient's malignancy is distinctive and different from one another. Therefore, the theory of "one drug fits all" that can treat an entire patient population is now considered outdated. The US National Cancer Institute defines "precision medicine" as an advancement from "personalized medicine" that represents a future that intends to enhance the quality of health through a personalized approach [141]. This entails having better control over individuals' genetic profiles, environmental factors, and lifestyle data to make choices for their medical applications. The main objective of precision medicine is to offer a more precise, organ-specific targeted release with on-demand targets of interventions for the prevention, diagnosis, risk assessment, and optimal treatment of disease. Furthermore, significant work has been put into modern oncology's introduction design, and efforts have been directed toward the future of a revolutionary approach in the realm of nanotechnology known as personalized nanomedicine [126, 142]. The process entails using nanosized carriers such as polymersomes to develop tailored therapy to administer "the right therapy to the right drug to the right patient at the right moment" for each patient [139].

Polymersomes exhibit an adaptive amphiphilic nature derived from the self-assembly of block copolymers containing hydrophobic and hydrophilic segments, which imparts significant structural versatility. This adaptability allows for the formation of nanostructures with tunable morphology and provides advantageous characteristics for personalized therapy, including precise drug delivery to targeted sites, sensitivity to changes in the tumor microenvironment, and stimulus-responsive controlled drug release. Thus, the future of personalized polymersomes is emerging in novel approaches, such as drug delivery and theragnostic and immunotherapeutic clinical applications.

Benito et al. demonstrated the potential of a tailored drug delivery system for cancer treatment using polymersomes formed through the self-assembly of a copolymer via solvent evaporation, resulting in vesicles approximately 130 nm in diameter. These polymersomes were efficiently loaded with doxorubicin hydrochloride (DOX), achieving a drug loading capacity of up to 69% and an encapsulation efficiency of 83%. The release of DOX from the polymersomes is triggered by glutathione, which induces cleavage of disulfide bonds within the polymer matrix. This cleavage causes destabilization and disassembly of the vesicular structure, facilitating drug release. Elevated concentrations of glutathione, characteristic of the tumor microenvironment, mimic these reductive conditions in vitro, resulting in a targeted and controlled release of up to 82% of the encapsulated DOX, thereby enhancing therapeutic efficacy [143].

Owing to the multicompartment nature of polymersomes, they are also appreciated for their potential to provide versatile vesicles combining tailor-made, targeted therapeutic agents and diagnostic features within a single unit. Using markers such as radio markers, inorganic compounds, and visible near-infrared fluorophores, substantial advances in the in vivo imaging of polymersomes have been achieved [144]. Cao J et al. prepared Iodine-Rich Polymersomes (IRP) (^{125}I and ^{131}I) using an isotopic exchange reaction that represented a versatile theragnostic single-photon emission radioisotope platform to evaluate dual-model SPECT or CT imaging and targeted therapy for cancer. For 125I-IRP, the radiolabeling efficiency reached 63.9 ± 6.3%, and for 131-IRP, it was comparable. The inherent characteristics of IRPs were unaffected by radiolabeling; 125-IRP remained 106 nm in size and had a spherical vesicular shape. For 4 T1 breast cancers in mice, 125-IRP allowed for high-resolution CT and high-sensitivity SPECT imaging. During 192 h, longitudinal SPECT/CT imaging revealed that 125-IRP exhibited a prolonged tumor retention [145]. In contrast, as compared to control IRP, therapeutic 131-IRP demonstrated strong antitumor activity, considerably reducing the development of 4 T1 breast tumors and increasing the duration of mice survival. Histological examination demonstrated that 131-IRP significantly reduced tumor cell viability.

Polymersomes for drug delivery monitoring provide real-time target site localization input, which may enable tailored medication techniques via preliminary screening and delivery system modification for specific individuals. Theragnostic magnetic polymersomes were developed by Bleul R et al. and their team using a continuous microfluidic method. Functionalized polymersomes loaded with cancer-targeting bombesin peptides were loaded with camptothecin and magnetite (SPION). It was determined that the medication load was between 2.4% and 2.6% of the total weight of the polymer. Research findings concluded that it has shown great promise for cancer diagnosis and treatment. It has improved anticancer action, encapsulates hydrophobic and hydrophilic components, targets drug administration, and enables real-time MRI tracking [146].

Recent studies have revealed that the therapeutic efficacy of personalized nano-vaccine cancer therapies has been significantly enhanced through improved vector delivery triggered by the immune response [147]. Personalized cancer vaccines have the potential to be improved using a variety of delivery vectors, such as synthetic materials, modified microbes, cells, and cell derivatives. Using these delivery vectors makes it possible to modify the distribution of antigens inside the body and facilitate the uptake, metabolism, and processing of antigens. As a result, the immune system response will gradually become more effective [148]. To efficiently produce robust and long-term antitumor immunity and cure melanoma in mice, Cui G et al. uniquely developed intelligent polymersome nano-vaccines. The nano-vaccines simultaneously transport tumor cell lysates and CpG adjuvants to lymph node dendritic cells. Research has revealed that the use of oxidation-sensitive polymersomes improved the maturation, antigen presentation, and T-cell activation capacity of dendritic cells derived from bone marrow (BMDCs). These polymersomes were used to deliver antigens and adjuvants to dendritic cells, specifically CpG/OVA. Further testing of CpG/TCL was conducted in live mice, resulting in the successful treatment of 85% of B16-F10 mice with melanoma. Additionally, this treatment has led to the development of sustainable anticancer immunological memory [149].

6 Conclusion

Cancer remains a highly complex and heterogeneous disease, ranking as the second leading cause of mortality worldwide. Stimuli-responsive materials, such as polymersomes, indicate that these approaches have demonstrated significant potential in cancer therapy. When exposed to specific physiological conditions and environmental triggers, these materials can undergo reversible changes in their properties, allowing for precise control of drug release. However, challenges remain to be addressed, including biocompatibility, toxicity, stability, manufacturing scalability, and regulatory and ethical considerations. Addressing these limitations through the integration of polymersomes with multi-responsive therapeutic strategies and the development of the next-generation stimuli-responsive materials may pave the way for more effective, safer, and personalized platforms in nanomedicine, with broad applications in drug delivery, theragnostics, and cancer immunotherapy.

Acknowledgments The authors would like to acknowledge the Ministry of Chemicals and Fertilizers, Govt. of India, and School of Pharmacy, The University of Mississippi, for the infrastructural facilities provided for the completion of work.

Author Contribution Amruta Prabhakar Padakanti, Prateek Uttreja: Conceptualization, Writing—original draft, Writing—review & editing; Indrajeet Karnik, Rasha M. Elkanayati, Gaurav Shanbhag: Data collection, Writing—original draft; Sateesh Kumar Vemula, Michael A. Repka, Naveen Chella: Conceptualization, Supervision, Writing—review & editing.

Declarations of Generative AI and AI-Assisted Technologies The authors declare that all sections of this chapter were written using human intellect. However, we have used a few AI-based tools for the grammatical and English language corrections.

References

1. Bertram JS. The molecular biology of cancer. Mol Asp Med. 2000 Dec;21(6):167–223.
2. Sherr CJ. Cancer cell cycles. Science. 1979;274(5293):1672–7.
3. Bailar JC, Gornik HL. Cancer undefeated. N Engl J Med. 1997;336(22):1569–74.
4. Mattiuzzi C, Lippi G. Current cancer epidemiology. J Epidemiol Glob Health. 2019;9(4):217.
5. Debela DT, Muzazu SG, Heraro KD, Ndalama MT, Mesele BW, Haile DC, et al. New approaches and procedures for cancer treatment: current perspectives. SAGE Open Med. 2021;12:9.
6. Trock BJ. Prostate cancer–specific survival following salvage radiotherapy vs observation in men with biochemical recurrence after radical prostatectomy. JAMA. 2008;299(23):2760.
7. Shilling V, Jenkins V, Fallowfield L, Howell T. The effects of hormone therapy on cognition in breast cancer. J Steroid Biochem Mol Biol. 2003;86(3–5):405–12.
8. Beral V, Peto R, Pirie K, Reeves G. Menopausal hormone therapy and 20-year breast cancer mortality. Lancet. 2019;394(10204):1139.
9. Col NF, Kim JA, Chlebowski RT. Menopausal hormone therapy after breast cancer: a meta-analysis and critical appraisal of the evidence. Breast Cancer Res. 2005;7(4):R535.
10. Whelan TJ, Pignol JP, Levine MN, Julian JA, MacKenzie R, Parpia S, et al. Long-term results of hypofractionated radiation therapy for breast cancer. N Engl J Med. 2010;362(6):513–20.
11. Hickman JA. Apoptosis and chemotherapy resistance. Eur J Cancer. 1996;32(6):921–6.
12. König J, Müller F, Fromm MF. Transporters and drug-drug interactions: important determinants of drug disposition and effects. Pharmacol Rev. 2013;65(3):944–66.
13. Xiao H, Zheng Y, Ma L, Tian L, Sun Q. Clinically-relevant ABC transporter for anti-cancer drug resistance. Front Pharmacol. 2021;19:12.
14. Levatić J, Ćurak J, Kralj M, Šmuc T, Osmak M, Supek F. Accurate models for P-gp drug recognition induced from a cancer cell line cytotoxicity screen. J Med Chem. 2013;56(14):5691–708.
15. Mohajer F, Mirhosseini-Eshkevari B, Ahmadi S, Ghasemzadeh MA, Mohammadi Ziarani G, Badiei A, et al. Advanced Nanosystems for cancer therapeutics: a review. ACS Appl Nano Mater. 2023;6(9):7123–49.
16. Song J, Lu C, Leszek J, Zhang J. Design and development of nanomaterial-based drug carriers to overcome the Blood–Brain Barrier by using different transport mechanisms. Int J Mol Sci. 2021;22(18):10118.

17. Dutta B, Nema A, Shetake NG, Gupta J, Barick KC, Lawande MA, et al. Glutamic acid-coated Fe3O4 nanoparticles for tumor-targeted imaging and therapeutics. Mater Sci Eng C. 2020;112:110915.
18. Pramanik A, Xu Z, Shamsuddin SH, Khaled YS, Ingram N, Maisey T, et al. Affimer tagged cubosomes: targeting of carcinoembryonic antigen expressing colorectal cancer cells using in vitro and in vivo models. ACS Appl Mater Interfaces. 2022;14(9):11078–91.
19. Urits I, Swanson D, Swett MC, Patel A, Berardino K, Amgalan A, et al. A review of Patisiran (ONPATTRO®) for the treatment of polyneuropathy in people with hereditary transthyretin amyloidosis. Neurol Ther. 2020;9(2):301–15.
20. Carvalho BG, Vit FF, Carvalho HF, Han SW, de la Torre LG. Recent advances in co-delivery nanosystems for synergistic action in cancer treatment. J Mater Chem B. 2021;9(5):1208–37.
21. Liang C, Yang Y, Ling Y, Huang Y, Li T, Li X. Improved therapeutic effect of folate-decorated PLGA–PEG nanoparticles for endometrial carcinoma. Bioorg Med Chem. 2011;19(13):4057–66.
22. Ying XY, Cui D, Yu L, Du YZ. Solid lipid nanoparticles modified with chitosan oligosaccharides for the controlled release of doxorubicin. Carbohydr Polym. 2011;84(4):1357–64.
23. Sapra P, Tyagi P, Allen T. Ligand-targeted liposomes for cancer treatment. Curr Drug Deliv. 2005;2(4):369–81.
24. Nishiyama N, Matsumura Y, Kataoka K. Development of polymeric micelles for targeting intractable cancers. Cancer Sci. 2016;107(7):867–74.
25. Gajbhiye KR, Salve R, Narwade M, Sheikh A, Kesharwani P, Gajbhiye V. Lipid polymer hybrid nanoparticles: a custom-tailored next-generation approach for cancer therapeutics. Mol Cancer. 2023;22(1):160.
26. Chen J, White SB, Harris KR, Li W, Yap JWT, Kim DH, et al. Poly(lactide-co-glycolide) microspheres for MRI-monitored delivery of sorafenib in a rabbit VX2 model. Biomaterials. 2015;61:299–306.
27. Lakhani PM, Rompicharla SVK, Ghosh B, Biswas S. An overview of synthetic strategies and current applications of gold nanorods in cancer treatment. Nanotechnology. 2015;26(43):432001.
28. Russo A, DeGraff W, Friedman N, Mitchell JB. Selective modulation of glutathione levels in human normal versus tumor cells and subsequent differential response to chemotherapy drugs. Cancer Res. 1986;46(6):2845–8.
29. Wu H, Zhu L, Torchilin VP. pH-sensitive poly(histidine)-PEG/DSPE-PEG co-polymer micelles for cytosolic drug delivery. Biomaterials. 2013;34(4):1213–22.
30. Zhu L, Kate P, Torchilin VP. Matrix metalloprotease 2-responsive multifunctional liposomal nanocarrier for enhanced tumor targeting. ACS Nano. 2012;6(4):3491–8.
31. Fleige E, Quadir MA, Haag R. Stimuli-responsive polymeric nanocarriers for the controlled transport of active compounds: concepts and applications. Adv Drug Deliv Rev. 2012;64(9):866–84.
32. Hu X, Li F, Wang S, Xia F, Ling D. Biological stimulus-driven assembly/disassembly of functional nanoparticles for targeted delivery, controlled activation, and bioelimination. Adv Healthc Mater. 2018;7(20)
33. Mura S, Nicolas J, Couvreur P. Stimuli-responsive nanocarriers for drug delivery. Nat Mater. 2013;12(11):991–1003.
34. Zhang J, Lin Y, Lin Z, Wei Q, Qian J, Ruan R, et al. Stimuli-responsive nanoparticles for controlled drug delivery in synergistic cancer immunotherapy. Adv Science. 2022;9(5)
35. Park W, Chen J, Cho S, Park S, Jung, Larson AC, Na K, et al. Acidic pH-triggered drug-eluting nanocomposites for magnetic resonance imaging-monitored intra-arterial drug delivery to hepatocellular carcinoma. ACS Appl Mater Interfaces. 2016;8(20):12711–9.
36. Deshpande PB, Dandagi P, Udupa N, Gopal SV, Jain SS, Vasanth SG. Controlled release polymeric ocular delivery of acyclovir. Pharm Dev Technol. 2010;15(4):369–78.
37. Karnik I, Youssef AAA, Joshi P, Munnangi SR, Narala S, Varner C, et al. Formulation development and characterization of dual drug loaded hot-melt extruded inserts for better ocular therapeutic outcomes: Sulfacetamide/prednisolone. J Drug Deliv Sci Technol. 2023;84:104558.

38. Uttreja P, Youssef AAA, Karnik I, Sanil K, Narala N, Wang H, et al. Formulation development of solid self-nanoemulsifying drug delivery systems of quetiapine fumarate via hot-melt extrusion technology: optimization using central composite design. Pharmaceutics. 2024;16(3):324.
39. Luk BT, Zhang L. Current advances in polymer-based nanotheranostics for cancer treatment and diagnosis. ACS Appl Mater Interfaces. 2014;6(24):21859–73.
40. Discher BM, Won YY, Ege DS, Lee JCM, Bates FS, Discher DE, et al. Polymersomes: tough vesicles made from Diblock copolymers. Science. 1979;284(5417):1143–6.
41. Anajafi T, Mallik S. Polymersome-based drug-delivery strategies for cancer therapeutics. Ther Deliv. 2015;6(4):521–34.
42. Villani S, Adami R, Reverchon E, Ferretti AM, Ponti A, Lepretti M, et al. pH-sensitive polymersomes: controlling swelling via copolymer structure and chemical composition. J Drug Target. 2017;25(9–10):899–909.
43. Broz P, Driamov S, Ziegler J, Ben-Haim N, Marsch S, Meier W, et al. Toward intelligent Nanosize bioreactors: a pH-switchable, channel-equipped, functional polymer nanocontainer. Nano Lett. 2006;6(10):2349–53.
44. Lale SV, Kumar A, Prasad S, Bharti AC, Koul V. Folic acid and trastuzumab functionalized redox responsive polymersomes for intracellular doxorubicin delivery in breast cancer. Biomacromolecules. 2015;16(6):1736–52.
45. Zhou Y, Chen R, Yang H, Bao C, Fan J, Wang C, et al. Light-responsive polymersomes with a charge-switch for targeted drug delivery. J Mater Chem B. 2020;8(4):727–35.
46. Sun J, Deng C, Chen X, Yu H, Tian H, Sun J, et al. Self-assembly of polypeptide-containing ABC-type triblock copolymers in aqueous solution and its pH dependence. Biomacromolecules. 2007;8(3):1013–7.
47. Sun J, Shi Q, Chen X, Guo J, Jing X. Self-assembly of a hydrophobic polypeptide containing a short hydrophilic middle segment: vesicles to large compound micelles. Macromol Chem Phys. 2008;209(11):1129–36.
48. Fetsch C, Gaitzsch J, Messager L, Battaglia G, Luxenhofer R. Self-assembly of amphiphilic block copolypeptoids—micelles, Worms and Polymersomes. Sci Rep. 2016;6(1):33491.
49. Houga C, Giermanska J, Lecommandoux S, Borsali R, Taton D, Gnanou Y, et al. Micelles and polymersomes obtained by self-assembly of dextran and polystyrene based block copolymers. Biomacromolecules. 2009;10(1):32–40.
50. Wong CK, Lai RY, Thordarson P. Polymersomes: fundamentals and shape transformation methods. In: Supramolecular nanotechnology. Wiley; 2023. p. 1027–54.
51. Lasic DD. The mechanism of vesicle formation. Biochem J. 1988;256(1):1–11.
52. Lasic DD, Martin FJ. On the mechanism of vesicle formation. J Memb Sci. 1990;50(2):215–22.
53. Fromherz P. Lipid-vesicle structure: size control by edge-active agents. Chem Phys Lett. 1983;94(3):259–66.
54. Grafmüller A, Shillcock J, Lipowsky R. The fusion of membranes and vesicles: pathway and energy barriers from dissipative particle dynamics. Biophys J. 2009;96(7):2658–75.
55. Ianiro A, Wu H, van Rijt MMJ, Vena MP, Keizer ADA, Esteves ACC, et al. Liquid–liquid phase separation during amphiphilic self-assembly. Nat Chem. 2019;11(4):320–8.
56. Lefley J, Waldron C, Becer CR. Macromolecular design and preparation of polymersomes. Polym Chem. 2020;11(45):7124–36.
57. Charleux B, Delaittre G, Rieger J, D'Agosto F. Polymerization-induced self-assembly: from soluble macromolecules to block copolymer Nano-objects in one step. Macromolecules. 2012;45(17):6753–65.
58. He J, Cao J, Chen Y, Zhang L, Tan J. Thermoresponsive block copolymer vesicles by visible light-initiated seeded polymerization-induced self-assembly for temperature-regulated enzymatic nanoreactors. ACS Macro Lett. 2020;9(4):533–9.
59. Blackman LD, Varlas S, Arno MC, Fayter A, Gibson MI, O'Reilly RK. Permeable protein-loaded polymersome cascade nanoreactors by polymerization-induced self-assembly. ACS Macro Lett. 2017;6(11):1263–7.

60. Contini C, Pearson R, Wang L, Messager L, Gaitzsch J, Rizzello L, et al. Bottom-up evolution of vesicles from disks to high-genus polymersomes. iScience. 2018;7:132–44.
61. Zhao X, Sun C, Xiong F, Wang T, Li S, Huo F, et al. Polymerization-induced self-assembly for efficient fabrication of biomedical nanoplatforms. Research. 2023 Jan;6
62. Rideau E, Wurm FR, Landfester K. Giant polymersomes from non-assisted film hydration of phosphate-based block copolymers. Polym Chem. 2018;9(44):5385–94.
63. Schlaad H. Solution properties of polypeptide-based copolymers. 2006. 53–73.
64. Kunzler C, Handschuh-Wang S, Roesener M, Schönherr H. Giant biodegradable poly(ethylene glycol)- block -poly(ε-caprolactone) polymersomes by electroformation. Macromol Biosci. 2020;20(6)
65. Vijayakrishna K, Mecerreyes D, Gnanou Y, Taton D. Polymeric vesicles and micelles obtained by self-assembly of ionic liquid-based block copolymers triggered by anion or solvent exchange. Macromolecules. 2009;42(14):5167–74.
66. Scheerstra JF, Wauters AC, Tel J, Abdelmohsen LKEA, van Hest JCM. Polymersomes as a potential platform for cancer immunotherapy. Mater Today Adv. 2022;13:100203.
67. Karayianni M, Pispas S. Block copolymer solution self-assembly: recent advances, emerging trends, and applications. J Polym Sci. 2021;59(17):1874–98.
68. Soomherun N, Kreua-ongarjnukool N, Chumnanvej S, Thumsing S. Encapsulation of nicardipine hydrochloride and release from biodegradable poly(D,L-lactic-co-glycolic acid) microparticles by double emulsion process: effect of emulsion stability and different parameters on drug entrapment. Int J Biomater. 2017;2017:1–13.
69. Danafar H, Rostamizadeh K, Davaran S, Hamidi M. PLA-PEG-PLA copolymer-based polymersomes as nanocarriers for delivery of hydrophilic and hydrophobic drugs: preparation and evaluation with atorvastatin and lisinopril. Drug Dev Ind Pharm. 2014;40(10):1411–20.
70. Moga A, Yandrapalli N, Dimova R, Robinson T. Optimization of the inverted emulsion method for high-yield production of biomimetic giant unilamellar vesicles. Chembiochem. 2019;20(20):2674–82.
71. Utada AS, Lorenceau E, Link DR, Kaplan PD, Stone HA, Weitz DA. Monodisperse double emulsions generated from a microcapillary device. Science (1979). 2005;308(5721):537–541.
72. Ahn J, Ko J, Lee S, Yu J, Kim Y, Jeon NL. Microfluidics in nanoparticle drug delivery; from synthesis to pre-clinical screening. Adv Drug Deliv Rev. 2018;128:29–53.
73. Roces CB, Christensen D, Perrie Y. Translating the fabrication of protein-loaded poly(lactic-co-glycolic acid) nanoparticles from bench to scale-independent production using microfluidics. Drug Deliv Transl Res. 2020;10(3):582–93.
74. Koide A, Kishimura A, Osada K, Jang WD, Yamasaki Y, Kataoka K. Semipermeable polymer vesicle (PICsome) self-assembled in aqueous medium from a pair of oppositely charged block copolymers: physiologically stable micro-/Nanocontainers of water-soluble macromolecules. J Am Chem Soc. 2006;128(18):5988–9.
75. Chenkual L, Lalchandani DS, Padakanti AP, Chella N, Porwal PK. Synthesis and self-assembly of block copolymers. In: Block co-polymeric nanocarriers: design, concept, and therapeutic applications. Singapore: Springer Nature Singapore; 2023. p. 75–119.
76. Ferji K, Nouvel C, Babin J, Li MH, Gaillard C, Nicol E, et al. Polymersomes from amphiphilic glycopolymers containing polymeric liquid crystal grafts. ACS Macro Lett. 2015;4(10):1119–22.
77. LoPresti C, Massignani M, Fernyhough C, Blanazs A, Ryan AJ, Madsen J, et al. Controlling polymersome surface topology at the nanoscale by membrane confined polymer/polymer phase separation. ACS Nano. 2011;5(3):1775–84.
78. Fonseca M, Jarak I, Victor F, Domingues C, Veiga F, Figueiras A. Polymersomes as the next attractive generation of drug delivery systems: definition, synthesis and applications. Materials. 2024;17(2):319.
79. Elmowafy M, Shalaby K, Elkomy MH, Alsaidan OA, Gomaa HAM, Abdelgawad MA, et al. Polymeric nanoparticles for delivery of natural bioactive agents: recent advances and challenges. Polymers (Basel). 2023;15(5):1123.

80. Shinde VR, Revi N, Murugappan S, Singh SP, Rengan AK. Enhanced permeability and retention effect: a key facilitator for solid tumor targeting by nanoparticles. Photodiagn Photodyn Ther. 2022;39:102915.
81. Dolai J, Mandal K, Jana NR. Nanoparticle size effects in biomedical applications. ACS Appl Nano Mater. 2021;4(7):6471–96.
82. Sharifi M, Cho WC, Ansariesfahani A, Tarharoudi R, Malekisarvar H, Sari S, et al. An updated review on EPR-based solid tumor targeting nanocarriers for cancer treatment. Cancers (Basel). 2022;14(12):2868.
83. Wojtynek NE, Mohs AM. Image-guided tumor surgery: the emerging role of nanotechnology. WIREs Nanomed Nanobiotechnol. 2020;12(4):e1624.
84. de Jong. Drug delivery and nanoparticles: applications and hazards. Int J Nanomedicine. 2008:133.
85. Wei M, Gao Y, Li X, Serpe MJ. Stimuli-responsive polymers and their applications. Polym Chem. 2017;8(1):127–43.
86. Felber AE, Dufresne MH, Leroux JC. pH-sensitive vesicles, polymeric micelles, and nanospheres prepared with polycarboxylates. Adv Drug Deliv Rev. 2012;64(11):979–92.
87. Bazban-Shotorbani S, Hasani-Sadrabadi MM, Karkhaneh A, Serpooshan V, Jacob KI, Moshaverinia A, et al. Revisiting structure-property relationship of pH-responsive polymers for drug delivery applications. J Control Release. 2017;253:46–63.
88. Hu X, Zhang Y, Xie Z, Jing X, Bellotti A, Gu Z. Stimuli-responsive polymersomes for biomedical applications. Biomacromolecules. 2017;18(3):649–73.
89. Yuan D, Zhou F, Niu Z, Shen P, Zhao M. Formation of mucus-permeable nanoparticles from soy protein isolate by partial enzymatic hydrolysis coupled with thermal and pH-shifting treatment. Food Chem. 2023;398:133851.
90. Aguirre-Pranzoni C, García MG, Ochoa NA. Structural and conformational changes on chitosan after green heterogeneous synthesis of phenyl derivatives. Carbohydr Polym. 2023;312:120843.
91. Yuan Y, Wang Y, Huang H, Tao S, Huang J. Fine-tuned polymer nanoassembly for codelivery of chemotherapeutic drug and siRNA. Macromol Biosci. 2023;23(4):e2200529.
92. Zhu L, Zhao L, Qu X, Yang Z. pH-sensitive polymeric vesicles from coassembly of amphiphilic cholate grafted poly(L-lysine) and acid-cleavable polymer–drug conjugate. Langmuir. 2012;28(33):11988–96.
93. Mukerabigwi JF, Yin W, Zha Z, Ke W, Wang Y, Chen W, et al. Polymersome nanoreactors with tumor pH-triggered selective membrane permeability for prodrug delivery, activation, and combined oxidation-chemotherapy. J Control Release. 2019;303:209–22.
94. Nan Y, Zhao C, Beaudoin G, Zhu XX. Synergistic approaches in the design and applications of UCST polymers. Macromol Rapid Commun. 2023;44(23):e2300261.
95. Li MH, Keller P. Stimuli-responsive polymer vesicles. Soft Matter. 2009;5(5):927.
96. Hajebi S, Abdollahi A, Roghani-Mamaqani H, Salami-Kalajahi M. Temperature-responsive poly(N-Isopropylacrylamide) nanogels: the role of hollow cavities and different Shell crosslinking densities on doxorubicin loading and release. Langmuir. 2020;36(10):2683–94.
97. Sánchez-Moreno P, De Vicente J, Nardecchia S, Marchal JA, Boulaiz H. Thermosensitive nanomaterials: recent advance in synthesis and biomedical applications. Nano. 2018;8(11):935.
98. Liu F, Kozlovskaya V, Medipelli S, Xue B, Ahmad F, Saeed M, et al. Temperature-sensitive polymersomes for controlled delivery of anticancer drugs. Chem Mater. 2015;27(23):7945–56.
99. Meng F, Hennink WE, Zhong Z. Reduction-sensitive polymers and bioconjugates for biomedical applications. Biomaterials. 2009;30(12):2180–98.
100. Cheng R, Feng F, Meng F, Deng C, Feijen J, Zhong Z. Glutathione-responsive nano-vehicles as a promising platform for targeted intracellular drug and gene delivery. J Control Release. 2011;152(1):2–12.
101. Li R, Peng F, Cai J, Yang D, Zhang P. Redox dual-stimuli responsive drug delivery systems for improving tumor-targeting ability and reducing adverse side effects. Asian J Pharm Sci. 2020;15(3):311–25.

102. Wu W, Chen M, Luo T, Fan Y, Zhang J, Zhang Y, et al. ROS and GSH-responsive S-nitrosoglutathione functionalized polymeric nanoparticles to overcome multidrug resistance in cancer. Acta Biomater. 2020;103:259–71.
103. Jia L, Cui D, Bignon J, Di Cicco A, Wdzieczak-Bakala J, Liu J, et al. Reduction-responsive cholesterol-based block copolymer vesicles for drug delivery. Biomacromolecules. 2014;15(6):2206–17.
104. Babin J, Pelletier M, Lepage M, Allard J, Morris D, Zhao Y. A new two-photon-sensitive block copolymer nanocarrier. Angew Chem Int Ed. 2009;48(18):3329–32.
105. Peters FB, Brock A, Wang J, Schultz PG. Photocleavage of the polypeptide backbone by 2-nitrophenylalanine. Chem Biol. 2009;16(2):148–52.
106. Lin YL, Chang HY, Sheng YJ, Tsao HK. Photoresponsive polymersomes formed by amphiphilic linear–dendritic block copolymers: generation-dependent aggregation behavior. Macromolecules. 2012;45(17):7143–56.
107. Mabrouk E, Cuvelier D, Brochard-Wyart F, Nassoy P, Li MH. Bursting of sensitive polymersomes induced by curling. Proc Natl Acad Sci. 2009;106(18):7294–8.
108. Hernández Becerra E, Quinchia J, Castro C, Orozco J. Light-triggered polymersome-based anticancer therapeutics delivery. Nanomaterials. 2022;12(5):836.
109. Wang M, Geilich B, Keidar M, Webster T. Killing malignant melanoma cells with protoporphyrin IX-loaded polymersome-mediated photodynamic therapy and cold atmospheric plasma. Int J Nanomedicine. 2017;12:4117–27.
110. Hou W, Liu R, Bi S, He Q, Wang H, Gu J. Photo-responsive polymersomes as drug delivery system for potential medical applications. Molecules. 2020;25(21):5147.
111. Leong J, Teo JY, Aakalu VK, Yang YY, Kong H. Engineering polymersomes for diagnostics and therapy. Adv Healthc Mater. 2018;7(8):e1701276.
112. Thambi T, Park JH, Lee DS. Stimuli-responsive polymersomes for cancer therapy. Biomater Sci. 2016;4(1):55–69.
113. Yao C, Li Y, Wang Z, Song C, Hu X, Liu S. Cytosolic NQO1 enzyme-activated near-infrared fluorescence imaging and photodynamic therapy with polymeric vesicles. ACS Nano. 2020;14(2):1919–35.
114. Fu J, Qiu L. Photo-crosslinked and esterase-sensitive Polymersome for improved antitumor effect of water-soluble chemotherapeutics. Nanomedicine. 2018;13(16):2051–66.
115. Huang H, Zheng Y, Chang M, Song J, Xia L, Wu C, et al. Ultrasound-based micro-/nanosystems for biomedical applications. Chem Rev. 2024;124(13):8307–472.
116. Chen W, Du J. Ultrasound and pH dually responsive polymer vesicles for anticancer drug delivery. Sci Rep. 2013;3(1):2162.
117. Araste F, Aliabadi A, Abnous K, Taghdisi SM, Ramezani M, Alibolandi M. Self-assembled polymeric vesicles: focus on polymersomes in cancer treatment. J Control Release. 2021;330:502–28.
118. Oliveira H, Pérez-Andrés E, Thevenot J, Sandre O, Berra E, Lecommandoux S. Magnetic field triggered drug release from polymersomes for cancer therapeutics. J Control Release. 2013;169(3):165–70.
119. He C, Zhang Z, Ding Y, Xue K, Wang X, Yang R, et al. LRP1-mediated pH-sensitive polymersomes facilitate combination therapy of glioblastoma in vitro and in vivo. J Nanobiotechnol. 2021;19(1):29.
120. Hasannia M, Abnous K, Taghdisi SM, Sazgarnia A, Nekooei S, Ramezani M, et al. Multifunctional polypeptide-based polymersome for chemophotothermal therapy of melanoma in vitro and in vivo. J Drug Deliv Sci Technol. 2024;94:105526.
121. Tang Q, Hu P, Peng H, Zhang N, Zheng Q, He Y. Near-infrared laser-triggered, self-Immolative smart polymersomes for in vivo cancer therapy. Int J Nanomedicine. 2020;15:137–49.
122. Zheng M, Du Q, Wang X, Zhou Y, Li J, Xia X, et al. Tuning the elasticity of polymersomes for brain tumor targeting. Adv Sci. 2021;8(20)
123. Albuquerque LJC, Sincari V, Jäger A, Kucka J, Humajova J, Pankrac J, et al. pH-responsive polymersome-mediated delivery of doxorubicin into tumor sites enhances the therapeutic efficacy and reduces cardiotoxic effects. J Control Release. 2021;332:529–38.

124. Wei Y, Gu X, Sun Y, Meng F, Storm G, Zhong Z. Transferrin-binding peptide functionalized polymersomes mediate targeted doxorubicin delivery to colorectal cancer in vivo. J Control Release. 2020;319:407–15.
125. Taléns-Visconti R, Díez-Sales O, de Julián-Ortiz JV, Nácher A. Nanoliposomes in cancer therapy: marketed products and current clinical trials. Int J Mol Sci. 2022;23(8):4249.
126. Negut I, Bita B. Polymersomes as innovative, stimuli-responsive platforms for cancer therapy. Pharmaceutics. 2024;16(4):463.
127. Matoori S, Leroux JC. Twenty-five years of polymersomes: lost in translation? Mater Horiz. 2020;7(5):1297–309.
128. Gaitzsch J, Canton I, Appelhans D, Battaglia G, Voit B. Cellular interactions with photocross-linked and pH-sensitive polymersomes: biocompatibility and uptake studies. Biomacromolecules. 2012;13(12):4188–95.
129. Gouveia MG, Wesseler JP, Ramaekers J, Weder C, Scholten PBV, Bruns N. Polymersome-based protein drug delivery—quo vadis? Chem Soc Rev. 2023;52(2):728–78.
130. Sharma AK, Prasher P, Aljabali AA, Mishra V, Gandhi H, Kumar S, et al. Emerging era of "somes": polymersomes as versatile drug delivery carrier for cancer diagnostics and therapy. Drug Deliv Transl Res. 2020;10(5):1171–90.
131. Sun H, Meng F, Cheng R, Deng C, Zhong Z. Reduction and pH dual-bioresponsive cross-linked polymersomes for efficient intracellular delivery of proteins and potent induction of cancer cell apoptosis. Acta Biomater. 2014;10(5):2159–68.
132. Khan RU, Yu H, Wang L, Zhang Q, Xiong W, Zain-ul-Abdin, et al. Synthesis of polyorganophosphazenes and preparation of their polymersomes for reductive/acidic dual-responsive anticancer drugs release. J Mater Sci. 2020;55(19):8264–84.
133. Oroojalian F, Babaei M, Taghdisi SM, Abnous K, Ramezani M, Alibolandi M. Encapsulation of thermo-responsive gel in pH-sensitive polymersomes as dual-responsive smart carriers for controlled release of doxorubicin. J Control Release. 2018;288:45–61.
134. DiazDuarte-Rodriguez M, Cortez-Lemus NA, Licea-Claverie A, Licea-Rodriguez J, Méndez ER. Dual responsive polymersomes for gold nanorod and doxorubicin encapsulation: nanomaterials with potential use as smart drug delivery systems. Polymers (Basel). 2019;11(6):939.
135. Zhou D, Fei Z, Jin L, Zhou P, Li C, Liu X, et al. Dual-responsive polymersomes as anticancer drug carriers for the co-delivery of doxorubicin and paclitaxel. J Mater Chem B. 2021;9(3):801–8.
136. Kozlovskaya V, Yang Y, Liu F, Ingle K, Ahmad A, Halade GV, et al. Dually responsive poly(N-vinylcaprolactam)-b-poly(dimethylsiloxane)-b-poly(N-vinylcaprolactam) polymersomes for controlled delivery. Molecules. 2022;27(11):3485.
137. Tsai MF, Lo YL, Soorni Y, Su CH, Sivasoorian SS, Yang JY, et al. Near-infrared light-triggered drug release from ultraviolet- and redox-responsive polymersome encapsulated with core–shell upconversion nanoparticles for cancer therapy. ACS Appl Bio Mater. 2021;4(4):3264–75.
138. Zhu D, Fan F, Huang C, Zhang Z, Qin Y, Lu L, et al. Bubble-generating polymersomes loaded with both indocyanine green and doxorubicin for effective chemotherapy combined with photothermal therapy. Acta Biomater. 2018;75:386–97.
139. Rana P, Singh C, Kaushik A, Saleem S, Kumar A. Recent advances in stimuli-responsive tailored nanogels for cancer therapy; from bench to personalized treatment. J Mater Chem B. 2024;12(2):382–412.
140. Kiyotani K, Toyoshima Y, Nakamura Y. Personalized immunotherapy in cancer precision medicine. Cancer Biol Med. 2021;18:10.
141. Krzyszczyk P, Acevedo A, Davidoff EJ, Timmins LM, Marrero-Berrios I, Patel M, et al. The growing role of precision and personalized medicine for cancer treatment. Technology (Singap World Sci). 2018;6(03n04):79.
142. Mura S, Couvreur P. Nanotheranostics for personalized medicine. Adv Drug Deliv Rev. 2012;64(13):1394–416.

143. Benito E, Romero-Azogil L, Galbis E, De-Paz MV, García-Martín MG. Structurally simple redox polymersomes for doxorubicin delivery. Eur Polym J. 2020;137:109952.
144. Thevenot J, Oliveira H, Lecommandoux S. Polymersomes for theranostics. J Drug Deliv Sci Technol. 2013;23(1):38–46.
145. Cao J, Wei Y, Zhang Y, Wang G, Ji X, Zhong Z. Iodine-rich polymersomes enable versatile SPECT/CT imaging and potent radioisotope therapy for tumor in vivo. ACS Appl Mater Interfaces. 2019;11(21):18953–9.
146. Bleul R, Thiermann R, Marten GU, House MJ, Pierre TG St, Häfeli UO, et al. Continuously manufactured magnetic polymersomes—a versatile tool (not only) for targeted cancer therapy. Nanoscale. 2013;5(23):11385.
147. Ye T, Li F, Ma G, Wei W. Enhancing therapeutic performance of personalized cancer vaccine via delivery vectors. Adv Drug Deliv Rev. 2021;177:113927.
148. Yang W, Zhu G, Wang S, Yu G, Yang Z, Lin L, et al. In situ dendritic cell vaccine for effective cancer immunotherapy. ACS Nano. 2019;13(3):3083–94.
149. Cui G, Sun Y, Qu L, Shen C, Sun Y, Meng F, et al. Uplifting antitumor immunotherapy with lymph-node-targeted and ratio-controlled codelivery of tumor cell lysate and adjuvant. Adv Healthc Mater. 2024;13(17):e2303690.

Stimuli-Responsive Dendrimers in Cancer Therapy

5

Ananya Goswami, Ritu Patel, Shagun Rangra, Dhwani Rana, and Derajram Benival

Abstract

Cancer remains a major global health burden, necessitating precise and effective therapeutic strategies. Nanotechnology-based interventions have enabled targeted drug delivery, with dendrimers emerging as promising nanocarriers due to their defined architecture, multivalency, and capacity for drug encapsulation or conjugation. Stimuli-responsive dendrimers allow controlled drug release in response to specific internal stimuli—such as pH shifts, redox conditions, and enzyme activity—or external triggers including light, temperature, and magnetic fields. These systems are designed to exploit the pathological characteristics of tumor microenvironments, such as acidic pH, elevated glutathione levels, and enzyme overexpression, to achieve selective drug release. Various chemical linkages and functional moieties are incorporated to render dendrimers responsive to individual or multiple stimuli. Thermo-responsive dendrimers utilize phase transitions near body temperature, while light-responsive systems enable on-demand release using photo-cleavable groups. Multi-stimuli-responsive designs further enhance targeting efficiency by addressing tumor heterogeneity. This chapter highlights the design principles, release mechanisms, and therapeutic outcomes

Ananya Goswami and Ritu Patel contributed equally with all other contributors.

A. Goswami · R. Patel · S. Rangra · D. Rana · D. Benival (✉)
Department of Pharmaceutics, National Institute of Pharmaceutical Education and Research - Ahmedabad (NIPER-A), An Institute of National Importance, Department of Pharmaceuticals, Ministry of Chemicals and Fertilizers, Government of India, Gandhinagar, Gujarat, India
e-mail: derajram@niperahm.res.in

P. V. Devarajan et al. (eds.), *Advances in Stimuli-Responsive Nanosystems for Cancer Therapy*, https://doi.org/10.1007/978-981-95-2744-1_5

of such systems, supported by in vitro and in vivo studies. Despite encouraging outcomes, challenges related to in vivo stability, biocompatibility, and clinical scalability persist. Continued research is essential to optimize dendrimer architecture and integrate diagnostic functionality for effective clinical translation in oncology.

Keywords

Cancer · Dendrimer · Stimuli-responsive · Drug delivery · Nanotechnology · Polymer

Abbreviations

5-ASA	5-amino salicylic acid
5-FU	5-fluorouracil
ATRA	All-trans-retinoic acid
AuNR	Gold nanorod
DOX	Doxorubicin
EDC	1-ethyl-3-(3-dimethylaminopropyl) carbodiimide
FA	Folic acid
FI	Fluorescein thiocyanate
GFLG	Glycylphenylalanylleucylglycine tetra-peptide
GSH	Glutathione
LCST	Lower critical solution temperature
MMPs	Matrix metalloproteinases
mPEG	Methyl polyethylene glycol
MTX	Methotrexate
NHS	N-hydroxysuccinimide
ONB	Ortho-nitrobenzyl
PAMAM	Polyamidoamine
PCBMA	Poly(carboxybetaine methacrylate)
PDEA	Poly(2-(N,N-diethylamino) ethyl methacrylate)
PEG	Polyethylene glycol
pHLIP	pH (low) insertion peptide
PPD	Poly(2-(N,N-diethylamino) ethyl methacrylate) with methoxy-poly(ethylene glycol)-poly(amidoamine)
PTX	Paclitaxel
TDN	Triple-responsive dendrimeric nanocages
TEM	Transmission electron microscope
VEGF	Vascular endothelial growth factor

1 Introduction

Cancer continues to be a major contributor to global mortality, presenting substantial hurdles to public health on a worldwide scale. Conventional cancer therapies, encompassing surgical intervention, chemotherapy, radiation therapy, and immunotherapy, have historically formed the foundation of cancer care [1, 2]. Nonetheless, these treatments often exhibit limitations such as severe adverse effects, non-specific targeting, and the emergence of drug resistance. The pursuit of more efficient and precise therapies has prompted the investigation of nanosystems, which present promising alternatives to traditional treatments [3, 4]. One of the main challenges in cancer therapy is the development of drug resistance. Cancer cells can adapt and become resistant to the effects of chemotherapy, making treatment less effective over time. Additionally, traditional cancer treatments are often linked to severe side effects that can significantly impact patients' quality of life [5]. The lack of specificity in current cancer treatments results in unintended damage to healthy cells, causing a variety of negative side effects. The challenge of developing cancer therapies lies in targeting cancer cells exclusively while preserving the integrity of healthy cells [6]. Nanotechnology has revolutionized cancer therapy by introducing nanosystems such as liposomes, nanoparticles, and dendrimers [7, 8]. These systems provide numerous advantages over conventional treatments, including improved delivery of therapeutic agents to cancer cells, enhanced drug solubility, and minimized side effects [9]. Among these, dendrimers have gained significant attention due to their unique structural properties and potential applications in drug delivery [10]. Dendrimers are complex, highly branched macromolecules with a well-defined, three-dimensional architecture. Their unique structure offers high surface area and multivalency, making them well-suited for drug delivery applications. Dendrimers can encapsulate or conjugate multiple drug molecules, enabling targeted delivery to cancer cells. Their capability to cross biological barriers and release drugs in a controlled manner enhances their potential in cancer therapy [11]. Stimuli-responsive systems are engineered to trigger the release of therapeutic agents in response to particular internal or external stimuli. These stimuli may encompass variations in pH, temperature, light, or magnetic fields. Stimuli-responsive dendrimers offer a means of precise and targeted drug release, thereby reducing side effects and enhancing therapeutic effectiveness. By adapting to the specific microenvironment of cancer cells, these systems can improve the accuracy of drug delivery [12]. Dendrimers were first developed in the late twentieth century, with initial research concentrating on their synthesis and structural characterization. Substantial progress has been achieved in understanding their properties and potential uses over the years. The idea of stimuli-responsive dendrimers has advanced, with researchers investigating different stimuli and mechanisms to achieve controlled drug release [13].

2 Dendrimers: Design and Targeting Pathways

Dendrimers are nano-sized, radially symmetric molecules with a well-defined, homogeneous, and monodisperse structure, resembling a tree-like architecture [14]. They are typically synthesized using two main methods: the divergent method, which starts from a central core and grows outward by successive addition of monomer units, and the convergent method, which involves the synthesis of dendritic fragments that are then attached to a core molecule. This synthesis approach provides precise control over the final structure and minimizes defects [11]. Dendrimers can be tailored to target specific cells or tissues by functionalizing their surface with various ligands, antibodies, or other targeting moieties. Their mechanism of action often relies on polyvalency, where the multiple functional groups on the surface enable strong and multivalent interactions with target molecules, electrostatic interactions with negatively charged cell membranes facilitating cellular uptake, and encapsulation of drugs within their structure, protecting the drug from degradation and enhancing its delivery to the target site [15]. Structurally, dendrimers are characterized by a core, branches extending from the core to form a tree-like structure, and surface groups that can be modified to enhance solubility, biocompatibility, or targeting ability. Due to these unique properties, dendrimers have a wide range of applications, including drug delivery, gene therapy, and diagnostic imaging. Their ability to encapsulate drugs and target specific cells makes them promising candidates for cancer therapy and other medical applications [16].

3 Stimuli-Responsive Dendrimers

Dendrimers do act as an interesting approach for drug delivery; however, their usage is limited due to a lot of factors, one of them being the difficulty in controlling the release of drug from the system [17]. Drugs are given with the aim of maximizing efficacy while reducing the side effects. This becomes especially important while dealing with anti-cancer drugs due to the high toxicity. Release of drugs at sites other than the site of action is not desirable. Stimuli-responsive systems offer a solution to this problem by promoting release only at the site of action at desired times. These systems are designed so that the drug payload is released in response to a trigger that is present only at the site of action. These triggers work by inducing some sort of a change in the delivery system that ultimately causes the drug release (Fig. 5.1).

These stimuli may be internal, such as enzymes, pH, or redox conditions, or external, such as light, temperature, or magnetic field. Internal triggers exploit differences between the microenvironments of diseased and normal tissues. For example, cancer tissues typically have a pH about 1 unit lower than normal tissues and exhibit an altered redox balance, leading to higher oxidative stress [18]. For external stimuli, we can control both the location and time of drug release [17].

Stimuli-responsive nanosystems offer many advantages over normal drug delivery. Combining them with those of dendrimers can result in the formation of an

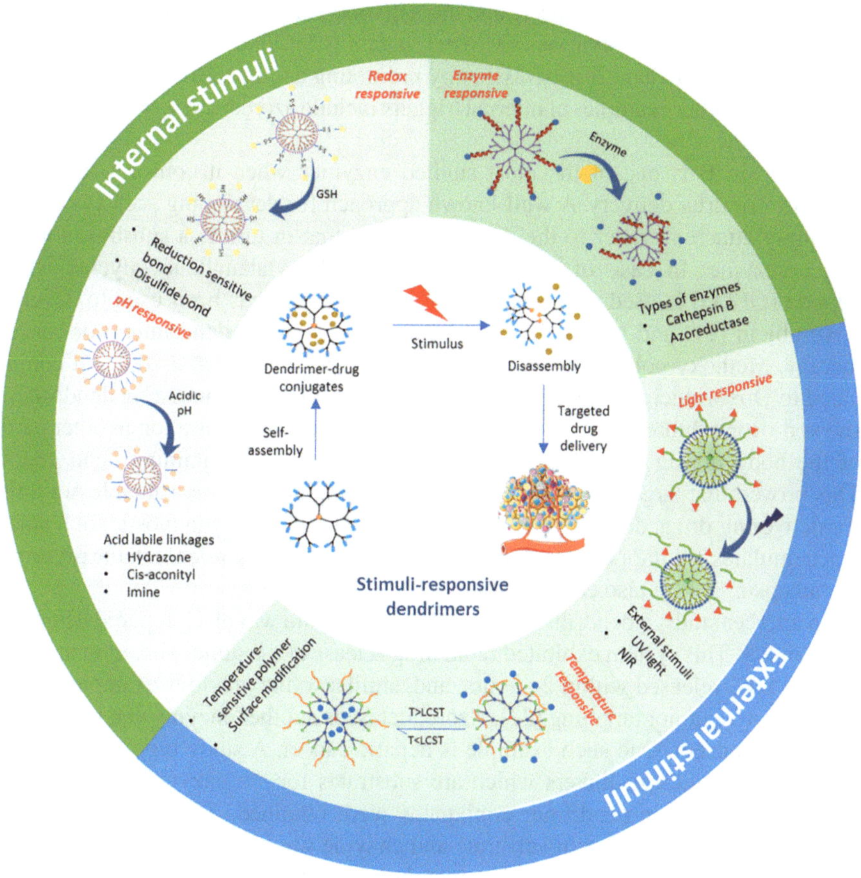

Fig. 5.1 Types of stimuli-responsive dendrimer systems and their basic mechanisms

excellent approach for drug delivery. There has been a lot of research conducted in the past and many more are still ongoing. This chapter focuses on covering a few of those to elaborate on the various kinds of stimuli that are available.

3.1 Enzyme-Responsive Dendrimers

Cancer is characterized by changes in the level and activity of various enzymes. Most commonly, in order to regulate the metabolic processes of the cancer tissue, the enzymes involved in the same are often upregulated [18]. Though this is meant to promote cancer growth, researchers have used it to their advantage as potential drug targets. This approach helps them achieve targeted delivery, thereby improving the efficacy and reducing side effects of the drugs. Certain enzymes that are upregulated during cancer include Cathepsin B, that is a lysosomal protease which is upregulated in most of the cancers and has a diverse role in promoting tumor growth [19]. Matrix metalloproteinases (MMPs), a family of zinc-dependent endopeptidases, contribute to cancer progression by promoting angiogenesis and other mechanisms [20]. Other examples of internal triggers include azoreductase, phospholipase, to name a few.

Cathepsin B is one of the most studied enzymes when it comes to targeting enzymes for drug delivery. A well-known approach for developing such systems is to simply attach a carrier to the delivery system that in turn is a substrate for the target enzyme. In case of Cathepsin B, glycylphenylalanylleucylglycine tetrapeptide (GFLG) is used as a linker. This very approach has been used to develop doxorubicin [21] and gemcitabine [22] enzyme-responsive dendrimers for cancer therapy. Methoxy poly (ethylene glycol) dendrimers were synthesized and conjugated to Doxorubicin using GFLG. It was noted that GFLG-containing dendrimers showed strong fluorescence at tumor sites with minimal concentration in other parts of the body during a comparison study with non-GFLG-containing dendrimers. This proved that targeting took place. This targeting approach is valuable not only for delivering drugs directly to the site of action but also as a diagnostic tool, since it accumulates in tumor tissues. Further, studies proving drug release in the presence of cathepsin B were also conducted [21].

In another study, a dendrimer-gemcitabine conjugate was also prepared using the same linker. This system exhibited rapid drug release at the tumor site, with approximately 80% released within 24 hours, and, similar to the previous example, facilitated effective tumor targeting [22]. Azoreductase is another enzyme that is elevated in certain cancers. One such example is hepatic cancer. A study reports the use of different aromatic azo-linkers which are substrates for the enzyme. Through this linkage, doxorubicin-dendrimer conjugates were obtained. The rationale behind choosing this combination of substrate and enzyme was to facilitate the movement of free doxorubicin across the nuclear membrane for it to show its action as it cannot penetrate in the bound form [23]. There are so many more studies conducted to understand the scope of enzyme-responsive dendrimers in treatment of cancer. However, these products are yet to reach the market.

3.2 pH-Responsive Dendrimers

The metabolic rate is high in cancer cells and it leads to the accumulation of metabolic byproducts that lowers the pH of the tumor cells compared to the physiological pH. This unique property of the cancer cells triggered the drug release from the pH-sensitive materials. To achieve this phenomenon, pH-responsive moieties are incorporated into the dendrimers [24–26]. Various pH-responsive functionalities have been used in multiple studies including hydrazone [27], acetal [28], and imine [29].

Zhang et al. developed a pH-sensitive PAMAM dendrimer of doxorubicin. Doxorubicin was covalently bound to partially acetylated and folic acid-modified G5 PAMAM dendrimers by the pH-responsive cis-aconityl linkage. This linkage led to the release of doxorubicin in the cancer cells due to the acidic pH of the tumor cells. However, folic acid (FA) was combined with PAMAM to provide the specific targeting of the dendrimers to cancer cells expressing the folate receptor. In this study, partially acetylated G5 dendrimers were formed by reacting the surface amino groups of the dendrimers with acetic anhydride, resulting in the formation of G5.NHAc. The formed product was further conjugated with FA and fluorescein isothiocyanate (FI) that ultimately formed G5.NHAc.FA and G5.NHAc-FA-FI dendrimers. They used cis-aconityl linkage as a pH-sensitive moiety to combine DOX with the dendrimer. The final product, G5.NHAc-FA-DOX, was characterized by UV–visible spectrophotometry and NMR spectroscopy. The UV–visible spectrum showed that the final product, G5.NHAc-FA-DOX, showed clear absorbance at 280 and 490 nm. This indicated the successful conjugation of FA and DOX. 1H NMR peaks further confirmed the conjugation. In vitro drug release study was performed in three different pH conditions, i.e., pH 7.4, pH 6.0, and pH 5.0. At pH 5.0, more amount of DOX was released from the dendrimers which indicated that in the acidic pH of the cancer cells, cis-aconityl linkage breaks down and promotes the release of drug in the tumor tissue. Cellular uptake by the KB cells showed high fluorescence intensity of the G5.NHAc-FA-DOX conjugates [30]. Similarly, in another study Jin et al. developed poly(2-(N,N-diethylamino)ethyl methacrylate) with methoxy-poly(ethylene glycol)-poly(amido amine) PPD dendrimers of 5-Flurouracil to cause drug release in the acidic environment. PEG was conjugated to increase the circulation half-life by protecting it from phagocytosis. PDEA chain was combined with the PAMAM to make it pH responsive. PPD is the derivative of PAMAM only. For the formulation of the dendrimeric conjugation, the first PPD was synthesized and it included three steps: the first step is a synthesis of the G4 PAMAM dendrimers by divergent approach, the second step involved the combination of mPEG and PAMAM to get mPEG-PAMAM, the third step includes the combination of the PDEA to the mPEG-PAMAM to form mPEG-PAMAM-PDEA (i.e., PPD). Synthesized PPD is then ready for the encapsulation of the 5-FU. Encapsulation was done at acidic pH near to pH. By doing so PDEA chain become hydrophilic and more space allows for higher encapsulation. After the diffusion of the 5-FU into the inner space pH was immediately adjusted to pH 8.0 to prevent drug leakage. TEM analysis was conducted to check the effect of pH on the size of the PPD nanocarrier. pH-dependent release was observed from the PPD dendrimers. The PDEA chain is pH sensitive and in acidic media it converted to its hydrophilic form and allows for the release of the drug by opening the inner space, so in acidic condition at pH 6.5 shows more amount of drug release as compared to the physiological pH at pH 7.4 [31].

One of the most widely used linkers is hydrazone [27]. One of the studies shows the use of hydrazone bond for specific drug release. A novel approach that combines photothermal therapy and chemotherapy by conjugating Doxorubicin PAMAM dendrimer with the mercaptohexadecanoic acid-functionalized gold nanorod (MHA-AuNR) has been developed by Li et al. PEGylation was done to improve the circulation. When the NIR laser is applied to the AuNR, it increases the temperature and kills the cancer cell. DOX itself is an anticancer drug, and when conjugated with PAMAM, it releases the drug into cancer tissue due to the acidic environment. The combination approach of both photothermal and chemotherapy provided a synergistic effect on the cancer cells. PEG-PAMAM is synthesized by modifying the amino group of PAMAM with monomethoxyl poly (ethylene glycol) 4-nitrophenyl carbonate (PEG-NPC). PEG-PAMAM-AuNR was synthesized by combining the PEG-PAMAM with MHA-AuNR using EDC/NHS as a coupling reagent. PAMAM conjugated to AuNR by the surface amino groups. Then hydrazon-DOX binds with PEG-PAMAM-AuNR via a covalent bond and forms PEG-DOX-PAMAM-AuNR. The final conjugate was characterized by the TEM and NMR spectroscopy. An in vitro drug release study was conducted at both physiological pH and pH 5.0, demonstrating pH-triggered release of DOX from the conjugate, attributable to the acid-labile hydrazone linkage. An in vitro cytotoxicity study was performed on HeLa cells, which suggested that the cell viability decreased more significantly when the NIR laser was applied in combination with PEG-DOX-PAMAM-AuNR, compared to either therapy alone. In addition, the IC50 of DOX approached that of free DOX [32].

Unimolecular structure of the dendrimers attracts its use in the drug delivery system. Poly(carboxybetaine methacrylate) (PCBMA)-modified PAMAM dendrimers loaded with DOX were developed to successfully release the drug at tumor site. Researchers used dissipative particle dynamic simulation to develop the conjugation and studied the effect of the number of polymerizations on the structure shape. Self-assembly between the hydrophobic PAMAM and hydrophilic PCBMA driven by the hydrophobic interaction leads to the formation of the core-shell unimolecular micelles. Here, hydrophilic PCBMA acts as a protective layer and forms a shell around the core of the PAMAM. Unimolecular micelles of the G5 PAMAM-PCBMA were prepared in the simulation box by adding water. Doxil load in the hydrophobic core of the micelles in the same simulation box. As the chain length of the zwitterion and PCBMA increases, the stability of the unimolecular micelles improves. Further higher polymerization number overcome the aggregation caused by the hydrophobic core. The drug release mechanism was studied in both pH 7.4 and pH 5.0. The release study suggested that the acidic condition caused the protonation of the PCBMA and amine group of the PAMAM and electrostatic repulsion leads to the release of the drug from the hydrophobic core. Another mechanism behind the drug release was observed during the simulation, where PAMAM-PCBMA interacted with water molecules and became protonated, thereby accelerating the swelling rate of micelles. At this step, Doxil begins to be exposed and dispersed into the aqueous solution, suggesting the drug release [33].

Further, a dual drug methotrexate (MTX) and *all-trans*-retinoic acid (ATRA) encapsulated G4 PAMAM dendrimers have been developed for cancerous cells targeted drug delivery in leukemia. The acidic pH of the tumor cells triggered the release of drugs from the dendrimers. Both drugs were loaded in the cavity of the dendrimers by the dialysis method. First, MTX was loaded and afterward, ATRA was loaded by dipping the dialysis bag containing the MTX-loaded dendrimers in the methanolic solution of the ATRA. Methanol facilitates the loading of the ATRA by generating the void in dendrimers. In vitro, drug release study was performed in three different pH conditions, i.e., pH 4.0, 7.4, and10.0. A higher amount of MTX and ATRA was released at pH 4.0. Release of drug in acidic pH attributed to the protonation of the both primary and tertiary amine group of the PAMAM [34].

3.3 Light-Responsive Dendrimers

Another external stimulus that has been studied for the purpose of targeted drug delivery is light. Triggers like ultraviolet radiation, visible radiation, and near-IR radiation have been the most explored so far. The basis of forming such systems is the presence of a moiety that can absorb the above-mentioned radiation and react to it, maybe through a conformational change that may be reversible or irreversible, or any other mechanism, that ultimately triggers the drug release [18].

One of the studies exploring the use of light-responsive dendrimers in cancer therapy employ the use of coumarin-conjugated dendrimers. Coumarin is a photo-responsive group that undergoes photodimerization when exposed to UV radiation above 300 nm. Therefore, cross-linked structures are developed that can act as cages for the purpose of drug encapsulation. When exposed to radiation of a lower wavelength, say around 254 nm, breaking of these cages takes place and hence the drug release. 5-Fluorouracil (5-FU) was used as the model anti-cancer drug in this study. It was found that the prepared system was less toxic than free 5-FU because it imparts a sustained release property to the drug while the system itself (coumarin–dendrimer conjugate) was not cytotoxic. The study was further extended to investigate the co-delivery of 5-FU and TRAIL, a gene with selective anti-cancer activity that is less potent on its own and therefore requires combination with other therapeutics. Successful co-delivery of the drug and gene was achieved. Further studies that can help avoid the use of UV radiation, because of it being phototoxic, are in progress [35]. Azobenzenes are another group of molecules that are studied for their photo-responsive properties. An advantage of using them in dendrimers is the flexibility they offer with regard to their placement in the dendrimer. They can be placed at the core, as the terminal group or in between the dendrimer. Their location also influences their photo-responsive property [36].

In another study, PAMAM dendrimers were synthesized whose surface was modified with the o-nitrobenzyl group that acts as the photo-responsive moiety in this case. These molecules cut off from the dendrimer when exposed to UV radiation of around 365 nm and hence cause drug release. Release of Salicylic acid and Adriamycin was explored in this study. Upon conducting a comparative study of drug release with and without the photo trigger, it was observed that the photo-responsive system exhibited superior performance [37].

3.4 Temperature-Responsive Dendrimers

Temperature is another stimulus that can be exploited to bring about targeted drug release. However, in this case, the trigger is external. We must know the exact time and location at which the trigger is to be applied for desired drug release. The theory of lower critical solution temperature (LCST) [38–40] becomes quite relevant in this scenario. It is the temperature at which rapid conformational changes take place in the dendrimer. Below the LCST, hydration and expansion of the dendrimers take place and above it, they collapse and form aggregates. Since dendrimers are known to have a very specific and organized structure, any trigger that causes conformational changes in the same is very much capable of causing drug release. There can be two main approaches to obtain temperature-sensitive dendrimers:

1. Using temperature-sensitive polymers for the synthesis of the dendrimers, for example, poly(N-isopropylacrylamide), is one of the most common polymers utilized for this purpose.
2. Modifying the dendrimer surface using groups that are known to have temperature-sensitive properties instead of using the specific polymer as a whole. For example, alkylamide groups from thermo-sensitive poly(N-alkylacrylamide) have been used to modify the surface of PAMAM and poly(propyleneimine) dendrimers. This one is the most preferred method.

However, when synthesizing temperature-responsive systems for application within the human body, an LCST close to physiological temperature is preferred to ensure that the applied trigger for drug release does not cause harm.

A study has been carried out using alkoxy diethylene glycol units as surface-modifying agents for PAMAM dendrimers to yield temperature-responsive systems. Another reason for choosing this particular surface modifier was also its ability to impart noncytotoxic property against normal cells to the entire system. The LCST was determined using transmittance studies, wherein the temperature at which a sudden decrease in transmittance was noted was taken as LCST. It was noted that, higher the dendrimer generation, lower was the LCST. The studies also proved the dependence of the LCST on the hydrophobicity of the surface-modifying groups, the ratio of hydrophobic to hydrophilic groups, etc. Therefore, by varying these parameters, the LCST can be modified bringing it close to the body temperature making it suitable for use in humans [38]. Phenylalanine residues are also known to impart thermo-sensitive properties to dendrimers. One study reports usage of the same to modify the surface of PAMAM dendrimers. Similar dependence of LCST on dendrimer generation, and hydrophobicity characteristics.

Stimuli-responsive dendrimers using temperature as the sole trigger for drug release are rather rare. Instead, studies on multi-stimuli-responsive dendrimers using temperature and pH as the stimuli for drug release are more common.

3.5 Multi-Responsive Dendrimers

In stimuli-responsive therapy, combining more than one stimulus has attained great attention due to specific targeting. Paclitaxel-loaded nanoparticles of multifunctional dendrimers were developed for combinational chemo and photodynamic therapy. Paclitaxel itself arrests the cell in the G2/M phase and shows anticancer action. The addition of the photosensitizer led to the formation of reactive oxygen species and caused cell death. Redox-responsive PTX released from the NPs in response to either GSH or H2O2 in the nucleus of the cell and accumulated in the nucleus causing increased photosensitizer concentration in the nucleus, thereby promoting DNA damage and eventually cell death. Together, the redox and photodynamic stimuli enhanced the cellular internalization of the NPs and overcame the PTX resistance. In vitro release studies showed that the maximum amount of drug was released in the presence of the GSH [41]. Nguyen et al. developed pH and redox dual-responsive PAMAM dendrimers of letrozole to target tumor cells. They used cystamine as a redox-responsive disulfide linker between PAMAMG3.5 and hydrophilic heparin (p-SS-Hep). PAMAM dendrimers were conjugated with heparin to reduce the side effects of the PAMAM due to its cationic amine groups.

Structure of the dendrimer was characterized by the 1H NMR. Cleavage of the disulfide linker was evaluated in various GSH and pH conditions. Drug release study data showed that at high concentrations of GSH and at low pH, the amount of drug released from dendrimers was high. The disulfide bond broke down in the reductive environment and due to the cleavage of the linker, heparin detached from the dendrimer and reduced the stearic hindrance allowing the drug release from the core to the medium [42].

3.5.1 Triple-Responsive Dendrimer

Tumor cells themselves have heterogeneity in their respective microenvironment. They are different in terms of pH gradient, and oxidation-reduction stress. Thus, one stimulus might be insufficient to provide effective treatment. For these reasons, multiple stimuli-responsive dendrimers target all biological microenvironments of the tumor cells and provide site-specific delivery within them. Zhong et al. synthesized triple-responsive dendrimeric nanocages (TDN) for site-specific drug delivery. The formed dendrimers responded to pH, temperature, and reductive conditions in the presence of GSH. The core of the dendrimers was made up of octa(3-aminopropyl) silsesquioxane and they were constructed with lysine by the divergent method. Furthermore, dendrimers were modified using lipoic acid to provide the disulfide bond. Co-assembly of dendrimers with PEG allows for prolonged blood circulation. Under acidic conditions, due to the enhanced hydrophilicity of PEG, dendrimers swelled. Under the high temperature, dehydration of the PEG caused shrinkage of the outer shell. In vitro drug release studies were performed under different conditions including various concentrations of GSH, at two different pH conditions (7.4, 5.0) and two different temperatures [20, 37]. Results show that the maximum amount of the drug was released at a lower pH (5.0), at a high concentration of GSH (10 mM), and high temperature [37]. The drug-loading capacity of TDN was twofold higher at lower temperatures and drug release was achieved through pH, reduction stimuli, and temperature [43]. Further, Hu et al. synthesized the redox and pH dual-responsive functional nanocarrier (His-PAMAM-ss-PEG-Tf) for targeted anticancer drug delivery to liver cancer. Generation 4 PAMAM was conjugated with histidine (His) on the surface to obtain the pH sensitivity and release of the drug at the pH of the cancer cells. Disulfide (s-s) linkage was introduced between PAMAM and PEG to allow drug release in the reductive environment. Transferrin (Tf) was introduced to provide active tumor targeting. Doxorubicin was loaded in the hydrophobic cavity of the dendrimers. The nanocarrier was characterized by 1H NMR and DLS. In vitro release by the dialysis method was done in three different buffers. Results showed that the drug release was influenced by pH. Faster drug release was achieved at the lower pH. pH-triggered release attributed to the protonation and deprotonation of imidazole of the histidine in the acidic condition. In the presence of the strong reducing agent, GSH breaks the disulfide bond between PAMAM and PEG and promotes the drug release by detaching the PEG chain [44].

pH and redox-responsive PAMAM dendrimers of DOX have been developed by Jinshan et al. for the treatment of gastric cancer. Hyperbranched PAMAM was PEGylated to reduce the toxicity of the same and increase the circulation half-life. DOX was attached to the PAMAM by the acid-sensitive cis-aconityl linkage. This conjugate self-assembled to form micelles. Further cross-linking with disulfide bonds made it redox-responsive. Data from drug release studies showed that acidic and reductive environments triggered the release of drug [45] (Table 5.1).

4 Conclusion and Future Scope

Stimuli-responsive dendrimers, designed to release therapeutic agents upon encountering specific triggers, offer significant potential for precision in drug delivery. By responding to stimuli such as pH changes, enzymes, redox conditions, and light, these dendrimers allow for controlled, targeted therapy with reduced side effects compared to traditional treatments. Their adaptability to the unique conditions of the tumor microenvironment enables enhanced therapeutic efficiency. Despite promising preclinical outcomes, the clinical application of stimuli-responsive dendrimers faces challenges. Issues such as biocompatibility, scalability, and the complexity of in vivo environments—where tumors exhibit heterogeneous levels of ROS and GSH—introduce unpredictability in drug release. These difficulties must be addressed to fully realize the potential of dendrimer-based systems. Moreover, research should focus on improving dendrimer stability and compatibility, while also advancing multi-responsive systems that can address multiple tumor-specific triggers simultaneously. Integrating these systems with imaging techniques will facilitate theranostic applications, enabling real-time monitoring of drug delivery and treatment progress. Additionally, thorough in vivo and clinical studies are essential to ensure that these dendrimers can safely and effectively scale for widespread use in cancer therapy.

Table 5.1 Summary of stimuli-responsive dendrimer systems: Polymers, mechanisms, and applications

Stimulus type	Case study	Employed polymers	Mechanism	Application	References
pH-responsive	pHLIP-conjugated polylysine dendrimers for siRNA targeting VEGF	Polylysine dendrimers	pHLIP inserts across cell membrane under acidic conditions, translocating siRNA	Targeting tumor microenvironment, gene delivery for cancer therapy	[46]
	PAMAM dendrimers modified with maleyl groups	PAMAM dendrimers with maleyl groups	Charge reversal in acidic vesicles for endosomal escape and nuclear entry of drugs	Gene and drug delivery to cancer cells	[47]
Redox-responsive	PAMAM dendrimer-valproic acid conjugates for GSH-triggered drug release	PAMAM dendrimers, valproic acid	GSH-induced cleavage of disulfide bonds, releasing valproic acid	Enhanced delivery of anticonvulsants with reduced systemic side effects	[48]
Enzyme-responsive	Doxorubicin-PAMAM conjugates using GFLG peptide linkers	PAMAM dendrimers, GFLG peptides	GFLG peptides cleaved by cathepsin B in tumor cells, releasing doxorubicin	Tumor-targeted drug release with minimal off-target effects	[49]
	PAMAM dendrimers conjugated to 5-aminosalicylic acid via azo linkers	PAMAM dendrimers, 5-ASA	Azoreductase in colon cleaves azo bond, releasing anti-inflammatory drugs	Targeted drug release for treating inflammatory bowel diseases	[50]

(continued)

Table 5.1 (continued)

Stimulus type	Case study	Employed polymers	Mechanism	Application	References
Light-responsive	Porphyrin-conjugated PAMAM dendrimers for photodynamic therapy	Porphyrin-cored PAMAM dendrimers	UV light activates ROS generation, leading to endosomal membrane disruption	Photodynamic therapy for improved cytoplasmic drug and gene delivery	[51]
	Doxorubicin-conjugated dendrimers with ortho-nitrobenzyl (ONB) linkers	PAMAM dendrimers, doxorubicin, ONB	UV light cleaves ONB linkers, triggering doxorubicin release	Controlled anticancer drug release under light exposure	[52]
Thermo-responsive	Isobutyl amide-terminated dendrimers forming temperature-sensitive capsules	Isobutyl amide-terminated dendrimers	LCST-induced micelle disassembly at elevated temperature	Temperature-controlled drug delivery	[53]
Multistimuli-responsive	Collagen-modified dendrimer conjugates responsive to both acid and enzymes	PAMAM dendrimers, collagen peptides	Hydrazone linkages and MMP-sensitive collagen peptides enable dual-responsive drug release	Improved drug delivery precision in tumor microenvironments	[54]

References

1. Rana D, Salave S, Longare S, Agarwal R, Kalia K, Benival D. Nanotherapeutics in tumour microenvironment for cancer therapy. Nanosci Nanotechnol Asia. 2021;12(1)
2. Rana D, Salave S, Perla A, Nadkarni A, Kolhe S, Jindal AB, et al. Bugs as drugs: understanding the linkage between gut microbiota and cancer treatment. Curr Drug Targets. 2022;23(9):869–88.
3. Burney I. Cancer chemotherapy and biotherapy: principles and practice. Sultan Qaboos Univ Med J. 2011;11:424–5.
4. Salave S, Rana D, Benival D. Encapsulation of anabolic peptide in lipid nano vesicles for osteoporosis. Curr Protein Pept Sci. 2022;23(7):495–503.
5. Wang X, Zhang H, Chen X. Drug resistance and combating drug resistance in cancer. Cancer Drug Resist. 2019;2(2):141–60.
6. Chehelgerdi M, Chehelgerdi M, Allela OQB, Pecho RDC, Jayasankar N, Rao DP, et al. Progressing nanotechnology to improve targeted cancer treatment: overcoming hurdles in its clinical implementation. Mol Cancer. 2023;22(1):169.
7. Salave S, Rana D, Benival D. Dual targeting anti-osteoporotic therapy through potential nanotherapeutic approaches. Pharm Nanotechnol. 2022;10(5):384–92.
8. Salave S, Rana D, Benival D. Peptide functionalised nanocarriers for bone specific delivery of PTH (1-34) in osteoporosis. Current Nanomed. 2021;11(3):142–8.
9. Xu M, Han X, Xiong H, Gao Y, Xu B, Zhu G, et al. Cancer Nanomedicine: emerging strategies and therapeutic potentials. Molecules. 2023;28(13)
10. Madaan K, Kumar S, Poonia N, Lather V, Pandita D. Dendrimers in drug delivery and targeting: drug-dendrimer interactions and toxicity issues. J Pharm Bioallied Sci. 2014;6(3):139–50.
11. Abbasi E, Aval SF, Akbarzadeh A, Milani M, Nasrabadi HT, Joo SW, et al. Dendrimers: synthesis, applications, and properties. Nanoscale Res Lett. 2014;9(1):247.
12. Mi P. Stimuli-responsive nanocarriers for drug delivery, tumor imaging, therapy and theranostics. Theranostics. 2020;10(10):4557–88.
13. Rahim MA, Jan N, Khan S, Shah H, Madni A, Khan A, et al. Recent advancements in stimuli responsive drug delivery platforms for active and passive cancer targeting. Cancers (Basel). 2021;13(4)
14. Bacha K, Chemotti C, Mbakidi JP, Deleu M, Bouquillon S. Dendrimers: synthesis, encapsulation applications and specific interaction with the stratum corneum—a review. Macromol. 2023;3:343–70.
15. Gaitsch H, Hersh AM, Alomari S, Tyler BM. Dendrimer technology in glioma: functional design and potential applications. Cancers (Basel). 2023;15(4)
16. Chis AA, Dobrea C, Morgovan C, Arseniu AM, Rus LL, Butuca A, et al. Applications and limitations of dendrimers in biomedicine. Molecules. 2020;25(17)
17. Wang H, Huang Q, Chang H, Xiao J, Cheng Y. Stimuli-responsive dendrimers in drug delivery. Biomater Sci. 2016;4(3):375–90.
18. Saluja V, Mankoo A, Saraogi GK, Tambuwala MM, Mishra V. Smart dendrimers: synergizing the targeting of anticancer bioactives. J Drug Deliv Sci Technol. Editions de Sante. 2019;52:15–26.
19. Mijanović O, Branković A, Panin AN, Savchuk S, Timashev P, Ulasov I, et al. Cathepsin B: a sellsword of cancer progression. Cancer Lett. 2019;(449):207–14.
20. Reunanen N, Kähäri VM. Madame curie bioscience database [Internet]. Matrix Metalloproteinases in Cancer Cell Invasion.
21. Lee S, Jeong Y, Park H, Kang D, Oh J, Lee S, et al. Enzyme-responsive doxorubicin release from dendrimer nanoparticles for anticancer drug delivery. Int J Nanomedicine. 2015;10(1):5489–503.
22. Zhang C, Pan D, Li J, Hu J, Bains A, Guys N, et al. Enzyme-responsive peptide dendrimer-gemcitabine conjugate as a controlled-release drug delivery vehicle with enhanced antitumor efficacy. Acta Biomaterialia. 2017;55:153–62.

23. Medina SH, Chevliakov MV, Tiruchinapally G, Durmaz YY, Kuruvilla SP, Elsayed MEH. Enzyme-activated nanoconjugates for tunable release of doxorubicin in hepatic cancer cells. Biomaterials. 2013;34(19):4655–66.
24. Hao G, Xu ZP, Li L. Manipulating extracellular tumour pH: an effective target for cancer therapy. RSC Adv. Royal Society of Chemistry;. 2018;8:22182–92.
25. Mu Y, Gong L, Peng T, Yao J, Lin Z. Advances in pH-responsive drug delivery systems. OpenNano. 2021;5
26. Bae Y, Nishiyama N, Kataoka K. In vivo antitumor activity of the folate-conjugated pH-sensitive polymeric micelle selectively releasing adriamycin in the intracellular acidic compartments. Bioconjug Chem. 2007;18(4):1131–9.
27. Sonawane SJ, Kalhapure RS, Govender T. Hydrazone linkages in pH responsive drug delivery systems. Eur J Pharm Sci. Elsevier B.V. 2017;99:45–65.
28. Chen W, Meng F, Cheng R, Zhong Z. pH-Sensitive degradable polymersomes for triggered release of anticancer drugs: a comparative study with micelles. J Control Release. 2010;142(1):40–6.
29. Han S, Lee J, Jung E, Park S, Sagawa A, Shibasaki Y, et al. Mechanochemical drug conjugation via pH-responsive imine linkage for polyether prodrug micelles. ACS Appl Bio Mater. 2021;4(3):2465–74.
30. Zhang M, Zhu J, Zheng Y, Guo R, Wang S, Mignani S, et al. Doxorubicin-conjugated PAMAM dendrimers for pH-responsive drug release and folic acid-targeted cancer therapy. Pharmaceutics. 2018;10(3)
31. Jin Y, Ren X, Wang W, Ke L, Ning E, Du L, et al. A 5-fluorouracil-loaded pH-responsive dendrimer nanocarrier for tumor targeting. Int J Pharm. 2011;420(2):378–84.
32. Li X, Takashima M, Yuba E, Harada A, Kono K. PEGylated PAMAM dendrimer-doxorubicin conjugate-hybridized gold nanorod for combined photothermal-chemotherapy. Biomaterials. 2014;35(24):6576–84.
33. Zeng S, Quan X, Zhu H, Sun D, Miao Z, Zhang L, et al. Computer simulations on a pH-responsive anticancer drug delivery system using zwitterion-grafted polyamidoamine dendrimer Unimolecular micelles. Langmuir. 2021;37(3):1225–34.
34. Tekade RK, Dutta T, Gajbhiye V, Jain NK. Exploring dendrimer towards dual drug delivery: PH responsive simultaneous drug-release kinetics. J Microencapsul. 2009;26(4):287–96.
35. Wang H, Miao W, Wang F, Cheng Y. A self-assembled coumarin-anchored dendrimer for efficient gene delivery and light-responsive drug delivery. Biomacromolecules. 2018;19(6):2194–201.
36. Deloncle R, Caminade AM. Stimuli-responsive dendritic structures: the case of light-driven azobenzene-containing dendrimers and dendrons. J Photochem Photobiol C: Photochem Rev. 2010;11(1):25–45.
37. Li Y, Jia X, Gao M, He H, Kuang G, Wei Y. Photoresponsive nanocarriers based on PAMAM dendrimers with a *o* -nitrobenzyl shell. J Polym Sci A Polym Chem. 2010;48(3):551–7.
38. Li X, Haba Y, Ochi K, Yuba E, Harada A, Kono K. PAMAM dendrimers with an oxyethylene unit-enriched surface as biocompatible temperature-sensitive dendrimers. Bioconjug Chem. 2013;24(2):282–90.
39. Zhang Y, Chapman WG. Modeling lower critical solution temperature behavior of associating dendrimers using density functional theory. Langmuir. 2019;35(33):10808–17.
40. Tono Y, Kojima C, Haba Y, Takahashi T, Harada A, Yagi S, et al. Thermosensitive properties of poly(amidoamine) dendrimers with peripheral phenylalanine residues. Langmuir. 2006;22(11):4920–2.
41. Zhong D, Wu H, Wu Y, Li Y, Yang J, Gong Q, et al. Redox dual-responsive dendrimeric nanoparticles for mutually synergistic chemo-photodynamic therapy to overcome drug resistance. J Control Release. 2021;329:1210–21.
42. Nguyen TL, Nguyen TH, Nguyen CK, Nguyen DH. Redox and pH responsive poly (amidoamine) dendrimer-heparin conjugates via disulfide linkages for letrozole delivery. Biomed Res Int. 2017;2017

43. Zhong D, Wu H, Wu Y, Li Y, Xu X, Yang J, et al. Rational design and facile fabrication of biocompatible triple responsive dendrimeric nanocages for targeted drug delivery. Nanoscale. 2019;11(32):15091–103.
44. Hu Q, Wang Y, Xu L, Chen D, Cheng L. Transferrin conjugated ph-and redox-responsive poly(Amidoamine) dendrimer conjugate as an efficient drug delivery carrier for cancer therapy. Int J Nanomedicine. 2020;15:2751–64.
45. Nie J, Wang Y, Wang W. In vitro and in vivo evaluation of stimuli-responsive vesicle from PEGylated hyperbranched PAMAM-doxorubicin conjugate for gastric cancer therapy. Int J Pharm. 2016;509(1–2):168–77.
46. Reshetnyak YK, Andreev OA, Lehnert U, Engelman DM. Translocation of molecules into cells by pH-dependent insertion of a transmembrane helix [Internet], vol. 103. PNAS; 2006. Available from: www.pnas.orgcgi. https://doi.org/10.1073/pnas.0601463103
47. Zhou Z, Shen Y, Tang J, Fan M, Van Kirk EA, Murdoch WJ, et al. Charge-reversal drug conjugate for targeted cancer cell nuclear drug delivery. Adv Funct Mater. 2009;19(22):3580–9.
48. Mishra MK, Beaty CA, Lesniak WG, Kambhampati SP, Zhang F, Wilson MA, et al. Dendrimer brain uptake and targeted therapy for brain injury in a large animal model of hypothermic circulatory arrest. ACS Nano. 2014;8(3):2134–47.
49. Zhang C, Pan D, Luo K, Li N, Guo C, Zheng X, et al. Dendrimer-doxorubicin conjugate as enzyme-sensitive and polymeric nanoscale drug delivery vehicle for ovarian cancer therapy. Polym Chem. 2014;5(18):5227–35.
50. Wiwattanapatapee R, Lomlim L, Saramunee K. Dendrimers conjugates for colonic delivery of 5-aminosalicylic acid. J Control Release. 2003;88(1):1–9.
51. Shieh MJ, Peng CL, Lou PJ, Chiu CH, Tsai TY, Hsu CY, et al. Non-toxic phototriggered gene transfection by PAMAM-porphyrin conjugates. J Control Release. 2008;129(3):200–6.
52. Ki Choi S, Thomas T, Li MH, Kotlyar A, Desai A, Baker JR. Light-controlled release of caged doxorubicin from folate receptor-targeting PAMAM dendrimer nanoconjugate. Chem Commun. 2010;46(15):2632–4.
53. Kono K, Murakami E, Hiranaka Y, Yuba E, Kojima C, Harada A, et al. Thermosensitive molecular assemblies from poly(amidoamine) dendron-based lipids. Angew Chem Int Ed. 2011;50(28):6332–6.
54. Kojima C, Suehiro T, Watanabe K, Ogawa M, Fukuhara A, Nishisaka E, et al. Doxorubicin-conjugated dendrimer/collagen hybrid gels for metastasis-associated drug delivery systems. Acta Biomater. 2013;9(3):5673–80.

Stimuli-Responsive Liposomes for Cancer Therapy

6

Namarta Thakur, Sunil Bharti, Shivani Sharma, Pemal Preet Kaur, and Amita Sarwal

Abstract

Liposomes have emerged as one of the most promising and versatile platforms for drug delivery, especially in the field of cancer. These nanoparticles are spherical vesicles composed of lipid bilayers, which exhibit biocompatibility and possess unique dual hydrophilic-hydrophobic properties. Liposomes effectively encapsulate a broad range of therapeutic agents, including both hydrophilic and lipophilic drugs, thereby offering a way to deliver both types of compounds. The primary benefit of liposomes is their ability to improve the safety profile by reducing the systemic toxicity of drugs by limiting their exposure to healthy tissues. Furthermore, liposomes can be designed with "smart" characteristics, allowing for controlled, stimulus-responsive drug release. These stimuli can be either internal (such as changes in pH, the presence of reducing agents, or enzymes) or external (such as light, magnetic fields, or ultrasound). Stimuli-responsive liposome is a precise approach for delivering chemotherapeutic agents at the tumor site in a controlled manner, thus reducing damage to normal tissues and preventing the side effects of the conventional chemotherapy regimen. This chapter looks at recent advances in liposome-based drug delivery methods, particularly in cancer therapy. It discusses several stimulus-responsive alterations, manufacturing procedures, characterization, and a brief literature review on their use in cancer treatment.

Keywords

Liposomes · Cancer · Targeted Delivery · Stimuli-response · Smart Drug Delivery

---

N. Thakur · S. Bharti · S. Sharma · P. P. Kaur · A. Sarwal (✉)
University Institute of Pharmaceutical Sciences and UGC Centre of Advanced Studies, Panjab University, Chandigarh, India

P. V. Devarajan et al. (eds.), *Advances in Stimuli-Responsive Nanosystems for Cancer Therapy*, https://doi.org/10.1007/978-981-95-2744-1_6

Abbreviations

AuNRs	Gold nanorods
CHEMS	Cholesteryl hemisuccinate
DLS	Dynamic light scattering
DOPE	Dioleoylphosphatidylethanolamine
GSH	Glutathione
HIV	Human immunodeficiency virus
HPLC	High-performance liquid chromatography
LUV	Large unilamellar vesicles
MLV	Multilamellar vesicles
MPS	Mononuclear phagocyte system
NIR	Near-infrared
PC	Phosphatidylcholine
PDI	Polydispersity index
PE	Phosphatidylethanolamine
PEG	Polyethylene glycol
PI	Phosphatidylinositol
PS	Phosphatidylserine
SEM	Scanning electron microscopy
SUV	Small unilamellar vesicles
TEM	Transmission electron microscopy

1 Introduction

Cancer has become one of the most prevalent diseases in India and worldwide, with many families having faced its wrath. This ailment isn't just a singular disease but a spectrum of conditions marked by uncontrolled cell growth. When cells in any part of our body grow uncontrollably, they can form a tumor. These tumors can be benign, which are generally harmless, or malignant, which are cancerous and can spread to other parts of the body. The World Health Organization's data paints a grim picture, placing cancer as a leading cause of death globally. In India, the surge in cancer cases has made the quest for effective treatments even more pressing [1]. The heterogeneity of cancer cells, resistance to drugs, and the adverse effects of conventional therapies make it imperative to develop targeted and efficient treatment modalities [2]. In the vast landscape of medical research, liposomes have emerged as a ray of hope. Liposomes are round structures made up of at least one lipid layer, primarily phospholipids. These lipids naturally fold into spherical shapes with a water-filled core surrounded by the lipid layers [3]. Figure 6.1 represents some of the challenges faced by current therapy options for cancer treatment.

Liposomes are microscopic, spherical vesicles formed by one or more lipid bilayers that encapsulate an aqueous phase both within and between these bilayers. They have gained immense popularity in drug delivery applications due to their ability to carry both hydrophilic (water-soluble) and hydrophobic (lipid-soluble)

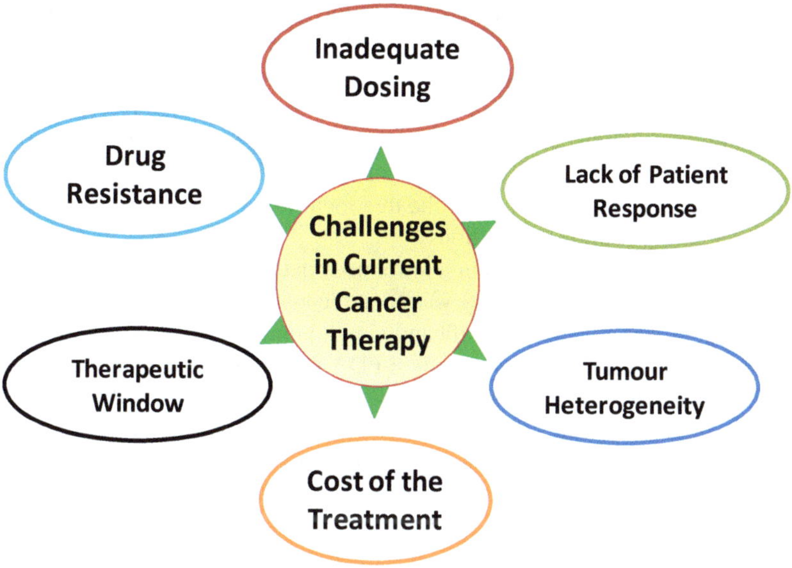

Fig. 6.1 Limitations of the current cancer therapy

substances. The structure of liposomes allows for the controlled release of drugs, improving their efficacy and reducing side effects. Different types of liposomes are tailored for specific uses: multilamellar vesicles (MLVs) consist of several concentric lipid bilayers and can carry a large volume of the aqueous phase, making them suitable for larger drug molecules or combinations of drugs. Small unilamellar vesicles (SUVs) are smaller, with a single bilayer, and are ideal for more precise drug delivery due to their uniform size. Large unilamellar vesicles (LUVs) are larger versions of SUVs and are often used in various therapeutic applications. Additionally, long-circulating liposomes, modified with polymers, stay in the bloodstream for extended periods, enhancing the pharmacokinetics of the drug, while immunoliposomes, which have antibodies on their surfaces, allow for targeted delivery of therapeutic agents to specific cells or tissues. This versatility makes liposomes invaluable carriers for a wide range of drugs, including cancer therapies and vaccines [4].

The primary component of liposomes is phospholipids, which are amphiphilic molecules with both hydrophilic heads and hydrophobic tails, allowing them to spontaneously form bilayer structures in aqueous environments. Phospholipids can be derived from natural sources, such as egg yolk or soybeans, or synthesized artificially. Cholesterol is often incorporated into liposomes to enhance their membrane rigidity and stability. Depending on the specific application, liposomes can be further modified with additional lipids, targeting ligands, or surface modifiers to optimize drug delivery. Effective characterization of liposomes is essential to ensure their quality, safety, and efficacy. Common techniques include dynamic light scattering (DLS) to measure particle size, zeta potential analysis to assess surface charge and stability, and electron microscopy to visualize liposomal morphology. The encapsulation efficiency—the percentage of drug successfully incorporated into the liposome—can be measured using methods like ultracentrifugation or gel filtration chromatography. The advantages of using liposomes in drug delivery include enhanced drug solubility, controlled and targeted drug release, reduced toxicity by minimizing direct contact with healthy tissues, and improved pharmacokinetics by prolonging the circulation time of drugs in the bloodstream. These benefits have made liposomes a promising tool in the treatment of various diseases, particularly in cancer and other chronic conditions.

The concept of liposomes was first introduced in the 1960s by Alec D. Bangham, and their potential for drug delivery was soon recognized. Over the decades, liposomal technology has advanced significantly, with numerous liposomal formulations being developed for clinical use. The first FDA-approved liposomal drug, Doxil®, a liposomal formulation of doxorubicin, was introduced in the 1990s for cancer treatment, marking a milestone in the use of liposomes in medicine. Since then, many liposomal drugs have been approved for clinical use, and a growing number are in various stages of research and development, with liposomal-based drug delivery systems playing a critical role in advancing personalized medicine and improving therapeutic outcomes across a range of medical conditions [5, 6]. Table 6.1 enlists the various approved products and some of the ongoing clinical trials.

Phospholipids have a unique dual nature, being both water-attracting and water-repelling. In water, their polar heads point outwards because their non-polar tails don't mix well with water, forming a protective layer. This setup is naturally preferred and gets a boost from hydrogen bonds, van der Waals forces, and other electric attractions. Liposomes, having a watery center and a surrounding fat layer, can hold both hydrophilic and hydrophobic molecules. How these molecules mix and act inside the body relies on the kind of liposomes used [7, 8].

2 Stimuli-Responsive Liposome

Stimuli-responsive systems, which release their therapeutic payload in response to specific triggers, offer a novel approach to targeted drug delivery. While they've shown promise in preclinical models, their clinical translation faces challenges, especially in controlling drug release. These systems can recognize and dynamically react to their microenvironment, mimicking living organisms' adaptability. The concept, introduced in the 1970s, has seen significant advancements. These systems can be sensitive to internal changes, such as pH shifts or enzyme concentrations, or external stimuli like temperature. Their adaptability allows for precise drug release in specific conditions, such as cancer or inflammation, and offers diverse administration methods, enhancing treatment efficacy [4]. By responding to specific stimuli present in the tumor microenvironment or applied externally, these systems ensure that the drug is released at the desired site, minimizing side effects and maximizing therapeutic efficacy [9].

Stimuli-responsive drug delivery offers several key advantages, particularly in enhancing the effectiveness of treatments. By ensuring that drugs are released

Table 6.1 Currently approved applications of liposomal drug formulations

Brand name	Manufacturer	Active ingredient	Medical indication
Approved products			
Myocet®	(Zeneus)	doxorubicin	Metastatic breast cancer
Doxil®	(Baxter)	doxorubicin	Kaposi's sarcoma, ovarian, and breast cancer
Lipodox®	(Sunpharmac Euticals Industries)	doxorubicin	Blood cancer, ovarian, and breast cancer
DaunoXome®	(Galen US Inc.)	daunorubicin	Dematological malignancy
Marqibo®	(Talon Therapeutics Inc.)	Vincristine sulfate	Acute lymphoblastic leukemia
Depocyt®	(Pacira Pharmaceuticals Inc.)	cytarabine	Neoplastic meningitis and lymphomatous meningitis
Clinical trials			
Carrier	Medical Indication	Description	Registration No.
Mitoxantrone hydrochloride liposome	Ovarian cancer (Phase I)	Evaluating potential in patients with cancer recurrence or resistance to platinum agents	NCT04718376
Bevacizumab and doxorubicin HCl liposome	Breast cancer (Phase II)	Use of this formulation for treatment of patients with metastasis or recurrence	NCT00445406
Doxorubicin HCl liposomes	Breast cancer (Phase II)	Examining potential compared to docetaxel in patients with recurrence or evaluating tolerance	NCT00193037
Mitoxantrone hydrochloride liposome	HER2-negative breast cancer (Phase II)	Evaluating safety and efficacy of nanovehicle in treatment of patients	NCT04927481
Paclitaxel liposomes	Pancreatic cancer (Phase IV)	Evaluating safety and efficacy on patients after paclitaxel treatment or combination therapy	NCT04217096

specifically at the desired site of action, it enables targeted delivery, which maximizes therapeutic effects and minimizes damage to healthy tissues [10]. This approach also helps reduce side effects by confining the drug's action to the target area. Moreover, controlled release systems maintain drug concentrations within the therapeutic window for extended periods, improving overall therapeutic efficacy [11]. Additionally, the reduction in side effects and the enhancement of treatment effectiveness contribute to better patient compliance, making stimuli-responsive drug delivery a promising strategy for improving patient outcomes in various medical conditions [12].

2.1 Stimuli-Responsive Mechanisms

Stimuli-responsive mechanisms refer to the ability of certain materials, in this context, liposomes, to change their properties or behavior in response to specific triggers or stimuli. In the realm of drug delivery, these mechanisms are of paramount importance as they allow for controlled and targeted release of therapeutic agents. By ensuring that the drug is released only in the presence of a specific stimulus, typically found in the pathological site (like a tumor), the therapeutic efficacy is enhanced while minimizing potential side effects [9].

3 Types of Stimuli

Stimuli can be broadly categorized into two types: internal stimuli and external stimuli.

3.1 Internal Stimuli

In today's medical landscape, there's a significant emphasis on advanced drug delivery systems. The objective is to ensure drugs are precisely targeted and to monitor their therapeutic release effectively. However, mastering drug distribution and pinpointing the exact release locations remains a substantial challenge. To address this, researchers are delving into the unique attributes of the tumor microenvironment, aiming to harness these internal stimuli to enhance drug delivery.

By understanding these distinct physiological differences, the goal is to develop a dynamic system that can adapt its drug release based on specific biological signals. Recent innovations have led to the creation of liposomes with special chemical modifications. These liposomes are tailored to respond to conditions commonly found within tumor environments, such as subtle pH changes, the presence of certain enzymes, or oxygen-deprived areas. This adaptability is crucial for enhancing the efficacy of chemotherapy treatments.

Imagine a system where drugs are released most effectively, either simultaneously or in stages, facilitated by penetration enhancers. These multifunctional systems promise improved therapeutic outcomes, ensuring the drug is delivered optimally to the desired sites in the body. With such advancements, the medical field is poised for a future where treatments could be more efficient and patient-centric, marking a promising era in healthcare [13]. These triggers are inherently present or generated within the biological system.

3.1.1 pH

The tumor microenvironment, characterized by its distinct physiological attributes, has become a focal point for targeted drug delivery. Many tumors exhibit a slightly acidic microenvironment compared to normal tissues [14]. Liposomes can be designed to release their payload in such acidic conditions. A significant hallmark of tumor progression is the deviation in pH levels from the physiological standard of 7.4. Due to their heightened glycolysis rate, many tumors manifest an acidic extracellular milieu, typically oscillating between pH 6.5–7.2. This unique acidity presents an opportunity to harness it for the site-specific drug release of liposomes, aiming for a more targeted and efficient cancer treatment. pH-sensitive liposomes are ingeniously designed with surface polymers that bear acid-sensitive bonds. When exposed to pH variations, these bonds dissociate, facilitating the release of therapeutic agents directly into the tumor tissue. This precision in drug delivery not only ensures the targeting of cancerous cells but also minimizes potential collateral damage to healthy cells. Some of these liposomes are even crafted to either destabilize their phospholipid bilayer or to fuse with cellular endosome or lysosome membranes, further optimizing drug delivery [15]. Delving deeper into the structural intricacies, a typical pH-sensitive liposome incorporates elements like phosphatidylamine derivatives, pNIPAM-based co-polymers, or weakly acidic amphiphiles such as cholesteryl hemisuccinate (CHEMS). In the presence of an acidic environment, these components undergo specific changes, leading to destabilization. This, in turn, triggers fusion with cellular membranes, culminating in drug release. Recent research has spotlighted certain formulations that demonstrate enhanced pH responsiveness. One such formulation emphasizes the potential of lipid diolein combined with CHEMS and other agents, including egg PC and Tween-80 [16]. However, the journey of harnessing the tumor's pH for drug delivery isn't without challenges. The acidic environment in perivascular regions might be remote from the bloodstream, or the pH variation might be minimal, potentially leading to an inadequate response from pH-sensitive liposomes. Moreover, the pH differential in tumors is often minimal, making it challenging for engineered liposomes to respond efficiently.

To circumvent these challenges and enhance the efficacy of pH-sensitive liposomes, researchers have incorporated site-specific ligands. When adorned with these targeted ligands, such as peptides or antibodies, pH-responsive carriers can bind to specific cells, leading to enhanced drug delivery. These ligands recognize and bind to specific receptors overexpressed on target cells, initiating processes like endocytosis and endosomal localization [17].

3.1.2 Enzymes

Certain enzymes are overexpressed in pathological conditions. Liposomes can be engineered to degrade and release their content in the presence of these enzymes [18]. Enzyme-responsive liposomes are emerging as a promising avenue in nanotherapeutics due to their precision and selectivity. These liposomes are designed to release their medicinal contents when they encounter specific enzymes. The release mechanisms include changes in the lipid bilayer, removal of protective polymers, or activation of a prodrug within the liposome. This approach targets enzymes that are typically overexpressed in diseased cells, ensuring targeted drug delivery [19].

The unique expression patterns of certain enzymes, like proteases, phospholipases, or glycosidases, seen in conditions like cancer or inflammation, offer a golden opportunity for targeted drug delivery. By harnessing these enzymes, drugs can be released directly at the intended biological site. A lot of the current research in this area focuses on enzymes present outside cells. For instance, recent studies have used short peptide sequences, which can be broken down by matrix metalloproteinases, as connectors between surface PEG chains and TAT-functionalized liposomes or other nanoparticles [20]. When these connectors are cleaved in the tumor environment, the surface ligands are exposed, leading to better penetration inside cells. Using this method, siRNA-loaded nanoparticles showed nearly 70% gene-silencing activity in tumor-bearing mice [21]. Similarly, there are designs for protease-sensitive polymer coatings or lipopeptides for controlled release from nanoparticles or liposomes. But it's not just about the external environment. Drugs can also be delivered inside cells using enzymes. For example, certain silica structures combined with polysaccharide derivatives have been used to deliver the drug doxorubicin after enzymes in the lysosome break down glycoside bonds [22]. Another enzyme, cathepsin B, which is found in many tumors, can release cargo by quickly breaking down certain polymers [23]. There are also innovative methods to deliver genes with high specificity using polymer systems that have a peptide which can be broken down by specific intracellular enzymes. This method has been tested in cells affected by HIV [24] or inflammation. Moreover, enzyme responsiveness isn't limited to treating just human diseases. It can also be used against bacterial infections. For instance, lipase-sensitive nanogels releasing the antibiotic vancomycin have been shown to effectively inhibit the growth of Staphylococcus aureus and even kill bacteria inside cells [25].

3.1.3 Redox

Utilizing disulfide bonds, which are susceptible to rapid breakdown by glutathione (GSH), provides a strategic approach for redox-sensitive drug delivery. The variance in GSH levels between the extracellular(~2–10 μM) and intracellular(~2–10 mM) spaces, and notably the higher concentrations in tumor tissues compared to normal ones, can be leveraged for targeted drug release. This principle has been actualized using micelles crafted from copolymers embedded with disulfide connections [26].

However, it's crucial to highlight that the journey of redox-responsive drug delivery systems hasn't been without setbacks. For example, Mylotrag, once green-lit for treating a particular leukemia variant, was eventually withdrawn from the market. This serves as a reminder of the complexities and challenges in harnessing redox mechanisms for drug delivery within the multifaceted human physiological landscape and release drug.

3.2 External Stimuli

These specialized liposomes are designed to release their therapeutic payload upon exposure to specific external triggers, such as temperature, light, magnetic fields, or ultrasound. The primary advantage of external stimulus-responsive liposomes is the precision they offer; the external triggers can be applied specifically to the target site, ensuring localized drug release and minimizing systemic side effects [4]. Such targeted approaches not only enhance the therapeutic efficacy but also reduce potential toxicities, paving the way for more effective and safer treatments, especially in challenging areas like oncology [27].

3.2.1 Temperature

In the fast-paced world of medical advancements, temperature-responsive liposomes are making quite a buzz. These aren't normal liposomes; they are a class apart, designed meticulously to undergo structural metamorphosis when exposed to specific temperature thresholds, typically oscillating between 40 and 45 °C. The foundational principle that drives the efficacy of these liposomes is their intrinsic ability to adapt their structure in response to thermal stimuli. This rapid transformation is not just a fascinating scientific phenomenon but holds profound the rapeutic implications. It ensures the timely and precise release of the encapsulated drug, optimizing its therapeutic potential and ensuring it reaches the intended target [28]. This precision becomes particularly pivotal in the context of oncological treatments. Tumors, by their very nature, can be tricky adversaries. Their treatment demands precision, ensuring that the therapeutic agents act selectively on malignant cells while sparing the healthy ones. By employing controlled hyperthermia techniques, tumors can be selectively heated, creating an environment conducive for these temperature-sensitive liposomes to spring into action. They release their medicinal arsenal right at the epicenter of the tumor, ensuring a targeted assault on cancerous cells. This not only amplifies the therapeutic impact but also significantly curtails potential collateral damage to surrounding healthy tissues, leading to a drastic reduction in side effects [29]. But the innovation doesn't stop here. In a bid to further refine their temperature sensitivity, some liposomes are infused with specific polymers, notably poly(N-isopropylacrylamide). These polymers are not mere passive passengers; they play an active role in the liposome's functionality. They undergo a phase transition at specific temperatures, instigating the liposome to modify its structure and initiate the drug release [30]. Temperature-responsive liposomes are not just another addition to the medical toolkit; they represent a paradigm shift. Their ability to deliver drugs with pinpoint accuracy, responding to finely calibrated temperature cues, heralds a new era in personalized medicine. As research continues and these liposomes evolve further, they hold the promise to redefine therapeutic interventions, making treatments more efficient, effective, and patient-centric.

3.2.2 Light

The core principle behind light-activated liposomes is the integration of photosensitive molecules or materials into the liposomal structure. Upon exposure to specific light wavelengths, especially near-infrared (NIR) light, these molecules undergo reactions or transformations. This process destabilizes the liposomal membrane, leading to the release of the encapsulated drug [31]. Gold nanorods (AuNRs) have emerged as a pivotal component in the design of light-responsive liposomes [32]. These nanorods can absorb NIR light and convert it into heat, a property known as photo-thermal conversion. The generated heat causes the liposomes to destabilize, releasing their contents. This mechanism ensures spatial and temporal control over drug delivery, enhancing therapeutic efficacy while minimizing side effects. The potential of light-activated liposomes extends beyond targeted drug delivery. In the realm of antimicrobial therapies, for instance, these liposomes can be used to combat drug-resistant bacteria. By controlling the release of antimicrobials, one can prevent overuse and the subsequent development of resistance [33]. Moreover, the ability to monitor and adjust the exact drug dose by tuning laser parameters adds another layer of precision to the treatment. Despite their promise, light-activated liposomes face challenges. Achieving uniform distribution of photosensitive molecules, controlling drug release rates, and ensuring deep tissue penetration are areas that warrant further exploration. Additionally, the long-term implications of repeated NIR light exposure and potential tissue damage need thorough investigation. Light-activated liposomes signify a paradigm shift in drug delivery. Their amalgamation of specificity, control, and safety positions them as a promising tool in modern therapeutics. As research progresses, it's anticipated that these liposomes will find broader applications, heralding a new era in personalized medicine. Light-sensitive liposomes can be triggered to release their payload upon exposure to specific wavelengths of light [27].

3.2.3 Magnetic Field

Magnetic-responsive liposomes are a revolutionary stride in the domain of drug delivery systems. These vesicles, embedded with magnetic nanoparticles, often composed of iron oxide, have the unique ability to respond to external magnetic fields. This responsiveness is not just a mere reaction; it's a meticulously designed mechanism that holds immense therapeutic potential. When these liposomes encounter an external magnetic field, the embedded nanoparticles heat up. This heat, in turn, destabilizes the liposomal membrane, leading to the release of the drug encapsulated within. This mechanism is not just about release; it's about control. By modulating the magnetic field's strength and duration, one can precisely dictate the rate and extent of drug release [34]. The implications of this are profound. In traditional drug delivery methods, the drug gets distributed throughout the body, which often leads to systemic side effects and reduced efficacy at the target site. Magnetic-responsive liposomes, however, offer a solution to this challenge. By applying a magnetic field to a specific location, one can ensure that the drug is released right where it's needed, ensuring maximum therapeutic effect and minimizing potential side effects [26]. But the potential of magnetic-responsive liposomes doesn't end at targeted drug delivery. Research has illuminated their utility in various therapeutic applications. For instance, in oncology, they present a promising strategy for delivering chemotherapeutic agents directly to tumor sites. This ensures that the drug acts effectively on the tumor, sparing the surrounding healthy tissues from potential damage [35]. However, the journey of magnetic-responsive liposomes is not devoid of challenges. Achieving a uniform distribution of magnetic nanoparticles within the liposome, ensuring deep tissue penetration, and understanding the long-term implications of magnetic fields on the body are areas that need further exploration. Moreover, the clinical translation of these drug delivery systems remains a topic of active research, with many studies still at the cell culture level [36].

3.2.4 Ultrasound Waves

Ultrasound-triggered liposomes, a pivotal innovation in the sphere of drug delivery, have garnered substantial attention in the scientific and medical communities. These liposomes, which are essentially vesicles encapsulating therapeutic agents, are embedded with specific agents that respond to ultrasound waves, thereby enabling controlled drug release. The mechanism of action is deeply rooted in the interaction between the liposomal membrane and the ultrasound waves. When exposed to ultrasound, the mechanical vibrations from the waves induce stress on the liposomal membrane, potentially causing it to become permeable or even rupture, which in turn facilitates the release of the encapsulated drug [37]. This mechanism is not merely about release; it's about control and precision. By modulating the strength and duration of the ultrasound waves, one can precisely control the rate and extent of drug release. This precision is not just a technological marvel; it's a therapeutic boon. Traditional drug delivery methods often result in the systemic distribution of the drug, which can lead to potential side effects and reduced efficacy at the target site. However, ultrasound-triggered liposomes offer a solution to this predicament. By focusing the ultrasound waves on a specific region, one can ensure localized drug release, thereby maximizing therapeutic effect and minimizing potential side effects. Moreover, the real-time monitoring capability offered by ultrasound provides a unique feedback mechanism, enabling on-the-fly adjustments to the treatment, thereby enhancing its efficacy [38]. The applications of these liposomes are vast and varied. In oncology, for instance, they present a promising strategy for delivering chemotherapeutic agents directly to tumor sites, ensuring maximum efficacy while minimizing damage to surrounding healthy tissues. However, the journey of ultrasound-triggered liposomes is not without challenges. Ensuring a uniform distribution of the agents within the liposome, achieving deep tissue penetration, and understanding the long-term effects of ultrasound on the body are areas that require further exploration and research. Moreover, the clinical translation of these drug delivery systems remains a topic of active research, with many studies still at the cell culture or preclinical stage. Table 6.2 provides a literature review of the various types of stimuli-responsive liposomes.

Table 6.2 Summary of liposomal carrier systems

Carrier	Drug	Cancer type	Description	References
pH-sensitive liposomes				
Folate-targeted pH-sensitive liposomes	Docetaxel doxycycline	Non-small cell lung cancer	Suppressing the proliferation rate of tumor cells, reducing side effects, exerting a synergistic impact, and selectively targeting tumor cells through folate modification	[39]
pH-sensitive liposomes	Docetaxel	Breast cancer	Decreasing tumor burden by as much as 70%, boosting the accumulation of cargo at the tumor site, and offering a platform for the combined delivery of genes and drugs to enhance tumor suppression efficacy	[40]
PEGylated pH-sensitive liposomes	Gemcitabine curcumin	Pancreatic cancer	The curcumin loading slightly reduced the pH sensitivity of the liposomes, but PEGylation had a more significant effect. The combination demonstrated a synergistic impact, exerting cytotoxicity in a concentration-dependent manner	[41]
pH-sensitive sialic acid (SA) conjugated liposomes	Topotecan	Sarcoma	At pH 4, 92.33% of the drug was released, leading to a 24-fold increase in apoptosis induction	[42]
Peptide-modified pH-sensitive liposomes	Doxorubicin	Breast cancer	Extended circulation time achieved through modification with a cell-penetrating peptide, resulting in a 1.9-fold increase in DOX accumulation at the tumor site. This enhanced accumulation promotes apoptosis induction and DNA damage	[43]

(continued)

Table 6.2 (continued)

Carrier	Drug	Cancer type	Description	References
Redox-responsive liposomes				
Redox-responsive liposomes	Indoleamine 2,3-dioxygenase (IDO) inhibito	Breast cancer	Blocking the IDO pathway, triggering immunogenic cell death, boosting tumor cell accumulation, prolonging circulation time, and preventing tumor growth	[44]
HA-functionalized redox-sensitive liposomes	Doxorubicin	Osteosarcoma	Redox-responsive characteristics were given by the SS-link, which specifically targeted tumor cells with overexpressed CD44. 60% of the burst release happened when GSH was present.	[45]
Redox-sensitive oligopeptide liposomes	Paclitaxel survivin-siRNA	Breast cancer	The oligopeptide lipid ($LHSSG_2C_{14}$) imparted a redox-sensitive property, while soybean phosphatidylcholine (SPC) facilitated proton sponge activity, promoting endosomal escape. This combination enabled a synergistic cancer therapy through the co-delivery of siRNA and anticancer drug	[46]
TAT-decorated redox-responsive liposomes	Paclitaxel	Melanoma	Functionalization with TAT, a cell-penetrating peptide, facilitated internalization in cancer cells. The conjugation of PEG via a redox-sensitive disulfide linker improved blood circulation time and boosted cancer cell accumulation	[47]

(continued)

Table 6.2 (continued)

Carrier	Drug	Cancer type	Description	References
Thermosensitive liposomes				
Thermosensitive exosome-liposome hybrid nanoparticles	Docetaxel GM-CSF	Peritoneal cancer	Cargo release triggered by hyperthermia at the tumor site, offering a combination of chemotherapy and immunotherapy	[48]
Thermosensitive liposomes	Parthenolide	Triple-negative breast cancer	Inhibiting the growth of tumors in vivo, Excellent biocompatibility and also the inhibition rate was 2.08-fold higher than Paclitaxel.	[49]
Image-guided thermosensitive liposomes	Carboplatin SN-38	Triple-negative breast cancer	SN-38 was embedded in the lipid bilayer, while carboplatin was loaded in the core of the liposomes. The cargo was released in a thermosensitive manner upon ultrasound exposure, resulting in the inhibition of tumor growth in vivo	[50]
Thermosensitive liposomes	Resiquimod	Breast cancer	Ultrasound-induced hyperthermia triggered the release of cargo from liposomes, enhancing the effectiveness of immunotherapy and promoting the infiltration of CD8+ T cells	[51]
Thermosensitive liposomal cerasome	Doxorubicin	Ovarian cancer	Surface modification of nanoparticles with WSG-peptide enhanced their targeting specificity towards cancer cells. The nanoparticles were prepared using thin-film hydration, ensuring high stability, and released their cargo in response to hypothermia	[52]

(continued)

Table 6.2 (continued)

Carrier	Drug	Cancer type	Description	References
Light-responsive liposomes				
Light-sensitive liposomal nanocarriers	Doxorubicin	Ovarian, breast and lung cancers	Surface modification of liposomes with the EGFR2 antibody enhanced their specificity for tumor cells, leading to a two-fold increase in antitumor drug accumulation at the tumor site. NIR laser radiation triggered drug release, resulting in a 92.7% suppression of tumor growth	[53]
Light-triggered theranostic liposomes	Tirapazamine	Breast cancer	Ce6-mediated photodynamic therapy (PDT) induces hypoxia, triggering the disassembly of liposomes and enabling controlled drug release for targeted cancer suppression	[54]
Light-responsive lipid-incorporated liposomes	Gemcitabine talaporfin sodium	Breast cancer	Drug release was triggered by NIR exposure, and the incorporation of DOPE into liposomes accelerated this release, resulting in high cytotoxicity toward tumor cells	[55]
NIR-sensitive liposomes	Calcein	Colon cancer	NIR irradiation triggers the generation of singlet oxygen, which facilitates the membrane permeabilization of liposomes and subsequent drug release	[56]

(continued)

Table 6.2 (continued)

Carrier	Drug	Cancer type	Description	References
NIR-activated IR780-loaded liposomes	Sunitinib	Breast cancer	The NIR dye IR780 was incorporated into the liposome phospholipid bilayer. NIR exposure disrupted the liposome membrane, causing drug release. Sunitinib targets VEGF receptors on the surface of nanoparticles to block angiogenesis	[57]
Magneto-responsive liposomes				
Magnetic-responsive liposome microbubbles	Doxorubicin	Pancreatic cancer	Nanoparticles exhibit magnetism and release DOX in response to ultrasonic, leading to apoptosis induction. Killing pancreatic cancer cells at low dosages	[58]
Magnetic liposomes	Doxorubicin	Liver cancer	The magnetic liposomes shows high stability and biocompatibility reducing almost 80% of the cell viability	[59]
Magnet-activatable nanoliposomes	Anethole trithione	Hepatocellular carcinoma	Magnet activalble nanoliposomes get penetrate easily in to the cell membrane suppresses the tumor by releasing the drug at specific location	[60]
EGFR peptide-conjugated magnetic polymeric liposomes	Paclitaxel	Liver cancer	These magnetic polymeric liposomes get accumulate in the liver with the help of magnetic force targeting tumor cells overexpressing EGFR having high cytotoxicity for cancer cells	[61]

(continued)

Table 6.2 (continued)

Carrier	Drug	Cancer type	Description	References
PEGylated magnetic liposomes	Curcumin	Breast cancer	Inductive magnetic hyperthermia concurrently promotes the release of medication from liposomes and reduces the viability of cancer cells. There is increased internalization in tumor cells	[62]
Multifunctional stimuli-responsive liposomes				
Multifunctional gold nanorod/docetaxel-loaded liposomes	Docetaxel	Prostate cancer	Growing intracellular buildup at the location of the tumor preventing the formation of tumors both in vivo and in vitro delivering thermotherapy and chemotherapy	[63]
Multifunctional temperature-sensitive liposomes	Tanespimycin	Breast cancer squamous cell carcinoma	By producing heat and ROS, NIR irradiation promotes medication release and tumor ablation.	[64]
Multifunctional traceable liposomes	Doxorubicin	Colon cancer	A tumor ablation technique that is minimally invasive. Due to surface modification with cRGD, there is a high concentration at the tumor site following intravenous injection. Local heating facilitates the release through liposomes.	[65]
Multifunctional liposomes	Oxaliplatin	Colorectal cancer	Unsaturated liposomes are exposed to light, porphyrin-phospholipid mediates photooxidation, which causes the liposomes to release their cargo. Liposome migration from the tumor site to nearby regions is revealed by MRI imaging.	[66]

4 Design and Preparation of Stimuli-Responsive Liposomes

4.1 Materials and Methods Used in Preparation

The design and preparation of stimuli-responsive liposomes require careful selection of materials and methods to ensure that the liposomes respond appropriately to the desired stimulus. Here's a detailed look at the materials and methods.

4.1.1 Choice of Lipids

The selection of lipids is a critical step in the design of liposomes, especially when crafting stimuli-responsive variants. The lipid composition dictates the liposome's stability, fluidity, and overall functionality.

Phospholipids

These are the primary components of liposomes. Phospholipids are the primary structural components of liposomes, playing a pivotal role in their formation and functionality. Structurally, liposomes are spherical vesicles formed by the self-assembly of diacyl-chain phospholipids, resulting in a lipid bilayer. This bilayer can encapsulate aqueous media, making liposomes ideal carriers for both hydrophilic and hydrophobic drugs [67]. The amphiphilic nature of phospholipids, possessing both hydrophilic (water-attracting) and hydrophobic (water-repelling) regions, allows them to spontaneously form bilayers in aqueous environments. Among the phospholipids, phosphatidylcholine (PC) is a commonly used type due to its biocompatibility and ability to form stable bilayers. However, other phospholipids like phosphatidylethanolamine (PE), phosphatidylserine (PS), and phosphatidylinositol (PI) are also employed in liposome formulations, each bringing unique characteristics to the table [68]. For instance, PE can enhance the fusion of liposomes with cell membranes, while PS has been explored for its potential in targeting specific cells due to its recognition by certain receptors. PI, on the other hand, can be used to introduce negative charges to the liposome surface. Phospholipids not only provide structural integrity to liposomes but also influence their stability, fluidity, and drug encapsulation efficiency. Their emulsifying properties further enhance the potential of liposomes in drug delivery applications [26].

Functional Lipids

Functional lipids are specialized lipids that are incorporated into liposomes to impart specific characteristics or functionalities. These lipids can enhance the liposome's stability, circulation time, targeting ability, and drug release properties, making liposomes smarter, more targeted, and more responsive [69]. One of the most popular functional lipids is the PEGylated lipid. By attaching a Polyethylene glycol (PEG) chain to lipids, these liposomes become like undercover agents, evading the immune system's radar and staying longer in the bloodstream. This ensures that the drug gets more time to act and doesn't get cleared out quickly. Then there are ligand-conjugated lipids. Imagine a missile with a specific target; that's what these lipids do. They have specific ligands, maybe peptides or antibodies, which can find and bind to particular receptors on cells. This ensures the drug is not just shot in the dark but reaches its intended target. For those applications where you need to deliver genes, cationic lipids come into play. These positively charged lipids are like magnets for negatively charged entities like DNA, making them perfect for gene delivery [67]. Now, when we talk about stimuli-responsive liposomes, thermosensitive lipids are the stars. These lipids respond to temperature changes. So, in areas where there's inflammation or a tumor, which might be slightly warmer, these lipids ensure the drug is released right there. For those who like to keep an eye on where the liposomes are going, fluorescent lipids are the answer. These lipids light up, allowing for real-time tracking, especially useful in diagnostic applications [70]. These lipids impart the stimuli-responsive property to the liposomes. For instance, pH-sensitive lipids like DOPE (dioleoylphosphatidylethanolamine) can be used for pH-responsive liposomes [71].

Cholesterol

It's often added to increase the rigidity and stability of the liposomal membrane [72]. Cholesterol, a naturally occurring sterol, has carved out a significant niche for itself in the domain of liposomal drug delivery. This molecule, commonly associated with biological membranes, has been harnessed for its unique properties in the realm of liposomes. At the heart of its utility is its ability to modulate the fluidity of the lipid bilayer. By seamlessly integrating between the fatty acid chains of phospholipids, cholesterol acts as a gatekeeper, reducing the permeability of the bilayer. This ensures that drugs encapsulated within the liposome are securely held, preventing premature release and ensuring that the therapeutic agent remains intact until it reaches its intended target [73]. But the role of cholesterol doesn't stop at modulating fluidity. It's also a guardian against liposome aggregation. In the complex environment of the bloodstream, liposomes face the risk of clumping together, which can compromise their efficacy. Cholesterol steps in here, ensuring that liposomes maintain their individuality, floating as discrete entities and maintaining a consistent size distribution [avantilipids.com]. This is pivotal, especially when considering the need for liposomes to navigate the intricate maze of the human circulatory system, passing through narrow capillaries and ensuring the drug is delivered uniformly [67]. Furthermore, cholesterol is not just a passive component; it actively enhances the resilience of the liposomal membrane. By imparting elasticity, it ensures that the liposomes are robust, capable of withstanding external stresses, be it mechanical forces or the challenges posed by the biological environment [74]. An intriguing aspect of cholesterol's role is its influence on the surface charge of the liposomal membrane. Depending on its concentration, cholesterol can tweak the liposome's interaction dynamics with cells and other biological entities, a factor that can be harnessed for targeted drug delivery [75].

4.1.2 Surface Modifiers

These can be added to enhance the circulation time or target specific cells or tissues. One of the primary surface modifiers used in liposomes is polysaccharides. These naturally occurring macromolecules, such as hyaluronic acid and chitosan, offer several advantages. They can improve the stability of liposomes, prolong their circulation time in the bloodstream, and facilitate targeted drug delivery by allowing specific interactions with cellular receptors. Polymers are another class of materials extensively used for liposomal surface modification. Polyethylene glycol (PEG) is a commonly used polymer that imparts a "stealth" characteristic to liposomes, allowing them to evade the immune system and achieve prolonged circulation times. PEGylation of liposomes reduces their recognition by the reticuloendothelial system, thereby increasing their half-life in the bloodstream [8]. Peptides and proteins can also be anchored to the liposomal surface to achieve targeted drug delivery. By recognizing and binding to specific receptors overexpressed on target cells, these bioactive molecules can guide liposomes to specific tissues or tumor sites [76]. Lipids with functional head groups are another category of surface modifiers. These lipids can interact with specific cell receptors or even respond to external stimuli, enabling controlled drug release or targeted delivery. PEGylated lipids are commonly used to increase the circulation time of liposomes in the bloodstream [77].

4.1.3 Drug Molecules

Liposomes, tiny vesicles made up of lipid bilayers, opened up new avenues for delivering a variety of drugs, from small molecules to large biologics. But, not every drug is suitable for liposomal encapsulation. Solubility plays a pivotal role. For a drug to be encapsulated efficiently, it should either be hydrophilic, meaning it dissolves in water, or lipophilic, meaning it dissolves in lipids. Hydrophilic drugs find their place in the aqueous core of the liposome, while lipophilic drugs get snugly fit within the lipid bilayer [78]. Next is the molecular weight of the drug. Traditionally, smaller drug molecules found it easier to get encapsulated within liposomes. But due to advancements in technology, even larger molecules, like proteins and peptides, are also encapsulated within the liposomes. The stability of a drug is crucial when considering it for liposomal encapsulation. If a drug is chemically unstable, it might degrade within the liposome, affecting its therapeutic efficacy. Physical changes, like crystallization, can hinder its release. Interactions between the drug and liposomal components can alter both the drug and the liposome's structure. Furthermore, for commercial viability, the drug must remain stable within the liposome during storage. Thus, a drug's stability directly influences its suitability for liposomal formulations [79]. The release profile of the drug from the liposome is equally important. Depending on the therapeutic need, some drugs should be released slowly over time, while others might need to be released quickly. Toxicity is a significant concern, especially for drugs that can cause side effects when they act on non-target tissues. By encapsulating toxic drugs in liposomes, their delivery can be targeted to specific tissues, reducing their side effects. The pharmacokinetics of a drug, which is the one to describes how it's absorbed, distributed, metabolized, and excreted in the body, plays a pivotal role in its selection for liposomal encapsulation. Drugs with rapid clearance or poor bioavailability can benefit from liposomal encapsulation, which can enhance absorption and prolong circulation time, ensuring effective drug delivery and therapeutic outcomes [80]. Lastly, the therapeutic index of the drug, which is the difference between its therapeutic and toxic doses, is considered. Drugs with a narrow therapeutic index are ideal candidates for liposomal encapsulation. This ensures that the drug acts precisely where it's needed, without causing harm elsewhere [81]. The selection of drugs for liposomal encapsulation is not a random choice. It's a well-thought-out decision, based on several criteria, ensuring that the final liposomal drug product is both safe and effective.

4.2 Preparation of Stimuli-Responsive Liposomes

The preparation of liposomes, particularly those that are stimuli-responsive, requires careful planning and execution to ensure that they exhibit the desired characteristics and functions. Below is a detailed overview of various techniques employed in the preparation of liposomes.

4.2.1 Thin Film Hydration Method

The thin film hydration (TFH) method is a widely used and straightforward technique for liposome preparation, appreciated for its simplicity, cost-effectiveness, and versatility. The process begins by dissolving lipids in a volatile organic solvent, such as chloroform or a chloroform-methanol mixture, which is then evaporated under reduced pressure using a rotary evaporator to form a thin lipid film on the flask walls. An aqueous solution, potentially containing a desired drug or therapeutic agent, is added to the lipid film, followed by gentle agitation and slight heating (above the lipid's transition temperature) to facilitate hydration and form multilamellar vesicles (MLVs). These MLVs can be further processed into small unilamellar vesicles (SUVs) through sonication or extrusion [82]. The method's main advantages include its simplicity and cost-effectiveness, as it doesn't require specialized equipment, making it accessible for both research and industrial applications. Additionally, it is versatile, capable of encapsulating both hydrophilic and hydrophobic drugs, with lipophilic drugs often achieving high loading efficiency [83]. However, the TFH method does have some challenges, including the risk of residual solvents in the final product, which can be toxic and necessitate thorough removal. Furthermore, the liposome size distribution can be heterogeneous, requiring additional processing for uniformity, and hydrophilic drugs may not be efficiently encapsulated within the aqueous core. Despite these challenges, TFH remains a highly effective and widely utilized method for liposome preparation due to its flexibility and ease of use.

4.2.2 Reverse Phase Evaporation

The reverse phase evaporation (REV) method is particularly effective for encapsulating both small molecules and large macromolecules, such as proteins and nucleic acids, offering high encapsulation efficiency. This method is especially advantageous for substances that are sensitive to sonication, which is commonly used in other liposome preparation techniques. The process begins by dissolving lipids in a volatile organic solvent, typically a mixture of diethyl ether and chloroform. The aqueous phase containing the substance to be encapsulated is then added to the lipid-solvent mixture, followed by sonication to form a water-in-oil emulsion. The organic solvent is removed under reduced pressure using a rotary evaporator, allowing the formation of liposomes, which are then hydrated with an aqueous phase to stabilize and form a liposome suspension. The REV method's primary advantages include its high encapsulation efficiency, particularly for hydrophilic substances, and its versatility in accommodating a wide range of substances, including proteins, nucleic acids, and small molecules. However, the method does come with challenges, such as the potential for residual organic solvents, which require thorough removal to ensure safety, and its complexity, requiring multiple steps and optimization [84]. Additionally, the method is limited to lipids that can form stable emulsions with organic solvents, potentially restricting the choice of lipids. In summary, while the REV method is a powerful tool for encapsulating macromolecules and hydrophilic drugs, it necessitates careful optimization and solvent management to achieve the best results.

4.2.3 Freeze-Thaw Method

The freeze-thaw method is a widely used technique in liposome preparation, particularly advantageous for encapsulating hydrophilic drugs, proteins, and other sensitive agents. This method is valued for its simplicity, cost-effectiveness, and high encapsulation efficiency. The process begins with dissolving lipids in an organic solvent, which is then evaporated to form a thin lipid film. This film is hydrated, resulting in the formation of multilamellar vesicles (MLVs), which are subjected to several freeze-thaw cycles—freezing in liquid nitrogen and thawing in a warm water bath. These cycles disrupt the MLVs and promote the formation of large unilamellar vesicles (LUVs). The method offers several advantages, including high encapsulation efficiency, especially for hydrophilic drugs, which enhances therapeutic efficacy. It is also simple and cost-effective, requiring minimal specialized equipment, and the freeze-thaw cycles contribute to improved stability and encapsulation properties [85]. However, the method does present some challenges, such as the potential for oxidative degradation of sensitive lipids during the freeze-thaw process, and the risk of size heterogeneity, which may necessitate additional processing steps like extrusion to achieve uniformity. Moreover, the multiple steps involved could increase the risk of contamination if not properly managed [86]. Despite these challenges, the freeze-thaw method remains a valuable tool in liposome preparation, especially for hydrophilic drug delivery systems, requiring careful optimization to mitigate potential issues like oxidative degradation and size distribution variability.

4.2.4 Microfludic Mixing

Microfluidic mixing is an advanced technique in liposome production that leverages microfluidic devices to precisely control the mixing of aqueous and organic phases, ensuring high precision and efficiency [87]. In this process, aqueous and organic phases containing dissolved lipids are introduced into a microfluidic device with tiny channels designed for precise fluid control. The two phases mix rapidly within these channels, leading to the spontaneous formation of liposomes. The flow rates, channel dimensions, and mixing conditions are carefully controlled to fine-tune liposome characteristics such as size, lamellarity, and encapsulation efficiency. This method offers several advantages, including exceptional control over liposome properties, the production of liposomes with narrow size distributions, scalability for industrial applications, and high reproducibility across different batches [88]. However, the technique is not without challenges. The initial setup cost of microfluidic devices can be high, and operating the system requires specialized knowledge. Additionally, the technique may be limited by the viscosity of the phases, which can affect the mixing process, and a deep understanding of fluid dynamics is essential to optimize liposome characteristics. Despite these challenges, Microfluidic Mixing holds great promise for drug delivery applications, providing highly controlled liposome properties but requiring an initial investment and expertise to ensure successful implementation.

Each liposome preparation method offers unique advantages and limitations. Thin Film Hydration (TFH) is simple and cost-effective, yet it can suffer from solvent residuals and size distribution issues. Reverse Phase Evaporation (REV) excels in encapsulating macromolecules and hydrophilic substances but requires careful optimization to avoid solvent toxicity and process complexity. The freeze-thaw method is effective for hydrophilic drug encapsulation but is susceptible to oxidative degradation and size heterogeneity. Microfluidic mixing offers exceptional control and scalability but involves high initial costs and specialized knowledge. Ultimately, the choice of method depends on the specific application, desired liposome characteristics, and the substances being encapsulated.

5 Characterization of Stimuli-Responsive Liposomes

Characterization techniques play a vital role in determining the stimuli-responsiveness of liposomes, helping to evaluate how they respond to specific internal or external triggers such as pH, temperature, light, enzymes, or magnetic fields. These techniques provide essential data on the liposome's physical, chemical, and biological properties, which are crucial for understanding how well the liposomes will function in targeted drug delivery. Below is an explanation of how various characterization methods can help in assessing the stimuli-responsiveness of liposomes:

5.1 Size and Morphology Analysis

Dynamic light scattering (DLS), transmission electron microscopy (TEM), and scanning electron microscopy (SEM) are essential characterization techniques for studying the stimuli-responsive behavior of liposomes. DLS is particularly useful for tracking changes in the hydrodynamic size of liposomes in response to environmental triggers like temperature or pH, which can indicate whether the liposomes are releasing their drug payload or undergoing structural changes. On the other hand, TEM and SEM provide high-resolution images that allow for direct observation of liposome morphology and structural integrity. These techniques can reveal changes in liposome shape, such as vesicle rupture or drug leakage, when exposed to specific stimuli. Together, these methods provide critical insights into the dynamics and performance of stimuli-responsive liposomes, aiding in the optimization of drug delivery systems [89, 90].

5.2 Drug Release Kinetics

In vitro release studies, UV–vis spectroscopy, and fluorescence spectroscopy play pivotal roles in evaluating the stimuli-responsive behavior of liposomes. In vitro release studies assess how liposomes release their drug payload in response to specific external or internal stimuli, such as variations in pH, temperature, or enzyme presence. By monitoring these conditions, researchers can determine how effectively liposomes respond to stimuli, and correlate release profiles with their material composition and design. UV–vis and fluorescence spectroscopy are used to quantify drug release by measuring the concentration of drug molecules in the release medium over time. Changes in absorbance or fluorescence intensity provide detailed insights into the dynamics of drug release, offering a clear understanding of the liposomes' responsiveness and the effectiveness of their stimuli-triggered behavior. Together, these techniques help optimize the design of stimuli-responsive liposomes for efficient drug delivery [91, 92].

5.3 Zeta Potential and Surface Charge

Zeta potential measurements and electrophoresis are crucial for understanding the behavior of stimuli-responsive liposomes. The surface charge of liposomes, as indicated by their zeta potential, influences their stability, colloidal properties, and interactions with biological membranes. Changes in zeta potential, triggered by factors like pH shifts or the presence of specific enzymes, may signal alterations in liposome structure or membrane disruption, which are often linked to drug release. Electrophoresis can complement this by assessing changes in charge distribution when liposomes are exposed to external triggers, such as electric fields or pH changes. Together, these techniques provide valuable insights into how stimuli-responsive liposomes react to environmental changes, affecting their drug release mechanisms and overall therapeutic effectiveness [93].

5.4 Encapsulation Efficiency and Drug Loading

High-performance liquid chromatography (HPLC) and ultracentrifugation are essential techniques for evaluating the encapsulation efficiency and drug loading of stimuli-responsive liposomes. HPLC allows for the precise quantification of the drug encapsulated within the liposomes, both before and after exposure to specific stimuli, helping to track changes in encapsulation efficiency and confirming whether the liposomes are releasing their cargo as intended. Ultracentrifugation, on the other hand, isolates liposomes from the surrounding medium, enabling the differentiation between free and encapsulated drug. By monitoring drug release upon exposure to various stimuli, these techniques offer valuable insights into the effectiveness of the stimulus-triggered release mechanism and the overall performance of liposomal drug delivery system [94, 95].

5.5 Stability and Long-Term Behavior

Storage Stability Testing: Liposomes may experience changes in their properties over time due to storage conditions. Stability studies under different environmental conditions (temperature, humidity) help assess whether the liposomes maintain their stimuli-responsive characteristics over time [96].

5.6 Magnetic or Light-Sensitive Liposomes

Magnetic resonance imaging (MRI) is a valuable tool for tracking magnetically responsive liposomes, enabling real-time visualization of liposome accumulation at the target site and monitoring drug release when exposed to a magnetic field. Similarly, fluorescent labeling and imaging techniques are employed for light-responsive liposomes, where fluorescent probes allow researchers to observe liposome behavior in real-time upon exposure to specific wavelengths of light. These imaging techniques provide critical insights into how liposomes respond to light stimuli and whether they effectively release their drug payload, aiding in the optimization of light-triggered drug delivery systems. These characterization methods are essential for optimizing liposome formulations for drug delivery applications.

6 Challenges and Limitations

Liposome-based drug delivery systems face several challenges and limitations that impact their effectiveness. One of the primary biological barriers is the immune response, where liposomes are often recognized and cleared by the mononuclear phagocyte system (MPS), reducing their circulation time and therapeutic efficacy. Additionally, liposomes larger than 100 nm are rapidly cleared by the liver and spleen, hindering their ability to reach target sites. From a technical perspective, reproducibility in achieving consistent size, charge, and drug encapsulation efficiency across different batches remains a challenge, and scaling up the production process for commercial use while maintaining quality is difficult. Stability concerns also pose significant issues, as liposomes can undergo fusion, aggregation, or leakage, particularly when stored at higher temperatures [97]. Moreover, interactions with serum proteins can destabilize liposomes and cause premature drug release. Further limitations include the depth of penetration for stimuli like light, which is often restricted, especially for visible wavelengths, as well as the difficulty in controlling stimuli precisely in clinical settings. Finally, potential side effects from stimuli, such as hyperthermia or specific light wavelengths, could cause damage to healthy tissues, limiting the overall therapeutic potential of liposome-based drug delivery systems [98].

7 Future Perspectives

The future of stimuli-responsive liposomes holds immense potential, with several innovations and advancements on the horizon. One key area of development is the design of multi-stimuli-responsive systems, where liposomes can be engineered to respond to multiple stimuli—either simultaneously or sequentially—providing enhanced control over drug release and improved specificity. The integration of smart materials, such as graphene or other 2D materials, may open new avenues for stimuli-responsive actions, enabling more precise and versatile drug delivery mechanisms [99]. Furthermore, the combination of stimuli-responsive liposomes with other therapeutic modalities, such as immunotherapy, gene therapy, or radiation, offers promising prospects for synergistic treatments with enhanced therapeutic outcomes [100]. These liposomes can also serve theranostic purposes, incorporating imaging agents that allow for real-time monitoring of therapeutic effects while delivering drugs to targeted sites.

Another exciting direction is the potential for personalized medicine. As our understanding of tumor biology deepens, liposomes can be specifically tailored to target the unique characteristics of individual tumors, improving treatment efficacy and reducing side effects. Additionally, the development of biomarker-responsive liposomes, which can react to specific biomarkers or enzymes overexpressed in a patient's tumor, offers a more precise approach to drug delivery [101]. These advances will not only improve the specificity of drug targeting but also open new pathways for treating a broader range of cancers and diseases.

In parallel, future research will focus on improving the safety profiles of liposome-based drug delivery systems, with an emphasis on minimizing off-target effects and reducing systemic toxicity. Enhanced penetration capabilities are another critical area, with efforts to design liposomes that can better navigate and penetrate the dense extracellular matrix of tumors, ensuring effective drug delivery to all tumor regions, including those that are harder to reach [102]. As these advancements unfold, stimuli-responsive liposomes could revolutionize cancer therapy and other areas of medicine, providing more personalized, efficient, and safer treatment options.

8 Conclusion

Stimuli-responsive liposomes represent a promising frontier in the realm of targeted drug delivery, particularly for cancer treatment. Their ability to release therapeutic agents in response to specific internal or external stimuli ensures precise drug delivery at the optimal time and location, maximizing therapeutic efficacy while minimizing systemic side effects. This innovative approach to drug delivery offers new hope in overcoming the challenges associated with conventional therapies, particularly in the context of cancer, where targeted treatment is essential for improving patient outcomes.

The evolution of stimuli-responsive liposomes, from their initial conceptualization to their current applications in clinical settings, highlights the transformative potential of nanotechnology in cancer therapy. The wide range of stimuli, such as pH, temperature, enzymes, light, and magnetic fields, provides a versatile toolkit that allows researchers and clinicians to fine-tune drug delivery systems tailored to the unique characteristics of each tumor. This adaptability enhances the precision and efficiency of cancer treatments, offering a targeted approach that could revolutionize patient care.

However, like all emerging technologies, the development of stimuli-responsive liposomes faces several challenges. Biological barriers, such as immune system recognition and rapid clearance, remain significant hurdles. Ensuring consistent production quality across batches and addressing stability concerns also present ongoing challenges. Despite these obstacles, the progress made in preclinical studies and clinical trials demonstrates the vast potential of this technology. The promising results so far suggest that with continued research and development, stimuli-responsive liposomes could play a pivotal role in the next generation of cancer therapies.

Looking ahead, the future of stimuli-responsive liposomes is filled with exciting possibilities. Advances in materials science, coupled with a deeper understanding of tumor biology and the integration of other therapeutic modalities, are expected to further propel the field. As more formulations enter clinical trials and achieve clinical success, stimuli-responsive liposomes may become a cornerstone of personalized cancer therapy, enabling more effective, targeted, and safer treatments.

In conclusion, while challenges remain, the progress made thus far is encouraging. The integration of nanotechnology with pharmacology through stimuli-responsive liposomes holds the promise of improving cancer treatment outcomes and providing a better quality of life for patients. As research continues and innovations unfold, there is great optimism that this approach will shape the future of cancer therapy, making it more precise, effective, and personalized. The journey of stimuli-responsive liposomes is just beginning, and it is an exciting time for researchers, clinicians, and stakeholders to invest in this groundbreaking field.

Acknowledgement The author would like to acknowledge Panjab University, Chandigarh, for providing necessary facilities and support.

Author Contributions Namarta Thakur—writing and original draft preparation.

Shivani Sharma—constructing an idea and reviewing the data collected.
Sunil Bharti—revision, writing, and editing of the manuscript.
Pemal Preet Kaur—revision of the work critically for important content.
Dr. Amita Sarwal—investigation, review, and supervision of the work.

Declaration The author used Quill-Bot during the preparation of this chapter to avoid similarity. The author reviewed and edited the work as necessary and takes full responsibility for the final version.

References

1. Bray F, Laversanne M, Sung H, Ferlay J, Siegel RL, Soerjomataram I, et al. Global cancer statistics 2022: GLOBOCAN estimates of incidence and mortality worldwide for 36 cancers in 185 countries. Cancer J Clin. 2024;74(3):229–63.
2. Vasan N, Baselga J, Hyman DM. A view on drug resistance in cancer. Nature. 2019;575(7782):299–309.
3. Allen TM, Cullis PR. Liposomal drug delivery systems: from concept to clinical applications. Adv Drug Deliv Rev. 2013;65(1):36–48.
4. Mura S, Nicolas J, Couvreur P. Stimuli-responsive nanocarriers for drug delivery. Nat Mater. 2013;12(11):991–1003.
5. Barenholz Y. (Chezy). Doxil® — the first FDA-approved nano-drug: lessons learned. J Control Release. 2012;160(2):117–34.
6. Akbarzadeh A, Rezaei-Sadabady R, Davaran S, Joo SW, Zarghami N, Hanifehpour Y, et al. Liposome: classification, preparation, and applications. Nanoscale Res Lett. 2013;8(1):102.
7. Pattni BS, Chupin VV, Torchilin VP. New developments in liposomal drug delivery. Chem Rev. 2015;115(19):10938–66.
8. Immordino ML, Dosio F, Cattel L. Stealth liposomes: review of the basic science, rationale, and clinical applications, existing and potential. Int J Nanomedicine. 2006;1(3):297–315.
9. Bae YH, Park K. Targeted drug delivery to tumors: myths, reality and possibility. J Control Release. 2011;153(3):198–205.
10. Lee ES, Gao Z, Bae YH. Recent progress in tumor pH targeting nanotechnology. J Control Release. 2008;132(3):164–70.
11. Du JZ, Du XJ, Mao CQ, Wang J. Tailor-made dual pH-sensitive polymer-doxorubicin nanoparticles for efficient anticancer drug delivery. J Am Chem Soc. 2011;133(44):17560–3.
12. Blanco E, Shen H, Ferrari M. Principles of nanoparticle design for overcoming biological barriers to drug delivery. Nat Biotechnol. 2015;33(9):941–51.
13. Nikolova MP, Kumar EM, Chavali MS. Updates on responsive drug delivery based on liposome vehicles for cancer treatment. Pharmaceutics. 2022;14(10):2195.
14. Gerweck LE, Seetharaman K. Cellular pH gradient in tumor versus normal tissue: potential exploitation for the treatment of cancer. Cancer Res. 1996;56(6):1194–8.
15. Abri Aghdam M, Bagheri R, Mosafer J, Baradaran B, Hashemzaei M, Baghbanzadeh A, et al. Recent advances on thermosensitive and pH-sensitive liposomes employed in controlled release. J Control Release. 2019;315:1–22.
16. Guo W, Gosselin MA, Lee RJ. Characterization of a novel diolein-based LPDII vector for gene delivery. J Control Release. 2002;83(1):121–32.
17. Zhao Y, Ren W, Zhong T, Zhang S, Huang D, Guo Y, et al. Tumor-specific pH-responsive peptide-modified pH-sensitive liposomes containing doxorubicin for enhancing glioma targeting and anti-tumor activity. J Control Release. 2016;222:56–66.
18. Miele E, Spinelli GP, Miele E, Tomao F, Tomao S. Albumin-bound formulation of paclitaxel (Abraxane ABI-007) in the treatment of breast cancer. Int J Nanomedicine. 2009;4:99–105.

19. Fouladi F, Steffen KJ, Mallik S. Enzyme-responsive liposomes for the delivery of anticancer drugs. Bioconjug Chem. 2017;28(4):857–68.
20. Zhu L, Kate P, Torchilin VP. Matrix metalloprotease 2-responsive multifunctional liposomal nanocarrier for enhanced tumor targeting. ACS Nano. 2012;6(4):3491–8.
21. Hatakeyama H, Akita H, Ito E, Hayashi Y, Oishi M, Nagasaki Y, et al. Systemic delivery of siRNA to tumors using a lipid nanoparticle containing a tumor-specific cleavable PEG-lipid. Biomaterials. 2011;32(18):4306–16.
22. Bernardos A, Mondragon L, Aznar E, Marcos MD, Martinez-Mañez R, Sancenon F, et al. Enzyme-responsive intracellular controlled release using nanometric silica mesoporous supports capped with "saccharides". ACS Nano. 2010;4(11):6353–68.
23. Lee JS, Groothuis T, Cusan C, Mink D, Feijen J. Lysosomally cleavable peptide-containing polymersomes modified with anti-EGFR antibody for systemic cancer chemotherapy. Biomaterials. 2011;32(34):9144–53.
24. Asai D, Kuramoto M, Shoji Y, Kang JH, Kodama KB, Kawamura K, et al. Specific transgene expression in HIV-infected cells using protease-cleavable transcription regulator. J Control Release. 2010;141(1):52–61.
25. Xiong MH, Bao Y, Yang XZ, Wang YC, Sun B, Wang J. Lipase-sensitive polymeric triple-layered nanogel for "on-demand" drug delivery. J Am Chem Soc. 2012;134(9):4355–62.
26. Li J, Huo M, Wang J, Zhou J, Mohammad JM, Zhang Y, et al. Redox-sensitive micelles self-assembled from amphiphilic hyaluronic acid-deoxycholic acid conjugates for targeted intracellular delivery of paclitaxel. Biomaterials. 2012;33(7):2310–20.
27. Yavlovich A, Smith B, Gupta K, Blumenthal R, Puri A. Light-sensitive lipid-based nanoparticles for drug delivery: design principles and future considerations for biological applications. Mol Membr Biol. 2010;27(7):364–81.
28. Abuwatfa WH, Awad NS, Pitt WG, Husseini GA. Thermosensitive polymers and thermo-responsive liposomal drug delivery systems. Polymers. 2022;14(5):925.
29. Ta T, Porter TM. Thermosensitive liposomes for localized delivery and triggered release of chemotherapy. J Control Release. 2013;169(1–2):112–25.
30. Nemoto R, Fujieda K, Hiruta Y, Hishida M, Ayano E, Maitani Y, et al. Liposomes with temperature-responsive reversible surface properties. Colloids Surf B Biointerfaces. 2019;176:309–16.
31. Yuan Z, Gottsacker C, He X, Waterkotte T, Park YC. Repetitive drug delivery using Light-Activated liposomes for potential antimicrobial therapies. Adv Drug Deliv Rev. 2022;187:114395.
32. Refaat A, Del Rosal B, Palasubramaniam J, Pietersz G, Wang X, Moulton SE, et al. Near-infrared light-responsive liposomes for protein delivery: towards bleeding-free photothermally-assisted thrombolysis. J Control Release. 2021;337:212–23.
33. Zangabad PS, Mirkiani S, Shahsavari S, Masoudi B, Masroor M, Hamed H, et al. Stimulus-responsive liposomes as smart nanoplatforms for drug delivery applications. Nanotechnol Rev. 2018;7(1):95–122.
34. Lee Y, Thompson DH. Stimuli-responsive liposomes for drug delivery. Wiley Interdiscip Rev Nanomed Nanobiotechnol. 2017;9(5):10.1002/wnan.1450.
35. TSA, Lu YJ, Chen JP. Optimization of the preparation of magnetic liposomes for the combined use of magnetic hyperthermia and photothermia in dual magneto-photothermal cancer therapy. Int J Mol Sci. 2020;21(15):5187.
36. Sun X, Tan A, Boyd BJ. Magnetically-activated lipid nanocarriers in biomedical applications: a review of current status and perspective. Wiley Interdiscip Rev Nanomed Nanobiotechnol. 2023;15(3):e1863.
37. Bahutair WN, Abuwatfa WH, Husseini GA. Ultrasound triggering of liposomal nanodrugs for cancer therapy: a review. Nanomaterials (Basel, Switzerland). 2022;12(17):3051.
38. de Matos MBC, Deckers R, van Elburg B, Lajoinie G, de Miranda BS, Versluis M, et al. Ultrasound-sensitive liposomes for triggered macromolecular drug delivery: formulation and in vitro characterization. Front Pharmacol. 2019;10:1463.

39. Park YI, Kwon SH, Lee G, Motoyama K, Kim MW, Lin M, et al. pH-sensitive multi-drug liposomes targeting folate receptor β for efficient treatment of non-small cell lung cancer. J Control Release. 2021;330:1–14.
40. Swami R, Kumar Y, Chaudhari D, Katiyar SS, Kuche K, Katare PB, et al. pH sensitive liposomes assisted specific and improved breast cancer therapy using co-delivery of SIRT1 shRNA and Docetaxel. Mater Sci Eng C Mater Biol Appl. 2021;120:111664.
41. Xu H, Li Y, Paxton JW, Wu Z. Co-delivery using pH-sensitive liposomes to pancreatic cancer cells: the effects of Curcumin on cellular concentration and pharmacokinetics of Gemcitabine. Pharm Res. 2021;38(7):1209–19.
42. Saraf S, Jain A, Tiwari A, Verma A, Jain SK. Engineered liposomes bearing camptothecin analogue for tumour targeting: in vitro and ex-vivo studies. J Liposome Res. 2021;31(4):326–41.
43. Ding Y, Cui W, Sun D, Wang GL, Hei Y, Meng S, et al. In vivo study of doxorubicin-loaded cell-penetrating peptide-modified pH-sensitive liposomes: biocompatibility, bio-distribution, and pharmacodynamics in BALB/c nude mice bearing human breast tumors. Drug Des Devel Ther. 2017;11:3105–17.
44. Liu D, Chen B, Mo Y, Wang Z, Qi T, Zhang Q, et al. Redox-activated Porphyrin-based liposome remote-loaded with Indoleamine 2,3-Dioxygenase (IDO) inhibitor for synergistic photoimmunotherapy through induction of immunogenic cell death and blockage of IDO pathway. Nano Lett. 2019;19(10):6964–76.
45. Chi Y, Yin X, Sun K, Feng S, Liu J, Chen D, et al. Redox-sensitive and hyaluronic acid functionalized liposomes for cytoplasmic drug delivery to osteosarcoma in animal models. J Control Release. 2017;261:113–25.
46. Chen X, Zhang Y, Tang C, Tian C, Sun Q, Su Z, et al. Co-delivery of paclitaxel and anti-survivin siRNA via redox-sensitive oligopeptide liposomes for the synergistic treatment of breast cancer and metastasis. Int J Pharm. 2017;529(1–2):102–15.
47. Fu H, Shi K, Hu G, Yang Y, Kuang Q, Lu L, et al. Tumor-targeted paclitaxel delivery and enhanced penetration using TAT-decorated liposomes comprising redox-responsive poly(ethylene glycol). J Pharm Sci. 2015;104(3):1160–73.
48. Lv Q, Cheng L, Lu Y, Zhang X, Wang Y, Deng J, et al. Thermosensitive exosome-liposome hybrid nanoparticle-mediated chemoimmunotherapy for improved treatment of metastatic peritoneal cancer. Adv Sci (Weinheim, Baden-Wurttemberg, Germany). 2020;7(18):2000515.
49. Jin X, Lu X, Zhang Z, Lv H. Indocyanine green-Parthenolide thermosensitive liposome combination treatment for triple-negative breast cancer. Int J Nanomedicine. 2020;15:3193–206.
50. Cressey P, Amrahli M, So PW, Gedroyc W, Wright M, Thanou M. Image-guided thermosensitive liposomes for focused ultrasound enhanced co-delivery of carboplatin and SN-38 against triple negative breast cancer in mice. Biomaterials. 2021;271:120758.
51. Zhang H, Tang WL, Kheirolomoom A, Fite BZ, Wu B, Lau K, et al. Development of thermosensitive resiquimod-loaded liposomes for enhanced cancer immunotherapy. J Control Release. 2021;330:1080–94.
52. Dai Y, Su J, Wu K, Ma W, Wang B, Li M, et al. Multifunctional thermosensitive liposomes based on natural phase-change material: near-infrared light-triggered drug release and multimodal imaging-guided cancer combination therapy. ACS Appl Mater Interfaces. 2019;11(11):10540–53.
53. Li Q, Tang Q, Zhang P, Wang Z, Zhao T, Zhou J, et al. Human epidermal growth factor receptor-2 antibodies enhance the specificity and anticancer activity of light-sensitive doxorubicin-labeled liposomes. Biomaterials. 2015;57:1–11.
54. Zhang K, Zhang Y, Meng X, Lu H, Chang H, Dong H, et al. Light-triggered theranostic liposomes for tumor diagnosis and combined photodynamic and hypoxia-activated prodrug therapy. Biomaterials. 2018;185:301–9.
55. Fuse T, Tagami T, Tane M, Ozeki T. Effective light-triggered contents release from helper lipid-incorporated liposomes co-encapsulating gemcitabine and a water-soluble photosensitizer. Int J Pharm. 2018;540(1–2):50–6.

56. Kono Y, Yokoyama K, Suzuki M, Takakura H, Ogawa M. Surface modification of liposomes using IR700 enables efficient controlled contents release triggered by near-IR light. Biol Pharm Bull. 2020;43(4):736–41.
57. Yang X, Li H, Qian C, Guo Y, Li C, Gao F, et al. Near-infrared light-activated IR780-loaded liposomes for anti-tumor angiogenesis and Photothermal therapy. Nanomed Nanotechnol Biol Med. 2018;14(7):2283–94.
58. Dwivedi P, Kiran S, Han S, Dwivedi M, Khatik R, Fan R, et al. Magnetic targeting and ultrasound activation of liposome-microbubble conjugate for enhanced delivery of anticancer therapies. ACS Appl Mater Interfaces. 2020;12(21):23737–51.
59. Redolfi Riva E, Sinibaldi E, Grillone AF, Del Turco S, Mondini A, Li T, et al. Enhanced in vitro magnetic cell targeting of Doxorubicin-loaded magnetic liposomes for localized cancer therapy. Nanomaterials (Basel, Switzerland). 2020;10(11):2104.
60. Liu Y, Li J, Chen H, Cai Y, Sheng T, Wang P, et al. Magnet-activatable nanoliposomes as intracellular bubble microreactors to enhance drug delivery efficacy and burst cancer cells. Nanoscale. 2019;11(40):18854–65.
61. Lin ZL, Ding J, Sun GP, Li D, He SS, Liang XF, et al. Application of Paclitaxel-loaded EGFR peptide-conjugated magnetic polymeric liposomes for liver cancer therapy. Curr Med Sci. 2020;40(1):145–54.
62. Hardiansyah A, Yang MC, Liu TY, Kuo CY, Huang LY, Chan TY. Hydrophobic drug-loaded PEGylated magnetic liposomes for drug-controlled release. Nanoscale Res Lett. 2017;12(1):355.
63. Hua H, Zhang N, Liu D, Song L, Liu T, Li S, et al. Multifunctional gold nanorods and docetaxel-encapsulated liposomes for combined thermo- and chemotherapy. Int J Nanomedicine. 2017;12:7869–84.
64. Tran TH, Nguyen HT, Van LN, Tran TTP, Lee JS, Ku SK, et al. Engineering of multifunctional temperature-sensitive liposomes for synergistic photothermal, photodynamic, and chemotherapeutic effects. Int J Pharm. 2017;528(1–2):692–704.
65. Yuba E, Takashima M, Hayashi T, Kokuryo D, Aoki I, Harada A, et al. Multifunctional traceable liposomes with temperature-triggered drug release and Neovasculature-targeting properties for improved cancer chemotherapy. Mol Pharm. 2021;18(9):3342–51.
66. Miranda D, Carter K, Luo D, Shao S, Geng J, Li C, et al. Multifunctional liposomes for image-guided Intratumoral chemo-phototherapy. Adv Healthc Mater. 2017;6(16):10.1002/adhm.201700253.
67. Nakhaei P, Margiana R, Bokov DO, Abdelbasset WK, Jadidi Kouhbanani MA, Varma RS, et al. Liposomes: structure, biomedical applications, and stability parameters with emphasis on cholesterol. Front Bioeng Biotechnol. 2021;9:705886.
68. Nsairat H, Khater D, Sayed U, Odeh F, Al Bawab A, Alshaer W. Liposomes: structure, composition, types, and clinical applications. Heliyon. 2022;8(5):e09394.
69. Li S, Wang W, Yu X. A perspective of engineered lipids and liposomes: chemical design and functional application based on therapeutic safety. Chem Mater. 2023;35(12):4587–97.
70. Kleusch C, Hersch N, Hoffmann B, Merkel R, Csiszár A. Fluorescent lipids: functional parts of fusogenic liposomes and tools for cell membrane labeling and visualization. Molecules (Basel, Switzerland). 2012;17(1):1055–73.
71. Kong G, Anyarambhatla G, Petros WP, Braun RD, Colvin OM, Needham D, et al. Efficacy of liposomes and hyperthermia in a human tumor xenograft model: importance of triggered drug release. Cancer Res. 2000;60(24):6950–7.
72. Manzoor AA, Lindner LH, Landon CD, Park JY, Simnick AJ, Dreher MR, et al. Overcoming limitations in nanoparticle drug delivery: triggered, intravascular release to improve drug penetration into tumors. Cancer Res. 2012;72(21):5566–75.
73. Briuglia ML, Rotella C, McFarlane A, Lamprou DA. Influence of cholesterol on liposome stability and on in vitro drug release. Drug Deliv Transl Res. 2015;5(3):231–42.
74. Kaddah S, Khreich N, Kaddah F, Charcosset C, Greige-Gerges H. Cholesterol modulates the liposome membrane fluidity and permeability for a hydrophilic molecule. Food Chem Toxicol. 2018;113:40–8.

75. Magarkar A, Dhawan V, Kallinteri P, Viitala T, Elmowafy M, Róg T, et al. Cholesterol level affects surface charge of lipid membranes in saline solution. Sci Rep. 2014;4:5005.
76. Aronson MR, Medina SH, Mitchell MJ. Peptide functionalized liposomes for receptor targeted cancer therapy. APL Bioeng. 2021;5(1):11501.
77. Klibanov AL, Maruyama K, Torchilin VP, Huang L. Amphipathic polyethyleneglycols effectively prolong the circulation time of liposomes. FEBS Lett. 1990;268(1):235–7.
78. Bulbake U, Doppalapudi S, Kommineni N, Khan W. Liposomal formulations in clinical use: an updated review. Pharmaceutics. 2017;9(2):12.
79. Pasarin D, Ghizdareanu AI, Enascuta CE, Matei CB, Bilbie C, Paraschiv-Palada L, et al. Coating materials to increase the stability of liposomes. Polymers. 2023;15(3):782.
80. Luo D, Carter KA, Molins EAG, Straubinger NL, Geng J, Shao S, et al. Pharmacokinetics and pharmacodynamics of liposomal chemophototherapy with short drug-light intervals. J Control Release. 2019;297:39–47.
81. Liu P, Chen G, Zhang J. A review of liposomes as a drug delivery system: current status of approved products, regulatory environments, and future perspectives. Molecules (Basel, Switzerland). 2022;27(4):1372.
82. Xiang B, Cao DY. Preparation of drug liposomes by thin-film hydration and homogenization. In: Lu WL, Qi XR, editors. Liposome-based drug delivery systems. Biomaterial engineering. Berlin/Heidelberg: Springer; 2018. https://doi.org/10.1007/978-3-662-49231-4_2-1.
83. Lombardo D, Kiselev MA. Methods of liposomes preparation: formation and control factors of versatile nanocarriers for biomedical and nanomedicine application. Pharmaceutics. 2022;14(3):543. https://doi.org/10.3390/pharmaceutics14030543.
84. Nsairat H, Khater D, Sayed U, Odeh F, Al Bawab A, Alshaer W. Liposomes: structure, composition, types, and clinical applications. Heliyon. 2022;8(5).
85. Koide H, Ochiai H, Suzuki H, Hirata S, Watanabe M, Yonezawa S, et al. Easy preparation of a liposome-mediated protein delivery system by freeze–thawing a liposome–protein complex. J Mater Chem B. 2022;10:6768–76. https://doi.org/10.1039/D2TB00271J.
86. Bernal-Chávez SA, Romero-Montero A, Hernández-Parra H, Peña-Corona SI, Del Prado-Audelo ML, Alcalá-Alcalá S, et al. Enhancing chemical and physical stability of pharmaceuticals using freeze-thaw method: challenges and opportunities for process optimization through quality by design approach. J Biol Eng. 2023;17:35. https://doi.org/10.1186/s13036-023-00353-9.
87. Ota A, Mochizuki A, Sou K, Takeoka S. Evaluation of a static mixer as a new microfluidic method for liposome formulation. Front Bioeng Biotechnol. 2023;22(11):1229829. https://doi.org/10.3389/fbioe.2023.1229829.
88. Zhang H, Tang WL, Kheirolomoom A, Fite BZ, Wu B, Lau K, et al. Development of thermosensitive resiquimod-loaded liposomes for enhanced cancer immunotherapy. J Control Release. 2021;330:1080–94. https://doi.org/10.1016/j.jconrel.2020.11.013.
89. Robson AL, Dastoor PC, Flynn J, Palmer W, Martin A, Smith DW, et al. Advantages and limitations of current imaging techniques for characterizing liposome morphology. Front Pharmacol. 2018;9:80.
90. Vorselen D, Piontek MC, Roos WH, Wuite GJL. Mechanical characterization of liposomes and extracellular vesicles, a protocol. Front Mol Biosci. 2020;7(July):1–14.
91. Fan Y, Marioli M, Zhang K. Analytical characterization of liposomes and other lipid nanoparticles for drug delivery. J Pharm Biomed Anal. 2021;192:113642.
92. Cipolla D, Wu H, Gonda I, Eastman S, Redelmeier T, Chan HK. Modifying the release properties of liposomes toward personalized medicine. J Pharm Sci. 2014;103(6):1851–62.
93. Mozafari MR. Nanoliposomes: preparation and analysis. Method Mol Biol (Clifton, NJ). 2010;605:29–50.
94. Bian J, Girotti J, Fan Y, Levy ES, Zang N, Sethuraman V, et al. Fast and versatile analysis of liposome encapsulation efficiency by nanoParticle exclusion chromatography. J Chromatogr A. 2022;1662:462688.
95. Panwar P, Pandey B, Lakhera PC, Singh KP. Preparation, characterization, and in vitro release study of albendazole-encapsulated nanosize liposomes. Int J Nanomedicine. 2010;5:101–8.

96. Bozzuto G, Molinari A. Liposomes as nanomedical devices. Int J Nanomedicine. 2015;10:975–99.
97. Wang S, Chen Y, Guo J, Huang Q. Liposomes for tumor targeted therapy: a review. Int J Mol Sci. 2023;24(3):2643.
98. Ashrafizadeh M, Delfi M, Zarrabi A, Bigham A, Sharifi E, Rabiee N, et al. Stimuli-responsive liposomal nanoformulations in cancer therapy: Pre-clinical clinical approaches. J Control Release. 2022;351:50–80.
99. Liu G, Wang J, Xia Q. Application of nanostructured lipid carrier in food for the improved bioavailability. Eur Food Res Technol. 2012;234:391.
100. van der Meel R, Sulheim E, Shi Y, Kiessling F, Mulder WJM, Lammers T. Smart cancer nanomedicine. Nat Nanotechnol. 2019;14(11):1007–17.
101. Sugahara KN, Teesalu T, Karmali PP, Kotamraju VR, Agemy L, Girard OM, et al. Tissue-penetrating delivery of compounds and nanoparticles into tumors. Cancer Cell. 2009;16(6):510–20.
102. Sindhwani S, Chan WCW. Nanotechnology for modern medicine: next step towards clinical translation. J Intern Med. 2021;290(3):486–98.

[illegible]

Stimuli-Responsive Lipid Nanoparticles for Cancer Therapy

7

Chukwuebuka Emmanuel Umeyor, Preeya Negi, Andrew Ezegbe, Ghanshamdas Bairagi, and Vandana B. Patravale

Abstract

Cancer is a fatal disease with increasing morbidities and mortalities worldwide. Considering the millions of deaths due to cancer, effective treatment interventions are urgently needed to halt the devastating mortality rates and disease burden. Current methods of therapy for cancer, such as phototherapy, radiotherapy, surgery, and chemotherapy, are limited due to their adverse and off-target effects. Available treatment for cancer patients has improved following a better understanding of the pathophysiology of cancer and advances in nanocarrier development. However, these traditional nanosystems lack innate targeting efficiencies, increased cellular internalization, and prolonged systemic circulation. Recent advances in the design of cutting-edge nanotechnology strategies like stimuli-responsive lipid nanoparticles (LNPs) have revolutionized clinical translation and use of LNPs in cancer treatment due to the efficiency of exogenous and endogenous factors to regulate drug release, increase cellular uptake, enhance systemic circulation, and promote tissue targeting. Beyond the brief introduction, this chapter provided an overview of challenges encountered in cancer therapy. It outlined the various lipid-based nanocarriers developed and deployed for

C. E. Umeyor
Department of Pharmaceutics and Pharmaceutical Technology, Faculty of Pharmaceutical Sciences, Nnamdi Azikiwe University, Awka, Anambra State, Nigeria

P. Negi · G. Bairagi · V. B. Patravale (✉)
Department of Pharmaceutical Sciences and Technology, Institute of Chemical Technology, Mumbai, Maharashtra, India
e-mail: vb.patravale@ictmumbai.edu.in

A. Ezegbe
Department of Pharmaceutical Technology and Industrial Pharmacy, University of Nigeria, Nsukka, Enugu State, Nigeria

Human and Natural Science Center, ABC Federal University, Santo Andre, Sao Paulo, Brazil

P. V. Devarajan et al. (eds.), *Advances in Stimuli-Responsive Nanosystems for Cancer Therapy*, https://doi.org/10.1007/978-981-95-2744-1_7

cancer therapy. It enumerated and discussed the progress made in using different types of stimuli-responsive LNPs in cancer treatment. While it discussed the benefits and limitations of stimuli-responsive LNPs for cancer, it also provided insights into the safety considerations for clinical translation of stimuli-responsive LNPs in cancer.

Keywords

Cancer · Lipid-based nanoparticles · Nanotechnology · Stimuli-responsive · Clinical translation · Tumour

Abbreviations

DMSO	Dimethyl sulphoxide
DOPE	Dioleoylphosphatidylethanolamine
DOX	Doxorubicin
EPR	Enhanced permeability and retention
FDA	Food and Drug Administration
GMS	Glycerine monostearate
GOx	Glucose oxidase
GRAS	Generally regarded as safe
HA	Hyaluronic acid
IA	Intra-arterial
IR	Infrared
IV	Intravenous
LCST	Low critical solution temperature
LDCs	Lipid–drug conjugates
LDL	Low-density lipoprotein
LNPs	Lipid nanoparticles
MMP-2	Matrix metalloproteinase-2
NLC	Nanostructured lipid carriers
NPAs	Nanoparticle aggregates
NSCLC	Non-small cell lung cancer
PDT	Photodynamic therapy
PEDDT	Poly-(3,4-ethylene dioxythiophene)
PLGA	Poly(lactic-co-glycolic acid)
PNIPAM	Poly-N-isopropyl acrylamide
ROS	Reactive oxygen species
SA	Sodium alginate
siRNA	Small interfering ribonucleic acid
SLNs	Solid lipid nanoparticles
SNEDDS	Self-nanoemulsifying drug delivery systems
TAMs	Tumour-associated macrophages
TME	Tumour microenvironment
UCST	Upper critical solution temperature
UV	Ultraviolet

1 Introduction

Globally, cancer remains an important fatal disease responsible for increasing mortalities. Before the emergence of nanotechnology, the efficacy of traditional treatment methods for cancer is limited by several factors, such as decreased bioavailability due to rapid renal clearance, poor aqueous solubility, irregular cellular uptake, and inability to target the therapeutics to desired sites of action. In the last few years, nanotechnology has redeemed and elevated clinical cancer therapy by revolutionizing drug development, enhancing the advantages offered by classical dosage forms, and producing notable improvements in drug biopharmaceutics [1]. Indeed, lipid nanoparticles (LNPs) is a viable nanotechnological platform that has made significant impact in nanomedicine research because LNPs are being developed and applied in cancer therapy, diagnosis, and sensing. Despite the potential for application of LNPs in cancer disease, the precise delivery and efficacy of the nanoparticles are hindered by limited penetration across several tissues and barriers [2, 3].

The development of nanoscale molecules that offer more precise, specific, targeted, controlled, and personalized effect and efficacy has become very paramount in cancer management. This is against the backdrop of the propensity of cancer cells to innovate ways of evading the effects of common therapies like radiotherapy, chemotherapy, photothermal therapy, and surgery, and continue replicating in their microenvironment [2]. Consequently, there is an unmet need to develop smart nanosystems which respond to stimuli like pH, light, temperature, ultrasound, enzymes, ultrasound, magnetic field, and redox potential, with improved efficacy. More so, stimuli-responsive lipid nanoparticles (LNPs) has gained considerable ground in drug delivery for disease therapy including cancer, due to their capacity to protect encapsulated drug molecules, improve anti-cancer activity, and produce controlled and targeted delivery in response to the triggering stimulus, whether exogenous or endogenous [1]. LNPs are designed and developed to specifically deliver bioactives into the microenvironment of cancers because functionalizing the LNPs by making them sensitive to exogenous or endogenous stimulus facilitates timely delivery and release response of the nanoparticles at the desired sites. Stimuli-responsive LNPs react in a certain complex but dynamic manner that mimics the response of a physiological tissue in the presence of a triggering stimulus. The smart nanomaterial is formulated using generally-regarded-as-safe (GRAS) materials that respond to physical triggers or that undergo chemically inspired molecular conformational, hydrolytic, and proton changes. In that way, it provides a versatile and convenient strategy for encapsulating nucleic acids, antibodies, imaging molecules, and drugs, for cancer therapy [4]. Like every human endeavour, developing stimuli-sensitive LNPs for cancer therapy may present some challenges that limit its clinical applications but they are surmountable so that maximum benefits could be obtained from stimuli-responsive LNPs. Some of these limitations will be discussed in the latter part of this chapter.

In addition, we discuss significant advances in lipid-based nanosystems applied in cancer therapy. We have provided fresh insights into the various classes of stimuli-responsive LNPs and also considered safety protocols that facilitate clinical translation of stimuli-responsive LNPs for cancer therapy.

2 Challenges in Clinical Management of Cancer

Clinical management of cancer is fraught with some challenges which threaten effective treatment of cancer and pose mortal risks to affected patients.

Cancer patients are immunocompromised and are susceptible to opportunistic infections with increased morbidity thereby making clinical treatment outcome unpredictable and oncologists tasks complicated. In this case, early and timely diagnosis is essential for clearer understanding of the developmental stage of the cancer cells in order to design the most effective treatment strategy and protect patients and caregivers from exposure to infections. However, during pandemics or outbreak of infectious diseases, clinical diagnosis of cancer might be difficult to access, especially for paediatric patients, because laboratory personnel and resources would be focused on the prognosis of the pandemic or epidemic [5–7].

Poor funding of cancer treatment programmes and lack of subsidy for cancer therapy in healthcare institutions imply that cancer patients must pay out-of-pocket for services rendered by oncologists. As a result, physicians charge patients exorbitant payments for a complete course of chemotherapy, and sometimes, patients overpay before they have access to treatment and care due to the absence or poor availability of healthcare insurance for cancer patients. In addition, incidences of overtreatment of cancer patients resulting in overpayment have been reported in some public healthcare facilities [8]. Many countries do not have structures for regulating the prices of drugs used for treating cancer nor negotiate with manufacturers of anti-cancer medicines for median prices that could be generally affordable nationwide. This scenario provides no compelling need for oncologists to prescribe effective medicines generally to patients irrespective of financial status. Further, it does not promote affordability and accessibility of cancer drugs even for patients who might be willing to pay for their therapy [9].

Another factor that diminishes the impact of effective clinical management of cancer is unrealistic expectation of patients about their disease condition or the effect expected from one clinical intervention or another. Cancer patients and their caregivers ask many questions about their illness, their likelihood of survival, disease prognosis, and whether the cancer is curable or not. Depending on the answer available to them or how it is communicated, the confidence of the patients is either diminished or boosted regarding the likely outcome of undergoing therapy. Indeed, it is important to communicate effectively to cancer patients with sufficient clarity to enable them understand possibilities in clinical management of their conditions, especially those who are treated with experimental drugs or trial procedures [10, 11].

Sometimes, cancer therapy and interventions might be misused and/or overused due to high cost of treatment or unrealistic expectations from treatment strategies. In advanced or terminal cases of cancer, there is increased need for remedies with the ability to reverse cancer progression and facilitate quick recovery. Thus, any therapy that has the potential to prolong life might be overused, especially if the drug is not expensive. Owing to curiosity, oncologists might misuse the medicine in different cancers to determine the levels of efficacy and safety provided by the therapy [12, 13].

Absence of a method for quality of care measurement is another vital challenge that impedes effective, efficacious, and safe cancer management. A good assessment method should identify gaps in performance of a treatment strategy and measure the extent to which improvements in the therapy can ameliorate these gaps. An effective and efficacious treatment is one that delivers precise effect on the illness, improves the survival chances of a patient, and avoids wasteful and valueless care. Hence, treatment evaluation protocol should obtain data about the response of a patient to an administered therapy and what could be done to enhance the quality of life of the patient [14].

3 Lipid-Based Nanoformulations for Cancer Treatment

The efficacy of existing treatment strategies (surgery, chemotherapy, radiotherapy, etc.) for malignancies is constrained by their adverse effects, narrow therapeutic index, brief half-life, and therapeutic resistance [15, 16]. To overcome these issues, nano-delivery-based systems like targeted distribution, diminished toxicity, controlled release, and enhanced bioavailability have been developed [15]. A recent study on breast cancer revealed that magnetic lipid-based nanosystems are capable of responding to magnetic stimuli and are especially designed for dipeptide L-carnosine delivery. Stimuli-responsive magnetic lipid-based nanosystems exhibited less systemic toxicity and enhanced chemotherapeutic effectiveness compared to non-targeted magnetic lipid-based nanosystems and L-carnosine solution [17]. Also, lipid-based nanosystems composed of naturally occurring biocompatible lipids (sphingomyelin and vitamin E) might influence the macrophage phenotype, resulting in the decrease of typical markers of tumour-associated macrophages (TAMs) [16]. This section explores a few such lipid-based nanosystems, including solid lipid nanoparticles (SLNs), nanostructured lipid carriers (NLCs), lipid nanoemulsions, self-nanoemulsifying drug delivery systems (SNEDDS), liposomes, ethosomes, and cubosomes, that are being utilized for the therapy of cancer.

3.1 Solid Lipid Nanoparticles

Solid lipid nanoparticles (SLNs) are first-generation lipid-based nanosystems. These nanoparticles are composed of solid lipids, which are biocompatible and capable of incorporating lipophilic drugs. The solid matrix allows for controlled drug release, enhancing the therapeutic efficacy while minimizing side effects. The particle size following drug encapsulation spans from nanoscale to submicron scale (50–1000 nm), and the production of nanoparticles does not need the use of organic solvents [18, 19]. SLNs can be utilized to deliver drugs that exhibit physicochemical incompatibility, possess a low pharmacokinetic profile, and are thermolabile. It can also deliver proteins and peptides with high efficacy and low toxicity [19].

3.1.1 Applications

- Curcumin-loaded SLNs showed higher anticancer activity than dimethyl sulfoxide-diluted curcumin in breast cancer cell line [19].
- Methotrexate loaded SLNs tagged with ferrous improved target specificity and anticancer activity in vitro and in vivo against breast cancer [19].
- Naringenin-loaded SLNs, when administered intratracheally, improved cellular absorption of drugs in vivo when compared to unloaded drug [19].
- Oxaliplatin loaded SLNs tagged with folic acid showed increased anticancer activity in colon cancer cell line when compared to free drug and non-functionalized SLNs [19].
- Andrographolide-loaded SLNs improved permeability across the blood–brain barrier in vitro and in vivo [19].
- Paclitaxel-loaded glyceryl tripalmitate SLNs showed increased anticancer activity against breast and lung cancer when compared to free drug [18].
- Temozolomide-loaded SLNs showed more anticancer activity against melanoma in vitro and in vivo [18].
- Docetaxel-loaded SLNs tagged with angiopep-2 for targeted therapy for glioblastoma at blood–brain barrier and the glioma cells [18].

3.2 Nanostructured Lipid Carriers

Nanostructured lipid carriers (NLCs) are a second-generation lipid-based nanocarrier system designed to overcome the limitations of SLNs. NLCs consist of a mixture of solid and liquid lipids in ratios from 70:30 to 99.9:0.1, providing a higher drug loading capacity and preventing drug expulsion during storage. The NLCs are solid at room temperature. More than one drug with different physicochemical properties can be encapsulated in NLCs and targeting of the NLCs can be improved by combination with amphiphilic copolymer [20].

3.2.1 Applications

- Encapsulation of docetaxel hydrochloride–oleic acid complexes in NLC showed stability at both neutral and acidic pH and improved anticancer activity when compared to unencapsulated against colon cancer [21, 22].
- Encapsulation of curcumin in NLC showed better anticancer activity when compared to free drug against human brain cancer cell lines [23].
- Encapsulation of artemisinin in NLC tagged with transferrin showed enhanced aqueous solubility, blood–brain barrier permeation, target specificity and hence

significantly higher anticancer activity when compared to artemisinin solution [23].

- Encapsulation of cytarabine in NLC coated with polysorbate 80 showed enhanced blood–brain barrier permeation and higher anticancer activity at lower dose when compared to free drug against meningeal leukaemia [23].
- Encapsulation of temozolomide and vincristine in NLC showed higher anticancer activity by 24% as compared to SLN of the same drugs at all concentrations against glioma tumour [23].
- Encapsulation of osthole in NLC system and tagged with chitosan conjugated folate showed higher target specificity and anticancer activity against colon cancer cell line [24].
- Encapsulating camptothecin and topotecan in NLC system showed higher cellular drug uptake and enhanced anticancer activity against melanoma and leukaemia [25].

3.3 Lipid Nanoemulsions

Lipid nanoemulsions are nanoscale emulsions composed of an oil phase dispersed in water, stabilized by surfactants. The size of lipid nanoemulsion is around 20–200 nm and by incorporation of tumour tags, target specificity can be improved [26].

3.3.1 Applications

- Paclitaxel–oleate nanoemulsion showed lower toxicity when compared to paclitaxel formulation containing cremophor EL [21].
- Docetaxel–linoleic acid nanoemulsion showed better anticancer activity against breast cancer when compared to free drug [22].
- Folic acid nanoemulsion for co-delivery of docosahexaenoic acid and paclitaxel for breast cancer showed controlled drug release, synergistic action, and enhanced permeability due to particle size less than 200 nm [27].
- Methotrexate–didodecyl nanoemulsion showed higher uptake by cancer cells and hence more anticancer activity against leukaemia when compared to free drug [28].
- Paclitaxel bind to low-density lipoprotein (LDL) receptors and hence showed less toxicity and higher anticancer activity. Also, simvastatin administered with paclitaxel nanoemulsions showed enhanced anticancer activity probably due to increased LDL receptor expression while the same did not happen with free paclitaxel [28, 29].
- Chlorambucil nanoemulsion showed enhanced pharmacokinetics and higher anticancer activity when compared to chlorambucil solution against colon adenocarcinoma in vivo [30].
- Cheliensisin A lipid nanoemulsion improved solubility and enhanced chemical stability of drug [30].
- Disulfiram–lipid emulsion enhanced stability of drug in blood [30].

3.4 Self-Nanoemulsifying Drug Delivery Systems

Self-nanoemulsifying drug delivery systems (SNEDDS) are isotropic mixtures of oils, surfactants, and co-surfactants that spontaneously form oil-in-water nanoemulsions upon contact with aqueous fluids with slight agitation [31]. SNEDDS can be explored for oral drug delivery as they prevent direct contact of drug with biological system during drug delivery and hence protects drug from the gastrointestinal fluid. Hence, SNEDDS have potential in oral delivery of peptides and proteins [32, 33].

3.4.1 Applications

- Apomorphine loaded SEDDS had low drug degradation when administered orally [21].
- Tamoxifen-loaded SNEDDS showed higher diffusion, more safety, higher anti-cancer efficiency compared to free drug in an ex-vivo permeation study [32].
- Paclitaxel-loaded SNEDDS reveals higher bioavailability due to faster release of drug when compared to drug suspension [34].
- Docetaxel-loaded SNEDDS demonstrated higher bioavailability and anti-cancer activity with low toxicity as compared to free drug [34].
- Brigatinib-loaded SNEDDS showed higher aqueous solubility, ex vivo penetration, in vitro release, and anticancer activity by two-fold when compared to free brigatinib [34].
- Resveratrol-loaded SNEDDS showed higher anticancer activity as it significantly inhibited cell migration in pancreatic cancer when compared to free drug [34].
- Piplartine-loaded SNEDDS demonstrated higher solubility, bioavailability, and anticancer activity when compared to free drug [34].
- Myricetin-loaded SNEDDS showed more than six times increase in oral bioavailability of drug [34].
- Solid SNEDDS co-loaded with docetaxel and carvacrol showed higher apoptosis and reduced toxicity as compared to free drugs [31].

3.5 Liposomes

Liposomes are spherical vesicles made up of one or more phospholipid layers with an aqueous cavity, capable of encapsulating both hydrophilic and hydrophobic drugs [35]. Targeting ligands can be attached to liposomes to improve target specificity [21]. But liposomes still have potential side effects and lack disposition [36].

3.5.1 Applications

- Liposome formed by mitomycin C–diacyl glycerol conjugate showed higher activity against ovarian cancer cells when compared to free drug [22].
- Liposome formed by mitomycin C–distearic acyl glycerol conjugate showed higher activity against colon carcinoma when compared to free drug [22].

- Liposome formed by gemcitabine–elaidic acid conjugate showed enhanced treatment of melanoma, lung carcinoma and adenocarcinoma, osteosarcoma, fibrous histiocytoma, breast cancer, prostate cancer, pancreas cancer, and colon cancer [22].
- Liposome formed by docetaxel–phosphatidyl choline conjugate prevented burst release of drug [33].
- Liposomes formed by the chlorambucil–phospholipid conjugate have stability in aqueous environment and better anticancer activity [33].
- Lipocisplatin (palmitoyl-2-glutaryl-sn-glycero-3-phosphocholine-cisplatin conjugate) have shown higher content stability when stored under 4 °C and lesser toxicity than free cisplatin [33].
- Doxil® (US)/Caelyx® (EU) a doxorubicin liposomes approved for AIDS-related Kaposi's sarcoma, recurrent ovarian cancer, multiple myeloma, and metastatic breast cancer showed reduced cardiotoxicity and myelotoxicity when compared to free doxorubicin [36].

3.6 Ethosomes

Ethosomes are lipid-based vesicular carriers containing high concentrations of ethanol, which enhances permeation and is a type of vesicular delivery system. As the ethosome bilayer is made up of water and lipid, both hydrophilic and hydrophobic drugs can be incorporated into it [37]. It can be used as a carrier for high molecular weight compounds [23]. Ethosomes serve as delivery systems for a wide range of medications [23], and are advantageous over conventional cancer therapies as it is non-invasive and hence show fewer side effects [38].

3.6.1 Applications

- Ethosome for thymoquinone encapsulation showed greater release and anticancer activity as compared to free thymoquinone [37].
- Ethosome for trihexyphenidyl hydrochloride encapsulation for Parkinson's disease management showed higher cutaneous permeation as compared to liposomes [38].
- Ethosomal gel containing mitoxantrone showed more permeability and notable cytotoxicity in vitro and in vivo [39].
- Ethosome for sulforaphane encapsulation enhanced anticancer activity when compared to the unencapsulated drug [39].
- Ethosome for cytarabine encapsulation for acute myeloid leukaemia showed increased percutaneous penetration [39].
- Ethosome for paclitaxel encapsulation for squamous cell carcinoma showed an increase in anti-apoptotic and anti-proliferative activity when compared to free drug [39].

3.7 Cubosomes

Cubosomes are self-structured lipid-based nanocarriers characterized by a cubic phase structure, consisting of bicontinuous lipid bilayers. The bicontinuous structure of cubosomes has higher entrapment efficiency for hydrophobic drugs as compared to liposomes [40]. Cubosomes have a broad surface area, in which hydrophobic drugs can homogeneously disperse. Coating cubosomes prevents the initial burst drug release, followed by a steady release [41].

3.7.1 Applications

- Cubosomes containing paclitaxel tagged with biotin, a targeted ligand, show higher anticancer activity and lower drug toxicity as compared to non-targeted therapy [42].
- Cubosomes containing hydrophobic drug AT101 showed an increase in solubility and hence bioavailability of the drug [42].
- Cubosomes containing cisplatin and metformin have higher anticancer activity than the free drug [42].
- Cubosomes containing cisplatin and paclitaxel are coated with a poly-ε-lysine layer to avoid an initial burst of anticancer medications followed by a steady release [42].
- Cubosomes containing piperine and quercetin exhibited enhanced activity on the HepG2 hepatocellular carcinoma cell line when compared to the standard, even at lower concentrations [43].
- Cubosomes of copper acetylacetonate with coupling of hyaluronic acid (HA) on the surface led to higher expression of CD44 receptors in cancer cells. HA tag improved the targeting of tumour cells [44].
- Cubosomes used for vaccine antigens as nanovehicles [41].

3.8 Lipid–Drug Conjugates

Lipid–drug conjugates (LDCs) are hybrid nanocarriers where the drug is covalently linked to a lipid molecule. Different types of lipids are used for conjugation with drugs, including fatty acids, fatty acid glycerides, phospholipid–drug conjugates, steroids, and fat-soluble endogenous molecules [22]. The anticancer drugs are conjugated to amphiphilic lipids, facilitating prolonged drug release and promoting the creation of self-assembled nanoparticles that may be precisely targeted to the tumour location [33]. The expression of low-density lipoprotein (LDL) receptors heightens in cancer cells and requires substantial quantities of cholesterol for their accelerated proliferation. Hence, cholesterol–drug conjugates enhance the anticancer drug incorporation into lipoproteins [21]. Tocopherol has also been used with anticancer drugs [22].

3.8.1 Applications

- Cholesterol conjugated 5-fluorouracil showed increased anticancer activity as compared to the unconjugated form [21].

- DHA-10-hydroxycamptothecin, a DHA drug conjugate, showed higher anticancer activity in human colorectal and lung cancer than the standard drug cisplatin [21].
- N-heptadecanoyl hydrazine conjugated doxorubicin exhibited better anticancer activity than the free drug [21].
- Squalene conjugated gemcitabine nanoassemblies showed a 3-fold increase in activity in drug-resistant murine and human leukemic cells [21].
- α-tocopherol conjugated siRNA resulted in enhanced cellular uptake, decreased degradation, and enhanced gene silencing activity [21] (Table 7.1).

Table 7.1 Advantages of different types of lipid-based nanosystems

S. No.	Lipid-based nanosystems	Advantages	References
1	Solid lipid nanoparticles	Biocompatible Less toxicity Higher stability Easy sterilization and scale-up process Delivery of hydrophobic and hydrophilic drugs	[19]
2	Nanostructured lipid Carriers	Biocompatible Less toxicity High physical stability Delivery of hydrophobic and hydrophilic drugs	[25]
3	Lipid nanoemulsions	Biocompatible Long-term stability Easy fabrication High drug loading Delivery of hydrophobic drugs	[26, 40]
4	Self-nanoemulsifying drug delivery systems	Protection against GI enzymes Reduces the first-pass metabolism Easy scale up Delivery of hydrophobic and hydrophilic drugs	[31, 32]
5	Liposomes	Biocompatible Biodegradable More target specificity Delivery of hydrophobic, hydrophilic, and amphiphilic drugs	[35, 36]
6	Ethosomes	Non-invasive drug delivery through skin Cost-effective Easy to apply Fewer side effects Delivery of hydrophobic, hydrophilic, and high molecular weight drugs	[37]
7	Cubosomes	Biocompatible Stable encapsulation Fewer side effects More target specificity High drug loading of hydrophobic or hydrophilic drugs	[40, 42, 44]
8	Lipid–drug conjugates	Improved oral bioavailability More target specificity Fewer side effects Potential to overcome drug resistance	[21]

These lipid-based nanosystems offer innovative solutions for cancer therapy, improving drug solubility, stability, bioavailability, and targeting. Their versatility and efficacy make them promising candidates for overcoming the challenges of traditional cancer treatments.

4 Overview of Stimuli-Sensitive LNPs for Cancer

Stimuli-sensitive or responsive LNPs are nanoformulations which when exposed to exogenous or endogenous triggers, undergo structural or conformational changes in size, shape, and surface characteristics, resulting in controlled release of encapsulated cargos. Essentially, structural modification of these nanoparticles due to the stimuli influences the guided delivery of loaded drugs in cancer tissues with enhanced efficiency, control and precision [45]. Endogenous stimuli are innate triggers generated as a result of extant conditions in the microenvironment of normal and cancer cells, such as enzyme activity, pH, and redox potential. Since the stimuli do not require the use of external equipment for their generation, LNPs which are responsive to endogenous stimuli can be easily applied in cancer therapy in clinical settings [46]. Conversely, exogenous stimuli are triggers including ultrasound, heat, light, and magnetic field, which are produced and transmitted externally in a short time following standard protocols for pulsatile drug release, tissue accumulation, and uptake. Due to the ease of manipulation of exogenous stimuli, the nanoparticles have the ability to permeate very deep into cancer cells for improved and regulated drug release [47, 48]. Stimuli-responsive LNPs are stable in physiological conditions but their stability is altered when they are exposed to changes in the physiological environment and this is prevalent in cancer. The microenvironment of cancer cells comprises factors which promote their growth and metastasis, such as hypoxia, increased amount of free radicals, high acidity, and enzymes over-secretion, giving rise to malignant and irregularly shaped structures which lack endothelial cells [49].

Owing to its considerable influence in lipid nanoformulations, many workers have developed stimuli-responsive LNPs for use in cancer therapy. In a report, cholesterol and phosphatidylethanolamine-based liposomes encapsulating magnetic cations were developed to be sensitive to magnetic field, an exogenous stimulus. Its use in gene therapy of human osteosarcoma cells recorded promising performance efficiency because in less than 1 h of magnetic induction, luciferase gene activity achieved in the cells was maximum and this effect was faster than that obtained without magnetic field [50]. Similarly, pH-responsive cationic lipid that undergoes self-assembly into multifunctional envelope-type nanodevice was synthesized for the encapsulation of anti-polo-like kinase 1 siRNA for cancer therapy. It was recorded that the cationic lipid nanodevice system produced efficient and reliable gene silencing compared to the nanodevice carrier system bearing cationic lipids [51]. These few examples highlight the propensity of stimuli-sensitive nanosystems to produce increased efficiency and efficacy by facilitating precise delivery, controlled release, and cellular localization of drug molecules. Though there are challenges encountered in clinical management of cancer, it is reassuring that stimuli-responsive nanoparticles have elicited immense interest and several studies have been commissioned in this regard. Moreover, the development of stimuli-sensitive LNPs has improved the search for better therapies for cancer affecting cardiovascular, immune, and genetic functions.

5 Classification of Stimuli-Sensitive LNPs for Cancer Treatment

LNPs applied in cancer therapy could be structured in such a way as to respond to various stimuli. LNPs are tuned to respond to exogenous and endogenous factors in order to deliver a better outcome in cancer treatment.

5.1 Exogenous Stimuli-Based Lipid Systems

Exogenous stimuli-based LNPs rely on exogenous factors which trigger the release and delivery of cargo loaded in the nanoparticles. Some Factors to be treated in this section include light, temperature, ultrasound, electric field and magnetic field.

5.1.1 Light-Triggered LNPs

The treatment of diseases including cancer, via the application of light has significantly improved over the years. Many diseases can be prevented and cured with the direct application of sunlight and artificial light sources [52]. Some of the light sources used in controlled drug delivery systems include infrared (IR), visible light and ultraviolet (UV). In drug delivery systems, some of the nanomaterials can be controlled by their structures and functions by using light intensity, wavelength and exposure time [53]. There are four types of light response mechanisms. They include: photo-cleavage group, photo-isomerization group, photo-cross-linking and photo-induced groups [54]. The photo-cleavage group contains a photo-fragmentation group that allows the reaction of the cleavage via chemical bond to occur under the IR irradiation and the UV light. When this is achieved, there will be a change in the delivery system that will cause the release of the drug [55]. Photo-isomerization groups consist of azobenzenes and spiropyran. Azobenzenes can easily be transformed from the cis- to the trans-form and vice versa under certain conditions. Under UV light, the cis- form predominates over the trans- conformation, while under the visible light, the trans- form predominates over cis- conformation [2]. The conformational change in the cis-trans form can cause a great change in the nanocarriers internal structure. In the light stimulation, the release of the drug is based on the swelling and disintegration of the nanoparticles [56]. Although different stimuli are employed in controlled drug release, a light stimulus is unique in the sense that high precision could easily trigger the drug release [48]. Light-triggered exogenous stimuli are normally used in photodynamic therapy (PDT) that require the utilization of three major parameters which include light, tissue oxygen and photoactivatable photosensitizer [57]. The photosensitizer principle of operation works on the ability of activation of light-mediated response which converts the

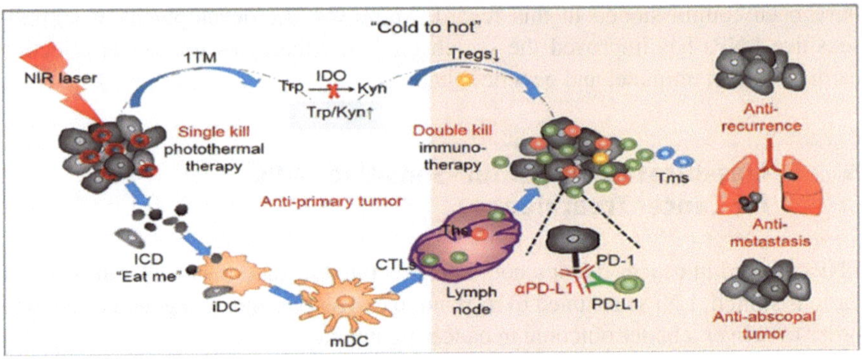

Fig. 7.1 Photo-mediated stimuli-responsive nanoparticles [62]. CC-BY license

tissue oxygen to radical oxygen species (ROS) causing cellular necrosis, thus making the photodynamic therapy to be an ideal candidate for cancer therapy [58]. Although photo-reactions can easily be induced by both visible light/ultraviolet and near-infrared, their applications differ based on their wavelengths. UV/visible light is used for topical treatments, while those treatments that require deeper light penetration go for the near-infrared due to their higher wavelengths between 650 and 900 nm [59]. According to Yang et al., the photodynamic therapy has shown to have destroyed tumour cells by generating oxygen that was produced through a photodynamic reaction at a specified wavelength [60]. There are two major strategies that can be used to alleviate tumour hypoxic microenvironment. The first could be by delivery of oxygen to the cancer site, while the second strategy could be by generation of the oxygen under cancer hypoxia [61]. This is shown in Fig. 7.1.

5.1.2 Temperature-Responsive Lipid Nanoparticles

Tissues of tumour cells are usually associated with higher temperatures (40–42 °C), when compared to normal tissues (37 °C) [63]. Although there are many exogenous stimulation factors used in cancer treatment, temperature medication delivery system has been widely used. Thermo-sensitive materials have two major properties that are put into consideration in thermally responsive nanosystems. They include the lower critical solution temperature (LCST) and the upper critical solution temperature (UCST) [64]. Thermally responsive polymers respond to changes based on the hydrophilic or hydrophobic nature at different temperatures. At an LCST, the polymer becomes insoluble while at a UCST, the polymer experiences a change in phase from hydrophilic to hydrophobic and causes drug release [65]. Materials used in cancer therapy include the polymeric nanogels, organic micelles, inorganic materials, liposomes and nanoparticles. The temperature-responsive polymer that is commonly used in cancer therapy is the poly N-isopropyl acrylamide (PNIPAM). One unique advantage associated with this polymer is its low LCST in aqueous solution at 32 °C. Above the LCST, the polymer becomes hydrophobic in nature [66]. Another polymer that is also used is the poly (N-alkylacrylamide).

In oncology, the thermo-responsive drug delivery has been in use over the years [38]. Since the tumour environment has different variation in temperature, drug release on this type of stimuli-response system is based on that. The drug is retained by the thermo-responsive carriers at the normal body temperature of 37 °C, while at high temperatures of the tumour environment between 40–42 °C, the drug is released [67]. In hyperthermia therapy which involves the microwave, ultrasound and radiofrequency, the body tissue is subsequently exposed to high temperature. This causes the cells to be killed [68]. One of the limitations of the hyperthermia therapy is its toxic effect on both the cancer and normal cells [69]. With this limitation, a thermo-responsive drug delivery system that is only toxic to the cancer cells and does not harm the normal cells needs to be developed [70]. The thermo-responsive liposomes work on their ability to minimize the metabolism, drug uptake and clearance which makes the drug to be highly concentrated at the heated tumour site [71]. Examples of thermo-responsive liposomes that have been used in hyperthermia treatment include: ThermoDox®, Doxil® and Myocet® [72]. Their activity is based on their changes in aqueous solubility. At low temperature (below the LCST), they dissolve in water, while at high temperature (above the LCST), they become insoluble, precipitates and causes the drug release via the delivery system [73, 74]. Other advantages associated with the temperature-responsive delivery system are its flexibility in design, easy regulation of the phase transition temperature and passive targeting compatibility [75]. Shen at al developed a temperature-responsive drug carrier using Fe_3O_4/PNIPAM/5-FU $mSiO_2$-chi/R6G [76]. A drug carrier that deployed the use of different temperatures in vitro was used for the study. Demirci et al. also developed a nanoparticle of iron oxide using dextran as a coating agent to destroy the retinoblastoma cells [77].

5.1.3 Electric-Field-Sensitive Lipid Nanoparticles

Drug treatment could be improved with the use of a non-stimulating factor called electric field. In the clinical aspect, electric field was used in the treatment of cancer. An example of electric field application is in the treatment of glioblastoma which was approved by the United States Food and Drug Administration (FDA) [78]. The electric field is one of the stimulatory methods used in cancer treatment. The permeability of cellular membranes can be affected by high intensity of the electric field which makes it useful as a stimulant in the administration of drugs [2]. Application of an electric potential causes the triggering of both the conductive polymer and implantable device. The drug release occurs by altering changes in the physical properties of the polymer. Some of the polymers commonly used in electric field stimulation include the polypyrrole (PPy), polyaniline and poly-(3,4-ethylene dioxythiophene) (PEDDT) [79, 80]. They have unique advantages such as biocompatibility and the conversion of electrical to mechanical energy [81]. Advantages associated with the electric field include: the ease of transportation and it does not require complicated equipment. When the electric signal is combined with a sensor or microchip system, it becomes easier to achieve a positive feedback from outside the body [82]. The response system of the electric field is made of a pair of electrodes and a polymer reservoir device. A change in the magnitude of the electric field, affects drug release [83].

5.1.4 Ultrasound-Sensitive Lipid Nanoparticles

The use of ultrasound-sensitive system has been recognized clinically. Advantages associated with the ultrasound system is their ability to focus on a target location, safety, and non-invasiveness and deep tissue penetration [2, 84]. The ultrasound system has some biological effects which include the cavitation, mechanical, thixotropic, thermal and acoustic effects. The release of carrier drugs is usually influenced by the effect of thermal, mechanical and radiation force which enhances tissue permeability [56]. The ultrasonic smart nanoparticles responsiveness is dependent on both the thermal and mechanical effects. The ultrasonic system can also be deployed in imaging diagnosis and in the promotion of image-guided drug delivery [85]. Another advantage associated with the use of ultrasound system is its dual response to both thermal and ultrasound. This can be harnessed in order to produce a nanosystem that could combine dual properties [86]. Other medical applications that have utilized the use of ultrasound include: the blood circulation analysis, tumour, lipectomy, kidney stone disruption and dentistry. In the past, vesicles were reduced in size from micro to nanoscale using low-frequency ultrasound, but currently, it has been used to target controlled drug release [87]. There are some physical effects that are usually produced whenever the ultrasound wave permeates the body and causes drug release. Some of the effects are made up of the acoustic fluid streaming, cavitation, hyperthermia and simple pressure variation [88, 89]. Among all the physical effects, the most pronounced of them is the cavitation. The transmission of an ultrasonic wave via the liquid medium causes the formation of a large number of microbubbles. Sonoporation is a process that enables the permeability of cell membrane to change in an acoustic cavitation [90, 91]. Many drug delivery systems such as emulsions, micelles, liposomes and nanobubbles can incorporate an ultrasound system in their design [92]. Both large and small macromolecules can be incorporated in ultrasound system. The energy vibration of the acoustic cavitation enables the ultrasound to produce hyperthermia [93]. The cavitation is what enables the acoustic wave to be applied as a release mechanism, which then causes the anticancer drugs accumulation at the site of tumour [94]. In 2014, Lu et al. reported the use of polymer-modified thermo-responsive liposomes for the treatment of cancer [48]. One of the most important factors on the use of ultrasound in local stimulus is its pressure waves that occur at frequencies of 20 KHz or greater [95]. Chen et al. developed a DOX-loaded titanium nanoparticles surface that was coated with poly(methyl vinyl ether-alt-maleic anhydride) PPBA via the generation of the ester bond which lead to the formation of pPBA@TNP-DOX nanoparticle. The PBA and sialylated epitope synergy, enhanced the tumour-targeting on the cancer cells [96].

In 2019, Li et al. was able to manufacture an injectable ultrasound-responsive PLGA nanoparticle aggregates (NPAs). The ultrasound that possessed low energy was irradiated which subsequently caused the NPAs loaded with DOX to be broken into PLGA nanoparticles. They entered the parenchyma of tumour by enhanced permeability and retention effect. This caused the drugs to be accumulated and released into the tumour site [97]. Ultrasound-responsive nanocarriers could be used as ultrasound sensitizers in the treatment of certain diseases [4, 98].

5.1.5 Magnetic Field Response Lipid Nanoparticles

The use of magnetic resonance imaging has been in use both in past and present in cancer treatment Advantages associated with the magnetic imaging include: biocompatibility, biodegradability, simplicity and ease of functionalization [99]. In tumour tissues, the magnetic nanoparticles play a major role as a transducer. They convert hysteresis loss into heat upon the application procedures associated with the magnetic response system. The first procedure involves the use of magnetic fields in drug targeting, while the second procedure involves the use of magnetic field to cause hyperthermia [100–102]. In magnetic-responsive drug delivery system, paramagnetic and superparamagnetic are usually assembled in polymer scaffolds. This helps to enhance the accumulation of anti-cancer drugs in tumours [103]. The superparamagnetic iron oxide nanoparticles have been used in combination with magnetite (Fe_3O_4) as a magnetic core [104]. The iron oxide nanoparticles possess outstanding qualities that distinguish it among others. They have good magnetic properties, biodegradability biocompatibility and low toxicity. The magnetic-responsive drug delivery system works on the principle on the ability of a drug to be entrapped in magnetic nanoparticles. The prepared drug/carrier complex is subsequently injected either intravenously (IV) or intra-arterial (IA) into the subject. The complex formed is then directed at the exact locations with the help of a high-magnitude magnetic field that is applied externally [104]. Magneto-liposomes are useful in the release of the liposome-based cargo under magnetic stimulus [105]. There are some properties associated with an ideal magnetic-responsive drug delivery system. They include: encapsulation of a wide variety of active pharmaceutical ingredients, adequate magnetic responsiveness, high biodegradability and biocompatibility, reduced antigenicity and toxicity effect [82].

5.2 Endogenous Stimuli-Based Lipid Systems

Drug release and transport from LNPs could also be activated by internal or physiological factors such as enzyme action, pH of the microenvironment, effect of glucose concentration, redox effect, and interaction with receptors, as presented in this section.

5.2.1 Enzyme-Responsive LNPs

In order to keep the human body functioning normally, enzymes are widely distributed throughout the tissues and organs. Since tumour cells develop more quickly than other normal organs and need more enzymes for functional support, the tumour microenvironment shows abnormal expression of enzymes such matrix phospholipases, cathepsins, metalloproteinases, and oxidoreductases. Researchers have been interested in the up- or down-regulation of enzymes. Another successful method for nanoparticle intelligence that could release the payload into the intended target is the creation of enzyme-responsive nanoparticles for targeted therapy of the tumour microenvironment [106]. Enzyme-responsive lipid nanoparticles (LNPs) demonstrate encouraging advancement in cancer treatment, leveraging the distinctive biochemical environment of tumours to improve drug delivery and efficacy. These nanoparticles are designed to react to specific enzymes that are highly expressed in tumour tissues, enabling targeted release of therapeutic agents. Enzymes are crucial in nanomedicine because of their outstanding catalytic activity and ability to recognize specific biological targets. Their highly efficient and selective reactions, which occur in various physiological, pathological, and metabolic pathways, make them ideal for use as responsive elements in the design of enzyme-sensitive stimulus-responsive nanoparticles (en-srNPs). Accordingly, enzyme-triggered en-srNPs for controlled drug delivery are classified based on the type of effector biomolecules involved, such as hydrolases, lipases, oxidoreductases, proteases, and glycosidases [107, 108].

This technology aims to reduce the systemic toxicity typically associated with conventional chemotherapy by ensuring that the drug is offered primarily to the cancerous cells. The mechanism behind enzyme-responsive lipid nanoparticles involves the use of substances that deteriorate or go through structural changes in the existence of specific enzymes. Upon entering the tumour microenvironment, these nanoparticles react to the localized enzyme activity (trigger the degradation of the core), resulting in the controlled release of the cargo (chemical drugs or therapeutic active agents) stored within the nanoparticles. This technique facilitates better accumulation of the drug at the tumour site while minimizing contact with healthy tissues, thus improving therapeutic effectiveness [109, 110]. The use of cleavable linkers within the drug and the nanoparticle matrix facilitates the drug's release once the linker is degraded by specific enzymes [106, 107]. The nanoparticles can be modified to include hydrophilic components that can detach upon reaching the tumour site, enabling the therapeutic drug encapsulated within to be delivered effectively. LNPs exploit the enhanced permeability and retention (EPR) effect, where leaky vasculature in tumours facilitates the aggregation of nanoparticles within the tumour interstitial space. This effect not only aids in passive targeting but also enhances the overall therapeutic concentration at the tumour site, further improving treatment outcomes [111].

An exemplary case of enzyme-responsive lipid nanoparticles utilized for cancer therapy involves the development of nanoparticles specifically designed to deliver the anticancer drug cisplatin within the framework of triple-negative breast cancer. In this innovative approach, researchers formulated nanoparticles utilizing peptide sequences sensitive to matrix metalloproteinase-2 (MMP-2), an enzyme that is often highly expressed in tumour microenvironments, especially in metastatic cancers. These nanoparticles were engineered to encapsulate cisplatin along with additional therapeutic agents, providing a dual-action approach against tumours. Upon administration, the nanoparticles circulate through the bloodstream before they reach the cancerous tissue, where MMP-2 facilitates the degradation of the peptides that form part of the nanoparticle structure. This enzymatic action triggers the release of cisplatin directly within the tumour site, enhancing the drug's local concentration while limiting systemic absorption and unwanted reactions. As a result, this targeted delivery not only improves the efficacy of cisplatin in killing cancer cells but also promotes better infiltration into the tumour masses. The success of this strategy highlights the efficacy of enzyme-responsive lipid nanoparticles to substantially enhance cancer treatment outputs by harnessing the unique biochemical environment of tumours for more effective drug delivery [106, 107, 110].

Despite the promising potential of enzyme-responsive lipid nanoparticles, challenges remain in their development. Issues such as stability in circulation (bloodstream), the variability of enzyme expression among different tumours and among patients and even among different stages of the same disease, exact regulation of the drug release kinetics, and the complexity of nanoparticle interactions with biological systems need to be resolved. Subsequent studies should concentrate on advancing the design of these nanoparticles to improve their performance, as well as conducting clinical trials to establish their efficacy and safety in cancer therapy. Developing scalable manufacturing processes for these sophisticated delivery systems remains a significant hurdle for widespread clinical adoption [49, 112].

5.2.2 pH-Sensitive LNPs

pH-sensitive lipid nanoparticles (LNPs) are an innovative approach in cancer therapy, designed to enhance drug delivery specifically to tumour sites. These nanoparticles exploit the unique acidic microenvironment of tumours, which is typically lower in pH compared to healthy tissues. This characteristic allows for targeted delivery of therapeutic compounds, reducing overall adverse effects and enhancing therapeutic effectiveness [113]. Cancer cells often exhibit a lower extracellular pH due to increased glycolysis (known as the Warburg effect), contributing to the accumulation of lactic acid and where cancer cells prefer anaerobic metabolism. This acidic environment can enhance pH-triggered drug delivery from nanoparticles when they reach the tumour site. This attribute allows pH-sensitive NPs to liberate the therapeutic agent in an acidic environment present within the tumour microenvironment, leading to improved therapeutic outcome with minimal systemic toxicity [114, 115]. The drug release from pH-sensitive lipid nanoparticles can occur through various mechanisms that are triggered by pH changes. At acidic pH levels, certain groups within the nanoparticles become protonated, altering their hydrophilic-hydrophobic balance. This destabilization leads to the disintegration of the nanoparticle architecture and followed drug release. Many pH-sensitive lipids are designed with acid-labile bonds, whichever fragmented under acidic conditions of tumours. This bond cleavage enhances drug liberation upon reaching the target site. Dioleoylphosphatidylethanolamine (DOPE) is a widely used lipid in creating pH-responsive nanoparticles, it transitions from a stable structure at physiological pH to a destabilized form in more acidic conditions, facilitating drug release [113, 116].

The primary advantage of pH-sensitive LNPs is potential to precisely deliver chemotherapeutics to tumour cells with minimal impact on healthy tissues. This specificity is crucial for reducing undesirable consequences of traditional chemotherapy. Recent research indicates that integrating pH-responsive nanoparticles with immunotherapy can enhance treatment outcomes by improving immune cell infiltration into tumours and reducing immunosuppressive factors [113, 117]. Studies have shown that these nanoparticles can significantly suppress tumour growth in vivo, demonstrating their suitability for targeted delivery systems for various anticancer drugs [118]. For instance, glycerin monostearate (GMS) and adipic acid dihydrazide (HZ) were used as lipid components to create an RGD-conjugated, pH-sensitive lipid known as RGD-HZ-GMS. To create an RGD-modified, DOX-loaded SLN (RGD-DOX-SLNs), DOX was encapsulated using RGD-HZ-GMS. The anti-cancer effect of RGD-DOX-SLNs was assessed using the MCF-7 breast cancer cell line and the MCF-7/ADR DOX-resistant cell line. Mice with the breast cancer model MCF-7/ADR cells were used to assess the cytotoxic effects and in vivo tumour suspension. The shape of RGD-DOX-SLNs was consistently spherical. The RGD-DOX-SLNs had a mean particle size of 96.3 nm and a zeta potential of 35.6 mV. Because the acid environment of tumour tissues differs from that of normal tissues, the pH-dependent release behaviour can decrease the drug release in normal tissues and increase the drug release at tumour sites, making it perfect for targeted tumour treatment [119, 120]. In pH 5.0, the release of DOX from RGD-DOX-SLNs exhibits a biphasic profile, with a fast initial burst and a slower release over the course of the study's 96 h. As a result, RGD-DOX-SLNs displayed an area under the plasma concentration–time curve (AUC) that was 5.58 times greater than that of DOX solution. For solid lipid nanoparticles, this two-phase release characteristic has been documented. While medications enclosed in lipid core continue to release continuously, drugs connected to the surface of shells and particles are linked to a burst release. Meanwhile, under pH 7.4 and pH 9.0, RGD-DOX-SLNs revealed no change and somewhat reduced release patterns. RGD-DOX-SLNs may be a promising new lipid carrier that could enhance breast cancer treatment, according to in vitro and in vivo antitumour outcomes [119–121].

Another study used a pH-responsive polycarboxybetain (PGlu(DET-Car))–LNPs to deliver small interfering RNA (siRNA) to malignant cells. The LNPs include an encapsulated siRNA. As a clever shell material for LNPs, the PGlu(DET-Car) shows switchable net charge in response to the ambient pH. In other words, the PGlu (DET-Car) had a cationic characteristic in acidic tumour microenvironments, which facilitated the transport of siRNA to cancer cells, but at physiological pH 7.4, it presented a neutral charge, making it stealthy during circulation. In malignant pH, PGlu(DET-Car)-modified LNPs may increase cellular uptake, which would promote endosomal escape and increase the effectiveness of gene knockdown. Following systemic injection, the PGlu(DET-Car)-modified LNPs demonstrated considerable tumour growth inhibition and achieved high tumour accumulation in the subcutaneous tumour models SKOV3-luc and CT 26 [122].

5.2.3 Glucose-Triggered LNPs

Cancer continues to be one of the primary causes of death globally, driving ongoing investigations into more efficient and targeted therapies. LNPs have established themselves as a promising Modality for cancer therapy, particularly when combined with glucose-triggered mechanisms. These nanoparticles exploit the unique metabolic characteristics of cancer cells, which often exhibit heightened glucose uptake due to altered metabolic pathways [111]. Cancer cells show modified metabolic activity compared to healthy cells, a phenomenon referred to as the Warburg effect, marked by elevated glucose consumption and enhanced glycolysis despite the availability of oxygen. This alteration in cellular metabolism provides a unique opportunity to design drug delivery platform that triggered to the elevated glucose levels inside the tumour microenvironment (TME). Glucose-triggered LNPs leverage this differential to achieve targeted drug release, enhancing specificity and reducing damage to surrounding normal tissues and also, it is innovative approach designed to enhance the precision and efficacy of cancer treatment [111, 123]. The mechanism of glucose-triggered lipid nanoparticles involves the alteration of glucose metabolic pathways within tumour cells. Cancer cells primarily exhibit the Warburg effect, characterized by a strong dependence on glycolysis for energy generation despite the availability of oxygen. GLNs are engineered to respond to elevated glucose levels within tumours, triggering drug delivery specifically at the location of malignancy. This targeted delivery not only improves therapeutic outcomes by accumulating the drug at the tumour place as well as minimizes systemic adverse effect associated with conventional therapies [124, 125].

This system ensures effective drug delivery predominantly within the tumour site, where glucose concentrations are significantly higher than in normal tissues. For example, glucose oxidase facilitates the conversion of glucose into hydrogen peroxide and gluconic acid through an oxidation reaction resulting in localized pH changes that destabilize the lipid bilayer and release the drug. LNPs utilize the EPR effect, which allows them to preferentially build up in tumour tissues as a result of the leaky blood vessels and poor lymphatic drainage typically seen in tumours. This selective accumulation enhances drug delivery as well as minimizing systemic adverse effects [111, 123–125]. Glucose-triggered lipid nanoparticles (LNPs) exemplify an innovative approach in cancer therapy, particularly through the use of glucose oxidase (GOx) to induce a starvation effect on tumours. One notable representative example is the design of dual-functional nanoparticle system that combines GOx with core–shell structured magnetic silica nanocarriers containing iron oxide for site-specific delivery in pancreatic cancer therapy. In this system, GOx accelerates the oxidative reaction of glucose, yielding hydrogen peroxide (H_2O_2) and gluconic acid and effectively depleting glucose levels in the tumour microenvironment and increasing oxidative stress [125, 126]. This depletion not only starves cancer cells of their primary energy source but also enhances the acidity of the tumour environment, facilitating the release of encapsulated chemotherapeutic agents, such as doxorubicin (DOX). The pH-sensitive properties of the nanoparticles ensure that drug release is optimized under acidic conditions typical of tumours. In vitro studies have shown that this combination significantly enhances cytotoxicity against cancer cells, while in vivo experiments demonstrated a marked inhibition of tumour growth in animal models. This dual-action strategy that encompasses more than improves the therapeutic effectiveness of DOX as well as utilizes the metabolic vulnerabilities of tumour cells, showcasing the efficacy of glucose-triggered LNPs as a valuable asset in modern cancer treatment [127, 128].

5.2.4 Redox-Sensitive LNPs

Redox-sensitive LNPs respond to the difference in redox potential of normal cells and malignant cells. The reduced glutathione (GSH) and glutathione disulfide

(GSSG) pair is one of the most prevalent redox pair. The GSH levels are elevated to four times in malignant cells that leads to cleavage of disulfide and ditelluride bonds causing drug release by carrier breakdown. The redox-sensitive LNPs are advantageous for the intracellular administration of medications [129]. Redox-sensitive LNPs were utilized to encapsulate mRNA and transfer it to the cytoplasm that successfully limit the growth of p53-null hepatocellular carcinoma and non-small cell lung cancer (NSCLC) cells through cell cycle arrest and triggering apoptosis in vitro and in vivo. A hybrid system of LNPs loaded in hydrogel was designed for the co-delivery of doxorubicin and cytochrome and disruption of the hydrogel structure by GSH resulted in the sequential release of the encapsulated molecules [129].

Doxorubicin conjugated with sodium alginate (SA) by disulfide linker was electrostatically combined with aminated superparamagnetic iron oxide nanoparticles for nanogel preparation. In vitro cytotoxicity indicated selective killing the human cervical cancer cells (HeLa) and considerably enhanced cytotoxicity of the African green monkey Sv40-transformed kidney fibroblast cells (COS-7). Thus, the nanogel expressed high targeted release but little systemic toxicity [130]. Higher concentration of reducing agent in malignant cells can be leveraged in the design of redox-sensitive LNPs with great specificity and efficiency. However, the intrinsic variability of malignant cells presents issues with the basic mechanism of redox-sensitive systems.

5.2.5 Receptor-Sensitive LNPs

Receptor-sensitive LNPs interact with specific cell surface receptors overexpressed in malignant cells. These LNPs are modified with ligands, antibodies, or peptides that bind to the receptors, permitting targeted cellular absorption. In order to achieve receptor-mediated uptake, unspecific interactions of the carrier with blood components have to be reduced by shielding. The ligand-receptor binding leads to receptor-mediated endocytosis of LNPs that is a highly selective process of LNP internalization, facilitating localized drug administration and decreasing systemic adverse effects [131]. Different sizes of LNPs can be taken up by different routes of internalization. The clathrin-mediated endocytosis of transferrin receptor, asialoglycoprotein receptor, or low-density lipoprotein receptor into early endosomes, followed by endosome maturation and fusion with lysosomes. The maximum size of 200 nm can be taken up by this route. The folate receptor or interleukin-2 receptor is internalized by caveolae-mediated endocytosis. The maximum size of 500 nm can be taken up by this route [131].

LNPs used for delivery of mRNA tagged with tyrosinase-related protein 2 showed enhanced overall survival of melanoma model mice by increasing anticancer activity [132]. Thus, endogenous stimuli-sensitive LNPs can be utilized for development of precision medicine for increased safety and efficacy. In addition, the novel concept of dual active and passive tumour targeting by dual cancer and gene therapy via multiple stimuli-responsive prodrugs could be viewed as a potential prospect for the management of malignancies.

5.3 Multi-Stimuli-Triggered LNPs

Multi-stimuli-triggered LNPs (MST LNPs) are developed to respond to several exogenous or endogenous stimuli, including as pH, temperature, redox potential, enzymes, light, or magnetic fields. Integrating multiple stimuli responses can compensate issues with single stimulus-responsive system, leading to more control over drug release, better absorption, enhanced therapeutic output, and decreasing side effects. However, problems like scale-up, stability, and regulatory hurdles must be resolved. There are different types of dual responsive systems like light/pH-responsive, pH/temperature-responsive, pH/redox-responsive, and light/temperature [129].

The LNPs responsive to endogenous stimuli have limitations like low flexibility and LNPs responsive to exogenous stimuli also have limitations like required power density and wavelength of light, or the increased cost of magnetic-responsive devices. Although the adoption of MST LNPs can overcome some of these constraints, the creation of more effective in vivo stimuli-sensitive platforms is required. The lignin-based poly(ionic liquid) hydrogel displayed stretchability to 1425%, toughness of 132 kPa, and cyclic stability due to the coordination of soft and hard segments, dynamic reversible coordination, hydrogen bonding, and electrostatic contact. Additionally, its high electrical conductivity permitted accurate monitoring of numerous stimuli like strain, pressure, and temperature. The polymer LNPs hydrogel acted as a flexible sensor for motion and pH monitoring [133]. Furthermore, the enhanced understanding of the cancer microenvironment's characteristics should be employed to develop safe, effective, and cost-efficient chemotherapy via advanced multiple stimuli-responsive nanomedicines and drug delivery systems [134].

6 Benefits and Limitations of Stimuli-Responsive LNPs for Cancer

Over the years, several researchers have documented evidence depicting the benefits and limitations of stimuli-sensitive LNPs. The benefits and limitations of these trigger-mediated nanocarriers would be presented in this section.

6.1 Benefits of Stimuli-Responsive LNPs for Cancer Management

(a) Stimuli-sensitive LNPs deliver encapsulated cargos to the intracellular compartments of cancer cells in a controlled, specific, and precise manner. This promotes improved efficacy, efficiency, and minimum toxicity of therapy.
(b) Depending on the desired treatment outcome, there is a high potential for controlling the influence of a stimulus on LNPs especially where LNPs are tuned

using external stimuli. An external stimulus could be incorporated or removed to suit the goal of therapy.

(c) In order to arrest cancer proliferation through multiple mechanisms, it is possible to design multifunction LNPs which are sensitive to multiple stimuli for maximum treatment outcome. This template could also be applied in cancer metastasis [4].

(d) Stimuli-responsive LNPs could be formulated and used for timely and non-invasive diagnosis and therapy (theranostics) of cancer. In this case, the LNPs would have the potential to diagnose cancers and initiate treatment to facilitate early remission of proliferating cells [135].

(e) LNPs which are responsive to ultrasound or magnetic field could initiate localization of cancers by directing the nanoparticles to the cancer site or disrupting the endothelial cells of the cancer with increased extravasation into the cancer microenvironment [136].

(f) They improve the outcomes of anti-cancer drug administration by promoting the simultaneous administration of both lipophilic and hydrophilic therapeutic agents like antibodies, proteins, small drug molecules, and nucleic acid [137].

(g) They leverage the biocompatibility property of the lipid moiety and the stimuli to produce increased permeation of the cancer vasculature resulting in accumulation and retention of therapeutic molecules in the microenvironment [138, 139].

6.2 Limitations of Stimuli-Responsive LNPs for Cancer Management

(a) Though LNPs could be functionalized to respond to exogenous or endogenous stimuli for enhanced anti-cancer efficacy, their performance and efficiency could be impeded by the destructive phagocytosing and clearance activities of the reticuloendothelial system [140]

(b) Very large particle size of the stimuli-responsive LNPs limits nanoparticles permeation of the cancer microvasculature thereby decreasing the efficiency of the therapy [141].

(c) In case of hypoxia-responsive LNPs, drug penetration, deposition, and accumulation in cancer tissues could be impaired due to poor modulation of the microenvironment of the hypoxic cancer and asphyxiating levels of oxygen [4].

(d) There could be hypoperfusion of anti-cancer agents in cancer tissues due to the cancer microvasculature, presence of fibroblasts, and the barrier functions of endothelial cells [142].

(e) Due to limited penetration of bones or tissues linked to bones by ultrasound, it might be difficult to achieve effective bone cancer remission using ultrasound-responsive LNPs, even at prolonged time and increased frequency of exposure [92, 143].

7 Safety Considerations for Clinical Translation of Stimuli-Responsive Lipid Nanoparticles

Stimuli-responsive LNPs encounter several physiological barriers when administered into a biological tissue which must be overcome to facilitate delivery of the carrier system to the desired location for the release of its payload. These obstacles including vascular endothelium, membrane pumps, immunological clearance, lysozyme, enzymatic degradation, the extracellular matrix, and renal clearance, might limit drug accumulation in cancer tissues leading to decreased efficacy [144]. Considering that stimuli-responsive LNPs used to combat cancer are in constant interaction with critical body tissues, it is imperative to ascertain their safety, suitability, and potential toxicity profiles before administration. Though it might not be necessary to compile entirely different test guidelines for stimuli-responsive LNPs compared to other micro- and/or nano-carrier systems; however, assessment of their safety which will facilitate clinical translation should take into account their beneficial interactions with systemic factors which boost their anti-cancer effects. Doxorubicin (Doxil®) liposomes, which were originally approved for use by AIDS patients, have been applied in multiple myeloma therapy due to its low cardiotoxicity compared to the free drug. This advantage could be linked to improved permeation and retention of the nanoparticles in the cancer microenvironment owing to the large fenestrations present in cancer vascular tissues which permit easy passage of nanoparticles and reduced cardiac interactions [145]. To foster easy approval and use of stimuli-dependent LNPs in cancer disease, the formulation should be designed and developed in a way that the nanoparticles serve as the solubilizing vehicle for drugs, especially lipophilic or hydrophobic drugs, without the addition of organic solvents like dimethyl sulphoxide (DMSO) to improve solubility. This is due to the toxicogenic characteristics of organic diluents which might require extemporaneous administration of corticosteroids or anti-histamines in order to resolve the toxicity reactions [144]. The size and shape of the nanoparticles are importance in the list of factors that affect clinical use of stimuli-responsive nanotherapeutics for cancer. Whereas nanoparticle size affects tailored drug transport across tight junctions and other physiological barriers, permeation, and deposition of anti-cancer drugs at the desired tumour site, acceptable particle shape, which defines morphological parameters, promotes the biocompatibility and structural suitability between the LNPs and the cancerous organelles while sparing surrounding healthy tissues from damage, demonstrating the importance of the physical properties of LNPs in effective cancer treatment [146, 147]. Functionalizing LNPs to be responsive to various stimuli could lead to functional and morphological changes in endogenous molecules such as cytochrome enzymes, plasma proteins like albumin, and other cellular peptides which interact with LNPs. These changes might result in biomolecular aggregation, especially proteins, leading to decreased anti-cancer efficacy or the development of newer pathological conditions [148–150]. Effective and safe stimuli-responsive LNPs for cancer therapy should not induce the destruction of cell membranes like lysozyme membrane, because it might lead to the leakage of cellular materials like enzymes and iron, and trigger mitochondrial dysfunction, and oxidative stress. This scenario would impede the efficacy of cargo-laden LNPs due to cell membrane damage and oxidative stress [151]. Development and manufacturing of LNPs for cancer treatment should be efficiently done to obtain stable nanocarriers because it has been shown that particle instability could result to premature nanocarrier degradation by serum proteins leading to very toxic premature release of encapsulated cargo. This scenario has been used to describe the mechanism of toxicity reactions affecting the bone marrow and gastrointestinal tract following in vivo administration of irinotecan liposomes [152]. These few illustration underscore the fact that efficient design of stimuli-responsive LNPs is as critical as the therapeutic response desired from the nanoparticles because any errors in formulation development would affect the final effect produced by the anti-cancer drug-loaded LNPs. Therefore, since clinical translation is the ultimate goal of stimuli-responsive LNPs, their application must indicate possible untoward effects derived from administration and the patient or the caregiver should be adequately informed, especially experimental preparations undergoing trial. In summary, any safety solution adopted should be applied on case-by-case basis and this process should be closely supervised by an oncologist, a pharmacist with specialty in cancer, and regulatory officer.

8 Conclusion and Future Research

A plethora of published reports indicate that stimuli-responsive LNPs have been accepted as an essential part of the armamentarium used by oncologists for the treatment of cancer. These nanocarriers are decorated to be sensitive to exogenous or endogenous stimuli which influence drug delivery and release in cancer microenvironment. The sensitivity of LNPs to these stimuli promotes targeting opportunities, biocompatibility, tumour accumulation, and enhanced drug release profile. Besides cancer treatment, stimuli-sensitive LNPs could be used in cancer diagnostics like probing, sensing, and molecular imaging, of cancer development. Available lines of evidences have shown that the structural design and optimization of stimuli-responsive LNPs for cancer therapy, safety concerns, and lack of streamlined regulatory frameworks to control their clinical application in cancer, are some of the challenges limiting translation. The perceived and identified obstacles should be adequately considered during the design and development of stimuli-responsive LNPs for effective cancer therapy. Further, there is a need to have clearly defined methods for assessing the biosafety of stimuli-responsive LNPs for effective clinical translation and application. Regulatory frameworks that encapsulate innovative strategies for designing stimuli-responsive LNPs and translating the nanocarriers for clinical applications should be articulated and implemented. There is a need to invest in extensive research and clinical studies to increase the number of stimuli-sensitive LNPs available for clinical application because, so far, only a few have recorded successful clinical translation. The outcomes of the studies and the data obtained would assist in formulating fit-for-purpose stimuli-responsive LNPs that would reach the market and are suitable for clinical utility.

References

1. Blum AP, Kammeyer JK, Rush AM, Callmann CE, Hahn ME, Gianneschi NC. Stimuli-responsive nanomaterials for biomedical applications. J Am Chem Soc. 2015;137(6):2140–54.
2. Pham SH, Choi Y, Choi J. Stimuli-responsive nanomaterials for application in antitumor therapy and drug delivery. Pharmaceutics. 2020;12:630.
3. Zhou H, Chen DS, Hu CJ, Hong X, Shi J, Xiao Y. Stimuli-responsive nanotechnology for RNA delivery. Adv Sci. 2023;10:2303597.
4. Mi P. Stimuli-responsive nanocarriers for drug delivery, tumor imaging, therapy and theranostics. Theranostics. 2020;10(10):4557–88.
5. Aden D, Zaheer S, Raj S. Challenges faced in the cancer diagnosis and management – COVID-19 pandemic and beyond – lessons for future. Heliyon. 2022;8:e12091.
6. Ahrensberg JM, Olesen F, Hansen RP, Schroder H, Vedsted P. Childhood cancer and factors related to prolonged diagnostic intervals: a Danish population-based study. Br J Cancer. 2013;108(6):1280–7.
7. Ding YY, Ramakrishna S, Long AH, Phillips CA, Montiel-Esparza R, Diorio CJ, Bailey LC, Maude SL, Aplenic R, Batra V, Reilly AF, Rheingold SR, Lacayo NJ, Sakamoto KM, Hunger SP. Delayed cancer diagnoses and high mortality in children during the COVID-19 pandemic. Paediatr Blood Cancer. 2020;67(9):e28427.
8. Sheets NC, Goldin GH, Meyer AM, Wu Y, Chang Y, Sturmer T, Holmes JA, Reeve BB, Godley PA, Carpenter WR, Chen RC. Intensity-modulated radiation therapy, proton therapy, or conformal radiation therapy and morbidity and disease control in localized prostate cancer. JAMA. 2012;307(15):1611–20.
9. National Cancer Policy Forum. Board on Health Care Services; Institute of Medicine. Delivering affordable cancer care in the 21st century: workshop summary. Washington (DC): National Academies Press (US); 2013.
10. Smith BD, Jiang J, McLaughlin SS, Hurria A, Smith GL, Giordano SH, Buchholz TA. Improvement in breast cancer outcomes over time: are older women missing out? J Clin Oncol. 2011;29(35):4647–53.
11. Temel JS, Greer JA, Admane S, Gallagher ER, Jackson VA, Lynch TJ, Lennes IT, Dahlin CM, Pirl WF. Longitudinal perceptions of prognosis and goals of therapy in patients with metastatic non-small-cell lung cancer: results of a randomized study of early palliative care. J Clin Oncol. 2011;29(17):2319–26.
12. Buys SS, Partridge E, Black A, Johnson CC, Lamerato L, Isaacs C, Reding DJ, Greenlee RT, Yokochi LA, Kessel B, Crawford ED, Church TR, Andriole GL, Weissfeld JL, Fouad MN, Chia D, O'Brien B, Ragard LR, Clapp JD, Rathmell JM, Riley TL, Hartge P, Pinsky PF, Zhu CS, Izmirlian G, Kramer BS, Miller AB, Xu JL, Prorok PC, Gohagan JK, Berg CD, PLCO Project Team. Effect of screening on ovarian cancer mortality: the prostate, lung, colorectal and ovarian (PLCO) cancer screening randomized controlled trial. JAMA. 2011;305(22):2295–303.
13. Anderson WF, Camargo MC, Fraumeni JF, Correa P, Rosenberg PS, Rabkin CS. Age-specific trends in incidence of noncardia gastric cancer in US adults. JAMA. 2010;303(17):1723–8.
14. Nwachukwu CR, Mudasiru O, Million L, Sheth S, Qamoos H, Onah JO, Okemini A, Rhodes M, Barry M, Banjo AA, Habeebu M, Olasinde TA, Bhatt AS. Evaluating barriers and opportunities in delivering high-quality oncology care in resource-limited setting using a comprehensive needs assessment tool. J Glob Oncol. 2018;4:1–9.
15. Keyvani V, Mollazadeh S, Riahi E, Mahmoudian RA, Tabari M, Lagzian E, Ghorbani E, Akbarzade H, Gholami AS, Gataa IS, Hassanian SM, Ferns GA, Khazaei M, Avan A, Anvari K. The application of nanotechnological therapeutic platforms against gynaecological cancers. Curr Pharm Des. 2024;30(13):975–87.
16. Bidan N, Dunsmore G, Ugrinic M, Bied M, Moreira M, Deloménie C, Ginhoux F, Blériot C, de la Fuente M, Mura S. Multicellular tumor spheroid model to study the multifaceted role of tumor-associated macrophages in PDAC. Drug Deliv Transl Res. 2024;14(8):2085–99.
17. Gaafar PME, Farid RM, Hazzah HA, AbouKilila HY, Helmy MW, Abdallah OY. Magnetic lipid-based hybrid nanosystems: a combined stimuli- responsive nanocarriers for enriched

chemotherapeutic potential of L-carnosine in induced breast Ehrlich ascites tumor model. Int J Pharm. 2024;655:124000.

18. German-Cortés J, Vilar-Hernández M, Rafael D, Abasolo I, Andrade F. Solid lipid nanoparticles: multitasking nano-carriers for cancer treatment. Pharmaceutics. 2023;15(3):831.
19. Sivadasan D, Ramakrishnan K, Mahendran J, Ranganathan H, Karuppaiah A, Rahman H. Solid lipid nanoparticles: applications and prospects in cancer treatment. Int J Mol Sci. 2023;24(7):6199.
20. Rizwanullah M, Ahmad MZ, Garg A, Ahmad J. Advancement in design of nanostructured lipid carriers for cancer targeting and theranostic application. Biochim Biophys Acta Gen Subj. 2021;1865(9):129936.
21. Irby D, Du C, Li F. Lipid–drug conjugate for enhancing drug delivery. Mol Pharm. 2017;14(5):1325.
22. Husni P, Lim C, Oh KT. Tumor microenvironment stimuli-responsive lipid-drug conjugates for cancer treatment. Int J Pharm. 2023;639:122942.
23. Garg J, Pathania K, Sah SP, Pawar SV. Nanostructured lipid carriers: a promising drug carrier for targeting brain tumours. Future J Pharm Sci. 2022;8(1):25.
24. Hosseini TG, Homayouni Tabrizi M, Karimi E, Younesi A, Larian Z. Designing nanostructured lipid carriers modified with folate-conjugated chitosan for targeted delivery of osthole to HT-29 colon cancer cells: investigation of anticancer, antioxidant, and antibacterial activities. Cancer Nanotechnol. 2024;15(1):7.
25. Rajoriya V, Gupta R, Vengurlekar S, Surendra SU. Nanostructured lipid carriers (NLCs): a promising candidate for lung cancer targeting. Int J Pharm. 2024;655:123986.
26. Kim B, Pena CD, Auguste DT. Targeted lipid nanoemulsions encapsulating epigenetic drugs exhibit selective cytotoxicity on CDH1-/FOXM1+ triple negative breast cancer cells. Mol Pharm. 2019;16(5):1813–26.
27. Giordano A, Provenza AC, Reverchon G, Baldino L, Reverchon E. Lipid-based nanocarriers: bridging diagnosis and cancer therapy. Pharmaceutics. 2024;16(9):1158.
28. Sánchez-López E, Guerra M, Dias-Ferreira J, Lopez-Machado A, Ettcheto M, Cano A, Espina M, Camins A, Garcia ML, Souto EB. Current applications of nanoemulsions in cancer therapeutics. Nanomaterials (Basel). 2019;9(6):821.
29. Kretzer IF, Maria DA, Guido MC, Contente TC, Maranhão RC. Simvastatin increases the antineoplastic actions of paclitaxel carried in lipid nanoemulsions in melanoma-bearing mice. Int J Nanomedicine. 2016;11:885–904.
30. Hörmann K, Zimmer A. Drug delivery and drug targeting with parenteral lipid nanoemulsions - a review. J Control Release. 2016;223:85–98.
31. Ateeq MAM, Aalhate M, Mahajan S, Kumar GS, Sen S, Singh H, Gupta U, Maji I, Dikundwar A, Guru SK, Singh PK. Self-nanoemulsifying drug delivery system (SNEDDS) of docetaxel and carvacrol synergizes the anticancer activity and enables safer toxicity profile: optimization, and in-vitro, ex-vivo and in-vivo pharmacokinetic evaluation. Drug Deliv Transl Res. 2023;13(10):2614–38.
32. Buya AB, Beloqui A, Memvanga PB, Préat V. Self-nanoemulsifying drug delivery systems: from the development to the current applications and challenges in oral drug delivery. Pharmaceutics. 2020;12(12):1194.
33. Sreekanth V, Bajaj A. Recent advances in engineering of lipid drug conjugates for cancer therapy. ACS Biomater Sci Eng. 2019;5(9):4148–66.
34. Parveen N, Sheikh A, Abourehab MAS, Karwasra R, Singh S, Kesharwani P. Self-nanoemulsifying drug delivery system for pancreatic cancer. Eur Polym J. 2023;190:111993.
35. Hamad I, Harb AA, Bustanji Y. Liposome-based drug delivery systems in cancer research: An analysis of global landscape efforts and achievements. Pharmaceutics. 2024;16(3):400.
36. Fulton MD, Najahi-Missaoui W. Liposomes in cancer therapy: how did we start and where are we now. Int J Mol Sci. 2023;24(7):6615.
37. Nasri S, Ebrahimi-Hosseinzadeh B, Rahaie M, Hatamian-Zarmi A, Sahraeian R. Thymoquinone-loaded ethosome with breast cancer potential: optimization, in vitro and biological assessment. J Nanostructure Chem. 2020;10(1):19–31.

38. Grossi LN, Braz WR, da Silva NP, Cazarim ELCC, Palmieri MGS, Tavares GD, Pittella F. Ethosomes as delivery system for treatment of melanoma: a mini-review. Oncologie. 2023;25(5):455–9.
39. Zhan B, Wang J, Li H, Xiao K, Fang X, Shi Y, Jia Y. Ethosomes: a promising drug delivery platform for transdermal application. Chemistry. 2024;6(5):993–1019.
40. Gowda BHJ, Ahmed MG, Alshehri SA, Wahab S, Vora LK, Singh Thakur RR, Kesharwani P. The cubosome-based nanoplatforms in cancer therapy: seeking new paradigms for cancer theranostics. Environ Res. 2023;237(Pt 1):116894.
41. Zhang L, Li J, Tian D, Sun L, Wang X, Tian M. Theranostic combinatorial drug-loaded coated cubosomes for enhanced targeting and efficacy against cancer cells. Cell Death Dis. 2020;11(1):1–12.
42. Umar H, Wahab HA, Gazzali AM, Tahir H, Ahmad W. Cubosomes: design, development, and tumor-targeted drug delivery applications. Polymer. 2022;14(15):3118.
43. Badiger P, Mannur VS, Koli R. Dual drug-loaded cubosome nanoparticles for hepatocellular carcinoma: a design of experiment approach for optimization and in vitro evaluation. Future J Pharm Sci. 2024;10(1):38.
44. Pramanik A, Xu Z, Ingram N, Coletta PL, Millner PA, Tyler AII, Hughes TA. Hyaluronic-acid-tagged cubosomes deliver cytotoxics specifically to CD44-positive cancer cells. Mol Pharm. 2022;19(12):4601–11.
45. Movahedi F, Hu RG, Becker DL, Xu C. Stimuli-responsive liposomes for the delivery of nucleic acid therapeutics. Nanomed Nanotechnol Biol Med. 2015;11:1575–84.
46. El Sayed ME, Hoffman AS, Stayton PS. Smart polymeric carriers for enhanced intracellular delivery of therapeutic macromolecules. Expert Opin Biol Ther. 2005;5:25–32.
47. Mura S, Nicolas J, Couvreur P. Stimuli-responsive nanocarriers for drug delivery. Nat Mater. 2013;12:991–1003 – Same as number 4.
48. Lu Y, Sun W, Gu Z. Stimuli-responsive nanomaterials for therapeutic protein delivery. J Control Release. 2014;194:1–19.
49. Zhou W, Jia Y, Liu Y, Chen Y, Zhao P. Tumor microenvironment-based stimuli-responsive nanoparticles for controlled release of drugs in cancer therapy. Pharmaceutics. 2022;14:2346.
50. Liu M, Huang L, Zhang W, Wang X, Geng Y, Zhang Y, Wang L, Zhang W, Zhang Y-J, Xiao S. A transistor-like pH-sensitive nanodetergent for selective cancer therapy. Nat Nanotechnol. 2022;17:541–51.
51. Sato Y, Hatakeyama H, Sakurai Y, Hyodo M, Akita H, Harashima H. A pH-sensitive cationic lipid facilitates the delivery of liposomal siRNA and gene silencing activity in vitro and in vivo. J Control Release. 2012;163:267–76.
52. Wanting H, Ruigi L, Siwei B, Qian H, Haibo W, Jun G. Photo-responsive polymersomes as drug delivery system for potential medical applications. Molecules. 2020;25(21):5147.
53. Chong L, Jiancheng W, Yiguang W, Huile G, Gang W, Yongzuo H, Hajun Y, Yong G, Yongjun W, Lin M, Huabing C, Hanjan H. Recent progress in drug delivery. Acta Pharm Sinica B. 2019;9(6):1145–62.
54. Zhang L, Qian M, Wang J. Progress in research of photo-controlled drug delivery systems. Acta Chim Sin. 2017;75:770–82.
55. Zhou LQ, Li P, Cui XW, Dietrich CF. Ultrasound nanotheranostics in fighting cancer: advances and prospects. Cancer Lett. 2020;470:204–19.
56. Greish K. Enhanced permeability and retention (EPR) effect for anticancer nanomedicine drug targeting. Cancer Nanotechnol. 2010;624:25–37.
57. Yin Q, Shen J, Zhang Z, Yu H, Li Y. Reversal of multidrug resistance by stimuli-responsive drug delivery systems for therapy of tumor. Adv Drug Deliv Rev. 2013;65:1699–715.
58. Alvarez-Lorenzo C, Bromberg L, Concheiro A. Light-sensitive intelligent drug delivery systems. Photochem Photobiol. 2009;85:848–60.
59. Xu F, Li H, Yao Q, Ge H, Fan J, Sun W, Wang J, Peng X. Hypoxia-activated NIR photosensitizer anchoring in the mitochondria for photodynamic therapy. Chem Sci. 2019;10:10586–94.
60. Yang B, Chen Y, Shi J. Reactive oxygen species (ROS)-based nanomedicine. Chem Rev. 2019;119(8):4881–985.

61. Yuan FA, Deng YB, Liu RR, Zheng XN, Rao XZ, Qiu DW, Zhang XY, Yu H, Li CSY. Adv Healthc Mater. 2021;10:2100198.
62. Jin Z, Yandai L, Zhe L, Qi W, Jiaqi Q, Renjie R, Xiancai J, Linxi H, Jibin S, Jianxun D, Huanghao Y. Stimuli-responsive nanoparticles for controlled drug delivery in synergistic cancer immunotherapy. Adv Sci. 2022;9:1–27.
63. Liu D, Yang F, Xiong FGN. The smart drug delivery system and its clinical potential. Theranostics. 2016;6:1306–23.
64. Kim YJ, Matsunaga YT. Thermo-responsive polymers and their application as smart biomaterials. J Mater Chem B. 2017;5:4307–21.
65. Guo X, Dan L, Guang Y, Chunli S, Zhaomn T, Jie W, Shaobing Z. Thermo-triggered drug release from actively targeting polymer micelles. ACS Appl Mater Interfaces. 2014;6(11):8549–59.
66. Tian H, Tang Z, Zhuang X, Chen X, Jing X. Biodegradable synthetic polymers: preparation, functionalization and biomedical application. Progr Poly Sci. 2012;37:237–80.
67. Tagami T, Foltz WD, Ernsting MJ, Lee CM, Tannock IF, May JP, Li S-D. MRI monitoring of intratumoral drug delivery and prediction of the therapeutic effect with a multifunctional thermosensitive liposome. Biomaterials. 2011;32:6570–8.
68. Gulzar A, Gai S, Yang P, Li C, Ansari MB, Lin J. Stimuli responsive drug delivery application of polymer and silica in biomedicine. J Mater Chem B. 2015;3:8599–622.
69. Torchilin VP. Multifunctional, stimuli-sensitive nanoparticulate systems for drug delivery. Nat Rev Drug Discov. 2014;13:813–27.
70. Chen KJ, Chaung EY, Wey SP, Lin KJ, Cheng F, Lin CC, Liu HL, Tseng HW, Liu CP, Wei MC. Hyper-thermia mediated local drug delivery by a bubble-generating liposomal system for tumor-specific chemotherapy. ACS Nano. 2014;8:5105–15.
71. Ganta S, Devalapally H, Shahiwala A, Amiji MA. Review of stimuli-responsive nanocarriers for drug and gene delivery. J Control Release. 2008;126:187–204.
72. Rapoport N. Physical stimuli-responsive polymeric micelles for anti-cancer drug delivery. Prog Polym Sci. 2007;32:962–90.
73. Sponchioni M, Palmiero UC, Moscatelli D. Thermo-responsive polymers: applications of smart materials in drug delivery and tissue engineering. Mater Sci Eng C. 2019;102:389–605.
74. Rahim MA, Jan N, Khan S, Shah H, Madni A, Khan A, Jabar A, Khan S, Elhissi A, Hussain Z. Recent advancements in stimuli responsive drug delivery platforms for active and passive cancer targeting. Cancer. 2021;13:670.
75. Roussakow S. The history of hyperthermia rise and decline. Conf Pap Med. 2013;2013:428027–40.
76. Shen B, Ma Y, Yu S, Ji C. Smart multifunctional magnetic nanoparticle-based drug delivery system for cancer thermo-chemotherapy and intracellular imaging. ACS Appl Mater Interfaces. 2016;8:24502–8.
77. Demirci H, Slimani N, Pawar M, Kumon RE, Vaishnava P, Besirli CG. Magnetic hyperthermia in Y79 retinoblastoma and ARPE-19 retinal epithelial cells: tumor selective apoptotic activity of iron oxide nanoparticle. Transl Vis Sci Technol. 2019;8:18.
78. Aguilar A, Michelle C, Edwin C, Kristen W, Arutselvan N, Tal M, Zeer B, Chirag B. Permeabilizing cell membranes with electric fields. Cancer. 2021;13(9):2283.
79. Paul G, David A, Burdick J, Chem C, Liang E, Langer R. Electrically controlled drug delivery from biotin-doped conductive polypyrrole. Adv Mater. 2006;18:577–81.
80. Shenghua L, Ying F, Guijun L, Li L, Helen K, Xianfeng C, Feng Y. Conjugated polymer for voltage-controlled release of molecules. Adv Mater. 2017;29:1701733.
81. Yue R, Xu J. Poly (3, 4-ethylenedioxythiophene) as promising organic thermoelectric materials: a mini-review. Synth Met. 2012;162:912–7.
82. Senyei A, Widder K, Czerlinski G. Magnetic guidance of drug-carrying microspheres. J Appl Phys. 1978;49:3578–83.
83. Pankhurst QA, Connolly J, Jones SK, Dobson J. Applications of magnetic nanoparticles in biomedicine. J Phys D Appl Phys. 2003;36:167.

84. Lening S, Hongmei L, Yangi Y, Yang L, Rehmat I, Sumin T, Rongsheng T, Yang B. Smart nanoparticles for cancer therapy. Cancer Lett. 2020;470:204–19.
85. Chandan R, Mehta S, Banerjee R. Ultrasound-responsive carriers for therapeutic applications. ACS Biomater Sci Eng. 2020;6:4731–47.
86. Alondra A, Michelle C, Edwin C, Kristen W, Arutselvan N, Tal M, Zecu B, Chirag B. Permeabilizing cell membranes with electric fields. Cancer. 2021;13:2283.
87. Schroeder A, Kost J, Barenholz Y. Ultrasound, liposomes, and drug delivery: principles for using ultrasound to control the release of drugs from liposomes. Chem Phys Lipids. 2009;162:1–16.
88. Di J, Kim J, Hu Q, Jiang X, Gu Z. Spatiotemporal drug delivery using laser-generated-focused ultrasound system. J Control Release. 2015;220:592–9.
89. Sirsi SR, Borden MA. State-of-the-art materials for ultrasound-triggered drug delivery. Adv Drug Deliv Rev. 2014;72:3–14.
90. Cheng W, Gu L, Ren W, Liu Y. Stimuli-responsive polymers for anti-cancer drug delivery. Mater Sci Eng C. 2014;45:600–8.
91. Xia H, Zhao Y, Tong R. Ultrasound-mediated polymeric micelle drug delivery. Adv Exp Med Biol. 2016;880:365–84.
92. Zhao YZ, Du LN, Lu CT, Jin YG, Ge SP. Potential and problems in ultrasound-responsive drug delivery systems. Int J Nanomedicine. 2013;8:1621.
93. Zardad AZ, Choonara YE, Toit LCD, Kumar P, Mabrouk M, Kondiah PD, Pillay VA. Review of thermo-and ultrasound-responsive polymeric systems for delivery of chemotherapeutic agents. Polymers. 2016;8:359.
94. Xi L, Li C, Wang Y, Gong Y, Su F, Li S. Novel thermosensitive polymer-modified liposomes as nano-carrier of hydrophobic antitumor drugs. J Pharm Sci. 2020;109(8):2544–52.
95. Yang W, Zhang F, Deng H, Lin L, Wang F, Kang G, Lau YJ, Tian R, Zhang M, Wang Z, He L, Ma Y, Niu G, Hu S, Chen X. ACS Nano. 2020;14:620.
96. Chen M, Liu D, Liu F, Wu Y, Peng F, Song J. Immunogenic cell death activates the tumor immune microenvironment to boost the immunotherapy efficiency. J Control Release. 2021;332:269.
97. Li J, Li X, Liu H, Ren T, Huang L, Deng Z. GSH and light dual stimuli-responsive supramolecular polymer drug carriers for cancer therapy. Polym Degrad Stab. 2019;168:108956.
98. Ko NR, Van SY, Hong SH, Kim S-Y, Kim M, Lee JS. Dual pH- and GSH-responsive degradable PEGylated graphene quantum dot-based nanoparticles for enhanced HER2-positive breast cancer therapy. Nano. 2020;10(1):91.
99. Yang HY, Li Y, Lee DS. Multifunctional and stimuli-responsive magnetic nanoparticle-based delivery systems for biomedical applications. Adv Ther. 2018;1:1800011.
100. Kobayashi T. Cancer hyperthermia using magnetic nanoparticles. Biotechnol J. 2011;6: 1342–7.
101. Shujie L, Lian L, Xi L, Cheng C, Chaohui L, Yuan H. Targeted inhibition of tumor inflammation and tumor-platelet crosstalk by nanoparticle-mediated drug delivery mitigates cancer metastasis. ACS Nano. 2022;16:50–67.
102. Yingting Y, Yi Y, Hao Y, Rui W, Zhenming Z, Xiaodan S, Lingyun Z, Xiang A, Zhen X, Qiong W. A tumor-specific MicroRNA recognition system facilitates the accurate targeting to tumor cells by magnetic nanoparticles. Mol Ther Nucleic Acids. 2016;5:e318.
103. Semkina A, Abakumor M, Grinenko N, Lipengoits A, Nukolova N, Chekhonin V. Magnetic resonance imaging of tumors with the use of iron oxide magnetic nanoparticles as a contrast agent. Bull Exp Biol Med. 2017;162:808–11.
104. Estelrich J, Escribano E, Queralt J, Busquets M. Iron oxide nanoparticles for magnetically-guided and magnetically-responsive drug delivery. Int J Mol Sci. 2015;16:8070–101.
105. De Cuyper M, Joniau M. Magneto-liposomes. Formation and structural characterization. Eur Biophys J. 1988;15:311–9.
106. Sun L, Liu H, Ye Y, Lei Y, Islam R, Tan S, Tong R, Miao YB, Cai L. Smart nanoparticles for cancer therapy. Signal Transduct Target Ther. 2023;8(1):418.

107. Li M, Zhao G, Su WK, Shuai Q. Enzyme-responsive nanoparticles for anti-tumor drug delivery. Front Chem. 2020;8:1–20.
108. Zhang J, Lin Y, Lin Z, Wei Q, Qian J, Ruan R, Jiang X, Hou L, Song J, Ding J, Yang H. Stimuli-responsive nanoparticles for controlled drug delivery in synergistic cancer immunotherapy. Adv Sci (Weinh). 2022;9(5):e2103444.
109. Tian H, Zhang T, Qin S, Huang Z, Zhou L, Shi J, Nice EC, Xie N, Huang C, Shen Z. Enhancing the therapeutic efficacy of nanoparticles for cancer treatment using versatile targeted strategies. J Hematol Oncol. 2022;15(1):132.
110. Gloria Kang GJ, Ewing-Nelson SR, Mackey L, Schlitt JT, Marathe A, Abbas KMSS. Enzyme-responsive nanomaterials for controlled drug delivery. Physiol Behav. 2018;176(1):139–48.
111. Waheed I, Ali A, Tabassum H, Khatoon N, Lai WF, Zhou X. Lipid-based nanoparticles as drug delivery carriers for cancer therapy. Front Oncol. 2024;14:1–17.
112. Kuang T, Liu Y, Gong T, Peng X, Hu X, Yu Z. Enzyme-responsive nanoparticles for anticancer drug delivery. Curr Nanosci. 2015;12(1):38–46.
113. AlSawaftah NM, Awad NS, Pitt WG, Husseini GA. pH-responsive nanocarriers in cancer therapy. Polymers (Basel). 2022;14(5):936.
114. Shinn J, Kwon N, Lee SA, Lee Y. Smart pH-responsive nanomedicines for disease therapy. J Pharm Investig. 2022;52(4):427–41.
115. Omidian H, Gill EJ, Cubeddu LX. Lipid nanoparticles in lung cancer therapy. Pharmaceutics. 2024;16(5):644.
116. Chu S, Shi X, Tian Y, Gao F. pH-responsive polymer nanomaterials for tumor therapy. Front Oncol. 2022;12:1–22.
117. Yan Y, Ding H. pH-responsive nanoparticles for cancer immunotherapy: a brief review. Nano. 2020;10(8):1–15.
118. Lv Y, Hao L, Hu W, Ran Y, Bai Y, Zhang L. Novel multifunctional pH-sensitive nanoparticles loaded into microbubbles as drug delivery vehicles for enhanced tumor targeting. Sci Rep. 2016;6:1–9.
119. Fan X, Zhao X, Qu X, Fang J. pH sensitive polymeric complex of cisplatin with hyaluronic acid exhibits tumor-targeted delivery and improved in vivo antitumor effect. Int J Pharm. 2015;496:644–53.
120. Zheng G, Zheng M, Yang B, Fu H, Li Y. Improving breast cancer therapy using doxorubicin loaded solid lipid nanoparticles: synthesis of a novel arginine-glycine-aspartic tripeptide conjugated, pH sensitive lipid and evaluation of the nanomedicine in vitro and in vivo. Biomed Pharmacother. 2019;116(440):109006.
121. Ma Y, Fan X, Li L. pH-sensitive polymeric micelles formed by doxorubicin conjugated prodrugs for co-delivery of doxorubicin and paclitaxel. Carbohydr Polym. 2016;137:19–29.
122. Sung YJ, Guo H, Ghasemizadeh A, Shen X, Chintrakulchai W, Kobayashi M, Toyoda M, Ogi K, Michinishi J, Ohtake T, Matsui M, Honda Y, Nomoto T, Takemoto H, Miura Y, Nishiyama N. Cancerous pH-responsive polycarboxybetaine-coated lipid nanoparticle for smart delivery of siRNA against subcutaneous tumor model in mice. Cancer Sci. 2022;113(12):4339–49.
123. Torres-Pérez SA, Torres-Pérez CE, Pedraza-Escalona M, Pérez-Tapia SM, Ramón-Gallegos E. Glycosylated nanoparticles for cancer-targeted drug delivery. Front Oncol. 2020;10:1–10.
124. Wang X, Wang L, Hao Q, Cai M, Wang X, An W. Harnessing glucose metabolism with nanomedicine for cancer treatment. Theranostics. 2024;14(17):6831–82.
125. Fu Y, Sun J, Wang Y, Li W. Glucose oxidase and metal catalysts combined tumor synergistic therapy: mechanism, advance and nanodelivery system. J Nanobiotechnol. 2023;21(1):1–23.
126. He T, Xu H, Zhang Y, Yi S, Cui R, Xing S, Wei C, Lin J, Huang P. Glucose oxidase-instructed traceable self-oxygenation/hyperthermia dually enhanced cancer starvation therapy. Theranostics. 2020;10(4):1544–54.
127. Fu LH, Qi C, Lin J, Huang P. Catalytic chemistry of glucose oxidase in cancer diagnosis and treatment. Chem Soc Rev. 2018;47(17):6454–72.
128. Li Z, Li X, Ai S, Liu S, Guan W. Glucose metabolism intervention-facilitated nanomedicine therapy. Int J Nanomedicine. 2022;17:2707–31.

129. Ahmadi S, Rabiee N, Bagherzadeh M, Elmi F, Fatahi Y, Farjadian F, Baheiraei N, Nasseri B, Rabiee M, Dastjerd NT, Valibeik A, Karimi M, Hamblin MR. Stimulus-responsive sequential release systems for drug and gene delivery. Nano Today. 2020;34:100914.
130. Chen L, Xue Y, Xia X, Song M, Huang J, Zhang H, Yu B, Long S, Liu Y, Liu L, Huang S, Yu F. A redox stimuli-responsive superparamagnetic nanogel with chemically anchored DOX for enhanced anticancer efficacy and low systemic adverse effects. J Mater Chem B. 2015;3(46):8949–62.
131. Steffens RC, Wagner E. Directing the way—receptor and chemical targeting strategies for nucleic acid delivery. Pharm Res. 2023;40(1):47–76.
132. Lee Y, Jeong M, Park J, Jung H, Lee H. Immunogenicity of lipid nanoparticles and its impact on the efficacy of mRNA vaccines and therapeutics. Exp Mol Med. 2023;55(10):2085–96.
133. Zheng Y, Moreno A, Zhang Y, Sipponen MH, Dai L. Harnessing chemical functionality of lignin towards stimuli-responsive materials. Trends Chem. 2024;6(2):62–78.
134. Rahim MA, Jan N, Khan S, Shah H, Madni A, Khan A, Jabar A, Khan S, Elhissi A, Hussain Z, Aziz HC, Sohail M, Khan M, Thu HE. Recent advancements in stimuli responsive drug delivery platforms for active and passive cancer targeting. Cancers (Basel). 2021;13(4):670.
135. Bukhari SZ, Zeth K, Iftikhar M, Rehman M, Munir MU, Khan WS, Ihsan A. Supramolecular lipid nanoparticles as delivery carriers for non-invasive cancer theranostics. Curr Res Pharmacol Drug Discov. 2021;2:100067.
136. Rosenblum D, Joshi N, Tao W, Karp JM, Peer D. Progress and challenges towards targeted delivery of cancer therapeutics. Nat Commun. 2018;9:1410.
137. Cheng Z, Fobian SF, Gurrieri E, Amin M, D'Agostino VG, Falahati M, Zalba S, Debets R, Garrido MJ, Saeed M, Seynhaeve ALB, Balcioglu HE, ten Hagen TLM. Lipid-based nanosystems: the next generation of cancer immune therapy. J Hematol Oncol. 2024;17:53.
138. Bobo D, Robinson KJ, Islam J, Thurecht KJ, Corrie SR. Nanoparticle-based medicines: a review of FDA-approved materials and clinical trials to date. Pharm Res. 2016;33:2373–87.
139. Lee MS, Dees EC, Wang AZ. Nanoparticle-delivered chemotherapy: old drugs in new packages. Oncology (Williston Park). 2017;31:198–208.
140. Amin M, Seynhaeve ALB, Sharifi M, Falahati M, Ten Hagen TLM. Liposomal drug delivery systems for cancer therapy: the Rotterdam experience. Pharmaceutics. 2022;14:2165.
141. Cabral H, Matsumoto Y, Mizuno K, Chen Q, Murakami M, Kimura M, Terada Y, Kano MR, Miyazono K, Uesaka M, Nishiyama N, Kataoka K. Accumulation of sub-100nm polymeric micelles in poorly permeable tumours depends on size. Nat Nanotechnol. 2011;6:815–23.
142. Whatcott CJ, Posner RG, Von Hoff DD, Han H. Desmoplasia and chemoresistance in pancreatic cancer. In: Grippo PJ, Munshi HG, editors. Pancreatic cancer and tumor microenvironment. Trivandrum: Transworld Research Network; 2012.
143. Heldin CH, Rubin K, Pietras K, Ostman A. High interstitial fluid pressure: an obstacle in cancer therapy. Nat Rev Cancer. 2004;4:806–13.
144. Wolfram J, Zhu M, Yang Y, Shen J, Gentile E, Paolino D, Fresta M, Nie G, Chen C, Shen H, Ferrari M, Zhao Y. Safety of nanoparticles in medicine. Curr Drug Targets. 2015;16(14):1671–81.
145. Barenholz Y. Doxil(R)--the first FDA-approved nano-drug: lessons learned. J Control Release. 2012;160(2):117–34.
146. Johnston HJ, Hutchison GR, Christensen FM, Peters S, Hankin S, Aschberger K, Stone V. A critical review of the biological mechanisms underlying the in vivo and in vitro toxicity of carbon nanotubes: the contribution of physicochemical characteristics. Nano. 2010;4(2):207–46.
147. Pan Y, Neuss S, Leifert A, Fischler M, Wen F, Simon U, Schmid G, Brandau W, Jahnen-Dechent W. Size-dependent cytotoxicity of gold nanoparticles. Small. 2007;3(11):1941–9.
148. Wolfram J, Yang Y, Shen J, Moten A, Chen C, Shen H, Ferrari M, Zhao Y. The nano-plasma interface: implications of the protein corona. Colloids Surf B Biointerf. 2014;0:17–24.
149. Zhu M, Perrett S, Nie G. Understanding the particokinetics of engineered nanomaterials for safe and effective therapeutic applications. Small. 2013;9(9–10):1619–34.

150. Pelaz B, Charron G, Pfeiffer C, Zhao Y, de la Fuente JM, Liang X-J, Parak WJ, del Pino P. Interfacing engineered nanoparticles with biological systems: anticipating adverse nano-bio interactions. Small. 2013;9(9–10):1573–84.
151. Stern ST, Adiseshaiah PP, Crist RM. Autophagy and lysosomal dysfunction as emerging mechanisms of nanomaterial toxicity. Part FibreToxicol. 2012;9:20.
152. Liu X, Tang I, Wainberg ZA, Meng H. Safety considerations of cancer nanomedicine – a key step towards translation. Small. 2020;16(36):e2000673.

Stimuli-Responsive Mesoporous Silica Nanoparticles for Cancer Therapy

8

Zahra Sifat Zaidi, Sai Veerabhadra Nikhil Teki, and Dhirendra S. Katti

Abstract

According to the World Health Organization, cancer accounted for 19.9 million incidences globally, in 2022, while 48.7% (9.7 million) of these cases resulted in deaths. These staggering statistics demand an immediate need for effective anti-cancer therapies. The current therapeutic options for cancer include surgery, chemotherapy, radiotherapy, and immunotherapy either alone or in combination. However, these therapies result in off-target interactions causing severe side

Zahra Sifat Zaidi and Sai Veerabhadra Nikhil Teki contributed equally.

Z. S. Zaidi
Department of Biological Sciences and Bioengineering, Indian Institute of Technology, Kanpur, Uttar Pradesh, India

The Mehta Family Centre for Engineering in Medicine, Indian Institute of Technology, Kanpur, Uttar Pradesh, India

S. V. N. Teki
Department of Biological Sciences and Bioengineering, Indian Institute of Technology, Kanpur, Uttar Pradesh, India

The Mehta Family Centre for Engineering in Medicine, Indian Institute of Technology, Kanpur, Uttar Pradesh, India

Department of Chemistry, Indian Institute of Technology, Kanpur, Uttar Pradesh, India

D. S. Katti (✉)
Department of Biological Sciences and Bioengineering, Indian Institute of Technology, Kanpur, Uttar Pradesh, India

The Mehta Family Centre for Engineering in Medicine, Indian Institute of Technology, Kanpur, Uttar Pradesh, India

School of Interdisciplinary Life Sciences, Indian Institute of Technology, Goa, India
e-mail: dsk@iitk.ac.in

P. V. Devarajan et al. (eds.), *Advances in Stimuli-Responsive Nanosystems for Cancer Therapy*, https://doi.org/10.1007/978-981-95-2744-1_8

effects. To circumvent this drawback, nanoparticles are used as a drug delivery system. Mesoporous silica nanoparticles (MSNs) are one of the classes of nanoparticles, commonly under investigation as an anti-cancer drug delivery system. MSNs offer a multitude of advantages, such as a uniform porous structure and a high surface area for improved drug loading, good biocompatibility, and ease of surface functionalization. To achieve spatio-temporal control over the release of anticancer therapeutics, MSNs are modified with stimuli-responsive moieties. As a result, these "smart" MSNs have garnered interest in the field of precision medicine. This chapter highlights the recent advancements of stimuli-responsive MSNs encompassing both endogenous and exogenous stimuli for cancer therapy. Strategies employed in the design of MSNs responsive to endogenous stimuli (such as pH, redox, and enzyme), and exogenous stimuli (such as light, heat, ultrasound, and magnetic field) are addressed. The chapter further discusses the strategies to develop hybrid-stimuli-responsive MSNs for cancer therapy. Moreover, their pre-clinical studies and clinical status have been summarized. Finally, the challenges and future perspective of stimuli-responsive MSNs have also been outlined.

Keywords

Mesoporous silica nanoparticles · Cancer therapy · Stimuli-responsive gatekeepers · Endogenous stimuli · Exogenous stimuli · Controlled drug delivery

Abbreviations

AMF	Alternating magnetic field
CatB	Cathepsin B
CD	Carbon nanodots
EPR	Enhanced permeation and retention
FDA	Food and Drug Administration
GRAS	Generally regarded as safe
GSH	Glutathione
Hyal	Hyaluronidase
LCST	Lower critical solution temperature
LDLo	Lethal doses
LOAEL	Lowest observed adverse effect levels
MAA	Methacrylic acid
MCM-41	Mobil Composition of Matter No. 41
MMP	Matrix metalloproteinase
MSN	Mesoporous silica nanoparticle
NIPAAm	N-isopropylacrylamide
NIR	Near-infrared
PET	Positron emission tomography

PLL(cit)	Citraconic anhydride-modified polylysine
PPz	Polyphosphazene
PTT	Photothermal therapy
PDT	Photodynamic therapy
SPION	Superparamagnetic iron oxide nanoparticle
THPMA	2-tetrahydropyranyl methacrylate
UCNP	Upconversion nanoparticle
US	Ultrasound
UV	Ultraviolet
β -CD	β-cyclodextrin

1 Introduction

Mesoporous silica nanoparticles (MSNs) are a distinctive class of inorganic nanomaterials characterized by an ordered porous structure with pore diameter in the range of 2–50 nm [1, 2]. The empty mesopores contribute to an overall high surface area and pore volume. This significantly improves the loading capacity of bioactive molecules when compared to solid non-porous nanoparticles [1, 3]. Besides high loading efficiency, various other characteristics of MSNs have garnered further interest for their application as a drug delivery vehicle. Some of these salient features of MSNs are briefly described below:

(i) *Low-toxicity profile*

MSNs constitute repeating silica units within their framework, similar to those used in the fabrication of FDA-approved colloidal silica and amorphous silica nanoparticles for the production of tablets and food additives, respectively [4]. Therefore, MSNs are expected to have a better safety profile compared to the other inorganic counterparts in a clinical setting. Sun et al. conducted an extensive study to estimate the human equivalent doses (HED) corresponding to the lowest observed adverse effect levels (LOAELs) and lethal doses (LDLo) of MSNs administered through different routes [5]. The LDLo values were determined to be as high as 4.1–81, 16–65, and 146 mg/kg HED for intravenous, intraperitoneal, and subcutaneous administrations, respectively. Meanwhile, the LOAELs were estimated as 0.45, 0.81, and 4.1 mg/kg HED for intravenous, intraperitoneal, and subcutaneous injections. This study gives an estimate of the concentration range of MSNs that can be safely administered for clinical studies.

(ii) *Ordered pore structure*

MSNs can exhibit a uniform and organized pore structure that provides precise control over drug loading and release kinetics, which are crucial for effective drug delivery [1, 6].

(iii) *Tunable size and morphology*

The size and morphology of MSNs can be customized by changing the reactants and modulating the reaction parameters [7, 8]. These adjustments have enabled the fabrication of nanoparticles of varying size (less than 50 nm to greater than 200 nm) and morphologies (such as spheres, rods, and cubes), thereby expanding their applications in drug delivery [9, 10].

(iv) *Ease of surface functionalization*

The silanol groups (Si-OH) on MSN surface serve as reactive substrates for grafting various silane coupling agents, drugs, polymers, ligands, proteins, antibodies, aptamers, nanovalves, and other inorganic nanoparticles. This wide array of MSN functionalization has enabled a versatility of MSNs for drug delivery [2, 4, 11].

(v) *Controlled release*

The chemically active mesoporous surface of MSNs provides a framework for achieving high drug loading with relative ease. A wide variety of small-molecule drugs and biologics can be encapsulated in the pores via physical adsorption [12]. However, this method poses a challenge due to the premature release of the cargo. This issue can be addressed by introducing stimuli-responsive gatekeepers on the MSN surface to enable controlled release [13]. These gatekeepers selectively expose the drug-loaded mesopores at the target site in response to stimuli such as pH, enzymes, temperature, or heat, thus allowing for the delivery of lower dosage with higher efficacy. Moreover, the introduction of these gatekeepers alters the release kinetics of the loaded drug, leading to a sustained release pattern instead of the burst release typically observed [14].

(vi) *Stable colloidal dispersion*

MSNs have a negatively charged surface due to the abundance of surface silanol groups [15]. This feature makes their outer surface hydrophilic by creating a hydration shell. As a result, MSNs exhibit excellent colloidal stability due to the repulsive surface charges that prevent aggregation, making them highly desirable for drug delivery applications [15].

(vii) *Thermal and chemical stability*

MSNs exhibit excellent chemical stability under physiological conditions, taking up to 15 days to completely degrade in simulated body fluids [16, 93]. Additionally, MSN matrices remain stable under extreme temperatures of up to 500 °C, without any changes in structure or porosity [17, 18]. This higher degree of thermal and chemical stability, compared to organic and polymeric nanocarriers, makes MSNs highly robust for biomedical applications.

The application of mesoporous silica as a drug delivery vehicle was, for the first time, reported by Regi et al. in 2001 [19]. They employed a material called MCM-41, with hexagonal honeycomb-like mesopores as a delivery system for the anti-inflammatory drug ibuprofen. This sets the groundwork for the development of MSNs as a drug delivery vehicle.

Interestingly, a PubMed search from 2001 to 2024 for the keywords "mesoporous silica nanoparticles in drug delivery" gave 3006 listings. Of these, 1579 articles (~50%) are published in the field of drug delivery in cancer. The substantial research on MSNs in cancer drug delivery, as highlighted by the PubMed search results, prompts an interest in understanding their role in cancer therapy.

2 Mesoporous Silica Nanoparticles and Cancer Therapy

A tumor consists of leaky blood vasculature with large fenestrations, thereby allowing macromolecules to enter the tumor more readily when compared to healthy tissue [20, 21]. This results in enhanced permeation of macromolecules in the tumor. Moreover, the impaired tumor lymphatics disallow drainage of the permeated macromolecules away from the tumor tissue, thus leading to enhanced retention. Taken together, this phenomenon coined as the "enhanced permeation and retention" or "EPR" effect, discovered by Matsumura and Maeda from Japan in 1986 [22], paved the foundation of nanoparticle-mediated drug delivery in cancer.

MSNs in cancer therapy attracted research interest due to their tunable physicochemical properties and high encapsulation efficiency. MSNs can be synthesized below 200 nm, which is essential for effective permeation via tumor fenestrations [21, 23]. Moreover, unlike other sub-200 nm non-porous particles, MSNs exhibit a significantly higher drug loading capacity [7]. Together, these properties confer a high bioavailability of chemotherapeutic drug in tumor when delivered via MSNs. The improved bioavailability in tumor further reduces the need of administrating high amount of drug to arrive at therapeutic concentration for cancer treatment. Consequently, the off-target effects of chemotherapeutic agents significantly reduce leading to lesser side effects. Moreover, the highly reactive surface of MSNs further allows for its modification with targeting moieties, which can potentially improve specificity to target tumor tissue and cancer cells. The tunable surface chemistry of MSNs also facilitates the engineering of particles with the ability to control drug delivery in response to a stimulus. These particles, also known as "stimuli-responsive" or "smart" MSNs [24], have gained momentum as they expand the scope of innovative nanoparticle designs for advancement in cancer therapy.

3 Stimuli-Responsive Mesoporous Silica Nanoparticles

The central concept of stimuli-responsive MSNs revolves around tailoring the surface of the particles with "gatekeepers," also called "capping," which responds to a stimulus [25, 26]. These gatekeepers can be polymers, peptides, inorganic nanoparticles, and lipids, among others. In the absence of stimulus, the gatekeeper inhibits the release of drug molecules encapsulated in the mesopores. However, when subjected to an appropriate stimulus, the gatekeeper undergoes physicochemical changes, thereby facilitating a controlled release of drug. The physicochemical properties of tumor tissue distinguishable from healthy tissue can be leveraged as endogenous stimuli (such as pH, redox, and enzyme) [27–29]. Additionally, stimuli can also be exogenously provided (such as light, heat, ultrasound and magnetic field) to achieve an on-demand release of the cargo [18, 30–32] . Moreover, gatekeepers on the surface of MSNs prevent immature leakage of anti-cancer drugs in circulation due to an absence of stimulus. This facilitates controlled drug delivery once the particles reach the tumor site and get exposed to an appropriate stimulus. Overall, stimuli-responsive MSNs further improve bioavailability of drug and mitigate the toxic side effects of chemotherapeutic agents. In this book chapter, we provide an overview of the recent advances in engineering stimuli-responsive MSNs sensitive to various endogenous and exogenous stimuli as drug delivery system for cancer therapy (Fig. 8.1).

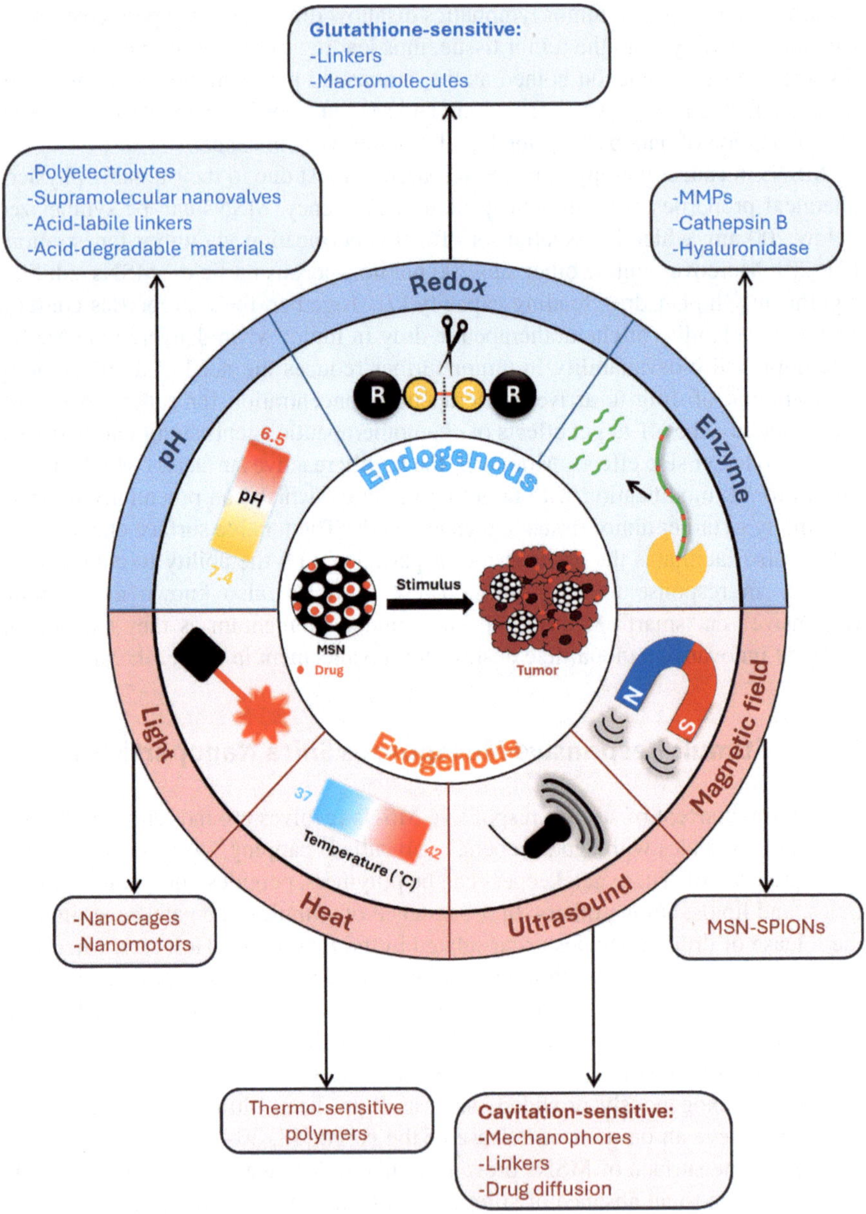

Fig. 8.1 Schematic depicting various strategies employed in the design of stimuli-responsiveness mesoporous silica nanoparticles for cancer therapy (SPION superparamagnetic iron oxide nanoparticles, MMP matrix metalloproteinase, MSN mesoporous silica nanoparticles)

3.1 Endogenous Stimuli

3.1.1 pH

The pH of the tumor microenvironment is an extensively explored stimulus for chemotherapeutic payload delivery using MSNs. The extracellular pH of tumor is mildly acidic (pH 6.5–6.8) compared to the healthy tissues (pH 7.4) [33]. This is a result of the rapid conversion of glucose into lactic acid by cancer cells, irrespective of the presence of oxygen- a phenomenon known as the Warburg effect [34, 35]. Consequently, the excessive lactate accumulation in the extracellular milieu causes a decrease in the pH of tumor microenvironment from a near neutral to a mildly acidic pH [33, 36, 37].

Strategies to Develop pH-Responsive MSNs

The notable difference in the pH of the microenvironment nanoparticle encounters, as it traverses from the bloodstream to tumor, has been leveraged to fabricate pH-responsive MSNs that facilitates tumor-specific drug delivery [38]. Researchers have employed various strategies to modify the surface of MSNs to achieve the same, some of which are:

(i) *pH-sensitive polyelectrolyte gatekeepers*

A relatively simple method to fabricate pH-responsive MSNs involves grafting polyelectrolytes onto the nanoparticle surface [39–44]. At physiological pH, these macromolecules tightly wrap around the nanoparticle surface, thereby covering a significant proportion of the mesopores. In the acidic pH of the tumor microenvironment, these polyelectrolytes either absorb or release H^+ ions, which affects the electrostatic, hydrophobic, and hydrophilic interactions between the polymeric chains and the surroundings. This results in conformational changes in the polymer chain, such as swelling (due to electrostatic repulsion among chain segments) [39, 44] or coiling (due to reduced solubility) [40–42], thereby exposing the mesopores and releasing the loaded drug.

(ii) *pH-sensitive supramolecular nanovalves*

The principles of supramolecular chemistry and assembly have been utilized to develop two-component pH-sensitive gatekeepers [28, 45–47]. This system consists of (a) a stationary stalk molecule (e.g., benzidine) immobilized on the silica surface, and (b) a mobile cyclic macromolecule (e.g., ß-cyclodextrin) surrounding the stalk [48, 49]. Non-covalent interactions between the immobilized stalk and cyclic cap effectively block the nanoparticle pores at physiological pH. However, the acidic pH reduces the binding affinity between the two components, causing the cyclic cap to slide off and expose the previously blocked mesopores, thereby releasing the drug [49].

(iii) *Acid-labile cross-linkers*

Another efficient strategy for pH-responsive drug delivery in MSNs is the use of acid-labile cross-linkers such as hydrazone [50], imine (Schiff-base)

[51], anhydride [52], and orthoesters [53]. These linkers are relatively stable at physiological pH but undergo acidic hydrolysis at tumor pH [54]. Researchers have utilized them to (a) cross-link gatekeepers and modify pore entrance [50, 51], or (b) to covalently conjugate drug molecules by modifying the inner pore walls [55, 56]. Acid-mediated cleavage of the linker results in removal of the gatekeeper and dissociation of the drug molecules, respectively, leading to drug release.

(iv) *Acid-decomposable materials*

A more recent strategy for engineering pH-responsive MSNs involves the use of acid- decomposable materials. Examples include self-immolative polymers [57], inorganic nanomaterials [58], and quantum dots [59], which act as gatekeepers at physiological pH but disintegrate into monomers or simpler molecules at acidic pH.

Despite extensive research in developing novel and efficient pH-responsive MSNs, current design strategies exhibit some shortcomings. Firstly, due to the similarity between the pH of wounded/ inflamed tissue with that of tumor tissue (i.e., mildly acidic), there is a possibility of off-target effects [60]. Secondly, the polyelectrolyte gatekeepers and the pH of the microenvironment are not the exclusive factors governing drug release kinetics. For example, the nature of the drug also plays a crucial role in the same. In cases where drug molecules possess an inherent negative or positive charge, the electrostatic interaction of the drug molecules with the polyelectrolyte gatekeeper may further influence their release kinetics. This interplay between the gatekeeping polymer, drug molecules, and ionic strength of the microenvironment may complicate the estimation of the drug release profile. Finally, the absence of a standardized design strategy hampers effective targeting of different cancer types and subtypes. This challenge primarily stems from the complex and highly heterogeneous tumor microenvironment, which exhibits significant diversity in extracellular pH depending on the cancer stage and type [61].

3.1.2 Redox

Glutathione (GSH), a ubiquitous tripeptide (glutamate-cysteine-glycine), is crucial for scavenging reactive oxygen species generated by cells [62, 63]. GSH, with concentration between 2 and 20 μM in healthy tissue, helps maintain redox homeostasis [64]. However, this balance is disrupted in cancerous tissues due to elevated glutathione levels, which can be nearly four times higher than healthy tissues [65, 66]. By exploiting these differences in the GSH concentrations between healthy and cancerous tissues, researchers have designed redox-responsive MSNs to achieve tumor-specific delivery of loaded drugs [67].

Strategies to Develop Redox-Responsive MSNs

To achieve redox-responsiveness, a wide variety of glutathione-sensitive gatekeepers, including redox-degradable cross-linkers [68, 69], polymers [70], and small inorganic nanoparticles [71–73], have been employed. These gatekeepers, in general, comprise a redox-sensitive moiety (like disulfide bonds) that undergoes a hydride exchange with glutathione. This results in the thiolation of both the sulfurs of the disulfide to eventually break the redox-sensitive bond [74]:

$$R - S - S - R' + 2GSH \rightarrow R - SH + R' - SH + G - S - S - G$$

On the basis of the aforementioned principle, various strategies to modify the surface of MSNs were employed to achieve the same, some of which are:

(i) *GSH-sensitive linkers*

A widely reported strategy to introduce redox-responsiveness involves the use of disulfide (-SS-) or diselenide (-SeSe-) linkers to attach gatekeepers to MSN surfaces [69, 75]. These gatekeepers seal the mesopores under low-GSH concentrations. However, at the tumor site, elevated GSH levels reduce these disulfides and diselenides, thereby separating the gatekeepers from the MSN surface and delivering the chemotherapeutic payload. Previously, bulky macromolecules (such as myoglobin and albumin) were used as gatekeepers [76]. However, in recent studies, other inorganic nanoparticles with size <10 nm (such as gold, iron oxide and quantum dots) were employed as gatekeepers [77]. These small nanoparticles offer an additional advantage to the drug delivery system by acting as contrast agents for imaging or photothermal agents to induce cancer cell death through tumor-localized hyperthermia.

(ii) *GSH-responsive macromolecules*

Some macromolecules like keratin, polydopamine, dextrin, and gelatin, exhibit redox-responsiveness [70, 78–81]. These macromolecules are used as gatekeepers to impart redox sensitivity to MSNs. Keratin, a biopolymer, contains cysteine residues with thiol functionalities that cross-link to form disulfides, creating a mesh over the MSN surface. These disulfides are cleavable under elevated glutathione levels, characteristic of tumor tissue, exposing the mesopores and releasing the chemotherapeutic agent [70, 78]. Dopamine, on the other hand, is first modified with cysteine residues. The thiol-modified dopamine monomers self-polymerize to form polydopamine, with disulfide cross-links [79]. Similar to keratin, the polydopamine network blocks the mesopores under normal conditions, whereas release the encapsulated drug in the presence of GSH. Additionally, polymers like dextrin and gelatin are first conjugated to the nanoparticle surface, followed by in situ cross-linking between polymer chains with cysteamine derivatives. The polymer mesh, thus formed, seals the drug-loaded mesopores under physiological conditions while providing a redox-sensitive release similar to the aforementioned mechanism [80, 81].

Redox-responsive MSNs utilize a simpler design and mechanism of action when compared to pH-responsive MSNs. While redox-responsive MSNs involve a mere reduction of disulfide linkages to deliver the drug, pH-responsive MSNs involve dynamic charge-based interactions between the gatekeeper, drug molecules, and tumor microenvironment. This simplicity in the mode of action confers more controlled release kinetics in redox-sensitive MSNs.

3.1.3 Enzyme

Enzymes play a myriad of functions in a biological system, crucial for maintaining tissue homeostasis. However, under pathological conditions such as cancer, specific enzymes like proteases, peptidases, phosphatases, and glycosidases exhibit aberrant expression due to altered metabolic pathways [82–85]. Drug delivery researchers have leveraged this altered enzyme expression in tumors to develop MSNs with gatekeepers that contain enzyme-cleavable motifs, enabling tumor-specific release of drugs [86, 87].

Strategies to Develop Enzyme-Responsive MSNs

Several enzymes were investigated to fabricate enzyme-responsive MSNs, wherein the nanoparticle surfaces are functionalized with enzyme-cleavable cross-linkers and gatekeepers. Some of these enzymes, along with their corresponding nanoparticle design, are as follows:

(i) *Matrix metalloproteinases (MMPs)*

MMPs are part of the zinc family of endopeptidases responsible for extracellular remodeling [88]. They are known to degrade proteins like collagen, elastin, and fibronectin [89, 90]. In cancer, specific MMPs (such as MMP-2 and -9) are overexpressed and play critical roles in cancer progression [91]. This has been leveraged by researchers to fabricate enzyme-responsive MSNs. These nanocarriers are either conjugated with (a) MMP-degradable macromolecules (like collagen or gelatin) on MSN surface that behave as stimuli-responsive gatekeepers [92–94], or (b) bulky gatekeepers linked via MMP-sensitive peptides (PLGVR: MMP-2; RSWMGLP: MMP-9) which are selectively cleaved in tumor microenvironment [95–98].

(ii) *Cathepsin B (CatB)*

CatB, part of the papain family of cysteine proteases, is aberrantly expressed both intracellularly and extracellularly in tumor [99–101]. Enzyme-responsive MSNs have been designed to display sensitivity to CatB by incorporating CatB-sensitive sequences such as GIVRAK, PGFK, GFLG, and CRRGGKKGGKKRK [102–104]. These sequences are used to link gatekeepers (such as polymers, peptides, and gold clusters) on the surface of MSNs, thereby facilitating drug release in response to elevated CatB levels in tumor.

(iii) *Hyaluronidase (Hyal)*

Hyal(s) are a type of endoglycosidases that degrade hyaluronic acid (HA), a glycosaminoglycan, which is an essential component of extracellular matrix [105, 106]. High molecular weight HA is degraded to low molecular weight fragments, which in turn, get internalized by cells via CD44 receptors [107]. Further, lysosomal Hyal degrades the low molecular weight fragments of HA into monosaccharides. By using low molecular weight HA as gatekeepers of MSNs, researchers have taken advantage of high expression of Hyal(s) and

CD44 receptor in cancer to develop a two-pronged approach: (a) cancer cell-specific drug release from MSNs due to lysosomal Hyal, and (b) HA-mediated targeted delivery of drug-loaded MSNs via CD44 receptor [108–111].

Enzyme-based stimuli-responsive MSNs offer notable advantages, particularly in their specificity and selectivity toward the enzymes within the tumor microenvironment. However, the efficiency of this strategy can be compromised by the heterogeneity of the tumor microenvironment, where the dysregulation of enzyme expression and activity is complex and dynamic across different cancer types and cancer stages [112].

3.2 Exogenous Stimuli

3.2.1 Light

Light-sensitive drug delivery via MSNs involves (a) the application of a light source focused on the target site and (b) the use of photosensitive molecules as gatekeepers to modify the surface of nanoparticles [113]. When these molecules absorb photons of enough energy to break their chemical bonds, they undergo photolysis, consequently, releasing the encapsulated drug [114, 115]. Photons spanning the ultraviolet/visible (UV/vis) spectra are widely reported for the application [115, 116]. However, UV/vis light presents a challenge in tissue penetration owing to its short wavelength. To address this drawback, researchers have employed near-infrared (NIR) radiation as it corresponds to a higher wavelength, thereby penetrating farther in tissues when compared to UV/vis light [117, 118]. NIR-responsive MSNs have demonstrated promising results in preclinical models for cancer therapy [119].

Strategies to Develop Light-Responsive MSNs

MSNs sensitive to NIR radiations have been designed by introducing lanthanides to the particles [120]. Lanthanide-doped materials exhibit upconversion-luminescence, wherein the material absorbs photons with wavelength in NIR region while emits photons in the UV/visible region of the electromagnetic spectrum [121, 122]. There are two major designs of upconversion mesoporous silica nanoparticles (UCNPs) employed in NIR-triggered controlled drug release, which are as follows:

(i) *Nanocages*

Nanocage UCNPs consist of mesoporous silica as the shell with lanthanide-doped core (e.g., Yb^{3+} and Tm^{3+}) [123, 124]. Polymer gatekeeping the mesopores are conjugated to the shell via photo-cleavable linkers sensitive to UV/vis light (e.g., nitrobenzyl derivates [114]), disallowing the release of drug molecules in the absence of an appropriate stimulus. Upon NIR irradiation, the core of the particle absorbs energy while emitting UV/vis radiation which, in turn, cleaves the linkers, resulting in the opening of pores and releasing of drug molecules [76]. This strategy circumvents the poor penetrability of short-wavelength light (i.e., UV/vis) in deep tissues, while taking advantage of the high energy of these radiations to cleave photo-sensitive linkers.

(ii) *Nanomotors*

While nanocages allow passive diffusion of the encapsulated drug molecules in the presence of a stimulus, nanomotors are designed to force the drug out of the mesopores via "photomechanical" thrust [125–127]. In this strategy, azobenzene molecules are linked in the mesopores, which undergo reversible isomerization between *cis* to *trans* forms in the presence of UV/blue radiation generated through the NIR upconversion [128, 129]. The change in the azobenzene structure physically propels the entrapped drug molecule out of the mesopores [130].

The utilization of UCNPs is not restricted to drug delivery, but its scope can be extended to in vivo imaging as well. UCNPs carry advantage over other NIR sensitive luminescent materials (e.g., organic dyes) as they exhibit an improved signal-to-noise ratio, reduced photo-bleaching, and deeper tissue penetration, with its excitation wavelength ranging over 800–1050 nm of the NIR spectrum [131]. This renders upconversion MSNs advantageous for both therapeutic and diagnostic purposes. Despite the advantages, UCNP-based therapeutics and diagnostics still face a major challenge. Although NIR penetrates deeper in the tissue when compared to UV/vis light, the depth of penetration may not be sufficient to reach more than a few centimeters below the skin [132]. This limits the usage of NIR-triggered drug delivery to subcutaneous tissues.

3.2.2 Heat

Thermo-responsive MSNs exhibit the potential to release their cargo in response to a difference in temperature. In this approach, thermo-responsive MSN delivery is coupled with an elevation of local tumor temperature, also known as hyperthermia therapy [133]. Hyperthermia therapy is an FDA-approved procedure for the treatment of all solid tumors, wherein an applicator emitting electromagnetic radiation (typically, microwaves and radio waves) is used to heat the tumor tissue up to 40–42 °C [134, 135]. This contrast in basal body temperature and tumor temperature is utilized to trigger drug release.

Strategy to Develop Heat-Responsive MSN

The central concept to incorporate heat-sensitivity to the MSNs is by incorporating thermo-responsive polymer as gatekeepers, as elaborated below:

(i) *Thermo-sensitive polymers as gatekeepers*

Commonly, materials exhibiting a temperature-dependent phase transition have been employed to modify the surface of MSNs. These materials exist in a water-soluble state below a temperature called lower critical solution temperature or LCST [136, 137]. Below LCST, there is an optimal formation of hydrogen bonds between water molecules and polymer chains. Thus, the polymer chains remain in an extended conformation which act as the gatekeepers of the mesopores of the nanoparticles. However, as the temperature is elevated beyond the LCST, these hydrogen bonds break, increasing the entropy of the polymer chains. This results in an enhanced intermolecular collision of the polymer chains, facilitating favorable interactions between these chains. Consequently, the polymer chains undergo aggregation and phase separation [136, 137]. The collapse of the gatekeeping chains results in the release of the encapsulated drug [138–140].

In the case of heat-triggered delivery of chemotherapeutic agents from MSNs, the role of hyperthermia is not restricted to acting as a stimulus alone but also as an adjuvant therapy [133, 141]. Moreover, the release kinetics of chemotherapeutic agents can be tuned by (a) changing the extent of surface modification, and (b) introducing additional chemical cross-links between the gatekeeping polymer chains on the MSN surface [142].

3.2.3 Ultrasound

Sound with a frequency greater than the upper human auditory limit (i.e., 20 kHz) is called ultrasound (US) [143]. US is widely used in clinics for biomedical imaging, physiotherapy, tumor ablation therapy, and echocardiography, among other applications [143]. In recent years, researchers have gained interest in using US as a stimulus for controlled drug delivery owing to its non-invasiveness, biosafety, and ability to penetrate deep tissue [144]. MSNs modified with US-sensitive components have exhibited potential as an on-demand drug delivery system for cancer therapy.

Strategies to Develop Ultrasound-Responsive MSNs

US results in "cavitation," a phenomenon wherein there is formation and collapse of gas bubbles (cavities) in a fluid as a result of rapid change in pressure [143, 145]. The collapse of the bubbles generates intense force resulting in a rapid increase of local temperature and pressure, which act on US-sensitive MSNs to release drugs. The major strategies to incorporate US sensitivity in MSNs are as follows:

(i) *Cavitation-sensitive mechanophores*

Certain chemical bonds undergo cleavage when subjected to US due to the generation of mechanical forces, as a consequence of cavitation [146]. The introduction of these molecules, called mechanophores, to the MSNs is one of the strategies employed to impart sensitivity to US [146]. One such example is that of 2-tetrahydropyranyl methacrylate (THPMA), a hydrophobic molecule consisting of US-labile acetal group [147, 148]. The cleavage of the acetal group results in the formation of methacrylic acid (MAA), a hydrophilic moiety. THPMA when conjugated with a polymer used for capping the MSN pores, can change the overall hydrophilicity of the polymeric network when subjected to US waves. Therefore, this transition of the gatekeeping polymer from an overall hydrophobic to hydrophilic state results in the release of the payload at the desired site.

(ii) *Cavitation-triggered diffusion*

The shock waves generated due to cavitation can also be employed to force the encapsulated drug out of the mesopores. In this strategy, MSNs consisting of hydrophobic internal channels are loaded with hydrophobic drug. Additionally, a hydrophilic polymer is used to cap MSNs. This particle design results in favorable interaction of the drug with the surface of the mesoporous channels while an unfavorable interaction with the gatekeeping polymer, facilitating retention of the drug molecules within the pores of the particle. The hydrophilic polymer readily absorbs water molecules on the surface of the particles, which consequently enhances cavitation upon exposure to US. The shock waves produced by enhanced cavitation forces the encapsulated drug molecules out of the mesopores resulting in an on-demand drug release [30, 31].

(iii) *Cavitation-labile gatekeepers*

Cavitation has also been used to disrupt MSN gatekeepers. For example, lipid membrane-coated MSNs, when exposed to shock waves produced by US, lose their integrity [149]. This leads to the removal of the gatekeeper and the release of the encapsulated drug from the pores of the particles.

Although US-responsive drug delivery displays promising results, a major caveat is the strong attenuation of US by calcified tissues such as bone. Therefore, US-responsive drug delivery systems may not be suitable for the treatment of head and lung cancers as cranial bones and rib cage act as barriers in the transmittance of sound waves, respectively [150, 151].

3.2.4 Magnetic Field

Magnetic field exhibits superior penetration in the body [30]. As a result, the application of magnetic fields, in the form of MRI scans, is a gold standard in diagnosis of pathologies concerning deep tissues [152–154]. The main strategy to develop magnetic-field-responsive MSNs is to incorporate iron oxide nanoparticles within the MSNs [155].

Strategy to Develop Magnetic-Field-Responsive MSNs

MSNs lack inherent sensitivity to magnetic field; therefore, incorporating magnetic materials is an essential strategy to develop magnetic-field-responsive MSNs. A widely reported strategy to achieve the same is introducing superparamagnetic iron oxide nanoparticles (SPIONs) to MSNs [155], the details of which are described below:

(i) *MSN-SPION hybrid particles*

SPIONs have gained interest as contrast agents in MRI [156]. However, they get heated in the presence of alternating magnetic field (AMF) [157]. This property of SPIONs has been leveraged by researchers to design MSN-SPION hybrid nanoparticles for magnetic field-induced, heat-triggered release of cargo [158]. The MSN-SPION hybrid nanoparticle system consists of SPION as core with mesoporous silica as shell, capped with thermo-responsive polymers [159–161]. In the presence of AMF, the temperature of SPION rises, facilitating the distribution of heat from core to shell of the nanoparticle. Consequently, in response to this increase in temperature, the thermosensitive cap (Table 8.2) renders the release of drug molecules encapsulated within the mesopores.

Alongside on-demand drug release, magnetic-field-responsive MSNs can also act as a contrast agent, aiding in biomedical imaging for diagnostic purposes [162]. Moreover, magnetic field can also direct magnetic nanoparticles to the target site, thereby further improving the enhanced permeation and retention of the delivery system in tumor [82].

3.3 Hybrid-Stimuli-Responsive MSNs

3.3.1 Dual-Stimuli-Responsive MSNs

The most commonly reported strategies for the fabrication of dual-stimuli-responsive MSNs entail MSNs responsive to pH along with either temperature or redox, as described below:

Strategies to Develop Dual-Stimuli-Responsive MSNs

(i) *pH-temperature*

The central principle of this approach is to modify the MSN surface with a thermo-responsive polymer and co-polymerize or modify it with proton-sensitive functional groups. An example of such a strategy is the incorporation of imidazole group (proton-sensitive) into NIPAAm (thermosensitive) [163]. Another example is to modify an appropriate gate-keeping polymer of MSNs with polyphosphazene (PPz), which undergoes conformational change with a change in both temperature and pH, resulting in faster drug release [164].

(ii) *pH-redox*

An interesting approach towards developing pH and redox-responsive MSNs is the doping of metal (e.g., calcium) as well as disulfide bonds (S-S) into the silica framework (Si-O-Si) [165]. The S-S bond reduces in the presence of elevated GSH levels in the tumor, resulting in the breakage of the bond. At the same time Ca-O bond breaks in acidic environment. Together, these modifications enhance the degradation of MSNs in the tumor microenvironment, resulting in the release of the payload. Another approach to achieve pH-redox dual sensitivity in MSNs is via host-guest chemistry. Here, the surface of MSNs is modified with benzene ring via benzoic imine bond (pH-sensitive) and disulfide bond (redox-sensitive) (MSN-SS-N=C-Ph). The benzene ring (host) interacts with β-cyclodextrin (guest) which caps the pores [166]. A relatively simplistic approach is the modification of MSNs with pH-sensitive polymers as gatekeepers (e.g., chitosan) via disulfide bonds. This approach is to accelerate drug release in the tumor when compared to healthy tissue [167].

3.3.2 Triple-Stimuli-Responsive MSNs

Triple-stimuli-responsive MSNs for cancer therapy is a relatively recent strategy. Most of the reported studies make use of light, tumor pH, and redox tumor microenvironment to design triple-stimuli-responsive MSNs. The strategy is described below:

Strategy to Develop Triple-Stimuli-Responsive MSNs

(i) *Light-pH-redox*

One of the reported strategies is the covalent attachment of NIR-responsive carbon nanodots (CD) on the surface of hollow MSNs via disulfide linkage [32]. The surface is also grafted with pH-sensitive polymer (such as citraconic anhydride-modified polylysine (PLL(cit)). PLL(cit) imparted pH-mediated charge conversion to the nanoparticles. This resulted in a prolonged systemic circulation of the nanoparticles (at pH 7.4) while enhanced cancer cell uptake at tumor pH (at pH 6.5). Upon internalization, the highly reducing environment would cleave the disulfide bond, thereby removing CDs, and subsequently releasing the encapsulated chemotherapeutic drug. The released CDs can be used as an imaging agent, as well as it presents the scope to be utilized in photo-thermal ablation of cancer cells [168]. Another strategy is to load NIR-responsive agents within the MSNs, while coating the surface with pH-responsive gatekeepers, via redox-sensitive linkages [169].

Examples of various stimuli-responsive mesoporous silica nanoparticles for cancer therapy are summarized in Table 8.1.

4 Translational Status and Challenges

The Food and Drug Administration (FDA) has classified silica as generally regarded as safe (GRAS) material [93]. Moreover, silica has a long history of being used in food, cosmetics, and pharmaceutical industries. Additionally, numerous studies have demonstrated biosafety of mesoporous silica nanoparticles in pre-clinical models. Studies in mouse models have revealed ~20 mg/kg of MSNs to be a safe dose [93]. However, nanoparticle size (32–142 nm) dependent investigations have revealed a higher accumulation of MSNs in liver and spleen of mouse models with an increase in size [173]. Besides, due to the slow degradation kinetics, MSNs pose a risk of bioaccumulation, which in turn can cause adverse effects in the long term [93]. Moreover, the incorporation of stimuli-responsive modifications further increases the size of MSNs [2], thereby increasing the risk of bioaccumulation. Because of easy clearance and low accumulation in reticuloendothelial organs,

Table 8.1 Examples of various stimuli-responsive mesoporous silica nanoparticles for cancer therapy

Endogenous stimuli-responsive mesoporous silica nanoparticles				
Stimulus; Strategy employed	*Cancer type; cell lines*	*Stimuli-responsive moiety*	*Application*	*References*
pH; Polyelectrolyte grafting	Breast Cancer; MCF-7 & MDA-MB-231	Mannosylated-poly (acrylic acid)	Acidic pH-triggered release of doxorubicin via protonation of polyacrylic acid	[42]
pH; Acid-sensitive linkages	Liver cancer; HepG2 cells Cervical cancer; HeLa cells	PEGMA-co-HEMA-g-DOX	Acidic pH-triggered release of doxorubicin grafted on the polymer via cleavage of acid-responsive aconitic anhydride	[52]
pH; Supramolecular nanovalves	Lung cancer; NCI-H460 cells Liver cancer; HEPG2 cells	β-cyclodextrin (mobile cyclic molecule) and benzimidazole (stationary stalk)	Acidic pH-triggered release of Quercetin due to a reduced binding constant between β-cyclodextrin and benzimidazole	[170]
pH; Acid decomposable materials	Cervical cancer; HeLa cells	ZnO quantum dots	Acid-sensitive dissolution of zinc oxide quantum dots capping the mesopores for controlled release of daunomycin	[59]
Redox; GSH-sensitive linkers	Cervical cancer; HeLa cells Breast cancer; SKBR3 cells	Disulfide bond bridged mesoporous organosilica	Glutathione-triggered disulfide cleavage in the nanoparticle framework for controlled release of doxorubicin	[75]
Redox; GSH-responsive macromolecules	Cervical cancer; HeLa cells	Disulfide cross-linked dialdehyde dextrin	Glutathione-triggered disulfide cleavage of gatekeeper to achieve controlled release of doxorubicin	[80]
Enzyme; MMP-responsive materials	Lung cancer; A549 cells	Collagen	MMP-2-triggered cleavage of collagen gatekeeper for controlled release of cisplatin	[92]
Enzyme; Hyaluronidase-responsive materials	Breast Cancer; MDA-MB-231 cells	Hyaluronic acid shell	Hyaluronidase-triggered cleavage of hyaluronic acid gatekeeper for controlled release of quercetin	[171]

(continued)

Table 8.1 (continued)

Enzyme; Cathepsin-B-responsive materials	Breast cancer; MCF-7 cells	Cathepsin-B cleavable peptide (DEGFLGED) linking resveratrol on MSN	Cathepsin-B-triggered cleavage of resveratrol-MSN complex in acidic pH for the controlled release of resveratrol	[172]
Exogenous stimuli-responsive mesoporous silica nanoparticles				
Stimulus: Strategy employed	*Cancer type; cell lines*	*Stimuli-responsive moiety*	*Application*	*References*
Light; Nanocages	Acute myeloid leukemia; MV-4-11 cells	Lanthanide-doping	Spatiotemporally controlled delivery of proteolysis targeting chimera for the degradation of BRD4 in tumors	[123]
Light; Nanomotors	Breast cancer, MCF-7 cells	Lanthanide-doping, photosensitizer Cys molecules	Nanomotors generated motion at 808 nm laser, facilitating deeper tissue penetration for an improved PDT/PTT	[126]
Heat; Thermo-sensitive polymers as gatekeepers	Breast cancer; MCF-7 cells	Poly (N-isopropyl acrylamide)	Temperature-sensitive delivery of doxorubicin at the tumor site above the LCST of the polymer.	[18]
Ultrasound; Cavitation-triggered diffusion	Prostate cancer; TRAMP cells	Perfluoro-decyltri-ethoxy-silane	Ultrasound-mediated doxorubicin release due to cavitation of superhydrophobic surface-coated MSNs	[31]
Ultrasound; Cavitation-labile gatekeepers	Breast cancer; MDA-MB-231 cells	Perfluoro-pentane	Ultrasound-mediated disruption of lipid coating MSNs for the triggered release of doxorubicin	[149]
Magnetic field; MSN-SPION hybrid particles	Breast cancer; MCF-7 cells	SPION	SPION-MSN core-shell nanoparticles coated with lipid, triggered doxorubicin release in the presence of magnetic field	[161]
Hybrid stimuli-responsive mesoporous silica nanoparticles				
Stimulus: Strategy employed	*Cancer type; cell lines*	*Stimuli-responsive moiety*	*Application*	*References*

(continued)

Table 8.1 (continued)

Dual stimuli; pH-temp	Lung cancer; A549 cells	Poly-phosphazene (PPz) (responsive to both pH and temperature)	Nanoparticle accumulation and release of camptothecin in the cells was significantly higher and faster in cancer cells compared to healthy cells in the dual-responsive MSN group	[164]
Dual stimuli; pH-redox	Breast cancer; MCF-7 and MDA-MB-231 cells	Chitosan conjugated; Disulfide linkages	Dual-stimuli-responsive MSNs exhibited better therapeutic outcomes in vitro in terms of migration assay and colony formation assay in cancer cells, as well as reduced cell viability.	[167]
Triple stimuli; Light-pH-redox	Cervical cancer, HeLa cells Breast cancer, 4T1 cells and MCF-7 cells	Citraconic anhydride-modified polylysine; Carbon dots; Disulfide linkages	Higher cellular uptake improved systemic circulation, and better therapeutic efficacy resulted in triple-stimuli-responsive MSNs. The particle system presents theranostic applications due NIR emitted by carbon dots.	[32]

GSH glutathione, *MMP* matrix metalloproteinase, *PDT* photodynamic therapy, *PTT* photothermal therapy, *SPION* superparamagnetic iron oxide, *MSN* mesoporous silica nanoparticles, *LCST* lower critical solution temperature

Cornell dots (ultra-small silica nanoparticles, ~20–25 nm) have received approval for clinical trials as a PET (positron emission tomography) tracer [174, 175].

Several other silica-based nanoparticles have progressed for clinical trials for various applications. For example, NANOM-FIM (silica-gold core-shell nanoparticles) for photothermal therapy in atherosclerosis patients [4, 174], Aura-Shell (silica-gold core-shell particles) for tumor ablation therapy of head and neck cancer [4, 174], and SiPore15™, mesoporous silica-based agent for the treatment of diabetes [4, 174]. However, MSNs are yet to progress to clinical trials as a drug delivery vehicle for cancer.

It is crucial to understand the challenges associated with MSN-based cancer therapy, particularly stimuli-responsive MSNs (in the interest of this chapter), to advance the drug delivery system to clinics. In the context of endogenous stimuli-triggered drug delivery, the system relies on the tumor microenvironment. However, the heterogeneous nature of cancer and similarities in the tumor microenvironment as compared to inflamed/injured tissue can result in compromised control over drug delivery and overall therapeutic efficacy of the system. On the other hand, a localized application of external stimuli renders greater precision and spatiotemporal control over drug delivery. However, it may pose a challenge in the treatment of advanced stages of cancer due to systemic dissemination of cancer cells.

Therefore, it is evident that advancements may still be needed for stimuli-responsive MSNs in cancer therapy to transition from bench to bedside.

A brief summary of pre-clinical studies of stimuli-responsive mesoporous silica nanoparticles for cancer therapy is presented in Table 8.2.

5 Conclusion and Future Perspectives

Mesoporous silica nanoparticles are inorganic biomaterials which are biocompatible, biodegradable, exhibit a high surface area to volume ratio, and can be synthesized to be monodispersed. Moreover, the MSN fabrication techniques can be highly reproducible and scalable. Additionally, they demonstrate tunable physicochemical properties, ease of surface functionalization, and high drug loading capacity. Taken together, these properties of MSNs have garnered interest for their use in cancer therapy. However, unmodified MSNs can result in premature drug release in circulation, thereby causing off-target effects. To overcome this challenge researchers have designed stimuli-responsive MSNs for an on-demand controlled release of the payload, thereby increasing the specificity and improving bioavailability of drug

Table 8.2 Summary of pre-clinical studies of stimuli-responsive mesoporous silica nanoparticles for cancer therapy

Endogenous stimuli-responsive mesoporous silica nanoparticles			
Stimulus	*Animal model*	*Result*	*Reference*
pH	Syngeneic BALB/c mice model (4T-1; Breast cancer)	pH-responsive MSNs resulted in the smallest tumor weight, volume, and highest growth inhibition. Nuclear deformation and atrophy observed after 17 days in treatment group.	[58]
Redox	Xenogeneic BALB/c nude mice model (Hep-G2; Liver cancer)	Redox-sensitive MSNs resulted in more than 65% tumor inhibition compared to 34% in free treatment. Reduced cell density and increased nuclear atrophy in tumor tissue in the treatment group.	[176]
Enzyme	Syngeneic Kunming mice model (H22; Liver cancer)	HA-coated MSNs demonstrated both active targeting and enzyme responsiveness. As a result, drug retention in tumor tissue lasted 72 hours with minimal distribution to other organs.	[177]

(continued)

Table 8.2 (continued)

Exogenous stimuli-responsive mesoporous silica nanoparticles			
Stimulus	*Animal model*	*Result*	*Reference*
Light	Xenogeneic nude mice model (MV-4-11; Acute myeloid leukemia)	Light-responsive MSNs resulted in a tumor growth inhibition value of 58.7%. Apoptosis and nuclear fragmentation were observed in the treatment group.	[124]
Heat	Xenogeneic nude mice model (SKBR-3; Breast cancer)	HER2+-targeted pNIPAm-coated MSNs loaded with doxorubicin resulted in significant tumor regression compared to both free drug and control.	[178]
Ultrasound	Syngeneic C57BL/6JNarl mice model (TRAMP; Prostate cancer)	Ultrasound-responsive MSNs exhibited improved tumor penetration, higher tumor regression as well as prolonged survival compared to the relevant control groups	[31]
Magnetic field	Syngeneic BALB/c mice model (C26; Colorectal cancer)	Higher tumor accumulation of SPION-MSNs at 6 h and 18 h post-injection; significant tumor regression in mice subjected to gene therapy via SPION-MSNs	[155]
Hybrid stimuli-responsive mesoporous silica nanoparticles			
Stimulus	*Animal model*	*Result*	*Reference*
Dual-stimuli	Syngeneic BALB/c mice model (4T-1; Breast cancer)	Dual-stimuli-responsive MSNs resulted in better survival outcomes when compared to stimuli-non-responsive MSNs.	[167]
Triple-stimuli	Syngeneic BALB/c mice model (EAC; Breast cancer)	Triple-stimuli-responsive MSNs resulted in better pre-clinical outcomes when compared to dual stimuli-responsive MSNs.	[32]

pNIPAm poly(N-isopropylacrylamide), *HA* hyaluronic acid, *SPION* superparamagnetic iron oxide, *MSN* mesoporous silica nanoparticles

in tumor. This chapter summarizes the strategies employed to design MSNs responsive to various endogenous and exogenous stimuli.

Continuous efforts in improving and standardizing the design of smart MSNs to achieve a better therapeutic outcome have resulted in promising pre-clinical findings. One approach in the improvement of the design of smart MSNs is the incorporation of dual or multi-stimuli-responsiveness in the system. This provides a more stringent approach for the release of drug specifically at the target site. However, literature constitutes limited pre-clinical studies validating the safety and therapeutic efficacy of hybrid systems compared to single-stimulus-responsive MSNs. The future holds the scope in evaluating the efficacy of stimuli-responsive MSNs for precision medicine in better pre-clinical models (such as patient-derived xenografts) and higher animal models (such as hamster, guinea pigs, and rabbits). This may hold the potential to significantly improve therapeutic outcomes and bring stimuli-responsive MSNs closer to clinical applications.

Acknowledgements The authors thank the Department of Biological Sciences and Bioengineering and the Mehta Family Centre for Engineering in Medicine at IIT Kanpur for their generous support. ZSZ and SVNT acknowledge the financial support from the Ministry of Education, Government of India. All the authors acknowledge the Department of Biotechnology for research funding. DSK acknowledges the Gireesh Jankinath Chair Professorship (2018–2021) and Rajeeva and Sangeeta Lahri Chair Professorship (2022–2024). DSK is also thankful to DBT for TATA Innovation Fellowship (2023–2028).

Author Contributions Writing, reviewing, and editing: ZSZ and SVNT

Reviewing and Supervision DSK

Declaration of Generative AI and AI-Assisted Technologies In the preparation of the manuscript Grammarly, Inc., Gemini, and ChatGPT were solely used for language correction. No generative AI was used in the writing of the manuscript, figures, or tables.

References

1. Manzano M, Vallet-Regí M. Mesoporous silica nanoparticles for drug delivery. Adv Funct Mater. 2020;30(2):3–5.
2. Lohiya G, Katti DS. Carboxylated chitosan mediated improved efficacy of mesoporous silica nanoparticle-based targeted drug delivery system for breast cancer therapy. Carbohydr Polym. 2021;277(2021):118822.
3. Vallet-Regí M. Our contributions to applications of mesoporous silica nanoparticles. Acta Biomater. 2022;137:44.
4. Xu B, Li S, Shi R, Liu H. Multifunctional mesoporous silica nanoparticles for biomedical applications. Signal Transduct Target Ther. 2023;8:435.
5. Sun L, Sogo Y, Wang X, Ito A. Biosafety of mesoporous silica nanoparticles: a combined experimental and literature study. J Mater Sci Mater Med [Internet]. 2021 [cited 2025 Feb 15];32(9):1–15. Available from: https://link.springer.com/article/10.1007/s10856-021-06582-y
6. Kankala RK, Han YH, Na J, Lee CH, Sun Z, Bin WS, et al. Nanoarchitectured structure and surface biofunctionality of mesoporous silica nanoparticles. Adv Mater. 2020;32:1907035.
7. Hong X, Zhong X, Du G, Hou Y, Zhang Y, Zhang Z, et al. The pore size of mesoporous silica nanoparticles regulates their antigen delivery efficiency. Sci Adv. 2020;6(25):eaaz4462.
8. Xu C, Lei C, Wang Y, Yu C. Dendritic mesoporous nanoparticles: structure, synthesis and properties. Angew Chem Int Edn. 2022;61:e202112752.
9. Madathiparambil Visalakshan R, González García LE, Benzigar MR, Ghazaryan A, Simon J, Mierczynska-Vasilev A, et al. The influence of nanoparticle shape on protein Corona formation. Small. 2020;16(25):e2000285.
10. Suteewong T, Sai H, Hovden R, Muller D, Bradbury MS, Gruner SM, et al. Multicompartment mesoporous silica nanoparticles with branched shapes: An epitaxial growth mechanism. Science (1979). 2013;340(6130):337.
11. Pham LM, Kim EC, Ou W, Phung CD, Nguyen TT, Pham TT, et al. Targeting and clearance of senescent foamy macrophages and senescent endothelial cells by antibody-functionalized mesoporous silica nanoparticles for alleviating aorta atherosclerosis. Biomaterials. 2021;269:120677.
12. Li Z, Xu K, Qin L, Zhao D, Yang N, Wang D, et al. Hollow nanomaterials in advanced drug delivery systems: from single- to multiple shells. Adv Mater [Internet]. 2023 [cited 2025 Feb 15];35(12):2203890. Available from: https://onlinelibrary.wiley.com/doi/full/10.1002/adma.202203890.

13. Thananukul K, Kaewsaneha C, Opaprakasit P, Lebaz N, Errachid A, Elaissari A. Smart gating porous particles as new carriers for drug delivery. Adv Drug Deliv Rev. 2021;174:425–46.
14. Frickenstein AN, Hagood JM, Britten CN, Abbott BS, McNally MW, Vopat CA, et al. Mesoporous silica nanoparticles: properties and strategies for enhancing clinical effect. Pharmaceutics. 2021 [cited 2025 Feb 15];13(4):570. Available from: https://www.mdpi.com/1999-4923/13/4/570/htm.
15. da Cruz SA, Albuquerque LJC, Mondo GB, Ceolin M, Picco AS, Cardoso MB. Colloidal stability and degradability of silica nanoparticles in biological fluids: a review. J Sol-Gel Sci Technol. 2022 [cited 2025 Feb 15];102(1):41–62. Available from: https://link.springer.com/article/10.1007/s10971-021-05695-8.
16. Moya SE, Hernández RR, Angelomé PC. Degradation of mesoporous silica materials in biological Milieu: the gateway for therapeutic applications. Adv Nanobiomed Res. 2024;4:2400005.
17. Khalbas AH, Albayati TM, Saady NMC, Zendehboudi S, Salih IK, Tofah ML. Insights into drug loading techniques with mesoporous silica nanoparticles: optimization of operating conditions and assessment of drug stability. J Drug Deliv Sci Technol. 2024;96:105698.
18. Porrang S, Rahemi N, Davaran S, Mahdavi M, Hassanzadeh B. Synthesis of temperature/pH dual-responsive mesoporous silica nanoparticles by surface modification and radical polymerization for anti-cancer drug delivery. Colloids Surf A Physicochem Eng Asp. 2021;623:126719.
19. Vallet-Regi M, Rámila A, Del Real RP, Pérez-Pariente J. A new property of MCM-41: drug delivery system. Chem Mater. 2001;13(2):308.
20. Maeda H, Wu J, Sawa T, Matsumura Y, Hori K. Tumor vascular permeability and the EPR effect in macromolecular therapeutics: a review. J Control Release. 2000;65(1–2):271–84.
21. Kang H, Rho S, Stiles WR, Hu S, Baek Y, Hwang DW, et al. Size-dependent EPR effect of polymeric nanoparticles on tumor targeting. Adv Healthc Mater. 2020;9(1):8–15.
22. Matsumura Y, Maeda H. A new concept for macromolecular therapeutics in cancer chemotherapy: mechanism of Tumoritropic accumulation of proteins and the antitumor agent Smancs. Cancer Res. 1986;46(8):6387.
23. Mitchell MJ, Billingsley MM, Haley RM, Wechsler ME, Peppas NA, Langer R. Engineering precision nanoparticles for drug delivery. Nat Rev Drug Discov. 2021;20:101.
24. Cui Y, Hong S, Jiang W, Li X, Zhou X, He X, et al. Engineering mesoporous bioactive glasses for emerging stimuli-responsive drug delivery and theranostic applications. Bioact Mater. 2024;34:436.
25. Mohanan S, Guan X, Liang M, Karakoti A, Vinu A. Stimuli-responsive silica Silanol conjugates: strategic nanoarchitectonics in targeted drug delivery. Small. 2023;20:2301113.
26. Feng Y, Liao Z, Li M, Zhang H, Li T, Qin X, et al. Mesoporous silica nanoparticles-based nanoplatforms: basic construction, current state, and emerging applications in anticancer therapeutics. Adv Healthc Mater. 2023;12:e2201884.
27. Wagner J, Gößl D, Ustyanovska N, Xiong M, Hauser D, Zhuzhgova O, et al. Mesoporous silica nanoparticles as pH-responsive carrier for the immune-activating drug Resiquimod enhance the local immune response in mice. ACS Nano. 2021;15(3):4450.
28. Cai Y, Deng T, Pan Y, Zink JI. Use of Ferritin capped mesoporous silica nanoparticles for Redox and pH triggered drug release in vitro and in vivo. Adv Funct Mater. 2020;30(39):2002043.
29. Hou J, Zhao Y, Sun L, Zou X. Enzyme/GSH/pH-responsive hyaluronic acid grafted porous silica nanocarriers bearing Ag2S QDs for fluorescence imaging and combined therapy. Carbohydr Polym. 2023;305:120547.
30. Lin FC, Xie Y, Deng T, Zink JI. Magnetism, ultrasound, and light-stimulated mesoporous silica nanocarriers for theranostics and beyond. J Am Chem Soc. 2021;143:6025.
31. Ho YJ, Wu CH, Jin Q, Feng LCY, Chiang PH, Wu N, et al. Superhydrophobic drug-loaded mesoporous silica nanoparticles capped with β-cyclodextrin for ultrasound image-guided combined antivascular and chemo-sonodynamic therapy. Biomaterials. 2020;232:119723.

32. Chen Z, Liao T, Wan L, Kuang Y, Liu C, Duan J, et al. Dual-stimuli responsive near-infrared emissive carbon dots/hollow mesoporous silica-based integrated theranostics platform for real-time visualized drug delivery. Nano Res. 2021;14(11):4264–73.
33. Boedtkjer E, Pedersen SF. The acidic tumor microenvironment as a driver of cancer. Annu Rev Physiol. 2020;82:103.
34. DeBerardinis RJ, Chandel NS. We need to talk about the Warburg effect. Nat Metab. 2020;2:127.
35. Gatenby RA, Gillies RJ. Why do cancers have high aerobic glycolysis? Nat Rev Cancer. 2004;4:891.
36. Brown TP, Ganapathy V. Lactate/GPR81 signaling and proton motive force in cancer: role in angiogenesis, immune escape, nutrition, and Warburg phenomenon. Pharmacol Ther. 2020;206:107451.
37. Li X, Yang Y, Zhang B, Lin X, Fu X, An Y, et al. Lactate metabolism in human health and disease. Signal Transduct Target Ther. 2022;7:305.
38. Ding H, Tan P, Fu S, Tian X, Zhang H, Ma X, et al. Preparation and application of pH-responsive drug delivery systems. J Control Release. 2022;348:206.
39. Narayan R, Gadag S, Cheruku SP, Raichur AM, Day CM, Garg S, et al. Chitosan-glucuronic acid conjugate coated mesoporous silica nanoparticles: a smart pH-responsive and receptor-targeted system for colorectal cancer therapy. Carbohydr Polym. 2021;261:117893.
40. Chen S, Huang R, Shen F, Wu Y, Lin Y, Yang X, et al. Enhancing antitumor immunity with stimulus-responsive mesoporous silicon in combination with chemotherapy and photothermal therapy. Biomater Sci. 2024;12:3826.
41. Edalatian Tavakoli S, Motavalizadehkakhky A, Homayouni Tabrizi M, Mehrzad J, Zhiani R. Study of the anti-cancer activity of a mesoporous silica nanoparticle surface coated with polydopamine loaded with umbelliprenin. Sci Rep. 2024;14(1):1–12.
42. Lee H, Choi M, Kim HE, Jin M, Jeon WJ, Jung M, et al. Mannosylated poly(acrylic acid)-coated mesoporous silica nanoparticles for anticancer therapy. J Control Release. 2022;349:241.
43. Zhang Y, Kim I, Lu Y, Xu Y, Yu DG, Song W. Intelligent poly(L-histidine)-based nanovehicles for controlled drug delivery. J Control Release. 2022;349:963.
44. Lahmadi S, Alamery S, Beagan A, Alotaibi K, Alswieleh A. Advanced hybrid silica nanoparticles with pH-responsive diblock copolymer brushes: optimized design for controlled doxorubicin loading and release in cancer therapy. RSC Adv. 2024;14(13):8819–28.
45. Lou XY, Yang YW. Pillar[n]arene-based supramolecular switches in solution and on surfaces. Adv Mater. 2020;32:e2003263.
46. Guo F, Li G, Zhou H, Ma S, Guo L, Liu X. Temperature and H2O2-operated nano-valves on mesoporous silica nanoparticles for controlled drug release and kinetics. Colloids Surf B Biointerfaces. 2020;187:110643.
47. Da Silva AFM, Da Costa NM, Fernandes TS, Bessa IAA, D'Amato DL, Senna CA, et al. Responsive supramolecular devices assembled from Pillar[5]arene nanogate and mesoporous silica for Cargo release. ACS Appl Nano Mater. 2022;5:13805.
48. Zhao T, Chen L, Liu M, Lin R, Cai W, Te HC, et al. Emulsion-oriented assembly for Janus double-spherical mesoporous nanoparticles as biological logic gates. Nat Chem. 2023;15(6):832.
49. Wang X, Li G, Li K, Shi Y, Lin W, Pan C, et al. Controlled-release of apatinib for targeted inhibition of osteosarcoma by supramolecular nanovalve-modified mesoporous silica. Front Bioeng Biotechnol. 2023;11:1135655.
50. Mladenović M, Morgan I, Ilić N, Saoud M, Pergal MV, Kaluđerović GN, et al. Ph-responsive release of ruthenium metallotherapeutics from mesoporous silica-based nanocarriers. Pharmaceutics. 2021;13(4):460.
51. Zhang Y, Ye Z, He R, Li Y, Xiong B, Yi M, et al. Bovine serum albumin-based and dual-responsive targeted hollow mesoporous silica nanoparticles for breast cancer therapy. Colloids Surf B Biointerfaces. 2023;224:113201.

52. Peng S, Zhang F, Huang B, Wang J, Zhang L. Mesoporous silica nanoprodrug encapsulated with near-infrared absorption dye for photothermal therapy combined with chemotherapy. ACS Appl Bio Mater. 2021;4(12):8225.
53. Li S, Hu L, Li D, Wang X, Zhang P, Wang J, et al. Carboxymethyl chitosan-based nanogels via acid-labile ortho ester linkages mediated enhanced drug delivery. Int J Biol Macromol. 2019;129:477.
54. Chu S, Shi X, Tian Y, Gao F. pH-responsive polymer nanomaterials for tumor therapy. Front Oncol. 2022;12:e1800917.
55. Saputra OA, Lestari WA, Kurniansyah V, Lestari WW, Sugiura T, Mukti RR, et al. Organically surface engineered mesoporous silica nanoparticles control the release of quercetin by pH stimuli. Sci Rep. 2022;12(1):20661.
56. Li T, Chen X, Liu Y, Fan L, Lin L, Xu Y, et al. pH-sensitive mesoporous silica nanoparticles anticancer prodrugs for sustained release of ursolic acid and the enhanced anti-cancer efficacy for hepatocellular carcinoma cancer. Eur J Pharm Sci. 2017;96:456.
57. Li Y, Liu J, He J, Dey A, Bui VD, Park JH. Recent advances in stimuli-responsive self-immolative polymers for drug delivery and molecular imaging. Chem Mater. 2024;36(9):4054–77.
58. Wang Y, Zhao K, Xie L, Li K, Zhang W, Xi Z, et al. Construction of calcium carbonate-liposome dual-film coated mesoporous silica as a delayed drug release system for antitumor therapy. Colloids Surf B Biointerfaces. 2022;212:112357.
59. Liu M, Sun X, Liao Z, Li Y, Qi X, Qian Y, et al. Zinc oxide end-capped Fe3O4@mSiO2 core-shell nanocarriers as targeted and responsive drug delivery system for chemo−/ions synergistic therapeutics. Drug Deliv. 2019;26(1):732.
60. Alsawaftah NM, Awad NS, Pitt WG, Husseini GA. pH-responsive nanocarriers in cancer therapy. Polymers. 2022;14:936.
61. Toft NJ, Axelsen TV, Pedersen HL, Mele M, Burton M, Balling E, et al. Acid-base transporters and ph dynamics in human breast carcinomas predict proliferative activity, metastasis, and survival. elife. 2021;10:e68447.
62. Wu K, El Zowalaty AE, Sayin VI, Papagiannakopoulos T. The pleiotropic functions of reactive oxygen species in cancer. Nat Cancer. 2024;5:384.
63. Hecht F, Zocchi M, Alimohammadi F, Harris IS. Regulation of antioxidants in cancer. Mol Cell. 2024;84:23.
64. Yoo HC, Yu YC, Sung Y, Han JM. Glutamine reliance in cell metabolism. Exp Mol Med. 2020;52:1496.
65. Cheung EC, Vousden KH. The role of ROS in tumour development and progression. Nat Rev Cancer. 2022;22:280.
66. Hayes JD, Dinkova-Kostova AT, Tew KD. Oxidative stress in cancer. Cancer Cell. 2020;38:167.
67. Ding Y, Dai Y, Wu M, Li L. Glutathione-mediated nanomedicines for cancer diagnosis and therapy. Chem Eng J. 2021;426:128880.
68. Zhu L, Li W, Liu C, Yue S, Qiao Y, Cui Y, et al. Glutathione-sensitive mesoporous nanoparticles loaded with cinnamaldehyde for chemodynamic and immunological therapy of cancer. J Mater Chem B. 2023;11(36):8717.
69. Yang G, Kim S, Oh JY, Kim D, Jin S, Choi E, et al. Surface protein-retractive and redox-degradable mesoporous organosilica nanoparticles for enhanced cancer therapy. J Colloid Interface Sci. 2023;649:1014.
70. Shang Y, Du J, Wang B, Lu P, Zhao Y, Yuan J. Preparation of MSNs@Keratin as pH/GSH dual responsive drug delivery system. J Biomater Sci Polym Ed. 2022;33(11):1369.
71. Yang Y, Wu H, Liu B, Liu Z. Tumor microenvironment-responsive dynamic inorganic nanoassemblies for cancer imaging and treatment. Adv Drug Deliv Rev. 2021;179:114004.
72. Liang J, Ling J, Zhang X, Ouyang X, Kun OAM, Yang G. pH/glutathione dual-responsive copper sulfide-coated organic mesoporous silica for synergistic chemo-photothermal therapy. J Colloid Interface Sci. 2024;657:1.

73. Li J, Tian H, Zhu F, Jiang S, He M, Li Y, et al. Amorphous ultra-small Fe-based nanocluster engineered and ICG loaded Organo-mesoporous silica for GSH depletion and photothermal-chemodynamic synergistic therapy. Adv Healthc Mater. 2022;11(21):e2201986.
74. Zhang P, Wu J, Xiao F, Zhao D, Luan Y. Disulfide bond based polymeric drug carriers for cancer chemotherapy and relevant redox environments in mammals. Med Res Rev. 2018;38:1485.
75. Zheng Q, Zou T, Wang W, Zhang C, Hu S, Cheng X, et al. Necroptosis-mediated synergistic photodynamic and glutamine-metabolic therapy enabled by a biomimetic targeting nanosystem for Cholangiocarcinoma. Adv Sci. 2024;2309203:1–18.
76. Yan H, Dong J, Huang X, Du X. Protein-gated upconversion nanoparticle-embedded mesoporous silica nanovehicles via Diselenide linkages for drug release tracking in real time and tumor chemotherapy. ACS Appl Mater Interfaces. 2021;13(24):29070.
77. Xu J, Shi R, Chen G, Dong S, Yang P, Zhang Z, et al. All-in-one theranostic nanomedicine with ultrabright second near-infrared emission for tumor-modulated bioimaging and chemodynamic/photodynamic therapy. ACS Nano. 2020;14(8):9613.
78. Du J, Wang L, Han X, Dou J, Jiang X, Yuan J. Polydopamine/keratin complexes as gatekeepers of mesoporous silica nanoparticles for pH and GSH dual responsive drug delivery. Mater Lett. 2021;293:129676.
79. Zhu D, Hu C, Liu Y, Chen F, Zheng Z, Wang X. Enzyme-/Redox-responsive mesoporous silica nanoparticles based on functionalized Dopamine as nanocarriers for cancer therapy. ACS Omega. 2019;4(4):6097.
80. Chen C, Sun W, Yao W, Wang Y, Ying H, Wang P. Functional polymeric dialdehyde dextrin network capped mesoporous silica nanoparticles for pH/GSH dual-controlled drug release. RSC Adv. 2018;8(37):20862.
81. Luo W, Xu X, Zhou B, He P, Li Y, Liu C. Formation of enzymatic/redox-switching nanogates on mesoporous silica nanoparticles for anticancer drug delivery. Mater Sci Eng C. 2019;100:855.
82. Moradi Kashkooli F, Soltani M, Souri M. Controlled anti-cancer drug release through advanced nano-drug delivery systems: static and dynamic targeting strategies. J Control Release. 2020;327:316.
83. Khaket TP, Kwon TK, Kang SC. Cathepsins: potent regulators in carcinogenesis. Pharmacol Ther. 2019;198:1.
84. Stanford SM, Bottini N. Targeting protein phosphatases in cancer immunotherapy and autoimmune disorders. Nat Rev Drug Discov. 2023;22:273.
85. Martin H, Lázaro LR, Gunnlaugsson T, Scanlan EM. Glycosidase activated prodrugs for targeted cancer therapy. Chem Soc Rev. 2022;51:9694.
86. Li Y, Mei T, Han S, Han T, Sun Y, Zhang H, et al. Cathepsin B-responsive nanodrug delivery systems for precise diagnosis and targeted therapy of malignant tumors. Chin Chem Lett. 2020;31(12):3027.
87. Naz S, Wang M, Han Y, Hu B, Teng L, Zhou J, et al. Enzyme-responsive mesoporous silica nanoparticles for tumor cells and mitochondria multistage-targeted drug delivery. Int J Nanomedicine. 2019;14:2533.
88. Monteiro P, Remy D, Lemerle E, Routet F, Macé AS, Guedj C, et al. A mechanosensitive caveolae–invadosome interplay drives matrix remodelling for cancer cell invasion. Nat Cell Biol. 2023;25(12):1787.
89. de Castro Brás LE, Frangogiannis NG. Extracellular matrix-derived peptides in tissue remodeling and fibrosis. Matrix Biol. 2020;91–92:176.
90. Laronha H, Caldeira J. Structure and function of human matrix metalloproteinases. Cells. 2020;9:1076.
91. Yuan Z, Li Y, Zhang S, Wang X, Dou H, Yu X, et al. Extracellular matrix remodeling in tumor progression and immune escape: from mechanisms to treatments. Mol Cancer. 2023;22:48.
92. Vaghasiya K, Ray E, Sharma A, Katare OP, Verma RK. Matrix Metalloproteinase-responsive mesoporous silica nanoparticles cloaked with cleavable protein for “self-actuating” on-demand controlled drug delivery for cancer therapy. ACS Appl Bio Mater. 2020;3(8):4987.

93. Lérida-Viso A, Estepa-Fernández A, García-Fernández A, Martí-Centelles V, Martínez-Máñez R. Biosafety of mesoporous silica nanoparticles; towards clinical translation. Adv Drug Deliv Rev. 2023;201:115049.
94. Zhou J, Wang M, Ying H, Su D, Zhang H, Lu G, et al. Extracellular matrix component shelled nanoparticles as dual enzyme-responsive drug delivery vehicles for cancer therapy. ACS Biomater Sci Eng. 2018;4(7):2404.
95. Liu L, Chu H, Yang J, Sun Y, Ma P, Song D. Construction of a magnetic-fluorescent-plasmonic nanosensor for the determination of MMP-2 activity based on SERS-fluorescence dual-mode signals. Biosens Bioelectron. 2022;212:114389.
96. Liang T, Zhang B, Xing Z, Dong Y, Xu H, Chen X, et al. Adapting and remolding: orchestrating tumor microenvironment normalization with photodynamic therapy by size transformable nanoframeworks. Angew Chem Int Edn. 2021;60(20):11464.
97. Vallet-Regí M, Schüth F, Lozano D, Colilla M, Manzano M. Engineering mesoporous silica nanoparticles for drug delivery: where are we after two decades? Chem Soc Rev. 2022;51:5365.
98. Gisbert-Garzarán M, Vallet-Regí M. Influence of the surface functionalization on the fate and performance of mesoporous silica nanoparticles. Nanomaterials. 2020;10:916.
99. Shi Q, Shen Q, Liu Y, Shi Y, Huang W, Wang X, et al. Increased glucose metabolism in TAMs fuels O-GlcNAcylation of lysosomal Cathepsin B to promote cancer metastasis and chemoresistance. Cancer Cell. 2022;40(10):1207.
100. Petruzzella A, Bruand M, Santamaria-Martínez A, Katanayeva N, Reymond L, Wehrle S, et al. Antibody–peptide conjugates deliver covalent inhibitors blocking oncogenic cathepsins. Nat Chem Biol. 2024;20:1188.
101. Biasizzo M, Javoršek U, Vidak E, Zarić M, Turk B. Cysteine cathepsins: a long and winding road towards clinics. Mol Asp Med. 2022;88:101150.
102. Yoon MC, Hook V, O'Donoghue AJ. Cathepsin B dipeptidyl carboxypeptidase and endopeptidase activities demonstrated across a broad pH range. Biochemistry. 2022;61(17):1904.
103. Zheng F, Zhang P, Xi Y, Huang K, Min Q, Zhu JJ. Peptide-mediated core/satellite/shell multifunctional nanovehicles for precise imaging of cathepsin B activity and dual-enzyme controlled drug release. NPG Asia Mater. 2017;9(3):e366.
104. Cheng YJ, Luo GF, Zhu JY, Xu XD, Zeng X, Cheng DB, et al. Enzyme-induced and tumor-targeted drug delivery system based on multifunctional mesoporous silica nanoparticles. ACS Appl Mater Interfaces. 2015;7(17):9078.
105. Martinez-Ordoñez A, Duran A, Ruiz-Martinez M, Cid-Diaz T, Zhang X, Han Q, et al. Hyaluronan driven by epithelial aPKC deficiency remodels the microenvironment and creates a vulnerability in mesenchymal colorectal cancer. Cancer Cell. 2023;41(2):252.
106. Winkler J, Abisoye-Ogunniyan A, Metcalf KJ, Werb Z. Concepts of extracellular matrix remodelling in tumour progression and metastasis. Nat Commun. 2020;11:5120.
107. Weng X, Maxwell-Warburton S, Hasib A, Ma L, Kang L. The membrane receptor CD44: novel insights into metabolism. Trends Endocrinol Metab. 2022;33:318.
108. Müller S, Sindikubwabo F, Cañeque T, Lafon A, Versini A, Lombard B, et al. CD44 regulates epigenetic plasticity by mediating iron endocytosis. Nat Chem. 2020;12(10):929.
109. Chen C, Zhao S, Zhao X, Cao L, Karnad A, Kumar AP, et al. Gemcitabine resistance of pancreatic cancer cells is mediated by IGF1R dependent upregulation of CD44 expression and isoform switching. Cell Death Dis. 2022;13(8):682.
110. Elakad O, Häupl B, Labitzky V, Yao S, Küffer S, von Hammerstein-Equord A, et al. Activation of CD44/PAK1/AKT signaling promotes resistance to FGFR1 inhibition in squamous-cell lung cancer. NPJ Precis Oncol. 2022;6(1):52.
111. Anjum S, Naseer F, Ahmad T, Jahan F, Qadir H, Gul R, et al. Enhancing therapeutic efficacy: sustained delivery of 5-fluorouracil (5-FU) via thiolated chitosan nanoparticles targeting CD44 in triple-negative breast cancer. Sci Rep. 2024;14(1):1–18.
112. Li M, Zhao G, Su WK, Shuai Q. Enzyme-responsive nanoparticles for anti-tumor drug delivery. Front Chem. 2020;8:647.

113. Tao Y, Chan HF, Shi B, Li M, Leong KW. Light: a magical tool for controlled drug delivery. Adv Funct Mater. 2020;30:2005029.
114. Levalley PJ, Neelarapu R, Sutherland BP, Dasgupta S, Kloxin CJ, Kloxin AM. Photolabile linkers: exploiting labile bond chemistry to control mode and rate of hydrogel degradation and protein release. J Am Chem Soc. 2020;142(10):4671.
115. Rivero-Buceta E, Encheva ME, Cech B, Fernandez E, Sastre G, Landry CC, et al. Light-activated controlled release of camptothecin by engineering porous materials: the ship in a bottle concept in drug delivery. Nanoscale. 2023;15(30):12506.
116. Yang Y, Chen F, Xu N, Yao Q, Wang R, Xie X, et al. Red-light-triggered self-destructive mesoporous silica nanoparticles for cascade-amplifying chemo-photodynamic therapy favoring antitumor immune responses. Biomaterials. 2022;281:121368.
117. Sana B, Finne-Wistrand A, Pappalardo D. Recent development in near infrared light-responsive polymeric materials for smart drug-delivery systems. Mater Today Chem. 2022;25:100963.
118. Yang X, Li M, Liang J, Hou X, He X, Wang K. NIR-controlled treatment of multidrug-resistant tumor cells by mesoporous silica capsules containing gold nanorods and Doxorubicin. ACS Appl Mater Interfaces. 2021;13(13):14894.
119. Karmakar A, Silswal A, Koner AL. Review of NIR-responsive '"Smart"' carriers for photothermal chemotherapy. J Mater Chem B. 2024;12:4785–808.
120. Yi Z, Luo Z, Qin X, Chen Q, Liu X. Lanthanide-activated nanoparticles: a toolbox for bioimaging, therapeutics, and neuromodulation. Acc Chem Res. 2020;53(11):2692.
121. Dong H, Sun LD, Yan CH. Local structure engineering in lanthanide-doped nanocrystals for tunable upconversion emissions. J Am Chem Soc. 2021;143(49):20546.
122. Zhang Z, Han Q, Lau JW, Xing B. Lanthanide-doped upconversion nanoparticles meet the needs for cutting-edge bioapplications: recent progress and perspectives. ACS Mater Lett. 2020;2:1516.
123. Tam V, Picchetti P, Liu Y, Skripka A, Carofiglio M, Tamboia G, et al. Upconverting nanoparticles coated with light-breakable mesoporous silica for NIR-triggered release of hydrophobic molecules. ACS Appl Mater Interfaces. 2024;16:29029–41.
124. He Q, Zhou L, Yu D, Zhu R, Chen Y, Song M, et al. Near-infrared-activatable PROTAC Nanocages for controllable target protein degradation and on-demand antitumor therapy. J Med Chem. 2023;66(15):10458.
125. Chen S, Sun X, Fu M, Liu X, Pang S, You Y, et al. Dual-source powered nanomotor with integrated functions for cancer photo-theranostics. Biomaterials. 2022;288:121744.
126. Liu H, Zhang J, Jia Y, Liu X, Chen X, Zhao W, et al. Theranostic nanomotors for tumor multimode imaging and photothermal/photodynamic synergistic therapy. Chem Eng J. 2022;442:135994.
127. Lv K, Hou M, Kou Y, Yu H, Liu M, Zhao T, et al. Black Titania Janus mesoporous nanomotor for enhanced tumor penetration and near-infrared light-triggered photodynamic therapy. ACS Nano. 2024;18:13910–23.
128. Kuntze K, Isokuortti J, Siiskonen A, Durandin N, Laaksonen T, Priimagi A. Azobenzene photoswitching with near-infrared light mediated by molecular oxygen. J Phys Chem B. 2021;125(45):12568.
129. Dudek M, Tarnowicz-Staniak N, Deiana M, Pokładek Z, Samoć M, Matczyszyn K. Two-photon absorption and two-photon-induced isomerization of azobenzene compounds. RSC Adv. 2020;10:40489.
130. Ye Z, Wang Y, Liu S, Xu D, Wang W, Ma X. Construction of nanomotors with replaceable engines by supramolecular machine-based host-guest assembly and disassembly. J Am Chem Soc. 2021;143(37):15063.
131. Schmidt EL, Zihao O, Ximendes E, Cui H, Keck CHC, DJ& GH. Near-infrared II fluorescence imaging. Nat Rev Methods Prim. 2024;4(1):23.
132. Chen G, Cao Y, Tang Y, Yang X, Liu Y, Huang D, et al. Advanced near-infrared light for monitoring and modulating the spatiotemporal dynamics of cell functions in living systems. Adv Sci. 2020;7:1903783.

133. Ribeiro TP, Moreira JA, Monterio FJ, Laranjeira MS. Nanomaterials in cancer: reviewing the combination of hyperthermia and triggered chemotherapy. J Control Release. 2022;347:89.
134. Stephen ZR, Zhang M. Recent progress in the synergistic combination of nanoparticle-mediated hyperthermia and immunotherapy for treatment of cancer. Adv Healthc Mater. 2021;10:e2001415.
135. Shen S, Qiu J, Huo D, Xia Y. Nanomaterial-enabled photothermal heating and its use for cancer therapy via localized hyperthermia. Small. 2024;20:e2305426.
136. Muttaqien SE, Nomoto T, Dou X, Takemoto H, Matsui M, Nishiyama N. Photodynamic therapy using LCST polymers exerting pH-responsive isothermal phase transition. J Control Release. 2020;328:608.
137. Badreldin M, Le Scouarnec R, Lecommandoux S, Harrisson S, Bonduelle C. Memory effect in thermoresponsive proline-based polymers. Angew Chem Int Ed Engl. 2022;61(46):e202209530.
138. Wei D, Sun Y, Zhu H, Fu Q. Stimuli-responsive polymer-based nanosystems for cancer theranostics. ACS Nano. 2023;17:23223.
139. Mishra S, Rawal A, Nebhani L. Imprinting the location of an in-built RAFT agent and selective grafting of polymer chains inside or outside the pores of mesoporous silica nanoparticles. Microporous Mesoporous Mater. 2020;294:109898.
140. Zhang C, Ye G, Wang W, Ding X, Feng Y, Wang R. Thermal, pH, and photo multistimuli responsive block copolymer self-assembly nanoparticles for drug delivery using visible-light-induced PET-RAFT. ACS Appl Nano Mater. 2024;7(1):370.
141. Lim MC, Chang SJ, Park B, Yoo HJ, Yoo CW, Nam BH, et al. Survival after Hyperthermic intraperitoneal chemotherapy and primary or interval Cytoreductive surgery in ovarian cancer: a randomized clinical trial. JAMA Surg. 2022;157(5):374.
142. Navarro L, Theune LE, Calderón M. Effect of crosslinking density on thermoresponsive nanogels: a study on the size control and the kinetics release of biomacromolecules. Eur Polym J. 2020;124:109478.
143. Huang D, Wang J, Song C, Zhao Y. Ultrasound-responsive matters for biomedical applications. Innovations. 2023;4:100421.
144. Zhao Z, Saiding Q, Cai Z, Cai M, Cui W. Ultrasound technology and biomaterials for precise drug therapy. Mater Today. 2023;63:210.
145. Liao M, Du J, Chen L, Huang J, Yang R, Bao W, et al. Sono-activated materials for enhancing focused ultrasound ablation: design and application in biomedicine. Acta Biomater. 2024;173:36.
146. Versaw BA, Zeng T, Hu X, Robb MJ. Harnessing the power of force: development of Mechanophores for molecular release. J Am Chem Soc. 2021;143(51):21461.
147. Lin S, Zhu L, Li Z, Yue S, Wang Z, Xu Y, et al. Ultrasound-responsive glycopolymer micelles for targeted dual drug delivery in cancer therapy. Biomater Sci. 2023;11(18):6149.
148. Baeza A, Vallet-Regí M. Mesoporous silica nanoparticles as theranostic antitumoral nanomedicines. Pharmaceutics. 2020;12:957.
149. Amin MU, Ali S, Tariq I, Ali MY, Pinnapreddy SR, Preis E, et al. Ultrasound-responsive smart drug delivery system of lipid coated mesoporous silica nanoparticles. Pharmaceutics. 2021;13(9):1396.
150. Guo J, Song X, Chen X, Xu M, Ming D. Mathematical model of ultrasound attenuation with skull thickness for transcranial-focused ultrasound. Front Neurosci. 2022;15:778616.
151. Dugar S, Fox S, Koratala A, Moghekar A, Mehta AC. Lung Ultrasonographic signs in pulmonary disease—a video review. J Intensive Care Med. 2023;38(2):220.
152. Saha A, Bosma JS, Twilt JJ, Van GB, Bjartell A, Padhani AR, et al. Artificial intelligence and radiologists in prostate cancer detection on MRI (PI-CAI): an international, paired, non- inferiority, confirmatory study. Lancet Oncol. 2024;2045(24):1–9.
153. Lehmann DH, Gomes B, Vetter N, Braun O, Amr A, Hilbel T, et al. Prediction of diagnosis and diastolic filling pressure by AI-enhanced cardiac MRI: a modelling study of hospital data. Lancet Digit Health. 2024;6(6):e407–17.

154. Cooley CZ, McDaniel PC, Stockmann JP, Srinivas SA, Cauley SF, Śliwiak M, et al. A portable scanner for magnetic resonance imaging of the brain. Nat Biomed Eng. 2021;5(3):229.
155. Abrishami A, Bahrami AR, Saljooghi AS, Matin MM. Enhanced theranostic efficacy of epirubicin-loaded SPION@MSN through co-delivery of an anti-miR-21-expressing plasmid and ZIF-8 hybridization to target colon adenocarcinoma. Nanoscale. 2024;16(12):6215.
156. Waddington DEJ, Boele T, Maschmeyer R, Kuncic Z, Rosen MS. High-sensitivity in vivo contrast for ultra-low field magnetic resonance imaging using superparamagnetic iron oxide nanoparticles. Sci Adv. 2020;6(29):eabb0998.
157. Torres-Herrero B, Armenia I, Alleva M, Asín L, Correa S, Ortiz C, et al. Remote activation of enzyme nanohybrids for cancer prodrug therapy controlled by magnetic heating. ACS Nano. 2023;17(13):12358.
158. Lin FC, Zink JI. Probing the local nanoscale heating mechanism of a magnetic core in mesoporous silica drug-delivery nanoparticles using fluorescence depolarization. J Am Chem Soc. 2020;142(11):5212.
159. Dadfar SMR, Pourmahdian S, Tehranchi MM, Dadfar SM. Design and fabrication of novel core-shell nanoparticles for theranostic applications. Colloid Polym Sci. 2020;298(10):1433.
160. Losito DW, de Araujo DR, Bezzon VDN, Oseliero Filho PL, Fonseca FLA, dos Chagas CS, et al. Mesoporous silica–Fe3O4 nanoparticle composites as potential drug carriers. ACS Appl Nano Mater. 2021;4(12):13363.
161. Patil-Sen Y, Torino E, De Sarno F, Ponsiglione AM, Chhabria V, Ahmed W, et al. Biocompatible superparamagnetic core-shell nanoparticles for potential use in hyperthermia-enabled drug release and as an enhanced contrast agent. Nanotechnology. 2020;31(37):375102.
162. Zheng S, Wang Y, Pan S, Ma E, Jin S, Jiao M, et al. Biocompatible nanomotors as active diagnostic imaging agents for enhanced magnetic resonance imaging of tumor tissues in vivo. Adv Funct Mater. 2021;31(24):2100936.
163. Kim HS, Kang JH, Jang JH, Lee EJ, Kim JH, Byun J, et al. Dual stimuli-responsive mesoporous silica nanoparticles for efficient loading and smart delivery of doxorubicin to cancer with RGD-integrin targeting. Eur J Pharm Sci. 2023;188:106525.
164. Salinas Y, Kneidinger M, Fornaguera C, Borrós S, Brüggemann O, Teasdale I. Dual stimuli-responsive polyphosphazene-based molecular gates for controlled drug delivery in lung cancer cells. RSC Adv. 2020;10(46):27305–14.
165. He Y, Shao L, Hu Y, Zhao F, Tan S, He D, et al. Redox and pH dual-responsive biodegradable mesoporous silica nanoparticle as a potential drug carrier for synergistic cancer therapy. Ceram Int. 2021;47(4):4572–8.
166. Lu J, Luo B, Chen Z, Yuan Y, Kuang Y, Wan L, et al. Host-guest fabrication of dual-responsive hyaluronic acid/mesoporous silica nanoparticle based drug delivery system for targeted cancer therapy. Int J Biol Macromol. 2020;146:363–73.
167. Bhavsar DB, Patel V, Sawant KK. Design and characterization of dual responsive mesoporous silica nanoparticles for breast cancer targeted therapy. Eur J Pharm Sci. 2020;152:105428.
168. Lu H, Zhao Q, Wang X, Mao Y, Chen C, Gao Y, et al. Multi-stimuli responsive mesoporous silica-coated carbon nanoparticles for chemo-photothermal therapy of tumor. Colloids Surf B Biointerfaces. 2020;190:110941.
169. Xu J, Chen M, Li M, Xu S, Liu H. Integration of chemotherapy and phototherapy based on a pH/ROS/NIR multi-responsive polymer-modified MSN drug delivery system for improved antitumor cells efficacy. Colloids Surf A Physicochem Eng Asp. 2023;663:131015.
170. Kawish M, Parveen S, Siddiqui NN, Jahan H, Elhissi A, Yasmeen S, et al. Highly functionalized pH-triggered supramolecular nanovalve for targeted cancer chemotherapy. Pharm Dev Technol [Internet]. 2024 [cited 2025 Feb 15];29(7):751–61. Available from: https://www.tandfonline.com/doi/abs/10.1080/10837450.2024.2392271.
171. Li Q, Lai S, Shang H, Qiao N, Sun X, Lu Y, et al. Construction and evaluation of biomass-modified mesoporous silica nanoparticles as enzyme-responsive and pH-responsive drug carriers for the controlled release of quercetin. J Drug Deliv Sci Technol. 2024;98:105852.

172. Li D, Song C, Zhang J, Zhao X. Targeted delivery and apoptosis induction activity of peptide-transferrin targeted mesoporous silica encapsulated resveratrol in MCF-7 cells. J Pharm Pharma [Internet]. 2023 [cited 2025 Feb 15];75(1):49–56. Available from: https://doi.org/10.1093/jpp/rgac028.
173. Dogra P, Adolphi NL, Wang Z, Lin YS, Butler KS, Durfee PN, et al. Establishing the effects of mesoporous silica nanoparticle properties on in vivo disposition using imaging-based pharmacokinetics. Nat Commun. 2018;9(1):4551.
174. Janjua TI, Cao Y, Yu C, Popat A. Clinical translation of silica nanoparticles. Nat Rev Mater. 2021;6(12):1072–4.
175. Phillips E, Penate-Medina O, Zanzonico PB, Carvajal RD, Mohan P, Ye Y, et al. Clinical translation of an ultrasmall inorganic optical-PET imaging nanoparticle probe. Sci Transl Med. 2014;6(260):1–9.
176. Xi X, Wang X, Xi Z, Fan C, Jiang Y, Li Z, et al. Facile GSH responsive glycyrrhetinic acid conjunction for liver targeting therapy. Chinese Chem Lett [Internet]. 2024 [cited 2025 Feb 15];110773. Available from: https://linkinghub.elsevier.com/retrieve/pii/S1001841724012890.
177. Wu Y, Wang X, Li L, Wang M, Tian S, Song J, et al. Preparation of Glutathione-regulated Sorafenib targeted nanodrug delivery system and its antihepatocellular carcinoma activity. ACS Appl Mater Interf [Internet]. 2024 [cited 2025 Feb 15];16:65131–41. Available from: https://pubs.acs.org/doi/full/10.1021/acsami.4c11076.
178. Zhuang J, Zhou L, Tang W, Ma T, Li H, Wang X, et al. Tumor targeting antibody-conjugated nanocarrier with pH/thermo dual-responsive macromolecular film layer for enhanced cancer chemotherapy. Mater Sci Eng C. 2021;118:111361.

Stimuli-Responsive Nanogels for Cancer Therapy

9

Raghuraj Singh, Krishna Jadhav, Agrim Jhilta, Swarnima Negi, and Rahul Kumar Verma

Abstract

The advancement of nanotechnology in drug delivery has revolutionized cancer therapy, particularly through the development of nanogels as promising drug carriers for targeted and "On-demand" controlled drug delivery. Nanogels are nano-sized hydrogel particles formed via physical or chemical cross-linking, display remarkable adaptability due to their ability to incorporate various functionalities. This adaptability enables surface modifications with targeting moieties, allowing for site-specific active cancer targeting. Furthermore, incorporating specific molecular functionalities transforms conventional nanogels into innovative smart stimuli-responsive systems. These surface-engineered nanogels can respond to internal biological stimuli such as pH, enzymes, temperature, ROS, and redox conditions, and external stimuli such as magnetic fields, light, and ultrasound. This stimuli-responsive behavior is mainly beneficial in cancer therapy, as it ensures site-specific controlled drug release, suppressing off-target drug toxicity and enhancing therapeutic efficacy. For instance, pH-responsive nanogels exploit the acidic microenvironment condition of tumors for selective drug release, while redox-sensitive systems leverage intracellular differences in oxidative states. Moreover, these carriers can encapsulate a variety of therapeutic agents, including chemotherapeutics, genetic molecules, and immunotherapeutics, offering a multimodal approach to cancer treatment.

R. Singh · R. K. Verma (✉)
Institute Nano Science and Technology (INST), Mohali, Punjab, India

Department of Pharmaceutics, Teerthanker Mahaveer College of Pharmacy (TMCOP), Teerthanker Mahaveer University (TMU), Moradabad, Uttar Pradesh, India
e-mail: rahulverma@inst.ac.in

K. Jadhav · A. Jhilta · S. Negi
Institute Nano Science and Technology (INST), Mohali, Punjab, India

P. V. Devarajan et al. (eds.), *Advances in Stimuli-Responsive Nanosystems for Cancer Therapy*, https://doi.org/10.1007/978-981-95-2744-1_9

Keywords
Nanogel · Cancer therapy · Nanotechnology · Drug delivery · Multifunctional nanogels

Abbreviations

ATP	Adenosine-5′-triphosphate
CPT	Camptothecin
DDS	Drug delivery systems
DOX	Doxorubicin
GSH	Glutathione
HA	Hyaluronic acid
LCST	Lower critical solution temperature
MMPs	Matrix metalloproteinases
PDEAEMA	Polymethacrylate 2-diethyl aminoketone
PNIPAM	Poly(N-isopropylacrylamide)
ROS	Reactive oxygen species

1 Introduction

Cancer remains one of the most persistent worldwide health challenges, characterized by the irregular and uncontrolled growth of cells capable of attacking adjacent tissues and metastasizing to distant organs [1]. This complex disease contributes significantly to global morbidity and mortality, with nearly 19.3 million new tumor diagnoses and ten million deaths reported in 2020 [2]. Projections estimate that by 2040, the worldwide cancer burden will rise to 28.4 million cases, an increase of almost 47% from 2020 [1–3]. Beyond the devastating health implications, cancer also poses a substantial economic burden, with the cost of care expected to exceed $240 billion by 2030 [1–3]. These challenges demand innovative therapeutic approaches capable of improving treatment outcomes while minimizing adverse effects and addressing the limitations of existing modalities.

Conventional treatments, such as chemotherapy, have been foundational in cancer management. However, these therapies are often associated with significant limitations, including non-specific drug delivery, systemic drug toxicity, and developing multidrug resistance by tumor cells [4, 5]. These drawbacks necessitate advanced drug delivery systems (DDS) that can enhance treatment precision, reduce side effects, and improve therapeutic efficacy. In this context, nanomedicine—an interdisciplinary field combining nanotechnology and medical science has developed as a transformative approach to revolutionizing cancer treatment [1, 6–8].

Among various nanomaterials explored, nanogels, also known as hydrogel nanoparticles or nano-hydrogels, have shown immense potential in oncology.

Nanogels are nanoscale particles, typically ranging from 1 to 1000 nm in size, composed of hydrophilic polymers that form 3D cross-linked networks capable of retaining water [4, 9]. This unique architecture imparts several advantageous properties, including high water content capacity, tunable size, large surface area for functionalization, biocompatibility, and stability. These features make nanogels highly versatile as carriers for therapeutic agents, enabling their use in targeted drug delivery, diagnostics, and combination therapies [9, 10].

An especially promising advancement in the field of nanogels is the development of stimuli-responsive, or environmentally sensitive, nanogels. These nanogels are engineered to undergo significant changes in their physicochemical properties, such as volume, network permeability, water content, refractive index, and hydrophilic-hydrophobic balance—in response to specific internal or external stimuli [9]. Internal stimuli, such as the acidic pH, upregulated enzymes, elevated redox potential, or ionic strength of the tumor microenvironment, allow these nanogels to deliver their payloads selectively at the site of action. External stimuli, including temperature shifts, light, magnetic fields, or ultrasound, provide additional avenues for controlled drug release [9, 11–13]. Importantly, these nanogels can often revert to their original state upon the removal or reversal of the stimulus, enhancing their versatility and reusability. The design and functionality of stimuli-responsive nanogels make them highly effective in addressing the challenges of cancer therapy. Their ability to respond dynamically to specific triggers allows for precise and controlled drug release, minimizing systemic toxicity and maximizing therapeutic efficacy. For example, pH-responsive nanogels can exploit the acidic environment of tumors to release chemotherapeutic agents directly at the site, while thermo-responsive systems release drugs in response to localized heating. Similarly, light- or magnetically responsive nanogels offer non-invasive control over drug delivery, making them suitable for applications requiring precision and minimal patient discomfort. Despite these advantages, several challenges hinder the clinical translation of stimuli-responsive nanogels. Issues such as complex fabrication processes, limited drug-loading capacity, rapid clearance from the body, and high production costs remain significant barriers. Additionally, ensuring scalability, stability, and reproducibility in large-scale manufacturing is essential for their broader clinical adoption [1, 10, 12, 14].

This chapter aims to provide an in-depth exploration of stimuli-responsive nanogels, emphasizing their design principles, mechanisms of action, and applications in cancer therapy. A comprehensive analysis of single, dual, and multi-stimuli-responsive systems, including those triggered by pH, enzyme, redox conditions, temperature, and light, is presented. Furthermore, the integration of advanced technologies, such as 3D printing, for fabricating nanogels with precise and customizable properties is discussed as a potential solution to existing challenges. By synthesizing recent advancements and highlighting future directions, this review seeks to demonstrate the transformative potential of stimuli-responsive nanogels in revolutionizing cancer therapy and other biomedical fields (Fig. 9.1).

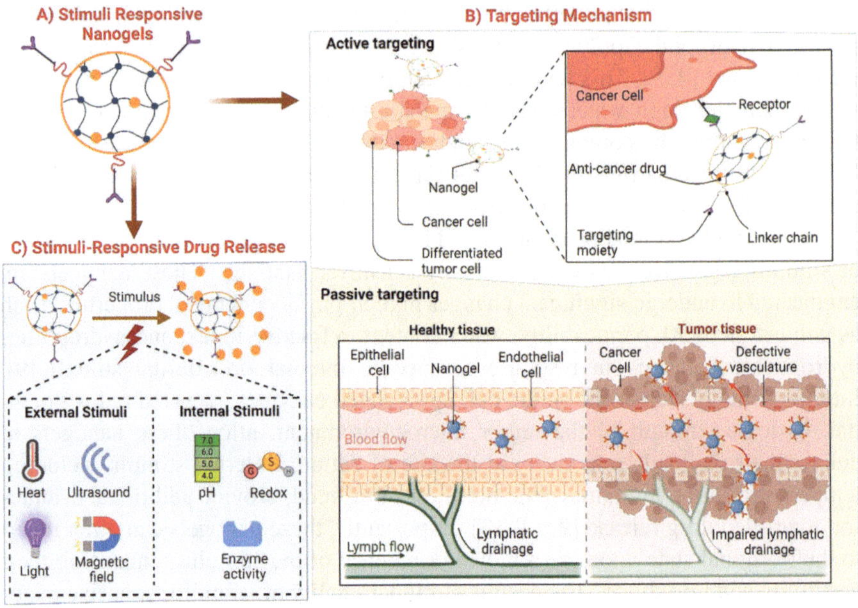

Fig. 9.1 Representation of stimuli-responsiveness of nanogels for cancer therapy with targeting mechanism and different stimuli-responsive release

2 Structure and Synthesis of Nanogels

Nanogels can be synthesized mainly through polymer precursors or the heterogeneous polymerization of different monomer units. Among these methods, the latter reveals superior efficiency. Cross-linking within nanogels can occur via two mechanisms: physical cross-linking, which commonly involves polymer precursors, and chemical cross-linking, which applies to both monomers and polymer precursors. These mechanisms are elaborated below [15].

2.1 Physical Cross-Linking

Physical cross-linking includes the formation of nanogels by non-covalent interactions, such as hydrophilic and hydrophobic polymer interactions, hydrogen bonding, ionic bonding, van der Waals forces, or electrostatic interactions. The size and other physiochemical properties of the nanogels can be fine-tuned by adjusting key parameters, such as polymer and concentration of cross-linking agent, pH, ionic strength, and temperature. This method is very efficient, as the formation of nanogels characteristically occurs in a very short time (within minutes) due to the self-assembly of amphiphilic copolymers or the complexation between the oppositely charged polymer chains. Nanogels resulting from polymeric micelles show substantial potential in drug delivery because their hydrophobic cores can encapsulate lipophilic drugs. Additionally, these delivery systems are biocompatible and biodegradable. Commonly used polymers for this purpose include PLGA, PLA, PCL, poly(propylene oxide) (PPO), poly(γ-benzyl L-aspartate), and poly(hydroxybutyrate) (PHB) [15].

2.2 Chemical Cross-Linking

This method is particularly appropriate for creating nanogels with larger pore sizes. The occurrence of a robust cross-linked polymeric network avoids dissolution in aqueous environments. Several cross-linking techniques have confirmed their effectiveness, such as click chemistry, photo-cross-linking of polymer, and quaternization of amino groups. Hydrophilic nanogels are often developed using chain-growth polymerization, employing an extensive range of monomers and cross-linkers, such as N,N′-methylenebisacrylamide (MBA), PEG diacrylate, diallyl phthalate, and divinylbenzene. These materials can be cross-linked through emulsion polymerization, reversible addition-fragmentation chain transfer (RAFT), click chemistry, and photoinduced cross-linking. Additionally, biodegradable nanogels based on amino acids can be synthesized using amino group cross-linking techniques [15].

3 Responsiveness and Properties of Smart Nanogels for Cancer Therapy

Nanogels are emerging as versatile and effective drug delivery systems in cancer therapy, characterized by their ability to respond to specific stimuli. These stimuli-responsive nanogels can enhance the precision of drug delivery, minimize side effects, and improve therapeutic outcomes. Their unique properties stem from their 3D cross-linked structure, which allows for significant tunability in terms of drug loading, release profiles, and targeting capabilities [9, 16]. Nanogels can be intended to respond to various internal or external stimuli. Many tumors exhibit an acidic microenvironment (pH 6.5–7.4), which can trigger drug release from pH-sensitive nanogels. For instance, studies have demonstrated that doxorubicin (DOX)-loaded xanthan gum nanogels show increased drug release at lower pH levels, simulating tumor conditions. Temperature-sensitive nanogels can undergo phase transitions at specific temperatures, facilitating drug release when they reach the target site. This property is particularly useful for localized hyperthermia treatments [9, 17]. Additionally, redox-responsive nanogels exploit the difference in redox potential between intracellular and extracellular microenvironments; disulfide bonds within the nanogel can be cleaved in the reducing environment of cancer cells, leading to controlled drug release [18, 19]. Some nanogels are also engineered to respond to specific enzymes that are overexpressed in tumor tissues, allowing for site-specific targeted drug delivery [12].

The properties of nanogels make them suitable for cancer therapy. They are generally composed of biocompatible materials that minimize adverse immune reactions and toxicity when administered in vivo. Their porous structure allows for significant amounts of therapeutic agents to be encapsulated, which is crucial for effective treatment regimens. The cross-linked network structure provides stability against degradation in physiological conditions, ensuring prolonged circulation time and bioavailability of the encapsulated drugs. Furthermore, by functionalizing the surface of nanogels with targeting ligands such as antibodies or peptides, they can selectively bind to cancer cells that overexpress specific receptors. This targeting capability is enhanced by the enhanced permeability and retention (EPR) effect commonly observed in tumors [12, 18]. Nanogels have shown promise in various applications within cancer treatment. They serve as carriers for chemotherapeutic agents like DOX and paclitaxel, allowing for controlled release at tumor sites while significantly minimizing exposure to healthy tissues. Additionally, nanogels are being explored as carriers for immunomodulatory drugs and vaccines, enhancing immune responses against tumors by delivering agents directly to immune cells within the tumor microenvironment. Their ability to co-deliver multiple therapeutic agents, such as chemotherapeutics and immunotherapeutics, can potentially overcome resistance mechanisms and improve treatment efficacy [18, 19]. Hybrid nanogels represent an advanced iteration of conventional nanogels, offering enhanced functionality through their ability to respond to both internal and external stimuli. By incorporating diverse nanomaterials, hybrid nanogels evolve into multifunctional, stimuli-responsive systems [19].

Commonly used inorganic materials include magnetic nanoparticles and noble metal nanoparticles, while organic components like carbon-based nanomaterials are also integrated into their structure. Unlike rigid nanoparticles, hybrid nanogels exhibit a distinctive softness and fluidity, significantly improving their cellular uptake and drug delivery capabilities. This softness not only aids in enhanced therapeutic delivery but also contributes to prolonged circulation times, as hybrid nanogels are less likely to be sequestered by the reticuloendothelial system, unlike their rigid, metal-based counterparts [20].

Furthermore, hybrid nanogels have demonstrated remarkable potential in tumor-targeted therapies by carrying therapeutic agents and releasing them specifically at tumor sites in response to endogenous stimuli, such as the acidic cancer microenvironment. Exogenous triggers, including near-infrared (NIR) light and magnetic fields, have also effectively achieved controlled drug release at targeted sites [14, 21]. These hybrid systems uniquely combine multiple desirable attributes, such as synergistic therapeutic approaches (e.g., dual heat- and chemotherapy) and imaging capabilities for image-guided therapy. The inherent softness and adaptability of hybrid nanogels improve their tumor-targeting ability and provide a significant edge over conventional nanogels and bare nanoparticles in delivering therapeutic payloads more effectively. This multi-functionality underscores their promise as an innovative platform for cancer therapy and diagnostics [22]. By seamlessly combining conventional nanogel advantages with these additional features, hybrid nanogels exemplify a significant advancement in stimuli-responsive drug delivery systems (Table 9.1).

Table 9.1 Types of stimuli and mechanisms of stimuli-responsive nanogel for cancer treatment

Types of stimuli	Mechanism	Examples	Ref.
Endo-stimuli-responsive			
pH-responsive	Polymer swelling upon protonation amine groups	Cationic hydrogels	[23]
Enzyme-responsive	Protease-mediated cleavage of the reactive peptide(s)	Protease-responsive Polymers or peptides	[24]
ROS-responsive	ROS-mediated cleavage of thioketal	Thioketal-linked polymers	[25]
Redox-responsive	GSH-mediated reduction and cleavage of the disulfide bond	Disulfide cross-linked polymers	[25]
Exo-stimuli-responsive			
Light-responsive	Photo-oxidation and photothermal release	Photosensitizer incorporated polymers plasmonic nanoparticle incorporated nanogel	[26]
Ultrasound-responsive	Ultrasound-induced cavitation and ultrasound-induced heat generation	Polymer-coated microbubbles	[27]
Temperature-responsive	Temperature-induced swelling	PNIPAM-based hydrogels	[28]
Magnetically responsive	Polymer phase transition due to heat generation by magnetic nanoparticles	Magnetic nanoparticle incorporated thermosensitive nanogel	[29]

4 Stimuli-Responsive Nanogels as Drug Carriers for Cancer Therapy

4.1 Enzyme-Responsive Nanogels

Enzyme-responsiveness has been given significant consideration in cancer therapy due to its ability to provide precise and controlled drug delivery tailored to the tumor microenvironment. These nanogels are designed to respond to specific enzymes overexpressed in cancerous tissues, such as matrix metalloproteinases (MMPs), hyaluronidases, and cathepsins, ensuring targeted therapeutic action. In a recent study, Wang et al. developed enzyme-responsive nanogels encapsulating the P-5 m peptide, synthesized using poly(N-isopropylacrylamide) (PNIPAM) through free radical precipitation polymerization. These nanogels are designed to undergo structural modifications when exposed to specific enzymes, facilitating the targeted release of the encapsulated drugs at the tumor site. This approach significantly reduces systemic toxicity and improves the therapeutic effectiveness of the treatment [30–32]. MMP-responsive nanogels, for instance, leverage the overexpression of MMPs in the tumor microenvironment. These nanogels are often cross-linked with MMP-sensitive peptides, which degrade upon enzyme interaction, releasing the payload. This approach has shown great promise in improving precision in chemotherapeutics such as cisplatin, DOX, and paclitaxel. Inflammation-responsive hydrogels with therapeutic factors are promising for bone regeneration [33]. In this study, an MMP-sensitive PEG-based hydrogel incorporating phosphatidylserine (PS) was developed, enabling controlled release to modulate inflammation, promote osteogenesis, and support bone repair through immunoregulation and osteoinduction. Similarly, hyaluronidase-sensitive nanogels take advantage of the abundant hyaluronidase in tumors. By integrating hyaluronic acid (HA) as a key component, these nanogels degrade in the presence of the enzyme, ensuring localized drug release. Dual-stimuli-responsive nanogels, combining enzyme sensitivity with other stimuli such as pH or temperature, have been developed to further refine control over drug release [34–36]. When triggered by salt, the nanogels exhibited a transition from hydrophobic to hydrophilic states. Additionally, adenosine-5′-triphosphate (ATP), a key molecule in energy transfer, serves as a unique triggering signal. Intracellular ATP concentrations (1–10 mM) are significantly higher than extracellular levels (<0.4 mM). This concentration gradient can be utilized to enable ATP-dependent drug release within the cytosol or cells [37, 38]. Enzyme-responsive nanogels offer versatile applications in cancer therapy, delivering RNA, immunotherapeutics, and photodynamic agents, though challenges in scalability, stability, and clinical translation persist.

4.2 pH-Responsive Nanogels

The pH-responsive nanogels have emerged as a highly efficient platform for cancer therapy, leveraging the acidic microenvironment of tumors for controlled drug delivery. Significant efforts have focused on designing pH-responsive nanogels for drug delivery, targeting the slightly acidic conditions of the tumor microenvironment (pH 6.5–7.2) and endosomal/lysosomal compartments (pH 4.5–6.5) relative to normal tissue pH (7.4) [39, 40]. PH-responsive nanogels exploit this physiological characteristic to enhance the specificity and efficacy of therapeutic agents. The pH-responsive nanogels can be broadly classified into three main types. The first type includes nanogels incorporating protonatable amino groups, such as primary, secondary, and tertiary amines. These nanogels exhibit swelling behavior in mildly acidic environments, which facilitates high drug release [41]. These nanogels are typically manufactured using polymers with ionizable functional groups, such as polyacrylic acid, polyhistidine, or chitosan, which respond to pH changes by swelling, degrading, or altering their conformation [23, 42]. For instance, nanogels composed of poly(N-isopropylacrylamide-co-acrylic acid) have been designed to deliver DOX, a widely used chemotherapeutic agent. These nanogels swell in response to the acidic extracellular pH, releasing the drug in a sustained manner. Similarly, chitosan-based nanogels have shown excellent biodegradability and pH-sensitive drug release profiles, making them suitable for various cancer therapies [43]. The degradation of nanogels, which influences the surrounding pH, can be utilized to regulate drug release rates or behavior. Additionally, the amino and carboxyl groups in nanogels play a dual role in controlling drug release and encapsulating small molecules, proteins, or nucleic acids. For example, Khaled et al. created hybrid silica nanogels with cross-linked cationic polymethacrylate 2-diethyl aminoketone

(PDEAEMA) for the targeted delivery and controlled release of siRNA [44]. Recent studies have also explored the use of pH-responsive nanogels in combination therapies. For example, these systems have been incorporated into photodynamic therapy, where the acidic microenvironment enhances the release of photosensitizers, improving light-activated tumor eradication [45]. The pH-responsive nanogels represent a versatile and potent strategy for targeted cancer therapy. Their ability to exploit tumor acidity, combined with multifunctional capabilities, positions them as a promising tool for improving the efficacy of cancer treatment.

4.3 Redox-Responsive Nanogels

Redox-responsive nanogels have gained prominence in cancer therapy for their ability to exploit the distinctive redox imbalance in tumor cells for targeted drug delivery. Tumor tissues exhibit elevated levels of reducing agents such as glutathione (GSH) compared to normal tissues. This disparity, with intracellular GSH levels in cancer cells often 100–1000 times higher than extracellular levels, serves as a key trigger for redox-responsive drug release [46, 47]. Redox-responsive nanogels are typically constructed using disulfide linkages, which are stable at normal physiological conditions but cleave in reductive environments. This structural property allows for precise, on-demand drug release in response to the high GSH concentrations within cancer cells. Upon cleavage of the disulfide bonds, the nanogels degrade or undergo structural rearrangements, releasing the encapsulated therapeutic agents directly at the tumor site [48–50]. Additionally, succinimide thioether is another type of redox-responsive chemical bond. However, research by Kiick and colleagues shows its cleavage is ten times slower than that of disulfide bonds [51]. Similarly, poly(acrylamide)-based nanogels incorporating disulfide cross-linkers have been utilized for the delivery of proteins and nucleic acids, offering a versatile platform for combinatorial cancer therapies. Redox-responsive nanogels have also been explored for co-delivery of drugs and genetic materials like siRNA, amplifying their therapeutic potential [52]. Additionally, they are being integrated into immunotherapies, where redox-triggered drug release enhances the activation of immune cells against tumors [25]. Despite their advantages, challenges persist, including stability during circulation, large-scale synthesis, and potential off-target activation in non-cancerous tissues with high GSH levels. Ongoing research focuses on improving the stability and specificity of these nanogels through advanced polymer chemistry and functional surface modifications.

4.4 Temperature-Responsive Nanogels

Researchers classify it as a response to exogenous stimuli. Temperature-responsive nanogels represent an innovative drug delivery approach in cancer therapy, utilizing the temperature variations between tumor tissues and normal tissues. Tumors often exhibit localized hyperthermia due to their abnormal vasculature and metabolic activity, creating an opportunity for targeted drug release using thermosensitive nanogels. These nanogels are typically designed with polymers that undergo phase transitions or structural changes at a specific temperature, known as the lower critical solution temperature (LCST) [53, 54]. Polymers such as PNIPAM are commonly used to construct temperature-responsive nanogels. Below their LCST, these nanogels remain swollen, encapsulating the drug securely. However, at temperatures above the LCST, which are often close to the hyperthermic tumor temperature of 40–42 °C, they collapse and release their payload, ensuring localized drug delivery. This mechanism reduces systemic toxicity and enhances therapeutic efficacy by concentrating the drug in the tumor region. PNIPAM-based nanogels have been widely studied for the delivery of chemotherapeutic agents like DOX [28, 55, 56]. These systems show efficient drug loading and rapid release in response to hyperthermic conditions. Furthermore, temperature-responsive nanogels can be combined with external heat-inducing techniques, such as photothermal or magnetic hyperthermia therapy. In addition to chemotherapy, temperature-responsive nanogels are being investigated for the co-delivery of immunotherapeutics, gene therapies, and imaging agents, enabling multifunctional cancer treatment approaches. However, challenges such as achieving precise LCST tuning, ensuring stability in circulation, and scaling up production for clinical applications remain key areas of ongoing research. In comparison to LCST-based nanogels, UCST-based nanogels have been less commonly developed and are not widely applied in drug delivery systems [57]. In contrast, the temperature-responsive nanogels represent a powerful strategy in cancer therapy, offering localized, on-demand drug release with minimal off-target effects. Their integration with other therapeutic modalities highlights their potential to advance the precision and efficacy of cancer treatment.

4.5 ROS-Responsive Nanogels

Reactive oxygen species (ROS)-responsive nanogels have appeared as a promising nano platform for targeted cancer therapy due to their ability to exploit the elevated ROS levels in tumor microenvironments. The ROS play a critical role in various pathophysiological processes and diseases, such as aging, atherosclerosis, and cancer. While maintaining ROS levels within a moderate range is essential, extreme ROS can lead to oxidative stress, causing cellular damage and altering vital components like lipids, DNA, and proteins. The overproduction of ROS and elevated oxidative stress observed in cancer cells serve as effective biological triggers for designing ROS-responsive drug delivery systems. Typically, these systems incorporate ROS-sensitive components, such as Se–Se bonds, to enable targeted drug release in cancer therapy [58]. ROS-responsive nanogels are typically designed with ROS-sensitive linkages, such as thioketal, selenium, or boronate ester bonds, which degrade in the presence of ROS. This degradation causes the nanogels to break down or release encapsulated drugs specifically at the tumor site, decreasing systemic drug toxicity and improving therapeutic efficacy. For example, nanogels incorporating boronate ester bonds have been used to deliver DOX, showing enhanced cytotoxicity in cancer cells while sparing healthy tissues [25, 59]. Deepagan et al. developed ROS-responsive nanogels based on PEG-polyglutamine derivatives, incorporating diselenide cross-links, for the delivery of DOX in cancer therapy. These nanogels demonstrated rapid DOX release in a simulated tumor microenvironment with H_2O_2 (10^{-4} M), exhibiting significant antitumor activity [60]. Additionally, these nanogels can be engineered to co-deliver multiple agents. Some designs include a combination of chemotherapeutics and ROS-scavenging molecules like antioxidants, which help to modulate the oxidative stress in cancer cells, enhancing the therapeutic outcome.

4.6 Light-Responsive Nanogels

Photo-responsive DDS can be classified into direct and indirect responsive types. In direct photo-responsive systems, the material's chemical structure is directly altered upon absorbing light of a specific wavelength. This type of the delivery system includes UV, visible light, and near-infrared responsive delivery system [61, 62]. In contrast, indirect photo-responsive systems involve an intermediate process, where light irradiation generates a secondary signal or molecule that triggers a series of reactions, ultimately leading to the transformation of the delivery vehicle or prodrug to release its cargo. Common intermediates in indirect photo-responsive systems include heat, hypoxia, and ROS, which interact with corresponding thermo-, ROS-, and hypoxia-sensitive components to facilitate the drug release process [61]. In this approach, light irradiation generates an internal signal within tissues or cells, which then activates downstream responses to the intermediary signals in a cascade-like manner.

Nanogels are typically hydrogels structured as nanoscale particles, which tend to respond quickly to stimuli such as variations in pH or temperature than bulk hydrogels. Numerous reports have shown that hydrogels are developed to target different pathological diseases. One of the most common ways to develop photo-responsive nanogels is to incorporate chromophore molecules or nanoparticles into the microgel. The absorption of light by these components generates heat, which triggers a volume transition by raising the temperature above the polymer's LCST [63]. The most exploited are the direct photo-responsive systems, where these nanogels use photo-responsive materials that can react to ultraviolet or near-infrared light. By employing light as a trigger, these hydrogels enable the controlled release of encapsulated drugs or other substances. The amount of drug released on demand is directly proportional to the intensity of the light source. This controlled release occurs as the hydrogel absorbs energy from the light, leading to polymer isomerization and subsequent degradation [64].

Chen et al. developed UV-degradable nano-capsules (HA-Azo/PDADMAC) by assembling photosensitive, azobenzene-functionalized HA with cationic polydiallyldimethylammonium chloride (PDADMAC). Upon exposure to 365 nm light, the nano-capsules undergo reversible cis–trans isomerization, causing disintegration into smaller polymer fragments. These nano-capsules exhibit prolonged circulation, high tumor accumulation, and CD44 receptor-targeting properties. When loaded with DOX, they significantly enhance cellular uptake and inhibit HepG2 cell proliferation. UV-induced dissociation of the capsules facilitates drug release and rapid tumor clearance [65]. In a separate study, Hang et al. formulated NIR- and UV-responsive nanogels using HA grafted with 7-N, N-diethylamino-4-hydroxymethylcoumarin (HA-CM). These nanogels actively target CD44 receptors, show efficient DOX loading, and enable controlled drug release upon NIR or UV irradiation. In vitro, the HA-CM nanogels demonstrated higher DOX release and cytotoxicity in CD44+ MCF-7 cells compared to CD44-U-87MG cells, highlighting their potential for tumor-targeted, photo-controlled chemotherapy [66].

4.7 Ultrasound-Responsive Nanogels

Ultrasound-responsive nanogels have emerged as a promising strategy in cancer treatment, offering precise control over drug delivery and improved therapeutic outcomes. These nanogels can be non-invasively triggered by incorporating ultrasound sensitivity to release therapeutic agents at targeted tumor sites. Ultrasound waves induce physical changes in the nanogel structure, such as swelling, destabilization, or degradation, allowing for controlled drug release and deeper tumor penetration. This approach combines the advantages of localized treatment with minimal side effects, making ultrasound-triggered nanogels a highly versatile and effective tool for enhancing the precision of cancer therapies [19]. Sun et al. have done extensive research to prolong the blood circulation and deep tumor penetration; an ultrasound-responsive feature was integrated into the nanocarriers. Polylysine and Pluronic F127 were initially co-assembled and then cross-linked with genipin to create a stable nanogel structure. The nanogel was further functionalized by attaching the ICAM-1 antibody (termed GenPLPFT) for targeted tumor delivery. When exposed to external ultrasound, the F127 component detached from the nanogel, triggering its swelling, which reduced stability and enhanced release of the encapsulated drug. Furthermore, sonication at the tumor site increased the deformability of GenPLPFT, enabling it to penetrate deeper into the tumor. In vivo experiments demonstrated that GenPLPFT effectively infiltrated xenografted tumors, improving antitumor efficacy and reducing systemic toxicity [67].

4.8 Magnetic-Field-Responsive Nanogels

These hybrid nanogels are highly responsive to magnetic fields, with their properties undergoing significant changes in response to even subtle variations in external magnetic stimuli. Common magnetic materials incorporated into these nanogels include metal particles such as iron (Fe), cobalt (Co), and nickel (Ni), as well as cobalt ferrite ($CoFe_2O_4$), iron oxides (Fe_2O_3 and Fe_3O_4), carbonyl iron (CI), nickel ferrite ($NiFe_2O_4$), and carbon nanotubes (CNTs). The concentration of magnetic nanoparticles (NPs) within the nanogel plays a pivotal role in influencing its responsiveness to magnetic fields and the amount of heat generated when subjected to alternating magnetic fields. This thermal response is particularly important in applying magnetic field-responsive nanogels for drug delivery, where localized heating can enhance targeted release and improve therapeutic efficacy [9, 19].

4.9 Multi-Stimuli-Responsive Nanogels

Multi-stimuli-responsive nanogels have emerged as a focal point in cancer therapeutics, driving significant advancements in their application within cancer research. One of the most compelling strategies is leveraging the tumor microenvironment's unique pH-sensitive characteristics in conjunction with other stimuli to enable precise, on-demand drug release. A notable example is the work by Qu et al., who developed a pH/redox dual-responsive nanogel system designed for the controlled release of camptothecin (CPT). This formulation exhibited robust CPT release under dual pH/redox-triggered conditions, facilitating targeted and controlled drug delivery specifically at tumor site [68]. Similarly, Gao et al. engineered polymeric nanogels with multi-responsive properties sensitive to temperature, pH, and redox changes to study the release kinetics of DOX. These nanogels demonstrated a highly effective, accelerated release profile under the combined stimuli, significantly enhancing the therapeutic efficacy of DOX against cancer cells. Such advancements highlight the potential of multi-stimuli-responsive nanogels in improving the precision and efficacy of cancer treatment [69]. A similar investigation was conducted by Huang et al., who engineered multi-stimuli-responsive nanogels sensitive to pH, redox conditions, and UV irradiation. These nanogels were based on polymer micelles self-assembled from a six-arm star copolymer, 6AS-PCL-PAAPPEGMA. Upon UV irradiation, the introduction of a reducing agent (glutathione, GSH), coupled with the pH-responsive behavior of the polyacrylic acid (PAA) block, triggered the de-cross-linking of the nanogel network, leading to a rapid release of DOX. Furthermore, these DOX-loaded nanogels demonstrated enhanced cellular uptake by HepG2 cells and effectively inhibited their proliferation, underscoring the therapeutic potential of this multi-stimuli-responsive drug delivery system [70].

5 Multifunctional Stimuli-Responsive Nanogels for Cancer Treatment

The field of cancer therapy is increasingly shifting toward multifunctional approaches that integrate therapeutic efficacy with precision and minimal side effects. In this context, hybrid nanogels have emerged as promising tools, offering a novel combination of features such as stimuli-responsiveness, biocompatibility, and multi-functionality [71]. These advanced nanocarriers are designed to target cancer cells with high specificity, either through their inherent cytotoxic properties or by delivering therapeutic agents such as drugs, enzymes, or genetic material.

Furthermore, hybrid nanogels contribute to enhanced imaging techniques, enabling simultaneous diagnosis and treatment, and they can selectively target tumor cells over normal tissues, reducing systemic toxicity [14, 71]. Their unique properties, including the ability to respond to specific environmental or externally applied stimuli, efficient encapsulation of therapeutic molecules, and capability to cross cellular barriers, make them versatile and effective platforms for addressing the complexities of cancer treatment. Multifunctional stimuli-responsive nanogels are cutting-edge platforms in cancer therapy, designed to address challenges such as drug resistance, systemic toxicity, and non-specific drug delivery [14, 71, 72].

Multifunctional hybrid nanogels combine the benefits of both organic and inorganic components, which allows for an array of applications in oncology. Organic polymers provide biocompatibility and tailored stimuli-responsiveness, while inorganic materials, such as gold or magnetic nanoparticles, enable functionalities like photothermal therapy or imaging enhancements. This synergistic integration equips hybrid nanogels to overcome various biological challenges, including poor drug solubility, multidrug resistance, and the need for spatiotemporal control of therapeutic delivery [71, 72]. However, while preclinical studies demonstrate significant potential, hybrid nanogels are yet to be widely adopted in clinical settings due to several unresolved challenges.

Both therapeutic and non-therapeutic factors hinder the clinical translation of hybrid nanogels. On the therapeutic side, there is a lack of comprehensive studies on their hemocompatibility, acute and chronic toxicity, interaction with physiological proteins, and degradation profiles. Additionally, understanding their behavior in the complex tumor microenvironment, including their ability to overcome tumor heterogeneity and navigate endocytosis and exocytosis pathways, remains limited. The toxicity and potential immunogenicity of their breakdown products also require thorough evaluation. The lack of standardized regulatory guidelines, reliable and reproducible testing protocols, and scalable manufacturing processes is a significant barrier on the non-therapeutic front. Moreover, limitations in the relevance of animal models to human cancer biology pose challenges for translating preclinical findings into clinical outcomes.

6 Future Prospects of Stimuli-Responsive Nanogel in Cancer Therapy

Stimuli-responsive nanogels are developing as a transformative technology in cancer treatment, offering precise and on-demand controllable drug delivery tailored to the unique characteristics of the cancer microenvironment. These nanogels are engineered to respond to specific internal and external stimuli. This selective responsiveness of nanogel enables localized controlled drug release, minimizing systemic drug toxicity and enhancing therapeutic efficacy. The acidic, enzyme-rich, and redox-imbalanced cancer microenvironment provides an ideal target for deploying such nanogels. The development of multi-stimuli-responsive nanogels marks a substantial advancement in this field. Combining multiple triggers, such as pH and

ROS, these systems attain superior specificity and adaptability, allowing more effective treatment of highly heterogeneous and drug-resistant cancer [64, 73].

Additionally, these platforms are highly appropriate for combination drug therapies, where the co-delivery of multiple drugs is crucial for overcoming cancer complications and resistance mechanisms. Another promising avenue is the incorporation of diagnostic and therapeutic functionalities within a single nanogel system, allowing superior theranostic applications. By incorporating imaging agents, stimuli-responsive nanogels allow real-time drug release and monitoring of therapeutic development. This competence improves treatment precision and aligns with personalized oncology goals. Despite their possible, significant challenges must be addressed to convert these systems into clinical use. These include ensuring reproducibility, large-scale production, regulatory compliance, and establishing biocompatibility and biodegradability [74]. Moreover, extensive preclinical and clinical evaluations are important to fully understand these innovative delivery systems' safety profiles and therapeutic consequences [64, 74].

Future research efforts will likely focus on optimizing the design and multi-functionality of nanogels. Strategies may include using hybrid materials, such as organic-inorganic composites, to enhance stability and responsiveness or leveraging machine learning techniques to modernize formulation development. By addressing these challenges, stimuli-responsive nanogels have the potential to transform cancer treatment, offering more precise, effective, and patient-specific therapies.

7 Conclusion

The Stimuli-responsive nanogels hold significant promise as efficient delivery systems for tumor targeting, leveraging the unique tumor microenvironment and their characteristic responsiveness to such conditions. Their biocompatibility, biodegradability, encapsulation efficiency, stability, and swelling behavior position them as attractive delivery candidates for anti-cancer therapy. Stimuli-responsive nanogels, capable of structural modifications in response to changes in pH, enzyme, temperature, and other environmental aspects, enable controlled and site-specific drug release. This functionality makes them particularly suitable for theranostic applications, where concurrent cancer diagnosis and treatment are attainable. The ability of nanogels to be smartly functionalized with various ligands enhances targeted delivery, ensuring specific interaction with cancer microenvironment and minimizing off-target toxicity. Multifunctional nanogels additionally increase their utility, offering superior efficacy in comparison to traditional drug delivery systems, particularly in targeting cancer stem cells. This competence not only contributes to tumor regression but also helps alleviate recurrence, addressing critical limitations of conventional chemotherapy, such as poor solubility, unfavorable pharmacokinetics, drug toxicity, systemic side effects, and instability. Despite these advantages, the industrial-scale production and clinical translation of stimuli-responsive nanogels remain hindered by several challenges. These include differences between in-vitro and in-vivo outcomes, potential immunogenicity, pharmacokinetic variability, toxicity, and biodistribution. Addressing these hurdles requires extensive research and solid clinical data to fully harness the potential of nanogels in cancer therapeutics. Future advancements should focus on refining stimuli-responsive nanogel design, enhancing their safety profiles, and improving the large-scale scalability of production processes to pave the way for their widespread application in cancer therapy.

References

1. Abedi F, Ghandforoushan P, Adeli F, Yousefnezhad M, Mohammadi A, Moghaddam S, Davaran S. Development of stimuli-responsive nanogels as drug carriers and their biomedical application in 3D printing. Mater Today Chem. 2023;29:101372.
2. Velazquez-Kronen R, Nelson J. World Cancer Day 2020–reflecting on a decade of NIOSH cancer research (2020).
3. Sung H, Ferlay J, Siegel RL, Laversanne M, Soerjomataram I, Jemal A, Bray F. Global cancer statistics 2020: GLOBOCAN estimates of incidence and mortality worldwide for 36 cancers in 185 countries. CA Cancer J Clin. 2021;71:209–49.
4. Cuggino JC, Blanco ERO, Gugliotta LM, Igarzabal CIA, Calderon M. Crossing biological barriers with nanogels to improve drug delivery performance. J Control Release. 2019;307:221–46.
5. Ray E, Jadhav K, Kadian M, Sharma G, Sharma K, Jhilta A, Singh R, Kumar A, Verma RK. Inhalable chitosan-coated nano-assemblies potentiate niclosamide for targeted abrogation of non-small-cell lung cancer through dual modulation of autophagy and apoptosis. Int J Biol Macromol. 2024;279:135411.
6. Sharma S, Chakraborty M, Yadav D, Dhullap A, Singh R, Verma RK, Bhattacharya S, Singh S. Strategic developments in polymer-functionalized liposomes for targeted colon cancer therapy: an updated review of clinical trial data and future horizons. Biomacromolecules. 2024;25:5650–69.
7. Kole E, Jadhav K, Singh R, Mandpe S, Abhang A, Verma RK, Naik J. Recent developments in tyrosine kinase inhibitor-based nanotherapeutics for EGFR-resistant non-small cell lung cancer. Crit Rev Ther Drug Carr Syst. 2024;42:6;83–123.
8. Ray E, Jadhav K, Kadian M, Sharma G, Sharma K, Jhilta A, Singh R, Kumar A, Verma RK. Targeted delivery of the metastasis-specific tumour homing TMTP1 peptide to non-small-cell lung cancer (NSCLC) using inhalable hybrid nano-assemblies. J Mater Chem B. 2024;12:9740–59.
9. Zha L, Banik B, Alexis F. Stimulus responsive nanogels for drug delivery. Soft Matter. 2011;7:5908–16.
10. Chakroborty S, Nath N, Mahal A, Barik A, Panda AR, Bal T, Obaidullah AJ, Elawady A. Stimuli-responsive nanogels: a smart material for biomedical applications. J Mol Liq. 2024;124828
11. Maya S, Sarmento B, Nair A, Rejinold NS, Nair SV, Jayakumar R. Smart stimuli sensitive nanogels in cancer drug delivery and imaging: a review. Curr Pharm Des. 2013;19:7203–18.
12. Attama AA, Nnamani PO, Onokala OB, Ugwu AA, Onugwu AL. Nanogels as target drug delivery systems in cancer therapy: a review of the last decade. Front Pharmacol. 2022;13:874510.
13. Vaghasiya K, Ray E, Singh R, Jadhav K, Sharma A, Khan R, Katare OP, Verma RK. Efficient, enzyme responsive and tumor receptor targeting gelatin nanoparticles decorated with concanavalin-A for site-specific and controlled drug delivery for cancer therapy. Mater Sci Eng C. 2021;123:112027.
14. Ali AA, Al-Othman A, Al-Sayah MH. Multifunctional stimuli-responsive hybrid nanogels for cancer therapy: current status and challenges. J Control Release. 2022;351:476–503.
15. Iyer S, Das A. Responsive nanogels for anti-cancer therapy. Mater Today Proc. 2021;44:2330–3.

16. García MC, Cuggino JC. Stimulus-responsive nanogels for drug delivery. In: Stimuli responsive polymeric nanocarriers for drug delivery applications, volume 1. Elsevier; 2018. p. 321–41.
17. Soni KS, Desale SS, Bronich TK. Nanogels: an overview of properties, biomedical applications and obstacles to clinical translation. J Control Release. 2016;240:109–26.
18. Ma X, Li S-J, Liu Y, Zhang T, Xue P, Kang Y, Sun Z-J, Xu Z. Bioengineered nanogels for cancer immunotherapy. Chem Soc Rev. 2022;51:5136–74.
19. Kumar N, Singh S, Sharma P, Kumar B, Kumar A. Single-, dual-, and multi-stimuli-responsive nanogels for biomedical applications. Gels. 2024;10:61.
20. Lim C-K, Singh A, Heo J, Kim D, Lee KE, Jeon H, Koh J, Kwon I-C, Kim S. Gadolinium-coordinated elastic nanogels for in vivo tumor targeting and imaging. Biomaterials. 2013;34:6846–52.
21. Wang H, Yi J, Mukherjee S, Banerjee P, Zhou S. Magnetic/NIR-thermally responsive hybrid nanogels for optical temperature sensing, tumor cell imaging and triggered drug release. Nanoscale. 2014;6:13001–11.
22. Peng Y, Gao Y, Yang C, Guo R, Shi X, Cao X. Low-molecular-weight poly (ethylenimine) nanogels loaded with ultrasmall iron oxide nanoparticles for T 1-weighted MR imaging-guided gene therapy of sarcoma. ACS Appl Mater Interfaces. 2021;13:27806–13.
23. Deen GR, Loh XJ. Stimuli-responsive cationic hydrogels in drug delivery applications. Gels. 2018;4:13.
24. Chandrawati R. Enzyme-responsive polymer hydrogels for therapeutic delivery. Exp Biol Med. 2016;241:972–9.
25. Rao NV, Ko H, Lee J, Park JH. Recent progress and advances in stimuli-responsive polymers for cancer therapy. Front Bioeng Biotechnol. 2018;6:110.
26. Zhu D, Roy S, Liu Z, Weller H, Parak WJ, Feliu N. Remotely controlled opening of delivery vehicles and release of cargo by external triggers. Adv Drug Deliv Rev. 2019;138:117–32.
27. Wei P, Cornel EJ, Du J. Ultrasound-responsive polymer-based drug delivery systems. Drug Deliv Transl Res. 2021;11:1323–39.
28. Xu X, Liu Y, Fu W, Yao M, Ding Z, Xuan J, Li D, Wang S, Xia Y, Cao M. Poly (N-isopropylacrylamide)-based thermoresponsive composite hydrogels for biomedical applications. Polymers. 2020;12:580.
29. Guisasola E, Vallet-Regí M, Baeza A. Magnetically responsive polymers for drug delivery applications. In: Stimuli responsive polymeric nanocarriers for drug delivery applications, volume 1. Elsevier; 2018. p. 143–68.
30. Huang D, Qian H, Qiao H, Chen W, Feijen J, Zhong Z. Bioresponsive functional nanogels as an emerging platform for cancer therapy. Expert Opin Drug Deliv. 2018;15:703–16.
31. Sharifzadeh G, Hosseinkhani H. Biomolecule-responsive hydrogels in medicine. Adv Healthc Mater. 2017;6:1700801.
32. Wang Y, Qiao S-L, Wang H. Facile synthesis of peptide cross-linked nanogels for tumor metastasis inhibition. ACS Appl Nano Mater. 2018;1:785–92.
33. Zhang M, Yu T, Li J, Yan H, Lyu L, Yu Y, Yang G, Zhang T, Zhou Y, Wang X. Matrix metalloproteinase-responsive hydrogel with on-demand release of phosphatidylserine promotes bone regeneration through immunomodulation. Adv Sci. 2024;11:2306924.
34. Vaghasiya K, Ray E, Sharma A, Singh R, Jadhav K, Khan R, Katare OP, Verma RK. Systematic development and optimization of spray-dried Quercetin-HP-β-cyclodextrin microparticles for DPI-based therapy of lung cancer. J Mater Sci. 2021;56:14700–16.
35. Vaghasiya K, Ray E, Sharma A, Katare OP, Verma RK. Matrix metalloproteinase-responsive mesoporous silica nanoparticles cloaked with cleavable protein for "self-actuating" on-demand controlled drug delivery for cancer therapy. ACS Appl Bio Mater. 2020;3:4987–99.
36. Gao D, Asghar S, Ye J, Zhang M, Hu R, Wang Y, Huang L, Yuan C, Chen Z, Xiao Y. Dual-targeted enzyme-sensitive hyaluronic acid nanogels loading paclitaxel for the therapy of breast cancer. Carbohydr Polym. 2022;294:119785.
37. Mo R, Jiang T, DiSanto R, Tai W, Gu Z. ATP-triggered anticancer drug delivery. Nat Commun. 2014;5:3364.

38. Sameiyan E, Bagheri E, Dehghani S, Ramezani M, Alibolandi M, Abnous K, Taghdisi SM. Aptamer-based ATP-responsive delivery systems for cancer diagnosis and treatment. Acta Biomater. 2021;123:110–22.
39. Binauld S, Stenzel MH. Acid-degradable polymers for drug delivery: a decade of innovation. Chem Commun. 2013;49:2082–102.
40. Li Z, Huang J, Wu J. pH-Sensitive nanogels for drug delivery in cancer therapy. Biomater Sci. 2021;9:574–89.
41. Zhu X, Sun Y, Chen D, Li J, Dong X, Wang J, Chen H, Wang Y, Zhang F, Dai J. Mastocarcinoma therapy synergistically promoted by lysosome dependent apoptosis specifically evoked by 5-Fu@ nanogel system with passive targeting and pH activatable dual function. J Control Release. 2017;254:107–18.
42. Pamfil D, Vasile C. Nanogels of natural polymers. In: Polymer gels: perspectives applications. Singapore: Springer; 2018. p. 71–110.
43. Lakshmanan V-K, Snima K, Bumgardner JD, Nair SV, Jayakumar R. Chitosan-based nanoparticles in cancer therapy. Chitosan Biomater. 2011;I:55–91.
44. Khaled SZ, Cevenini A, Yazdi IK, Parodi A, Evangelopoulos M, Corbo C, Scaria S, Hu Y, Haddix SG, Corradetti B. One-pot synthesis of pH-responsive hybrid nanogel particles for the intracellular delivery of small interfering RNA. Biomaterials. 2016;87:57–68.
45. Chu S, Shi X, Tian Y, Gao F. pH-responsive polymer nanomaterials for tumor therapy. Front Oncol. 2022;12:855019.
46. Ghassami E, Varshosaz J, Taymouri S. Redox sensitive polysaccharide based nanoparticles for improved cancer treatment: a comprehensive review. Curr Pharm Des. 2018;24:3303–19.
47. Meng F, Hennink WE, Zhong Z. Reduction-sensitive polymers and bioconjugates for biomedical applications. Biomaterials. 2009;30:2180–98.
48. Raza A, Hayat U, Rasheed T, Bilal M, Iqbal HM. Redox-responsive nano-carriers as tumor-targeted drug delivery systems. Eur J Med Chem. 2018;157:705–15.
49. Wang Y, Zu M, Ma X, Jia D, Lu Y, Zhang T, Xue P, Kang Y, Xu Z. Glutathione-responsive multifunctional "Trojan Horse" nanogel as a nanotheranostic for combined chemotherapy and photodynamic anticancer therapy. ACS Appl Mater Interfaces. 2020;12:50896–908.
50. Morgulchik N, Kamaly N. Meta-analysis of in vitro drug-release parameters reveals predictable and robust kinetics for redox-responsive drug-conjugated therapeutic nanogels. ACS Appl Nano Mater. 2021;4:4256–68.
51. Baldwin AD, Kiick KL. Tunable degradation of maleimide–thiol adducts in reducing environments. Bioconjug Chem. 2011;22:1946–53.
52. Notabi MK, Arnspang EC, Peppas NA, Andersen M. siRNA delivery mediated by pH and redox responsive p (DEAEMA-co-HEMA-g-PEGMA) nanogels. J Drug Deliv Sci Technol. 2023;86:104510.
53. Wang Y, Xu H, Ma L. Recent advances of thermally responsive nanogels for cancer therapy. Ther Deliv. 2015;6:1157–69.
54. Cheng W, Gu L, Ren W, Liu Y. Stimuli-responsive polymers for anti-cancer drug delivery. Mater Sci Eng C. 2014;45:600–8.
55. Vihola H, Laukkanen A, Valtola L, Tenhu H, Hirvonen JJB. Cytotoxicity of thermosensitive polymers poly (N-isopropylacrylamide), poly (N-vinylcaprolactam) and amphiphilically modified poly (N-vinylcaprolactam). Biomaterials. 2005;26:3055–64.
56. Pelton R. Preparation of aqueous latices with N-isopropylacrylamide. Colloids Surf. 1986;20:247–56.
57. Bansal K, Upadhyay P, Saraogi G, Rosling A, Rosenholm J. Advances in thermo-responsive polymers exhibiting upper critical solution temperature (UCST), eXPRESS. Polym Lett. 2019;13
58. Zhang Y, Ma C, Zhang S, Wei C, Xu Y, Lu W. ROS-responsive selenium-containing polyphosphoester nanogels for activated anticancer drug release. Mater Today Chem. 2018;9:34–42.
59. Li Q, Wen Y, Wen J, Zhang Y-P, Xu X-D, Victorious A, Zavitz R, Xu X. A new biosafe reactive oxygen species (ROS)-responsive nanoplatform for drug delivery. RSC Adv. 2016;6:38984–9.

60. Deepagan V, Kwon S, You DG, Um W, Ko H, Lee H, Jo D-G, Kang YM, Park JH. In situ diselenide-crosslinked polymeric micelles for ROS-mediated anticancer drug delivery. Biomaterials. 2016;103:56–66.
61. Zhou Y, Ye H, Chen Y, Zhu R, Yin L. Photoresponsive drug/gene delivery systems. Biomacromolecules. 2018;19:1840–57.
62. Chen Q, Wen J, Li H, Xu Y, Liu F, Sun S. Recent advances in different modal imaging-guided photothermal therapy. Biomaterials. 2016;106:144–66.
63. He J, Tong X, Zhao Y. Photoresponsive nanogels based on photocontrollable cross-links. Macromolecules. 2009;42:4845–52.
64. Yu J, Liu Y, Zhang Y, Ran R, Kong Z, Zhao D, Liu M, Zhao W, Cui Y, Hua Y. Smart nanogels for cancer treatment from the perspective of functional groups. Front Bioeng Biotechnol. 2024;11:1329311.
65. Chen Q, Li X, Xie Y, Hu W, Cheng Z, Zhong H, Zhu H. Azo modified hyaluronic acid based nanocapsules: CD44 targeted, UV-responsive decomposition and drug release in liver cancer cells. Carbohydr Polym. 2021;267:118152.
66. Hang C, Zou Y, Zhong Y, Zhong Z, Meng F. NIR and UV-responsive degradable hyaluronic acid nanogels for CD44-targeted and remotely triggered intracellular doxorubicin delivery. Colloids Surf B: Biointerfaces. 2017;158:547–55.
67. Sun M, Yue T, Wang C, Fan Z, Gazit E, Du J. Ultrasound-responsive peptide nanogels to balance conflicting requirements for deep tumor penetration and prolonged blood circulation. ACS Nano. 2022;16:9183–94.
68. Qu Y, Chu B, Wei X, Lei M, Hu D, Zha R, Zhong L, Wang M, Wang F, Qian Z. Redox/pH dual-stimuli responsive camptothecin prodrug nanogels for "on-demand" drug delivery. J Control Release. 2019;296:93–106.
69. Gao D, Duan L, Wu M, Wang X, Sun Z, Zhang Y, Li Y, He P. Preparation of thermo/redox/pH-stimulative poly (N-isopropylacrylamide-co-N, N′-dimethylaminoethyl methacrylate) nanogels and their DOX release behaviors. J Biomed Mater Res A. 2019;107:1195–203.
70. Huang Y, Tang Z, Peng S, Zhang J, Wang W, Wang Q, Lin W, Lin X, Zu X, Luo H. pH/redox/UV irradiation multi-stimuli responsive nanogels from star copolymer micelles and Fe3+ complexation for "on-demand" anticancer drug delivery. React Funct Polym. 2020;149:104532.
71. Motornov M, Roiter Y, Tokarev I, Minko S. Stimuli-responsive nanoparticles, nanogels and capsules for integrated multifunctional intelligent systems. Prog Polym Sci. 2010;35:174–211.
72. Ashwani P, Gopika G, Arun Krishna K, Jose J, John F, George J. Stimuli-responsive and multifunctional nanogels in drug delivery. Chem Biodivers. 2023;20:e202301009.
73. Rana P, Singh C, Kaushik A, Saleem S, Kumar A. Recent advances in stimuli-responsive tailored nanogels for cancer therapy; from bench to personalized treatment. J Mater Chem B. 2024;12:382.
74. Patra J. Theranostic Application of Nanoparticulated System: Present and Future Prospects, Materials Research Foundations, 47.

Stimuli-Responsive Metal–Organic Frameworks for Cancer Therapy

10

Shipra Shukla, Sanchit Arora, Vaibhav Sadana, Abhishek Saxena, and Dalapathi Gugulothu

Abstract

Metal–organic frameworks (MOFs) have become a potential material in cancer therapy because of their unique porosity and customisable features. These materials provide novel approaches to targeted treatment, imaging, and drug administration. The potential of MOFs has been further boosted by the addition of stimuli-responsive components, which allow for the regulated release of medicinal compounds in response to particular biological or environmental triggers. This chapter explores the principles of stimulus-responsive MOFs, explaining their basic structure, processes of response, and the range of stimuli, including pH, temperature, enzymes, and light, that can be used to treat cancer. It demonstrates how these cutting-edge materials are being used to create more effective and focused drug delivery systems, enhance imaging methods, and achieve accurate tumour targeting with negligible side effects through a thorough examination of their uses. It also discusses important elements of toxicity and biocompatibility, offering information from studies conducted in vivo and in vitro as well as regulatory considerations. The potential for stimuli-responsive MOFs to transform cancer therapy is highlighted in the chapter's discussion of emerging trends, future opportunities for innovation, and current obstacles facing the field.

S. Shukla
Department of Pharmaceutics, Indo-Soviet Friendship College of Pharmacy, Moga, Punjab, India

S. Arora · V. Sadana · A. Saxena · D. Gugulothu (✉)
Department of Pharmaceutics, Delhi Institute of Pharmaceutical Sciences and Research (DIPSAR), Delhi Pharmaceutical Sciences and Research University (DPSRU), New Delhi, India
e-mail: dalapathig@dpsru.edu.in

P. V. Devarajan et al. (eds.), *Advances in Stimuli-Responsive Nanosystems for Cancer Therapy*, https://doi.org/10.1007/978-981-95-2744-1_10

Keywords

Stimuli-responsive MOFs · Cancer therapy · Targeted drug delivery · Biocompatibility · Theranostics

Abbreviations

3-MA	3-methyladenine
5-FU	5-fluorouracil
ATP	Adenosine triphosphate
CDT	Chemodynamic therapy
CMC	Carboxymethylcellulose
DOX	Doxorubicin
ECM	Extracellular matrix
GSH	Glutathione
ICA	Imidazole-2-carboxyaldehyde
ICG	Indocyanine green
MIL	Materials of Institute Lavoisier
MOFs	Metal–organic frameworks
MRI	Magnetic resonance imaging
NCP	Nanoscale coordination polymer
NMOF	Nanoscale metal–organic framework
NPs	Nanoparticles
PEG	Polyethylene glycol
PMOCol	Protein-metal ion-organic ligand coordination
ROS	Reactive oxygen species
SDT	Sonodynamic therapy
TEG	Triethylene glycol monomethyl ether
THF	Tetrahydrofuran
TME	Tumour microenvironment
ZIF-8	Zeolitic imidazolate framework-8

1 Introduction

Cancer is one of the main causes of death for individuals globally [1]. Currently, the three main cancer treatment modalities are radiation therapy, open surgery, and chemotherapy [2]. However, the traditional direct delivery of therapeutic drugs to patients is no longer applicable due to inherent limitations such as poor biodistribution, poor pharmacokinetics, and undesired side effects [3]. Chemotherapy, however, primarily targets cancerous tissues selectively and has severe adverse effects on healthy, normal tissues as a side effect. As a result, not all cancer cells are effectively removed from the body. In a similar way, patients receiving radiotherapy also experience significant physiological dysfunctions [4, 5]. While these treatments can slow down the growth and spread of tumours, they are largely ineffective against malignant tumours that cause significant harm to healthy human organs. Due to the differences in tumour growth from normal somatic cell growth and the lack of targeting in anti-cancer medications, once they enter the systemic circulation, they destroy cells randomly. Therefore, materials have been researched as carriers to deliver drugs directly to tumour areas in targeted drug therapy by leveraging the differences between normal somatic cells and the tumour microenvironment. As a result, numerous initiatives have been undertaken to create drug delivery systems with controlled release that produce fewer side effects and higher therapeutic efficacy [6].

Multifaceted therapeutic agents have drawn increased interest as a result of the rapid advancements in nanoscience and technology. Over the past few decades, a variety of nanomaterials, including liposomes, metal nanoparticles (NPs), organic micelles, and inorganic mesoporous silica, have been created as therapeutic carriers [7–10]. However, the majority of these nanocarriers have drawbacks, including low biocompatibility, cytotoxicity, and restricted loading capacity. Drug delivery systems that work well should be non-toxic, biocompatible, and biodegradable, with a high drug loading capacity. Reliability of the therapeutic index is guaranteed by controlled therapeutic release, which results in multimodal systems for cancer therapy and diagnostic uses [11].

In this context, a great deal of research has been done on MOFs, a multipurpose platform. Organic molecules function as bridges between metallic clusters or ions to construct three-dimensional networks with high porosity and organisation, resulting in MOFs, an absorbing class of hybrid materials [12, 13]. A significant external surface area, porous structures, programmable sizes, customisable frameworks, and biocompatibility are only a few of the amazing qualities of MOFs [14–16]. Because of their perfect characteristics, which include their potency for loading drugs, porous frameworks, and variable pore volume, they are significant possibilities for biomedical uses. As a smart nanomedicine, MOFs' external surface can be readily altered to enhance their biodistribution, biodegradability, and biocompatibility. It can also be used to reduce burst uncontrolled release of the drug and provide intelligent drug release [17]. However, the main obstacles to their biomedical application include in vivo retention and ineffective drug release of the MOF at the tumour site. Numerous stimuli, including pH, glutathione (GSH), glucose, ATP, ions, light, H_2S, magnetic, thermal, pressure, and others, have been reported to help overcome these challenges [18].

Tumour microenvironments differ from healthy tissues in that they have certain properties, such as higher temperatures and lower pH levels. These properties can be used to create internal drug delivery systems that respond to cues like ATP, redox, and pH. Systems that react to external stimuli (e.g. magnetic field, temperature, light, etc.) distinguish themselves from intrinsic systems by visualising the NPs accumulated at the tumour site and then activating the nanocarriers through external triggers at a predetermined time [19]. Stimuli-responsive devices have shown higher promise for clinical application and offer improved spatiotemporal medication release. Therefore, MOF-based systems have garnered significant attention as a stimuli-responsive drug delivery method because they can achieve a controlled release of the medication [20–22]. For this reason, the current review focuses on the latest developments in smart MOFs for cancer treatment.

2 Fundamentals of Stimuli-Responsive MOFs

2.1 Basics of MOFs

With a rapid advancement of nanotechnology and nanomedicine several attempts have been made to design a nanoplatform for controlled release of medicines in order to increase therapeutic efficacy and reduce adverse effects. Among them organic materials like polymeric particles, polymers, dendrimers, cyclodextrin, and so on and inorganic materials such as quantum dots, iron oxide, graphene, and so on have all been combined for drug delivery up to this point [23]. Although organic materials often have qualities like low toxicity, biodegradability, chemical modification, and so on, they are insufficient for regulating drug release. On the other hand, inorganic materials have strong magnetic, optical, and photon output characteristics, but biocompatibility and low degradability is sometimes an issue [24–26]. A type of organic-inorganic hybrid material that has received a lot of interest since it was initially proposed is called MOF [27]. Thus far, a variety of synthetic approaches have been employed to design MOFs, which have then been explored for a wide range of applications. These include gas storage and separations, sensing, nonlinear optics, catalysis, and a growing array of potential biomedical applications [28].

When compared to the previously described typical drug carriers, MOFs have the following clear advantages [29, 30]. First of all, because of their adaptable structures (Fig. 10.1), MOFs have a wide range of morphological characteristics, compositions, sizes, and chemical characteristics. Furthermore, it is noteworthy that these materials' desired physicochemical features remain largely unaffected by modification; that is, MOF-based materials continue to have excellent homogeneity,

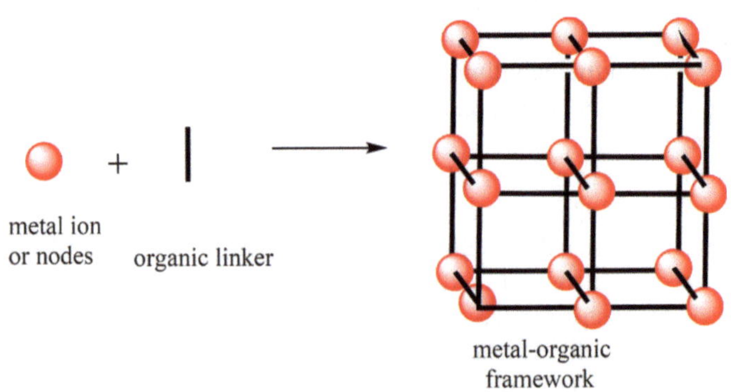

Fig. 10.1 Illustration of MOF's building blocks and resulting porous, crystalline structure. (Reprinted from [31] under CC-BY license)

controllable size, and shape [31]. Second, MOFs with high loading capacities are made possible by wide surface areas and high porosities. Various interactions, including covalent bonds, hydrogen bonds, Van der Waals forces, the π–π effect between aromatic rings, electrostatic contacts, coordination bonds, and so on, can physically or chemically combine drugs within the carriers [32, 33]. Furthermore, biomolecules are commonly integrated into MOFs by diverse techniques, including pore encapsulation [34], surface attachment, covalent interaction, in situ encapsulation, resulting in bio-MOFs (metal-biomolecule frameworks). Weak coordination links also ensure the biodegradability of MOFs [23–36]. However, in vivo retention and inadequate drug dissolution of the MOF at the tumour site are the primary barriers to their biological utilisation. Therefore, to overcome such obstacles, numerous stimuli-responsive drug delivery systems have been documented, which include pH, GSH, glucose, ATP, ions, light, H_2S, magnetic, heat, pressure, etc.

2.2 Stimuli-Responsiveness in MOFs

Driven by the natural world, the creation of artificial systems with the ability to perceive and/or react to outside stimuli in a predictable and regulated way has attracted a lot of attention due to the growing need for intelligent systems [37]. These artificial systems or smart materials are a diverse group of materials that include photonic crystals sensitive to changes in temperature, light, or magnetic field, surfactants that can be stabilised in a controlled way; membranes that can be switched on and off for transportation abilities; actuators reacting to changes in temperature; solvents with polarity that can be switched; and thin films that can be transparent or hydrophilic [22]. The body's spatially and temporally controlled qualities are provided by both external (such as stimuli) and internal (such as signals from within the cancer microenvironment) triggers, which enable the release of the drugs contained in these systems [38, 39]. These stimuli-responsive gated micro/nanocarriers have a lot of potential uses in personalised and nanomedicine in the future. Additionally, as described earlier that MOFs' loading capabilities and the variety of ways in which they can be chemically modified present prospects for the design of stimuli-responsive MOFs, a large class of nanomaterials with potential applications in controlled drug delivery and nanomedicine [19, 40]. Therefore, MOFs that can change their intrinsic physical and/or chemical properties in response to one or more stimuli; as a result, the overall structure may change (becomes dynamic) or stay the same (becomes rigid) are called smart MOFs. In this way, they can enhance their therapeutic efficacy by releasing their cargo on demand at the tumour sites [41]. The characteristics of the MOF may change as a result of these modifications, either irreversibly or reversibly. While the irreversible alterations are irreversible and cannot be recovered by stopping the stimulus or administering an antagonistic stimulus, the reversible changes can be recovered by stopping the stimulus or delivering an antagonistic stimulus.

2.3 Mechanism of Stimuli-Responsiveness

Stimuli-responsive MOFs undergo a specific transition in their physical structure or chemical makeup upon activation by particular triggers, leading to a notable alteration in their physicochemical properties [23, 42]. This modification causes a controlled release of the medication at the targeted location. These triggers are typically classified as intrinsic or extrinsic. The target microenvironment, which has particular characteristics like pH and temperature, contains intrinsic stimuli. These stimuli have been used to create internal (e.g. ATP, redox, and pH) stimuli-responsive media for MOFs. Conversely, systems that are sensitive to external stimuli (e.g. magnetic fields, temperatures, ions, pressure, and light) utilise the properties of the target microenvironment to accumulate MOFs at the target site, which are then triggered externally when the timing is right. Systems that respond to external stimuli have shown greater promise for clinical use and offer greater spatiotemporal release of drugs. A few smart MOFs, with an emphasis on cancer treatments, are outlined and explained in depth in the sections that follow.

3 Types of Stimuli for Cancer Therapy

A novel approach to cancer treatment is the use of stimuli-responsive MOFs, which use their special capacity to react to a range of biological and environmental stimuli to deliver drugs precisely where they are needed. These include temperature-responsive MOFs that use small changes in temperature to trigger drug release, pH-responsive MOFs that use the acidic tumour microenvironment to release therapeutic agents selectively at the tumour site, and light-responsive MOFs that allow controlled drug delivery by applying particular light wavelengths (Table 10.1). This responsiveness's adaptability improves the accuracy and efficacy of cancer treatments while reducing side effects and enhancing patient outcomes.

3.1 pH-Responsive MOFs

The pH of the tumour microenvironment (5.7–7.8), endosome (5.5–6.0), and lysosome (4.5–5.0) is lower than that found in normal blood and tissues. Many acid-labile bonds, such as amide, oxime, orthoester bonds, imine, polyacetal, polyketal, sulfonates, etc., or acid-sensitive MOFs, such as the Materials of Institute Lavoisier (MIL-n), Zeolitic imidazolate framework (ZIF-n), or other pH-sensitive materials, like chitosan, gelatin polymer, and carboxymethylcellulose (CMC), may degrade when exposed to low pH. This acidic pH instability causes structural dissociation, releasing the loaded medicines partially or completely (Fig. 10.2) [51–53]. Therefore, utilising this strategy, MOF-based pH-responsive drug delivery systems were developed for the administration of drugs both intracellularly and extracellularly. In this instance, we outlined the processes of pH-responsive drug delivery systems' drug release as seen from the four different angles below: (a) coordination

Table 10.1 Overview of Stimuli-responsive MOFs for cancer therapy

Stimuli type	MOF example	Mechanism of action	Application in cancer therapy	Ref.
pH-responsive	ZIF-8	Protonation/deprotonation triggers structural changes or drug release	Targeted drug delivery to acidic tumour microenvironments	[43]
Temperature-responsive	UiO-66	Thermal expansion/contraction influences pore size and drug release rates	Controlled drug release in response to body or external temperature changes	[44]
Solvent-switching	NMOF	Swelling or shrinking in different solvents alters pore accessibility	Solvent-induced drug release for specific delivery sites	[45]
Light-responsive	UiO-66	Photo-induced isomerization or degradation triggers drug release	Precise control over drug release timing using external light sources	[46]
Magnetically responsive	Fe_3O_4@MOF	External magnetic fields guide MOF to the target site or trigger drug release	Targeted therapy and imaging with magnetic guidance	[47]
Redox-responsive	Mn-MOF	Redox reactions within the tumour microenvironment induce structural changes	Specific drug release in redox-active tumour sites	[48]
Enzyme-responsive	Polymer-coated MOF	Enzyme-specific cleavage triggers structural changes	Site-specific drug delivery exploiting tumour-associated enzyme overexpression	[49]
Multiple-stimuli-responsive	Hybrid MOF systems	Combination of above mechanisms for enhanced control and specificity	Advanced drug delivery systems capable of responsible to multiple tumour markers	[50]

bond breakdown brought on by protonation, (b) bonds that can be broken by pH, (c) materials that are sensitive to pH, and (d) exchanges between hosts and guests.

Creating pH-induced responsivity is a common technique that involves protonation-induced coordination breakdown. MOFs comprised of organic ligands with ionizable chemical groups, such as imidazolyl (-C3H3N2) [54], daunosamine and amino groups (-NH2) [55], COOH group [56], pyridine groups [57], phenolic hydroxyl groups [58], etc., are included in this technique. Under physiological pH, the aforementioned groups remain deprotonated/deionized; however, when subjected to acidic tumour microenvironments, they are prone to being protonated for charge reversal, which can lead to structural changes or the breaking down of the coordination bonds among metal clusters and organic ligands, causing the MOFs to collapse and release encapsulated drugs at the tumour site. Members of the widely used acid-labile MOF family include the imidazolyl-rich ZIF-n family, the carboxyl-rich MIL-n family, and the University of Oslo-n (UiO-n) family.

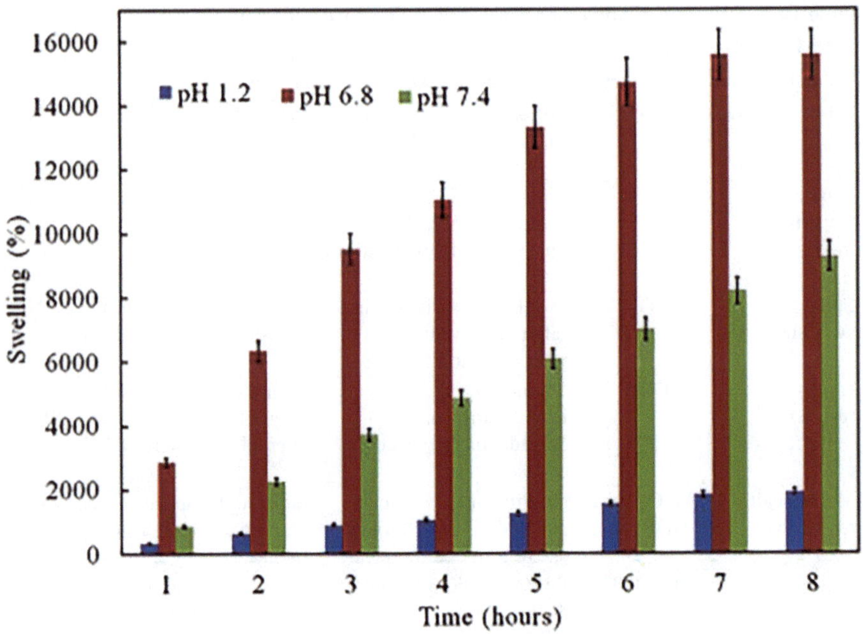

Fig. 10.2 Swelling behaviour of CMC/Cu-MOF@IBU at pH values of 1.2, 6.8 and 7.4. (Adapted from [51], Copyright, Elsevier, 2018)

Constructed from 2-methylimidazole and Zn^{2+}, ZIF-8 is a type of acid-responsive MOF that collapsed at lower pH (5.4–6) while demonstrating "zero premature release" under neutral conditions (pH 7.4) [43]. Because of its superior acid-labile properties, ZIF-8 is a highly regarded carrier material that is frequently used in drug delivery systems. Furthermore, ZIF-8's large cavity (11.6 Å) enhances its drug encapsulation efficiency, while its tiny window size (~3.4 Å) inhibits early guest molecule leakage. The imidazolyl N atom in the ligand becomes negatively charged in alkaline or neutral environments as a result of deprotonation, and it combines with positively charged Zn^{2+} to develop a ZIF-8 structure. The 2-methylimidazole, on the other hand, is protonated in an acidic environment, weakly coordinating with Zn^{2+} to hasten ZIF-8's collapse, unleashing the guest molecules [54]. In ZIF-8-based drug delivery systems, the mechanism of protonation-induced breakdown of coordination bonds is extensively observed. Chen et al. [59] synthesised a pH-responsive 3-MA@ZIF-8 to give high loading (19.798 wt%) of 3-methyladenine (3-MA). A rapid release of 3-MA (40%) in PBS at pH 6.0 was seen within the first 4 h, and this was closely linked to the protonation-induced degradation of ZIF-8 NPs. Concurrently, the study found that ZIF-8 enhanced the inhibition of 3-MA, which stopped the development of autophagosomes. Different transition metal elements are combined with dicarboxylic acid ligands, like succinic and glutaric acids, to generate materials in the MIL-n class. The ligand's carboxyl group is protonated in an acidic environment to improve electrostatic contact, which breaks the coordination connection between the organic ligand and the metal cluster [60]. Generally speaking, Lajevardi's group developed Fe3O4@silica@MIL-100(Fe)/β-CD, a pH-responsive drug nanocarrier. The drug delivery system was constructed using beta cyclodextrin (β-CD) polymer, which has acid groups and is sensitive to pH and ionic strength. It also helps with stability. Remarkably, the medication released more quickly in an acidic buffer solution (pH = 5) than it did in a physiological condition (pH = 7.4), suggesting that pH adjustment may be able to affect the release behaviour [61].

Another strategy uses covalent bonds to connect medications to MOFs, producing a highly tailored and selective reaction. By hydrolyzing the acid-labile bonds between MOF and the cargos at a low pH, this strategy releases the medications [62]. By using the right linkers, one may control the rate of pH-triggered bond cleavage. pH-sensitive linkers, such as ether, hydrazine, benzoic imimine bond, epoxy bond, Schiff-base bond, amide, and others, are usually modified by chemical processes [23]. In addition, various acid-labile bonds had been compiled in earlier research [63], including oxime, orthoester linkages, imine, polyacetal, and polyketal. These covalent bonds start to hydrolyze in an acidic environment, releasing the medication from the carrier. A few amino-rich antitumour drugs, such as DOX, can covalently bind to ZIF-90's surface through the Schiff base reaction of the aldehyde group (-CHO) in the imidazole-2-carboxaldehyde (ICA) ligand [62]. Only 20% of DOX was released at pH 7.4, which may have been caused by covalent fixing. A minor quantity was also partially ascribed to the hydrolysis of the Schiff-base bond, C=N.

Adding pH-sensitive components to the mix is another way to create pH-responsive MOFs. For this kind of application, chitosan is an appropriate substance. Here, Sivakumar et al. synthesised chitosan-capped Fe-MIL-88B-NH_2 as a pH-responsive delivery system that effectively distributes DOX. Under basic pH settings, the chitosan capping layer prevents the Fe-MIL-88B-NH2 MOFs structure from disintegrating; but, under acidic pH conditions, it induces disintegration by forming an insoluble gel-like form. To release DOX into the tumour microenvironment, chitosan's physiochemical property was used [64].

Another method for making pH-responsive MOFs involves host-guest interactions, which can include electrostatic, hydrogen-bonding, π–π stacking, and other interactions [65]. Many functional groups that aid in the protonation of the medicines and ligands are necessary for electrostatic interactions to take place. This could result in charge reversal and electrostatic repulsion, which would speed up the pace at which the drug releases due to pH changes [66]. Some drugs with aromatic rings can form many π-interactions and hydrogen bonds with the MOF benzene rings in a favourable way, improving the drug's loading and enabling controlled release. For example, the π-conjugated 1,3,6,8-tetrakis (p-benzoic acid) pyrene (H4TBAPy) linker was used to construct Mg (H2TBAPy) (H2O)3·C4H8O2 (TDL-Mg) MOF. After that, it was loaded with 5-FU (28.2 wt%) in order to deliver the medication gradually. In 72 hours, the loaded 5-FU (76%) from 5-FU@TDL-Mg was released at a rate that is suitable for medical applications, thanks to the synergistic effect caused by the poly-directional hydrogen bonding interactions, multiple π-interactions, and constrained channels [67].

3.2 Temperature-Responsive MOFs

A change in the ambient temperature causes the thermoresponsive materials employed in temperature-responsive drug delivery systems to alter, controlling the release of the drug. One recognised thermoresponsive polymer with a low critical solution temperature is poly (N-isopropyl acrylamide) (PNIPAM). Because of its hydrophilicity, it dissolves easily in water at temperatures below its cloud point (Tc, or around 31 °C); at higher temperatures, it condenses to aggregate. PNIPAM can therefore be a building block for drug delivery systems that are thermoresponsive [68]. For example, Nagata et al. reported on a PNIPAM-modified MOF (UiO-66-PNIPAM) nanocarrier that demonstrated a thermally regulated release triggered by just increasing the temperature [44]. The quick release of the encapsulated medicines (resorufin, caffeine, and procainamide) at a lower temperature (25 °C) and its cessation at a higher temperature (40 °C) above its Tc (31 °C) indicated the drug-loaded UiO-66-PNIPAM's temperature responsiveness. Moreover, the Zr-based MOF ZJU-801 was synthesised using the organic linker H2NPDA, or (2E,2′E)-3,3′-(naphthalene-1,4-diyl) diacrylic acid, exhibiting thermoresponsive characteristics [69].

3.3 Solvent-Switching MOFs

A regulated, reversed transition between two or more stable molecular states that produces distinct properties is referred to as a "molecular switch." A regulated, reversed transition between two or more stable molecular states that produces distinct properties is referred to as a "molecular switch." When exposed to outside solvent-responsive MOFs have the ability to change their own structures or characteristics. Samanta et al. synthesised a dynamic nanoscale MOF Zn(1a) $(H_2O)_2$ responsive to solvents by self-assembling ZnII and asymmetric linker 1a (bola-amphiphilic oligo-(p-phenylene ethynylene) (OPE) dicarboxylate) utilising dodecyl and triethylene glycol monomethyl ether. NMOF was present in tetrahydrofuran (THF) in the NMOF-1 form, which is a nano-vesicular shape with dodecyl chains on the outside. The shape of NMOF altered to an inverse-nanovesicle (NMOF-2) in water and a nano-scroll (NMOF-3) in methanol when TEG chains were exposed to the exterior of the compound. The produced hybrid MOFs in the nano-vesicular form (NMOF-1) have TEG chains oriented towards the interior and dodecyl chains on the exterior, creating a hydrophilic cavity that can facilitate H-bonding interactions for the encapsulation of DOX and cisplatin. When the solvent changes, the medicines are released via this morphologically flexible NMOF. It was possible to load 14.4 weight percent cisplatin and 4.1 weight percent DOX into NMOF-1. The drug-filled NMOFs demonstrated micromolar cytotoxicity, and both in vitro and in vivo drug release analyses were performed [45].

3.4 Light-Responsive MOFs

Light is a typical response option because it can be precisely controlled in terms of both penetration depth and intensity. As a result, drug delivery systems made of materials that are light-sensitive are being investigated more and more. Photoactive substances like porphyrin, azobenzene dicarboxylate (AZB), anthracene and its derivatives, and indocyanine green (ICG) have been employed as ligands in the MOF framework to construct light-responsive drug delivery systems [70]. UiO-AZB is a UiO-type MOF that breaks down when exposed to light because the photoisomerization of the ligand AZB causes it to become photoresponsive, together with Zr^{4+} nodes. Kristina and colleagues [46] built PEG-NH_2@5-FU-UiO-AZB for light-responsible release of 5-FU based on this feature. The experimental findings demonstrated that, in the presence of UV light, AZB is broken down into its constituent ions (Zr^{4+} and AZB^{2-})· which cleaves the MOF structure and releases the medication. With the help of external light stimulation, this work offers good control over drug release and offers a novel method for designing a photo-responsive drug delivery system.

3.5 Magnetically Responsive MOFs

A novel method known as "magnetic-responsive drug delivery" makes use of an outside magnetic field to both precisely guide drug-loaded magnetic NPs to the desired location and act as an external stimulus to cause a controlled release of the medicament, as illustrated in Fig. 10.3 [71]. Above all, the MOF-based magnetic framework may be applied as an imaging-guided therapy and MRI theranostic system. Core–shell architectures are frequently used in MOF-based nanocarriers for magnetically responsive drug release. Magnetic NPs are frequently used in these nanocarriers as a magnetic core, with the MOF serving as the shell. Fe_3O_4@$Cu_3(BTC)_2$ (BTC, benzene-1,3,5-tricarboxylate) is a MOF-based magnetic nanocomposite that was made by integrating Fe_3O_4 nanorods into $Cu_3(BTC)_2$ nanocrystals [47]. These magnetic composites demonstrated the high porosity and desired magnetic performance of a potential tailored medication delivery device. Fe3O4@$Cu_3(BTC)_2$ was soaked in a NIM trichloromethane solution to introduce the anticancer drug Nimesulide (NIM), which has a 0.2 g g^{-1} loading capacity, into these nanocomposite materials. Over the course of 11 days at 37 °C in physiological saline, the loaded NIM was fully released. Similarly, an iron carboxylate NMOF (F-NMOF) with high magnetic saturation was reported by Sethi et al. [72] for use in magnet-assisted drug delivery. In PBS, the DOX-loaded F-NMOF showed consistent DOX release, releasing 88% in 15 days. An exterior magnetic force could further increase the drug-mediated toxicity of the DOX-loaded F-NMOF, according to the cell

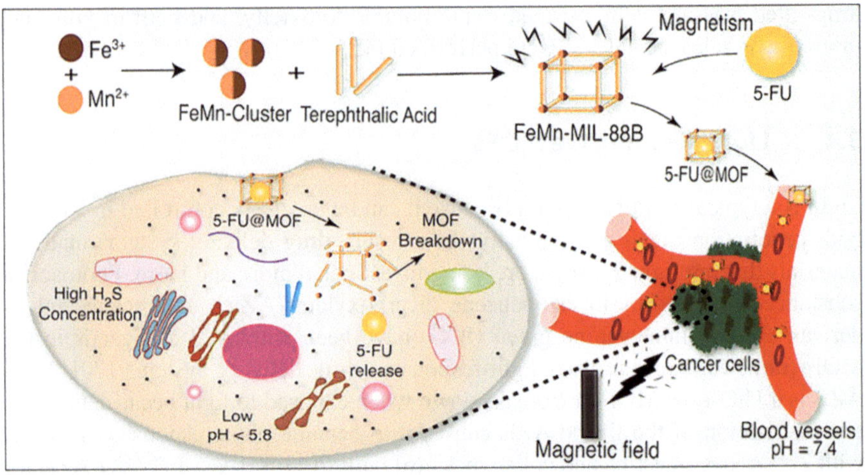

Fig. 10.3 Stimuli-responsive magnetic MOFs. (Reprinted with permission from [71], CC-BY NCND)

viability assay, although very little magnetic targeting would be accomplished with free drug (DOX).

3.6 Redox-Responsive MOFs

The fundamental basis for delivering redox-responsive medications is the significantly different redox concentrations present in both normal tissues and tumours. For instance, reducing substances like GSH are found in much higher amounts in human cancer tissues in comparison to normal tissues [73]. In the building of redox-sensitive drug delivery systems, the disulfide bond (SS), a redox-responsive group, is an attractive receptor site because it is easily broken in the presence of GSH. Zhao et al. synthesised GSH-responsive NCPs (nanoscale coordination polymers) with dithiodiglycolic acid and manganese ions (Mn^{2+}) as the organic linker containing disulfide (SS) for delivery of drugs [48]. These Mn–SS NCPs were loaded with DOX to produce Mn–SS/DOX NPs through hydrophobic interactions. Additionally, at lower pH levels, the release of DOX was accelerated due to the protonation of the amino group. Moreover, these NCPs' Mn^{2+} ions produced a significant T1 contrast for MR imaging. These Mn-SS/DOX@PDA-PEG NPs also show enhanced in vivo therapeutic efficacy as compared to free DOX.

Similar to this, hydrogen peroxide (H_2O_2), the most common reactive oxygen species (ROS) molecule produced during ischemia/reperfusion, usually induces apoptosis and inflammation, which results in oxidative tissue damage. Therefore, using H_2O_2 as a target for treatment and diagnosis offers enormous promise for cancer treatment. An NCP-based H_2O_2-responsive drug delivery system was described by Liu et al. [74]. The following is how this system was put together.

These NPs have potential use in both radiation (using Hf ions) and chemotherapy (using cisplatin released in a reducing environment). Through the breakdown of endogenous H_2O_2 generated in the cancer cells, the MnO_2 core in these NPs initiates in situ O_2 production. The investigation revealed that although the addition of these NPs to H_2O_2 solutions caused the H_2O_2 to degrade and produce O_2, the H_2O_2 solution remained stable and did not undergo much spontaneous breakdown in the absence of the NPs. Moreover, the induction of O_2 production was not possible for NPs lacking the MnO_2 core. The O_2 created here could be useful in alleviating radioresistance associated with hypoxia. Furthermore, improved T1 contrast for in vivo tumour MR imaging may result from the MnO_2 breakdown. With remarkable efficacy, this biodegradable theranostic system exhibits potential as an in vivo combinational chemo-radiotherapy. Similarly, glucose-responsive materials have drawn a lot of interest because of their capacity to competitively blend with glucose, indicating their enormous promise for developing drug delivery systems.

3.7 Enzyme-Responsive MOFs

Enzyme-responsive MOF-based drug delivery systems were created by taking advantage of the susceptibility of materials to different enzymes. Many enzyme-responsive components, such as pectin/pectinase and hyaluronic acid (HA)/hyaluronidase (HAase) [23], have been used for the majority of MOF-based enzyme-responsive drug delivery systems. Only certain places can release the medicine due to the ability of specific enzymes to break down these materials through redox reactions inside the enzyme and the associated substance. As an illustration, Kim et al. [49] described an enzyme-responsive polymer-coated MOF gatekeeper system with PCN-224 nano MOF and HA, which is HAdase-responsive. Noncovalent interactions, such as hydrogen bonds and van der Waals interactions, mediate the encapsulation process primarily, and the nanocarrier exhibited a high loading capacity (108%). In addition, electrostatic interaction as well as simple single and polyvalent coordination bonds may be responsible for the HA-coating on nano MOF. The addition of HAdase to the HA-Dox-PCN system led to a progressive increase in Dox fluorescence intensity after 3 h of incubation, indicating that HAdase broke down the HA gatekeeper that was covering the PCN-224 surface to release the drug. In this case, HAdase ought to be held responsible for the release of DOX. Overall, enzyme-responsive release offers fresh perspectives for creating stimuli-responsive drug delivery systems; nevertheless, it has several drawbacks as well, including a restricted range of enzyme types and ease of inactivation.

3.8 Multiple-Stimuli-Responsive MOFs for Drug Delivery

The complexity of the human body's environment limits the precision with which drugs can be delivered using MOF drug carriers that react to single inputs. To increase loading capacities and chemotherapeutic efficiency, multiple-stimuli-responsive MOFs may be a preferable substitute.

Wang et al. described a pH- and GSH-sensitive drug release mechanism based on MIL-101(Fe) [50]. DOX was used to load MIL-101(Fe), which was further modified by K(ad)RGDS-PEG1900 (a targeted peptide functionalized polymer) and β-CD-SS-BCN (a β-CD derivative). The GSH-responsive disulfide bond that formed after joining β-CD to the MOF's surface and the pH-responsive benzoic imine bond in K(ad) RGDS-PEG1900 are the reasons why this system demonstrated both GSH- and pH-responsive drug release behaviour. Tumour growth was significantly inhibited by this method, which demonstrated increased cellular absorption in the acidic tumour environment and tumour intracellular GSH-responsive drug release. Similarly, pH/temperature-responsive MOFs, pH/redox-responsive MOFs, NIR/H_2O_2-responsive MOFs, etc., are developed to achieve more precise cancer therapies.

4 Applications of Stimuli-Responsive MOFs in Cancer Therapy

In the sections that follow, several smart MOFs—with a focus on cancer diagnosis, treatment and therapy are detailed and explained. Lie and coworkers [75] used iron, aluminium, or zirconium as metal nodes and 4,4′-dithiobisbenzoic acid (4,4′-DTBA) as the organic ligand to create a novel redox-responsive MOF carrier, MOFM(DTBA) (M = Fe, Al, or Zr). GSH, which is frequently overexpressed in cancerous cells, has the ability to cleave the disulfide link in 4,4′-DTBA. To make CCM@MOF-Zr (DTBA) NPs, curcumin (CCM) was added to MOF-Zr (DTBA). Compared to free CCM, CCM@MOF-Zr (DTBA) NPs released CCM more quickly in vitro and accelerated cell death.

Zhang and coworkers [76] present here a unique copper MOF NP (Cu-MOF NPs) that responds to hypoxia and can be used in sonodynamic therapy (SDT) and chemodynamic therapy (CDT). Under normal oxygen partial pressure, large-sized Cu-MOF NPs exhibit good stability and promote tumour accumulation. However, when exposed to the hypoxic tumour microenvironment (TME), they quickly decomposed and produced Cu^{2+} and Ce6, which significantly strengthened the intratumoral penetration. Internalised Cu^{2+} combines with surrounding GSH to reduce Cu^{2+} to Cu^{+}, which then combines with native H_2O_2 in a Fenton-like reaction for CDT to form harmful hydroxyl radicals (•OH). Under US radiation, the released Ce6 further mediated SDT. GSH depletion significantly increased the synergistic efficacy of SDT/CDT, which caused MCF-7 to die in a minimally invasive, selective, and efficient manner. The unique hypoxia-responsive MOF nanosystem with intrinsic CDT features shown in this work can be easily integrated with different therapeutic modalities, primarily because of its flexibility.

Increased temperatures have the potential to reduce host-guest interactions and accelerate the release of drugs. Temperature readily destroys the π–π connection, as demonstrated by DS@ZJU-801 [69], also mentioned earlier. The synergistic efficacy of SDT/CDT was greatly enhanced by GSH depletion, leading to the least invasive, efficient, and selective death of MCF-7. The larger ligand of ZJU-801 and the π–π association of ZJU-801 and diclofenac sodium (DS) allowed for on-command heating-activated release of drug. An on-demand release occurs when the temperature rises because the π–π interaction becomes weaker. On the other hand, human tissue can burn in high temperatures. Silva and colleagues developed a thermo-responsive adsorbent-heater-thermometer core-shell nanomaterial (ZIF-8, EuxTby) @AuNP to overcome this problem and allow for controlled drug release below physiological temperatures (about 37 °C). The lanthanide ions and the ligand's luminescence intensities were used to monitor the temperature in this system, while the AuNPs served as heaters and ZIF-8 as the drug's adsorbent. Because AuNPs absorb visible light, these adaptable materials release the model medicines, 5FU and caffeine (CAF), in response to increased temperatures. Furthermore, the drug release rate may be increased by the polymer's shrinking at high temperatures [77].

In this instance, MOF-encapsulated AuNCs (AuNCs@MOF) were loaded with DOX to produce pH-responsive particles (AuNCs@MOF-DOX) with controlled release for enhanced PDT/chemotherapy. This indicates a move towards pH-responsive MOFs. The structure of ZIF-8 deteriorated in an acidic tumour microenvironment, causing the tumour cells to release AuNCs and DOX more quickly and improving the efficacy of PDT/chemotherapy. Significant amounts of singlet oxygen were created when subjected to a 670 nm laser, and the release rate of DOX increased to 77.1% at a pH of 5.5. The tumours were only somewhat inhibited by a single PDT treatment and a single chemotherapy treatment; however, when PDT and chemotherapy were combined, the tumours vanished entirely. Excellent PDT/chemotherapy performance was demonstrated by the constructed pH-responsive AuNCs@MOF-DOX nanoprobes with regulated release [78]. Similarly, Zeng et al. [79] developed pH-sensitive and redox-responsive folic acid-modified MOFs (FA-MOF/Buf). Additionally, research on the anticancer effects of nanomedicines was done in vivo as well as in vitro. In vitro, the FA-MOF/Buf NPs showed increased cytotoxicity in breast cancer cells, improved intracellular absorption, and enhanced water solubility and stability when compared to free Buf. In vivo, it also showed better accumulation at the tumour location, increased anticancer effectiveness, and less side effects. According to our research, FA-MOF could be developed as a Buf delivery method to boost the drug's anticancer effects when treating breast cancer.

NIR-responsive DOX release was also demonstrated by the DOX-loaded PPy@MIL-100(Fe) described by Zhu et al. [80]. In the absence of NIR irradiation, the loaded DOX released 31.9% (pH 7.4) and 49.1% (pH 5) after two hours; in the presence of NIR irradiation, however, the DOX released up to 53.2% (pH 7.4) and 70.4% (pH 5.0) because of the elevated local temperature brought on by the photothermal influence of PPy under NIR laser irradiation. Due to the drug's temperature-dependent molecular mobility and thermal movement within the carrier's crystalline lattice, the local heating caused molecular desorption. The combination exhibits remarkable potential in the context of synergistic cancer therapy. A similar multi-stimuli responsive photodynamic treatment, i.e. NIR/H_2O_2-responsive one, was developed against hypoxic cancer cells using a core-shell nanocomposite called

UCNPs/MB@ZIF-8@catalase (MB stands for methylene blue; UCNPs for upconversion NPs) [81]. In this nanocomposite, MB has a loading capacity of 1.97 weight percent. This nanocomposite's ability to produce singlet oxygen ($^1O_2$) extracellularly was evaluated using DPBF (1,3-diphenylisobenzofuran; absorption peak at 410 nm) as a probe. The ZIF-8 shell's high porosity offered an effective platform for adsorbing oxygen molecules and encouraging the production of $^1O_2$. It also prevented MB molecules from aggregating and leaking, and it facilitated the 17.9% efficient transfer of fluorescence resonance energy from the UCNPs to the MB. These findings suggest that this nanocomposite could be used for both photodynamic therapy and bioimaging.

Here is another example where a noncovalent interaction between the protein and the MOF makes MOF a highly promising carrier for encasing protein drugs. In order to improve deep tissue pancreatic cancer photoimmunotherapy, a tumour targeting collagenase-encapsulating MOF by protein-metal ion-organic ligand coordination (PMOCol) is created here [82]. When PMOCol reaches the tumour microenvironment (TME), it releases collagenase with strong enzymatic activity for the control of the extracellular matrix (ECM). Furthermore, compared to the ZIF-8 biomineralized protein, PMOCol shows significantly less toxicity. Due to extracellular matrix disintegration that enhances suppressive TME, the pancreatic tumour with extensive ECM exhibits greater immunocyte infiltration following therapy. PMOCol is able to produce systemic immunological memory and cause acute immunogenicity to activate host immunity in order to stop tumour development and recurrence. This is achieved by combining a hyperthermia agent with significant near-infrared absorption at approximately 1064 nm.

Another work where Xiang and coworkers [83] created innovative porous Fe_3O_4@C nanocomposites derived from MOFs (Fe_3O_4@C-PVP@DOX) for integrated cancer therapy using magnetic resonance imaging (MRI) guided magnetically driven hyperthermia and chemotherapy. This smart nanoplatform's magnetic behaviour, porous nature, and excellent surface modification provided it with highly efficient MRI imaging, magnetically triggered immediate DOX release function, favourable biocompatibility, and synergistic chemotherapy and magnetic hyperthermia therapy, which suggested a single platform for cancer therapy. Furthermore, the substantial inhibition of malignant tumour growth with few adverse effects was confirmed by in vivo animal tests, and this was linked to the sequential 13-day combination therapy of chemotherapy and magnetic hyperthermia in one. Therefore, this smart nanoplatform bears tremendous potential for use in the precise treatment of magnetically triggered hyperthermia due to its highly efficient combination of chemotherapy and hyperthermia.

5 Biocompatibility and Toxicity Considerations

One of the primary prerequisites for a material to be accepted for use in biomedical applications is its biocompatibility. Biocompatibility is the ability of a material to perform its intended function inside the human body without negatively affecting neighbouring tissue or changing homeostasis [84]. We are only beginning to explore the potential toxicological effects of MOFs. However, a number of well-known factors have the ability to increase MOF toxicity. They include the kinds of metals and cross-linkers, the size of the particles, ligands, functionalized groups, and the solvent system utilised to create the MOF. Furthermore, the surface functional groups' type, quantity, rate of degradation, and shape all have a significant impact on the body's cellular absorption, biodistribution, translocation, and excretion [85]. In order to create MOFs, metal ions that are primarily nonbiodegradable and of nanoscale size are utilised. Due to the toxicity of certain MOF-forming elements, well-known toxic metals including lead, arsenic, chromium, and cadmium may result in serious health problems. So, the least harmful metals that the body normally needs as nutrients, such as iron and zinc, should be utilised to create the MOFs intended for medication delivery or other therapeutic uses. As a result, non-toxic substitutes like Fe, Zn, and Zr have replaced metals like Cr that were once used in MOFs [86]. Carboxylates, phenolates, sulfonates, amines, and phosphonates are the most often employed organic linkers in the synthesis of MOFs. Because of these linkers' properties, MOFs are anticipated to break down into these constituent elements and may pose serious health hazards [87].

Furthermore, there is still work to be done to maximise the therapeutic effects of chemotherapy medications while minimising their harmful side effects. Acidic environments cause the protonation of ligands to degrade coordination bonds, which results in weak coordination reactions among metal ions and ligands for drug release. However, this process also leads to premature drug release, which can lead to unwanted toxicity. The degree of protonation varies amongst MOFs, as does the drug release profile. It was not feasible to accurately control the reaction since the in vivo physical and chemical properties, such as pH and GSH expression levels, differed in different organs and at different times. Using protective shells, such as glucose [88], pH gelatin polymer [89], PEG [90], pectin and chitosan [91], GSH polymer [92], hydrogel [93], and so forth, coating the surface of the vehicle is the most common method to extend circulation, prohibit premature ejaculation, and enhance biocompatibility. By decreasing metabolism and preventing NMOFs from being removed from the bloodstream too soon, these coatings may increase the effectiveness of medications with short half-lives.

Thus far, MOF-based nanocarriers' cytotoxicity has been evaluated using in vitro toxicological investigations using different cell lines as a means of determining cell survival [94]. However, in addition to monitoring body and organ weights, more work needs to be done to assess the toxicity of these nanocarriers in vivo by looking at their metabolic pathways [19, 95, 96]. The stability and biodegradability of MOFs are crucial issues that must be highlighted for bio-applications. In order to prevent endogenous build-up, it is advised that MOF nanocarriers break down at a specific location. There have been several published investigations on the stabilities of different MOF materials in physiologically simulated situations. For instance, MIL-101 (Fe) degraded after several days in water, although MIL-100 (Fe) was stable in phosphate buffer [19, 97]. Therefore, in order to fully comprehend the stability and degradation process of various MOF carriers both in vitro and in vivo, further extensive research is needed.

Even while MOFs currently have a number of obstacles to overcome before they can be considered for clinical usage, such as inconsistent quality and an unknown path throughout the body, we are confident that with unwavering work and time, researchers will be able to overcome these problems very soon. Consequently, it seems that intelligent MOF-based systems will be beneficial in biomedical applications for delivering medicinal compounds to the site of action.

6 Conclusion

Considerable progress has been made in the creation of supports with caged therapeutic ligands as components or gated, stimuli-responsive MOFs as drug carriers for medical applications. MOFs are the perfect high-capacity carriers for medical applications because of their high porosity, biocompatibility, versatility in functionalizing the NPs, and effective cell absorption. The opportunities presented by smart MOFs have been examined with reference to possible uses in sensing, optical, biological, controlled release, catalysis, clean energy, and sensing. Despite these developments, much more work needs to be done and there is still room for improvement. The next urgent goals might be to understand the synergistic interactions between MOFs and stimuli or to improve the current performance of smart MOFs, such as specificity towards a particular stimulus in the presence of other stimuli for improved control over material performance. Moreover, issues including enhanced stimulation and characterization methods, hybridization with different materials without sacrificing overall efficiency, and biocompatibility must be addressed. We believe that additional advancements in the field of smart MOFs will be crucial in resolving major human concerns.

References

1. Siegel RL, Jemal A, Wender RC, Gansler T, Ma J, Brawley OW. An assessment of progress in cancer control. CA Cancer J Clin. 2018;68(5):329–39.
2. Tran S, DeGiovanni PJ, Piel B, Rai P. Cancer nanomedicine: a review of recent success in drug delivery. Clin Transl Med. 2017;6:1–21.
3. Cho K, Wang XU, Nie S, Chen Z, Shin DM. Therapeutic NPs for drug delivery in cancer. Clin Cancer Res. 2008;14(5):1310–6.
4. Tohme S, Simmons RL, Tsung A. Surgery for cancer: a trigger for metastases. Cancer Res. 2017;77(7):1548–52.
5. Chen HH, Kuo MT. Improving radiotherapy in cancer treatment: promises and challenges. Oncotarget. 2017;8(37):62742.
6. Doane TL, Burda C. The unique role of NPs in nanomedicine: imaging, drug delivery and therapy. Chem Soc Rev. 2012;41(7):2885–911.
7. Zahiri M, Babaei M, Abnous K, Taghdisi SM, Ramezani M, Alibolandi M. Hybrid nanoreservoirs based on dextran-capped dendritic mesoporous silica NPs for CD133-targeted drug delivery. J Cell Physiol. 2020;235(2):1036–50.

8. Mohammadi M, Taghavi S, Abnous K, Taghdisi SM, Ramezani M, Alibolandi M. Hybrid vesicular drug delivery systems for cancer therapeutics. Adv Funct Mater. 2018;28(36):1802136.
9. Palazzolo S, Bayda S, Hadla M, Caligiuri I, Corona G, Toffoli G, Rizzolio F. The clinical translation of organic nanomaterials for cancer therapy: a focus on polymeric NPs, micelles, liposomes and exosomes. Curr Med Chem. 2018;25(34):4224–68.
10. Belfiore L, Saunders DN, Ranson M, Thurecht KJ, Storm G, Vine KL. Towards clinical translation of ligand-functionalized liposomes in targeted cancer therapy: challenges and opportunities. J Control Release. 2018;277:1–3.
11. He L, Liu Y, Lau J, Fan W, Li Q, Zhang C, Huang P, Chen X. Recent progress in nanoscale metal-organic frameworks for drug release and cancer therapy. Nanomedicine. 2019;14(10):1343–65.
12. Zhang X, Chang L, Yang Z, Shi Y, Long C, Han J, Zhang B, Qiu X, Li G, Tang Z. Facile synthesis of ultrathin metal-organic framework nanosheets for Lewis acid catalysis. Nano Res. 2019;12:437–40.
13. Wang LJ, Deng H, Furukawa H, Gándara F, Cordova KE, Peri D, Yaghi OM. Synthesis and characterization of metal–organic framework-74 containing 2, 4, 6, 8, and 10 different metals. Inorg Chem. 2014;53(12):5881–3.
14. Liu Y, Zhao Y, Chen X. Bioengineering of metal-organic frameworks for nanomedicine. Theranostics. 2019;9(11):3122.
15. Beg S, Rahman M, Jain A, Saini S, Midoux P, Pichon C, Ahmad FJ, Akhter S. Nanoporous metal organic frameworks as hybrid polymer–metal composites for drug delivery and biomedical applications. Drug Discov Today. 2017;22(4):625–37.
16. Simon-Yarza T, Mielcarek A, Couvreur P, Serre C. NPs of metal-organic frameworks: on the road to in vivo efficacy in biomedicine. Adv Mater. 2018;30(37):1707365.
17. Rojas S, Arenas-Vivo A, Horcajada P. Metal-organic frameworks: a novel platform for combined advanced therapies. Coord Chem Rev. 2019;388:202–26.
18. Yao J, Liu Y, Wang J, Jiang Q, She D, Guo H, Sun N, Pang Z, Deng C, Yang W, Shen S. On-demand CO release for amplification of chemotherapy by MOF functionalized magnetic carbon NPs with NIR irradiation. Biomaterials. 2019;195:51–62.
19. Horcajada P, Chalati T, Serre C, Gillet B, Sebrie C, Baati T, Eubank JF, Heurtaux D, Clayette P, Kreuz C, Chang JS. Porous metal–organic-framework nanoscale carriers as a potential platform for drug delivery and imaging. Nat Mater. 2010;9(2):172–8.
20. Horcajada P, Gref R, Baati T, Allan PK, Maurin G, Couvreur P, Férey G, Morris RE, Serre C. Metal–organic frameworks in biomedicine. Chem Rev. 2012;112(2):1232–68.
21. Cai W, Chu CC, Liu G, Wáng YX. Metal–organic framework-based nanomedicine platforms for drug delivery and molecular imaging. Small. 2015;11(37):4806–22.
22. Mandel K, Granath T, Wehner T, Rey M, Stracke W, Vogel N, Sextl G, Müller-Buschbaum K. Smart Optical Composite Materials: Dispersions of Metal–Organic Framework@Superparamagnetic Microrods for Switchable Isotropic–Anisotropic Optical Properties. ACS Nano. 2017;11(1):779–87.
23. Wang Y, Yan J, Wen N, Xiong H, Cai S, He Q, Hu Y, Peng D, Liu Z, Liu Y. Metal-organic frameworks for stimuli-responsive drug delivery. Biomaterials. 2020;230:119619.
24. Greenwald RB, Choe YH, McGuire J, Conover CD. Effective drug delivery by PEGylated drug conjugates. Adv Drug Deliv Rev. 2003;55(2):217–50.
25. Tong R, Cheng J. Paclitaxel-initiated, controlled polymerization of lactide for the formulation of polymeric nanoparticulate delivery vehicles. Angew Chem Int Ed. 2008;47(26):4830–4.
26. Tang S, Huang X, Chen X, Zheng N. Hollow mesoporous zirconia nanocapsules for drug delivery. Adv Funct Mater. 2010;20(15):2442–7.
27. Cui Y, Li B, He H, Zhou W, Chen B, Qian G. Metal–organic frameworks as platforms for functional materials. Acc Chem Res. 2016;49(3):483–93.
28. Wu MX, Yang YW. Metal–organic framework (MOF)-based drug/cargo delivery and cancer therapy. Adv Mater. 2017;29(23):1606134.
29. Della Rocca J, Liu D, Lin W. Nanoscale metal–organic frameworks for biomedical imaging and drug delivery. Acc Chem Res. 2011;44(10):957–68.

30. Zhou HC, Long JR, Yaghi OM. Introduction to metal–organic frameworks. Chem Rev. 2012;112(2):673–4.
31. Singh C, Mukhopadhyay S, Hod I. Metal–organic framework derived nanomaterials for electrocatalysis: recent developments for CO2 and N2 reduction. Nano Convergence. 2021;8(1):1–0.
32. Lin SX, Pan WL, Niu RJ, Liu Y, Chen JX, Zhang WH, Lang JP, Young DJ. Effective loading of cisplatin into a nanoscale UiO-66 metal–organic framework with preformed defects. Dalton Trans. 2019;48(16):5308–14.
33. Nezhad-Mokhtari P, Arsalani N, Javanbakht S, Shaabani A. Development of gelatin microsphere encapsulated cu-based metal-organic framework nanohybrid for the methotrexate delivery. J Drug Deliv Sci Technol. 2019;50:174–80.
34. Ke X, Song X, Qin N, Cai Y, Ke F. Rational synthesis of magnetic Fe3O4@ MOF NPs for sustained drug delivery. J Porous Mater. 2019;26:813–8.
35. Zhang W, Ma YB, Li YA, Wang HP. A low cytotoxic porous zinc-adeninate metal-organic framework carrier: pH-triggered drug release and anti-breast cancer study. J Iran Chem Soc. 2019;16:65–71.
36. Au-Duong AN, Lee CK. Iodine-loaded metal organic framework as growth-triggered antimicrobial agent. Mater Sci Eng C. 2017;76:477–82.
37. Nagarkar SS, Desai AV, Ghosh SK. Stimulus-responsive metal–organic frameworks. Chem Asian J. 2014;9(9):2358–76.
38. Chu C, Lin H, Liu H, Wang X, Wang J, Zhang P, Gao H, Huang C, Zeng Y, Tan Y, Liu G. Tumor microenvironment-triggered supramolecular system as an in situ nanotheranostic generator for cancer phototherapy. Adv Mater. 2017;29(23):1605928.
39. Mura S, Nicolas J, Couvreur P. Stimuli-responsive nanocarriers for drug delivery. Nat Mater. 2013;12(11):991–1003.
40. Peng S, Bie B, Sun Y, Liu M, Cong H, Zhou W, Xia Y, Tang H, Deng H, Zhou X. Metal-organic frameworks for precise inclusion of single-stranded DNA and transfection in immune cells. Nat Commun. 2018;9(1):1293.
41. Wang D, Zhou J, Chen R, Shi R, Wang C, Lu J, Zhao G, Xia G, Zhou S, Liu Z, Wang H. Core–shell metal-organic frameworks as Fe2+ suppliers for Fe2+-mediated cancer therapy under multimodality imaging. Chem Mater. 2017;29(8):3477–89.
42. Cai W, Wang J, Chu C, Chen W, Wu C, Liu G. Metal–organic framework-based stimuli-responsive systems for drug delivery. Adv Sci. 2019;6(1):1801526.
43. Zheng H, Zhang Y, Liu L, Wan W, Guo P, Nyström AM, Zou X. One-pot synthesis of metal–organic frameworks with encapsulated target molecules and their applications for controlled drug delivery. J Am Chem Soc. 2016;138(3):962–8.
44. Nagata S, Kokado K, Sada K. Metal–organic framework tethering PNIPAM for ON–OFF controlled release in solution. Chem Commun. 2015;51(41):8614–7.
45. Samanta D, Roy S, Sasmal R, Saha ND, Viswanatha R, Agasti SS, Maji TK. Solvent adaptive dynamic metal-organic soft hybrid for imaging and biological delivery. Angew Chem. 2019;131(15):5062–6.
46. Stefaniak KR, Epley CC, Novak JJ, McAndrew ML, Cornell HD, Zhu J, McDaniel DK, Davis JL, Allen IC, Morris AJ, Grove TZ. Photo-triggered release of 5-fluorouracil from a MOF drug delivery vehicle. Chem Commun. 2018;54(55):7617–20.
47. Ke F, Yuan YP, Qiu LG, Shen YH, Xie AJ, Zhu JF, Tian XY, Zhang LD. Facile fabrication of magnetic metal–organic framework nanocomposites for potential targeted drug delivery. J Mater Chem. 2011;21(11):3843–8.
48. Zhao J, Yang Y, Han X, Liang C, Liu J, Song X, Ge Z, Liu Z. Redox-sensitive nanoscale coordination polymers for drug delivery and cancer theranostics. ACS Appl Mater Interfaces. 2017;9(28):23555–63.
49. Kim K, Lee S, Jin E, Palanikumar L, Lee JH, Kim JC, Nam JS, Jana B, Kwon TH, Kwak SK, Choe W. MOF× biopolymer: collaborative combination of metal–organic framework and biopolymer for advanced anticancer therapy. ACS Appl Mater Interfaces. 2019;11(31):27512–20.

50. Wang XG, Dong ZY, Cheng H, Wan SS, Chen WH, Zou MZ, Huo JW, Deng HX, Zhang XZ. A multifunctional metal–organic framework based tumor targeting drug delivery system for cancer therapy. Nanoscale. 2015;7(38):16061–70.
51. Javanbakht S, Pooresmaeil M, Hashemi H, Namazi H. Carboxymethylcellulose capsulated Cu-based metal-organic framework-drug nanohybrid as a pH-sensitive nanocomposite for ibuprofen oral delivery. Int J Biol Macromol. 2018;119:588–96.
52. Yang K, Yang K, Chao S, Wen J, Pei Y, Pei Z. A supramolecular hybrid material constructed from pillar (6) arene-based host–guest complexation and ZIF-8 for targeted drug delivery. Chem Commun. 2018;54(70):9817–20.
53. Karakeçili A, Topuz B, Korpayev S, Erdek M. Metal-organic frameworks for on-demand pH controlled delivery of vancomycin from chitosan scaffolds. Mater Sci Eng C. 2019;105:110098.
54. Zhao H, Shu G, Zhu J, Fu Y, Gu Z, Yang D. Persistent luminescent metal-organic frameworks with long-lasting near infrared emission for tumor site activated imaging and drug delivery. Biomaterials. 2019;217:119332.
55. Xue Z, Zhu M, Dong Y, Feng T, Chen Z, Feng Y, Shan Z, Xu J, Meng S. An integrated targeting drug delivery system based on the hybridization of graphdiyne and MOFs for visualized cancer therapy. Nanoscale. 2019;11(24):11709–18.
56. Adhikari C, Mishra A, Nayak D, Chakraborty A. Metal organic frameworks modified mesoporous silica NPs (MSN): a nano-composite system to inhibit uncontrolled chemotherapeutic drug delivery from bare-msn. J Drug Deliv Sci Technol. 2018;47:1–1.
57. Lago AB, Pino-Cuevas A, Carballo R, Vázquez-López EM. A new metal–organic polymeric system capable of stimuli-responsive controlled release of the drug ibuprofen. Dalton Trans. 2016;45(4):1614–21.
58. Ke F, Zhang M, Qin N, Zhao G, Chu J, Wan X. Synergistic antioxidant activity and anticancer effect of green tea catechin stabilized on nanoscale cyclodextrin-based metal–organic frameworks. J Mater Sci. 2019;54:10420–9.
59. Chen X, Tong R, Shi Z, Yang B, Liu H, Ding S, Wang X, Lei Q, Wu J, Fang W. MOF NPs with encapsulated autophagy inhibitor in controlled drug delivery system for antitumor. ACS Appl Mater Interfaces. 2018;10(3):2328–37.
60. Naeimi S, Faghihian H. Controlled release of doxycycline by magnetized microporous MIL53 (Fe); focus on magnetization and drug loading. Curr Drug Deliv. 2019;16(1):42–50.
61. Lajevardi A, Sadr MH, Yaraki MT, Badiei A, Armaghan M. A pH-responsive and magnetic Fe3O4@ silica@ MIL-100 (Fe)/β-CD nanocomposite as a drug nanocarrier: loading and release study of cephalexin. New J Chem. 2018;42(12):9690–701.
62. Zhang FM, Dong H, Zhang X, Sun XJ, Liu M, Yang DD, Liu X, Wei JZ. Postsynthetic modification of ZIF-90 for potential targeted codelivery of two anticancer drugs. ACS Appl Mater Interfaces. 2017;9(32):27332–7.
63. Kanamala M, Wilson WR, Yang M, Palmer BD, Wu Z. Mechanisms and biomaterials in pH-responsive tumour targeted drug delivery: a review. Biomaterials. 2016;85:152–67.
64. Sivakumar P, Priyatharshni S, Nagashanmugam KB, Thanigaivelan A, Kumar K. Chitosan capped nanoscale Fe-MIL-88B-NH2 metal-organic framework as drug carrier material for the pH responsive delivery of doxorubicin. Mater Res Express. 2017;4(8):085023.
65. Millange F, Guillou N, Medina ME, Férey G, Carlin-Sinclair A, Golden KM, Walton RI. Selective sorption of organic molecules by the flexible porous hybrid metal– organic framework MIL-53 (Fe) controlled by various host– guest interactions. Chem Mater. 2010;22(14):4237–45.
66. Zhao H, Hou S, Zhao X, Liu D. Adsorption and pH-responsive release of tinidazole on metal–organic framework CAU-1. J Chem Eng Data. 2019;64(4):1851–8.
67. Hu Z, Qiao C, Xia Z, Li F, Han J, Wei Q, Yang Q, Xie G, Chen S, Gao S. A luminescent mg-metal–organic framework for sustained release of 5-fluorouracil: appropriate host–guest interaction and satisfied acid–base resistance. ACS Appl Mater Interfaces. 2020;12(13):14914–23.
68. Roy D, Brooks WL, Sumerlin BS. New directions in thermoresponsive polymers. Chem Soc Rev. 2013;42(17):7214–43.

69. Jiang K, Zhang L, Hu Q, Zhang Q, Lin W, Cui Y, Yang Y, Qian G. Thermal stimuli-triggered drug release from a biocompatible porous metal–organic framework. Chem A Euro J. 2017;23(42):10215–21.
70. Falsafi M, Saljooghi AS, Abnous K, Taghdisi SM, Ramezani M, Alibolandi M. Smart metal organic frameworks: focus on cancer treatment. Biomater Sci. 2021;9(5):1503–29.
71. Akbar MU, Badar M, Zaheer M. Programmable drug release from a dual-stimuli responsive magnetic metal–organic framework. ACS Omega. 2022;7(36):32588–98.
72. Sethi K, Sharma S, Roy I. Nanoscale iron carboxylate metal organic frameworks as drug carriers for magnetically aided intracellular delivery. RSC Adv. 2016;6(80):76861–6.
73. Meng F, Hennink WE, Zhong Z. Reduction-sensitive polymers and bioconjugates for biomedical applications. Biomaterials. 2009;30(12):2180–98.
74. Liu J, Chen Q, Zhu W, Yi X, Yang Y, Dong Z, Liu Z. Nanoscale-coordination-polymer-shelled manganese dioxide composite NPs: a multistage redox/pH/H2O2-responsive cancer theranostic nanoplatform. Adv Funct Mater. 2017;27(10):1605926.
75. Lei B, Wang M, Jiang Z, Qi W, Su R, He Z. Constructing redox-responsive metal–organic framework nanocarriers for anticancer drug delivery. ACS Appl Mater Interfaces. 2018;10(19):16698–706.
76. Zhang K, Meng X, Yang Z, Dong H, Zhang X. Enhanced cancer therapy by hypoxia-responsive copper metal-organic frameworks nanosystem. Biomaterials. 2020;258:120278.
77. Silva JY, Proenza YG, da Luz LL, de Sousa Araújo S, Gomes Filho MA, Junior SA, Soares TA, Longo RL. A thermo-responsive adsorbent-heater-thermometer nanomaterial for controlled drug release:(ZIF-8, EuxTby)@ AuNP core-shell. Mater Sci Eng C. 2019;102:578–88.
78. Zhang L, Gao Y, Sun S, Li Z, Wu A, Zeng L. pH-responsive metal–organic framework encapsulated gold nanoclusters with modulated release to enhance photodynamic therapy/chemotherapy in breast cancer. J Mater Chem B. 2020;8(8):1739–47.
79. Zeng H, Xia C, Zhao B, Zhu M, Zhang H, Zhang D, Rui X, Li H, Yuan Y. Folic acid–functionalized metal-organic framework NPs as drug carriers improved bufalin antitumor activity against breast cancer. Front Pharmacol. 2022;12:747992.
80. Zhu YD, Chen SP, Zhao H, Yang Y, Chen XQ, Sun J, Fan HS, Zhang XD. PPy@ MIL-100 NPs as a pH-and near-IR-irradiation-responsive drug carrier for simultaneous photothermal therapy and chemotherapy of cancer cells. ACS Appl Mater Interfaces. 2016;8(50):34209–17.
81. Cai HJ, Shen TT, Zhang J, Shan CF, Jia JG, Li X, Liu WS, Tang Y. A core–shell metal–organic-framework (MOF)-based smart nanocomposite for efficient NIR/H2O2-responsive photodynamic therapy against hypoxic tumor cells. J Mater Chem B. 2017;5(13):2390–4.
82. Liu Q, Wang L, Su Y, Dong W, Wang H, Liu Y, Liu H, Liu L, Wang Y. Ultrahigh enzyme loading metal–organic frameworks for deep tissue pancreatic cancer photoimmunotherapy. Small. 2023:2305131.
83. Xiang Z, Qi Y, Lu Y, Hu Z, Wang X, Jia W, Hu J, Ji J, Lu W. MOF-derived novel porous Fe3O4@ C nanocomposites as smart nanomedical platforms for combined cancer therapy: magnetic-triggered synergistic hyperthermia and chemotherapy. J Mater Chem B. 2020;8(37):8671–83.
84. Williams DF. On the mechanisms of biocompatibility. Biomaterials. 2008;29(20):2941–53.
85. Al Sharabati M, Sabouni R, Husseini GA. Biomedical applications of metal– organic frameworks for disease diagnosis and drug delivery: a review. Nano. 2022;12(2):277.
86. Tibbetts I, Kostakis GE. Recent bio-advances in metal-organic frameworks. Molecules. 2020;25(6):1291.
87. Kumar P, Anand B, Tsang YF, Kim KH, Khullar S, Wang B. Regeneration, degradation, and toxicity effect of MOFs: opportunities and challenges. Environ Res. 2019;176:108488.
88. Zhang H, Shang Y, Li YH, Sun SK, Yin XB. Smart metal–organic framework-based nanoplatforms for imaging-guided precise chemotherapy. ACS Appl Mater Interfaces. 2018;11(2):1886–95.
89. Javanbakht S, Nezhad-Mokhtari P, Shaabani A, Arsalani N, Ghorbani M. Incorporating Cu-based metal-organic framework/drug nanohybrids into gelatin microsphere for ibuprofen oral delivery. Mater Sci Eng C. 2019;96:302–9.

90. Gupta V, Mohiyuddin S, Sachdev A, Soni PK, Gopinath P, Tyagi S. PEG functionalized zirconium dicarboxylate MOFs for docetaxel drug delivery in vitro. J Drug Deliv Sci Technol. 2019;52:846–55.
91. Tang J, Ding G, Niu J, Zhang W, Tang G, Liang Y, Fan C, Dong H, Yang J, Li J, Cao Y. Preparation and characterization of tebuconazole metal-organic framework-based microcapsules with dual-microbicidal activity. Chem Eng J. 2019;359:225–32.
92. Liu Y, Gong CS, Dai Y, Yang Z, Yu G, Liu Y, Zhang M, Lin L, Tang W, Zhou Z, Zhu G. In situ polymerization on nanoscale metal-organic frameworks for enhanced physiological stability and stimulus-responsive intracellular drug delivery. Biomaterials. 2019;218:119365.
93. Chen WH, Liao WC, Sohn YS, Fadeev M, Cecconello A, Nechushtai R, Willner I. Stimuli-responsive nucleic acid-based polyacrylamide hydrogel-coated metal–organic framework NPs for controlled drug release. Adv Funct Mater. 2018;28(8):1705137.
94. Ruyra À, Yazdi A, Espín J, Carné-Sánchez A, Roher N, Lorenzo J, Imaz I, Maspoch D. Synthesis, culture medium stability, and in vitro and in vivo zebrafish embryo toxicity of metal–organic framework NPs. Chem A Euro J. 2015;21(6):2508–18.
95. Baati T, Njim L, Neffati F, Kerkeni A, Bouttemi M, Gref R, Najjar MF, Zakhama A, Couvreur P, Serre C, Horcajada P. In depth analysis of the in vivo toxicity of NPs of porous iron (III) metal–organic frameworks. Chem Sci. 2013;4(4):1597–607.
96. Tamames-Tabar C, Cunha D, Imbuluzqueta E, Ragon F, Serre C, Blanco-Prieto MJ, Horcajada P. Cytotoxicity of nanoscaled metal–organic frameworks. J Mater Chem B. 2014;2(3):262–71.
97. Taylor-Pashow KM, Della Rocca J, Xie Z, Tran S, Lin W. Postsynthetic modifications of iron-carboxylate nanoscale metal– organic frameworks for imaging and drug delivery. J Am Chem Soc. 2009;131(40):14261–3.

Stimuli-Responsive Hydrogels for Cancer Therapy

11

Shivani Sharma, Sunil Bharti, Namarta Thakur,
Pemal Preet Kaur, and Amita Sarwal

Abstract

Cancer, one of the most lethal diseases, kills millions of people globally and has a significant economic impact. Although immunotherapy, chemotherapy, and surgery have improved, the intricate nature of the tumor microenvironment, the rise of cancer recurrence, their severe adverse effects, and multidrug resistance call for novel therapeutic options. Hydrogels, particularly stimuli-responsive hydrogels (SRH), in light of their high biocompatibility, flexible characteristics, and ability to encapsulate drugs it are gaining popularity as a drug delivery option. SRH has shown considerable potential in enhancing the safety and efficacy in cancer treatment by delivering site-specific drug delivery and reducing the adverse effects. These hydrogels release drugs in reaction to particular biological or environmental stimuli, including light, pH, magnetic fields, temperature, enzymes, and so on. This chapter provides an overview of SRH for treating cancer, its classification, properties, and mechanisms of drug release, reviews literature and recent advancements in hydrogel-based cancer therapy, explores patented technologies and clinically approved treatments, and discusses future prospects, emphasizing their potential in personalized medicine, combinatorial therapies, and nanotechnology integration. By harnessing the unique characteristics of hydrogels, researchers are opening the door for more accurate and effective therapies in the future.

Keywords

Hydrogel · Smart-hydrogels · Cancer · Stimuli-response · Drug delivery

S. Sharma · S. Bharti · N. Thakur · P. P. Kaur · A. Sarwal (✉)
University Institute of Pharmaceutical Sciences and UGC Centre of Advanced Studies, Panjab University, Chandigarh, India

P. V. Devarajan et al. (eds.), *Advances in Stimuli-Responsive Nanosystems for Cancer Therapy*, https://doi.org/10.1007/978-981-95-2744-1_11

Abbreviations

SRH	Stimuli responsive hydrogel
5-Fu	Fluorouracil
CMC	Carboxymethylcellulose
DOX	Doxorubicin
EG	Ethylene glycol
GP	Glycerophosphate
HPMA	N-(2-Hydroxypropyl)methacrylamide
MSH	Magnetic supramolecular hydrogel
NCI	National Cancer Institute
NIR	Near-infrared
PAA	Poly(acrylic acid)
PBA	Phenylboronic acid
PEG	Poly(ethylene glycol)
PMAA	Poly(methacrylic acid)
PTX	Paclitaxel
ROS	Reactive oxygen species
SBMA	Sulfobetaine methacrylate
STZ	Streptozocin
TME	Tumor microenvironment
TMZ	Temozolomide
VAc	Vinyl acetate

1 Introduction

Millions of people worldwide are impacted by the immensely diverse disease known as cancer, which continues to be a global health concern. The word cancer refers to a broad category of diseases that can impact any body part. Neoplasm and malignant tumors are alternate terms used for cancer. Cancer involves the rapid growth of aberrant cells that proliferate beyond their normal limits and can invade surrounding tissues, eventually spreading to other organs is known as metastasis. Metastasis is the major cause of cancer-related deaths.

The National Cancer Institute (NCI) states that the term "cancer" refers to disorders that show abnormal and uncontrollable cell division. These diseases can infect surrounding tissues and may or may not migrate to other body parts through the lymphatic and systemic circulation [1]. This condition is caused by the unchecked growth of altered cells that are subject to natural selection [2]. One of the deadliest groups of diseases that affects people is cancer, which causes millions of deaths worldwide each year and has a wide range of distinctive clinical characteristics [3].

Cancer was initially identified as a tumor, which was a swelling mass of tissue and was known to be lethal and occasionally appeared to propagate to the surrounding tissues of the body. Advancements in molecular biology have revealed that cancer cells undergo a series of driver mutations, often known as "hits," which result in uncontrolled growth, inability to differentiate, and effective invasion of healthy tissue. This understanding revealed that cancer was "a disease of the genes."

1.1 Epidemiology

Globally, there were 19.3 million new cases of cancer and approximately 10 million cancer-related deaths in the year 2020 [4]. Cancer is now a major public health concern. More than 27,000 people die from cancer every day, and more than 52,900 people receive a cancer diagnosis. By 2040, 16.2 million deaths and around 28 million new cases are expected. The broad use of targeted, tailored treatment and greater funding for the development of cancer medications are the best ways to consistently lower the death rate from cancer worldwide [5].

1.2 Cancer Targeting Strategies

Globally, one out of every six deaths is because of cancer, which makes it a major health issue all over the world. The procedure of treating cancer has proven to be quite difficult. Although chemotherapeutic agents decreased morbidity and mortality, nearly all of them damage normal cells, especially proliferating and developing cells. Radiation therapy also may cause damage to healthy cells and organs [4]. While there have been significant advancements in recent years, including targeted therapy, nanoparticles, radionics, chemodynamic therapy, stem cell therapy, natural antioxidants, sonodynamic therapy, ablation therapy, and ferroptosis-based therapies, these have been used along with the traditional approaches [6]. Stimuli-responsive delivery systems can accurately target cancer cells by releasing the medicaments only when exposed to unique conditions in the tumor microenvironment. Therefore, they are thought to be a better option for cancer patients, because they can minimize damage to uninfected tissues, potentially reduce the adverse effects related to traditional chemotherapy, and maximize efficacy [7].

2 Hydrogels

Hydrogels emerged as successful materials with diverse applications over the last century. Hydrogels are ideal biomaterials because of their special network structure, which allows for higher hydrophilicity and biocompatibility, as well as displaying the soft physical characteristics of live tissue. Hydrogels can be termed as a 3D (three-dimensional) network that has hydrophilic groups like-CONH, -COOH, NH_2, -OH, $-SO_3H$, and $-CONH_2$ that allow them to absorb a lot of water and swell in it. Usually, crosslinked polymer chains, which can occasionally be created by cross-linked colloidal clusters form its network. Their ability to absorb water allows them to be flexible & soft [8]. Technically, a polymeric network that swells and holds a higher amount of water in its structure and does not dissolve in the aqueous medium is known as a hydrogel. The word "hydrogel" comes from the combination of the words "hydro," which means water, and "gel," which denotes a substance that is solid and resembles jelly [9, 10].

Hydrogels are a fascinating class of materials that have gained immense importance across various scientific, industrial, and biomedical fields due to their unique properties and versatile applications. These three-dimensional networks of polymers, which are mainly made of hydrophilic polymers, are remarkably good at absorbing and holding onto a large amount of water without compromising their structural integrity [11, 12].

Hydrogels differ from conventional polymers because of their higher aqueous content, making them to resemble to a natural tissues and highly biocompatible. Hydrogels belong to a flexible class of materials with a wide range of characteristics and functions. Various synthetic and natural polymers can be used to formulate hydrogels [13].

Hydrogels are utilized for targeted delivery of anticancer drugs, enabling precise drug release to tumor sites while minimizing exposure to healthy tissues. It has been demonstrated that hydrogel-based drug delivery systems increase drug half-life, enable regulated and customizable drug release, and therefore reduce non-targeted exposure, thereby improving chemotherapy outcomes and gene therapy efficacy. Hydrogels have also been extensively studied in the fields of tumor imaging, permanent tissue replacement, and temporary prosthetics [14].

For the first time, the biological use of these polymers was given by Wichterle and Lim in 1960s; later on, their use in drug delivery, nanotechnology, and tissue engineering was identified and researched. In the last two decades, hydrogels have been extensively used and researched for their application in medical fields like as smart drug delivery systems, proteins and gene delivery systems, and as artificial organs like the pancreas. Despite a large number of research, the major application areas of hydrogels in the medical field, especially in drug delivery, include contact lenses, wound dressings, transdermal patches, etc. They have many properties and characteristics which are discussed in the following text, but the main ones are the sol–gel phase transition properties, swelling and deswelling properties which can be utilized and improved to make them suitable for the required action. Based on various stimuli applied, the properties of these molecules can be changed and modified to fit the need, which makes them really interesting and valuable in drug delivery [15]. Hydrogels are broadly used to deliver the medicaments due to their qualities like easy administration, higher drug loading, and controlled release. These are useful in drug delivery to different organs, be it the skin, eyes, colon, lungs, kidneys, or the central nervous system.

3 Applications of Hydrogels in Cancer Treatment

Hydrogels are among those emerging therapies that address the current medical challenges, especially associated with chemotherapeutic drugs. Excellent controlled release, biocompatibility, drug loading, and biodegradability are all provided by hydrogels, which are extensively utilized in immunotherapy, photodynamic treatment, cancer radiation, chemotherapy, and photothermal therapy. Hydrogels have potential for delivering hydrophobic entities because of its ability to improve aqueous compatibility. Hydrogels are transforming cancer treatment by enabling tailored and localized delivery. They can release chemotherapeutic or targeted agents directly at tumor sites, enhancing efficacy while minimizing side effects. Additionally, hydrogels facilitate gene delivery, allowing for gene therapies that can alter tumor growth and support immunotherapy by encapsulating cancer vaccines or boosting immune responses. They are also used in photodynamic and photothermal therapies, delivering photosensitizers or nanoparticles for light-based treatments that destroy cancer cells. Hydrogels can be implanted for sustained drug release, aid in wound healing post-surgery or radiation, and assist in cancer detection through functionalized imaging agents. Furthermore, they enable combination therapies, delivering multiple therapeutic agents simultaneously for more effective cancer therapy.

Shukla et al. formulated a biocompatible parenteral hydrogel that controls drug release using cyclodextrin assembly, following non-Fickian kinetics. It enhances cancer treatment efficacy, achieving 75% cell-killing efficiency and complete tumor regression in albino mice within a month, without chemotherapy-related side effects. This hydrogel offers a regulated delivery system for safer and more effective cancer treatment [16].

Capanema et al. developed a hybrid hydrogel incorporating silver nanoparticles within carboxy-methyl cellulose (CMC) which electrostatically bound to doxorubicin (DOX). These vesicle-like nanosystems exhibited controlled DOX release, synergistic anticancer effects against melanoma cells, and antibacterial activity. This nanoparticle-polysaccharide-drug hybrid hydrogel presents a promising platform for topical chemotherapy, offering both anticancer and antibacterial benefits for skin cancer treatment [17].

Gonsalves et al. developed a pH-responsive nanocomposite hydrogel incorporating nanoparticles fabricated using PLGA-carboxymethyl chitosan, aiming to treat skin cancer and promote ulcer healing. It exhibited rapid gelation, high swelling, mechanical strength, controlled drug release, and antibacterial activity. This multifunctional hydrogel provides an effective platform for skin cancer therapy and wound healing, offering regulated drug delivery and antibacterial benefits [18].

Amatya et al. developed BSA/silver nanoparticle (AgNP)-incorporated hydrogel layers with excellent bioadhesive, mechanical, and photothermal properties for external photothermal therapy (PTT) of skin cancer. These hydrogels efficiently converted light into heat, maintaining their photothermal performance in gel form. In vivo studies demonstrated effective tumor ablation, achieving temperatures above 45 °C with a low laser power of 0.6 W, leading to significant tumor inhibition after a single treatment. Overall, BSA/AgNP-loaded hydrogels provide a promising and safe platform for minimally invasive skin cancer therapy [19].

Minhas et al. developed a thermally stable, colon-specific pectin hydrogel system for effective 5-fluorouracil (5-FU) delivery in colon cancer treatment. Formed by cross-linking polymerization of EGDMA, MAA, & pectin, it remains stable in the upper gastrointestinal tract, which is acidic in nature, ensuring targeted drug release. When used for lung cancer, microgel enables uniform drug distribution, prolonged release, low toxicity, high solubility, reduced first-pass metabolism, and biodegradability. This composite hydrogel system offers a promising approach for targeted and efficient cancer therapy [20].

Xu Y. et al. developed a PF127-based hydrogel system having PEG-PLGA nanoparticles for co-delivery of paclitaxel and Timozolomide, ensuring controlled drug release through hydrogel degradation. Optimized with Pluronic® F68, sodium alginate, and HPMC, the gel demonstrated superior apoptosis against C6 and U87 MG cell lines, with apoptosis rates of 32.5% and 39.2%, respectively. This system offers a promising approach for enhanced and controlled cancer therapy [21].

4 Stimuli-Responsive Hydrogels or Smart Hydrogels

Hydrogels have a unique property of modifying their structure and pore size as a result of different stimuli, i.e., physical, chemical, or biochemical, and according to changes in environment, therefore called smart hydrogels. Stimuli-responsive hydrogels are a class of materials that have garnered significant attention for their ability to adapt and respond to changes in their surrounding environment. Physical stimuli like temperature, pressure, mechanical stress, electric fields, magnetic fields, and light can change the molecular interactions within the polymer. Chemical stimuli such as ionic factors, chemical agents and pH, can also change the interactions among polymer chains at the molecular level and in-between polymer and solvents. Biochemical stimuli include the response to various enzymes, proteins, antigens etc. Another type of stimulus-responsive system includes the dual or multi-responsive hydrogel system, which results from the combination of multiple types of stimulus-responsive polymers in a single hydrogel [13, 22, 23].

Therefore, smart hydrogels are the molecules of interest for various fields like pharmaceutical, biochemical, and biotechnological fields and can be utilized for their role in drug delivery to obtain different rates of drug release and also targeted delivery. Figure 11.1 represents the different types of stimuli to which hydrogels respond.

This chapter explores stimuli-responsive hydrogels for cancer therapy, highlighting their responsiveness to light, pH, temperature, chemicals, electric, and magnetic fields. It reviews recent advancements, granted patents, clinical trials, and future prospects, emphasizing their potential in next-generation cancer treatments.

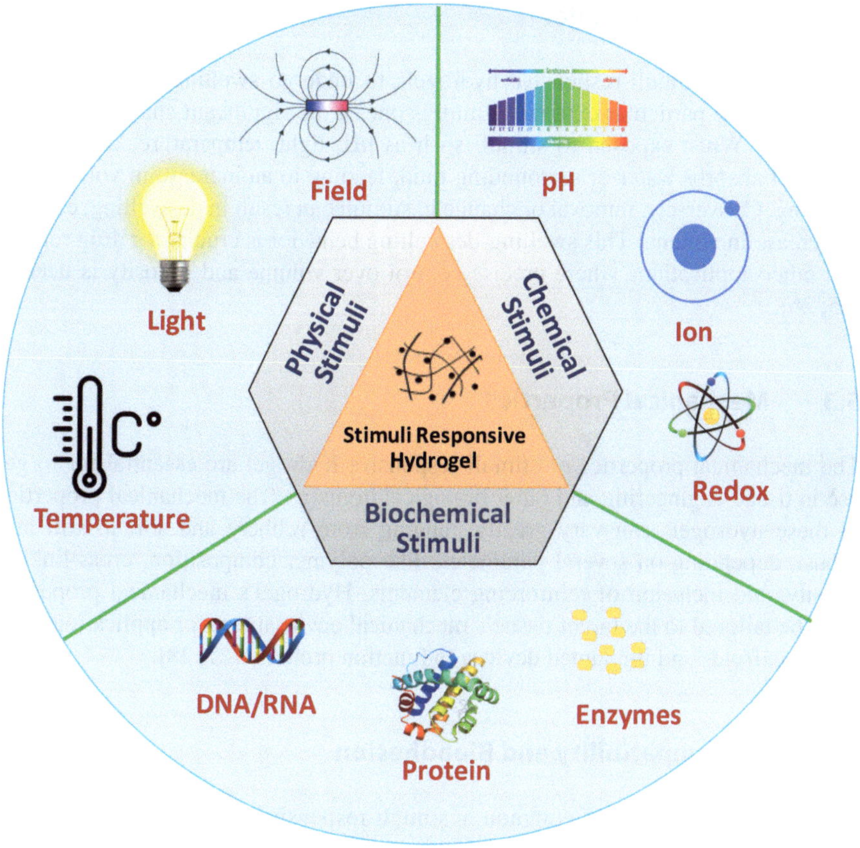

Fig. 11.1 Stimuli-based classification of hydrogels

5 Properties of Stimuli-Responsive Hydrogels

SRH exhibit a wide range of characteristics that make them highly versatile and selective for several applications, from development of delivery to tissue engineering and beyond following are some of them are explained.

5.1 Hydrophilic Nature

As we all know hydrophilic means the water-loving moieties and molecules, one essential property of hydrogels is their hydrophilic nature, i.e., their water-loving nature. They have the capacity to attract water molecules towards themselves, resulting in the absorption of water molecules and forming age age-like structure. In simple words, the water molecules are drawn to and interact with hydrophilic polymers, which make it easier for water to enter their structure. Hydrogels have a characteristic, inflated, gel-like look due to this feature [24].

5.2 Swelling Behavior

The ability of stimuli-responsive hydrogels to undergo swelling and deswelling because of the particular external stimuli is one of the significant characteristics of hydrogels. When exposed to stimuli such as pH, light, temperature, or ions, the hydrogel absorbs water or surrounding fluid, leading to an increase in volume and swelling. Conversely, removal or change in stimuli can result in deswelling, causing a decrease in volume. This swelling-deswelling behavior is crucial for drug release and other applications where precise control over volume and porosity is needed [25, 26].

5.3 Mechanical Properties

The mechanical properties of stimuli-responsive hydrogel are essential for to get use in tissue engineering and other biological domains. The mechanical properties of these hydrogels can vary greatly, ranging from rubbery and soft to stiff and robust, depending on several parameters like polymer composition, cross-linking density, and inclusion of reinforcing elements. Hydrogel's mechanical properties must be tailored to the target tissue's mechanical environment for applications like tissue scaffolds and implanted devices to function properly [27, 28].

5.4 Biocompatibility and Bioadhesion

Biocompatible designs are common in stimuli-responsive hydrogels, which makes them appropriate for a broad range of biomedical uses. The hydrogel's ability to interact with biological systems without having negative consequences is ensured by its biocompatibility. hydrogels may also have bioadhesive qualities, which enable them to stick to biological tissues and support tissue regeneration or targeted medication delivery [29].

5.5 Drug Encapsulation and Release Kinetics

Drugs varying from tiny molecules to macromolecules like proteins and nucleic acids can be efficiently encapsulated in stimuli-responsive hydrogels. The composition and structure of the hydrogel have an impact on the encapsulation process. And also, the hydrogels can regulate the release kinetics of the medications they encapsulate, providing a gradual, regulated release. The hydrogel's composition, cross-linking density, and stimulus sensitivity can all be altered to control the release rates [30, 31].

Therefore, stimuli-responsive hydrogels possess a fascinating array of properties that make them a vital class of smart materials in the realm of biomaterials and beyond. Their capacity to respond to environmental stimuli, along with characteristics like swelling behavior, mechanical properties, and biocompatibility, positions them at the forefront of cutting-edge research and innovation.

6 Stimuli-Responsive Hydrogels in Cancer Treatment

Stimuli-responsive hydrogels are intelligent materials that detect certain environmental changes and react predictably to modifications in the material's characteristics brought on by external stimuli. For cancer therapy, stimuli-responsive hydrogels offer a viable method for medication administration and therapeutic approaches. Drug delivery can be made more precise and controlled by modifying these hydrogels so that they can release drugs in response to particular stimuli like temperature, pH, light, or enzyme activity seen in tumors. They are useful for improving the efficacy and lowering the negative effects of cancer therapies because of their capacity to react with the tumor microenvironment as well as the ability to interact with biological systems [32]. A new class of smart biomaterials called stimuli-responsive hydrogels is designed for drug delivery applications that use chemical and physical stimuli, like redox reactions & pH, biochemical signals (like enzymes & glucose), that can respond to two or more stimuli. Because of their stimuli-responsiveness, these hydrogels allow for regulated manipulation of therapeutic drug release. This feature makes it easier to implement precise and regulated drug distribution plans, which improves treatment outcomes and reduces side effects in medical settings. Over time, a number of hydrogel-based formulations for the treatment, detection, and prevention of cancer have been developed; several are currently undergoing clinical studies [33]. The efficacy of the incorporated drugs is significantly raised by these responsive hydrogels, which also retarded the damage to the unaffected cells by identifying and releasing the drugs on demand based on the specific variables inside the tumor microenvironment. The presence of specific chemicals, variations in temperature, and pH are just a few of the stimuli that these smart materials react to. For example, redox-pH hydrogels react to changes in glutathione in tumor cell cytoplasm, and the acidic tumor microenvironment causes the release of the medicine; temperature-pH responsive hydrogels that react to the high temperature and acidic surroundings of infected cells. This focused strategy guarantees efficient medication administration and regulated in situ release, improving therapeutic results and reducing adverse effects on healthy tissues.

7 Stimuli-Based Hydrogel Types

Smart hydrogels or the SRH can be categorized based on the nature of stimulus that imparts a different character into them, i.e., physical and chemical responsive hydrogels, and further into more specific categories based on the specific stimulus, i.e., pH responsive, temperature responsive, etc. In this chapter, some of these categories, along with their examples, are being discussed [13].

7.1 Chemical Stimuli-Responsive Hydrogels

7.1.1 pH-Responsive Hydrogels

pH-responsive hydrogels, also called pH-sensitive or pH-dependent hydrogels. A smart delivery system that undergoes significant changes in its properties, such as its structure, swelling behavior, and reaction to the variations in the pH (acidity or alkalinity) of its surroundings. Polymers like (PAA) poly(acrylic acid), (PMAA) poly(methacrylic acid), and chitosan are commonly used in the preparation of pH-responsive hydrogels [13].

Salem, Heba F et al. prepared *in situ* pH-responsive hydrogel incorporated with niosomes loaded with TZO, for delivering and targeting TZO to its site of activity. This study successfully showed that niosomes loaded with *in situ* gel exhibited stronger anticancer activity than tamoxifen. It can be said that *in situ* gel loaded with TZO may be helpful in breast cancer therapy [34].

Kaushik, Priyanka et al. formulated pH-sensitive pectin hydrogel incorporated with zein nanoparticles loaded with doxorubucin. The hydrogel facilitates pH-responsive release of drug, particularly targeting the acidic environment of HeLa cells. The hydrogel offers controlled drug delivery, improved stability, and a protective microenvironment to the loaded drug. These findings demonstrate the potential of SRH as an injectable anticancer platform, combining with pH-sensitivity targeting with retarded off-target damage and précised therapy [35].

Ning, Like et al. formulated a hydrogel using surface-modifying nanocellulose having pH-responsive properties. This hydrogel demonstrated a significant potential for delivering controlled release of paclitaxel. The system enhanced intracellular drug distribution in cancer cells, showing notable improvement in its antitumor efficacy. The pH-responsive nature of the hydrogel enabled more efficient targeted release, demonstrating its ability as a targeted anticancer drug delivery platform [36].

7.1.2 Redox-Responsive Hydrogels

Redox-active compounds, such as disulfide-cross-linked hydrogels, show potential in drug delivery along with biomedical applications. Thiols, present in synthetic polymers and protein-based compounds, can be used for quick hydrogel synthesis. Divinyl sulphone and maleimide serve as effective crosslinking agents for developing thiol-based hydrogel systems. The crosslinking efficiency can be notably improved by oxidation using periodate, ferricyanide & hydrogen peroxide. For redox responsive systems, disulfide-containing compounds are especially valuable as they undergo reduction of the silfhyfryl group in the tumor microenvironment, allowing targeted drug release. In normal tissue, disulfide bonds are stable, but reduction in tumor environments can cause structural disruptions, leading to abrupt drug release. Disulfide bonds can be introduced into a polymer through direct or indirect synthetic approaches, where the indirect method is being more straightforward and widely preferred [37–39].

Di Huang et al. developed a smart microgel that was redox sensitive using an emulsion polymerization method, by combining (DSDMA) bis(2-methacryloyloxyethyl) disulfide, (DPA) 2-(diisopropylamino) ethyl methacrylate, and (PEG) polyethylene glycol. These hydrogels showed promising results in *in vitro* as well as preclinical studies by inhibiting proliferation and suppressing tumor growth [40].

Akhmad Irhas Robby et al. reported that, in a tumor-bearing mice, a tumor microenvironment-selective PAA-MnO_2 hydrogel showed potent ROS-scavenging activity at a tumor site. These results highlight the ability of redox-responsive hydrogels in cancer therapy, as their selective formation in tumor environments supports targeted and effective treatment [41].

7.2 Biochemical/Biological-Responsive Hydrogel

Systems having biological functional groups or moieties that are formed to interact with the biological environment for their action are known as bio-responsive hydrogels. The main biological entities used are the peptide, proteins and biological enzymes. The main concept is to use the biomolecules, such as the peptide sequences, which are cleavable in response to particular enzymes, which help hydrogels in the cross-linking. The enzymes effect on its substrate modifies the gel's swelling behavior. Due to their precise reactivity and critical roles in numerous biological processes, enzymes can guide the customization of materials to react specifically by including the particular substrate. The main advantage of these systems is the *in vivo* application [42].

7.2.1 Enzyme-Responsive Hydrogel

Enzyme-responsive nanocarriers have attracted significant interest in developing functional biomaterials, particularly for delivering drugs. Given their crucial role in biochemical processes and high biocompatibility, enzymes serve as effective triggers in hydrogel-based platforms. Enzyme-responsive hydrogels offer a versatile and dynamic system for site-specific delivery of drugs, biosensing, regenerative medicine, and tissue engineering. These hydrogels replicate the dynamic properties of uninfected tissues by permitting spatiotemporal regulation over the release of the drug, resulting in a biomimetic matrix that promotes tissue regeneration and cell proliferation [43, 44].

Su T. et al. developed an enzyme-responsive hydrogel using glucose oxidase (GOx), β-D-glucose, and an N-hydroxyimide–heparin conjugate. This system enables regulated drug release by exploiting the overexpression of heparanase in cancer cells, which particularly cleaves the heparin component. As a result, the hydrogel minimizes premature drug release in healthy tissues while ensuring targeted activation in tumor environments [45].

Zhao, Zongren et al. prepared Enzyme-responsive hydrogels that can be injected after surgery. They loaded temozolomide (TMZ), a first-line therapeutic medication and O6-benzylguanine (an MGMT inhibitor) for a therapy that targets gliomas that are resistant to TMZ. Because the hydrogels loaded with TMZ decreased *in vivo* MGMT expression, glioma cells that were resistant to TMZ became more receptive to TMZ therapy. Additionally, these hydrogels decreased TMZ-resistant gliomas recurrence and dramatically increased the efficacy of TMZ in glioma growth inhibition after surgery, indicating their promise as targeted drug delivery for glioma recurrence prevention [46].

7.2.2 Glucose-Responsive Hydrogel

Glucose-responsive systems have a significant promise for cancer and diabetes treatment, as they dynamically adjust drug release in response to glucose levels. This property enables personalized and efficient therapies, improving efficacy. Studies on glucose-responsive systems are typically classified based on mechanism and application. Mechanistically, these systems are classified into three primary categories: glucose oxidase (GOx)-driven, lectin-based, and those modified with phenylboronic acid (PBA). Based on applications, they are primarily used for insulin release facilitation or glucose concentration diagnostics. Extensive research continues to refine these elements, further advancing their therapeutic and diagnostic potential [47–49].

Li X. et al. formulated innovative, glucose-responsive hydrogels with insulin, catalase, and GOx in peptide IA-0. STZ-induced diabetic mice model, the preclinical & *in vitro* investigations showed that the hydrogels that were created controlled blood glucose levels. Because diabetes is more common and patients with diabetes require accurate insulin delivery immediately, glucose-responsive hydrogels are used more frequently in the management of diabetic disease than in cancer therapy [50].

7.3 Physical Stimuli-Responsive Hydrogels

7.3.1 Thermo-Responsive Hydrogels

Thermal-responsive hydrogels are a captivating class of smart materials having remarkable and reversible variations in their chemical and physical properties in response to variations in temperature. This characteristic makes them a focal point in various scientific and technological applications, ranging from biomedical engineering to environmental sensing. Widely used temperature-responsive hydrogel contains poly(N-isopropylacrylamide) (PNIPAAm) and others consist of chitosan-based copolymers such as methylcellulose and hydroxyl-propyl methylcellulose, etc [51].

Farjadian, Fatemeh, and colleagues created a smart pH and temperature-responsive nano-hydrogel with a well-defined structure. This involved using RAFT polymerization to create a structure that was cross-linked using poly (ethylene glycol) diacrylate and then ring-opening to bind it to lysine. Through the Schiff-base reaction, the lysine moiety of the obtained molecule was connected to the anticancer medication doxorubicin. Higher temperatures and an acidic pH are associated with increased release rates. Additionally, it found that the doxorubicin-conjugated nanoparticle causes increased cytotoxicity and cellular uptake [52].

Yue, Zhen, et al. formulated a thermo-sensitive hydrogel for regulated drug administration by easily copolymerizing PEGMA and AAc using free aqueous radicals without the need for crosslinkers. Findings showed that 5-Fu incorporated hydrogel exhibits higher cytotoxicity against LO2 & HepG-2 cells, and the blank hydrogel exhibited very low toxicity. The study demonstrates the promise of pH and thermo-sensitive hydrogel for accurate and controlled delivery of 5-Fu for cancer therapy [53].

Pang, Xiuping, et al. utilized ring-opening polymerization to develop thermosensitive hydrogels based on diblock copolymers (mPEG-b-PELG). GDMH suppressed tumor growth in both the post-surgical recurrence model and the subcutaneous tumor model, according to an in vivo antitumor effectiveness study [54].

Matthew C. Garrett et al. developed an injectable, hydrogel-based system of paclitaxel using L-alanine and L-lysine-based diblockcopolypeptide. Study highlighted hydrogel loaded with paclitaxel resulted in less reactive astrocytes, tissue damage, and cellular inflammation. Studies conducted in vivo revealed better survival and local tumor control [55].

7.3.2 Light-Responsive Hydrogels

Hydrogels responding to light stimulus, i.e., photo/light responsive systems, which exhibit alterations in their properties as exposed to light, specifically ultraviolet (UV) or visible light. Photo-reactive groups or entities, such as spiropyran or azobenzene derivatives, are incorporated into the hydrogel structure. Upon exposure to light, these groups undergo isomerization or other photoreactions, leading to variations in swelling nature, mechanical properties, and sol-gel transformation [56].

As explained by Qiu and Park (2001), light-responsive gel systems are of major interest as a delivery system. They have the advantage over the application and accuracy, as light stimulus can be exposed rapidly to the system. An example of such a system includes the azobenzene or cyclodextrin modified dextran hydrogel, which was studied as a regulated delivery platform for proteins and leads to the conclusion that SRH has towards a particular molecule is of great interest for delivery systems. But the major drawback of these platforms is their lower response time. Therefore, they require modifications to be made faster-acting [57].

Wang, Cheng, and colleagues formulated multicompartment hydrogels that are responsive to both light and heat. The triblock copolymers are utilized for formulation development poly(N-isopropylacrylamide)-b-poly(4-acryloylmorpholine)-b-poly(2-(((2-nitrobenzyl)oxy)carbonyl) amino) ethyl methacrylate) (PNBOC-b-PNAM-b-PNIPAM). When exposed to higher concentrations and temperature, it forms PNBOC-cored micelles and hydrogels. It contained doxorubicin and gemcitabine due to its hydrophilic and hydrophobic domains. Both medications are retained at the administration site when the gel transforms from sol-to-gel phase following administration. Applying UV radiation at the intended target of action results in a considerable drug release because of the sol-to-gel transition, which is caused by the crosslinking of micellar cores and the hydrophobic to hydrophilic transition [58].

Hou M. et al., formulated a photo-responsive hydrogel containing doxorubicin, from natural material, i.e., agarose/humic acid. Doxorubicin and SH (Sodium humate) were encapsulated in this hydrogel, which produced combination chemotherapeutic and photothermal effects. As a strong absorber, SH can efficiently transform light energy into thermal energy from near-infrared (NIR) light, leading to localized hyperthermia. This enhances the absorption of therapeutic medicines by cells by favoring continuous release of drug from the gel complex through a typical gel-phase to sol-phase conversion. Furthermore, when the solid tumors were exposed to radiation (NIR laser) hydrogel-based intratumoral injections responded to combined, i.e., chemical and photothermal stimuli, potentially preventing the overall tumor recurrence. Preclinical studies showed improved anticancer activity [59].

7.3.3 Electro-Responsive Hydrogels

Electro-responsive hydrogels, also called as electrically conductive hydrogels, are the systems in which the electric stimuli alters the structure and properties of the material. They can respond to electrical stimulation through various mechanisms, including electrophoresis, electroosmosis, and changes in ionic concentration within the hydrogel [60, 61]. Electrically responsive hydrogels typically contain electro-active polymers, like PPy (polypyrrole) and PEDOT (poly(3,4-ethylenedioxythiophene), that can undergo redox reactions and changes in conductivity in response to electrical stimulation [62]. Such systems can be used not only for the delivery of drugs but also in various fields like biosensors, smart medical devices, etc [63].

Servant, Ania et al. developed electro-responsive polymer-carbon nanotube hydrogel hybrids. As the electric field was induced ON/OFF, the gel hybrids produce a regulated drug release, resulting in pulsatile release profiles of the system evaluated in both *in vivo* as well as *in* vitro characterization [64].

Rahimi et al. fabricated electro-responsive hydrogels using polyacrylic acid and fibrin, stimulated cell growth and tissue formation, and are also biocompatible, making them suitable for use [65].

Zheng et al. formulated conductive *in-situ* double-crosslinked gel via ionic interactions & chemical covalent bondings that have favorable handling characteristics, improved mechanical properties & self-healing capabilities, which reported significant recovery and generation of the structural and functional attributes of myocardium [66].

7.3.4 Magnetic Field-Responsive Hydrogels

Magnetically-responsive hydrogels respond to a magnetic field stimulus by changing their structure or properties. This type of hydrogels were formulated by dispersing micro/nanoparticles with magnetic properties in the polymeric network, and later those were controlled by applying a magnetic field externally [67]. This allows the maintenance of an optimal concentration at the intended site while reducing unwanted effects [68]. By toggling the external magnetic field, the release of the drug and porosity of the hydrogels can be controlled. Hydrogels sensitive to magnetic fields have been designed to enhance scaffold functionality, including tissue function restoration, maintenance, and stimulation of cellular responses after application of the magnetic field externally [69].

Gao, Fei et al., formulated a hydrogel having a magnetic stimulus-responsive nature for the prevention of recurrence of breast cancer. This study highlights the enhanced efficacy of FVIO-based hydrogel, supported by notable suppression in the recurrence of local tumor in comparison with hyperthermia and chemotherapy alone. as evidenced by the significant suppression of the local tumor recurrences compared to chemotherapy or hyperthermia alone. With the optimally adaptive function, this magnetic hydrogel shows great potential in preventing the recurrence of various cancer types [70].

Wu, Haoan et al. Designed a self-assembled magnetic supramolecular hydrogel (MSH) which allows the sol to gel transition in response to temperature and shear thinning injection via complexation associated with the threading of α-CD on the copolymer moieties on the surface of the PEGylated iron oxide (Fe_3O_4) nanoparticles. After surgical removal, MSH works in concert to eradicate the tumor and prevent the local recurrence of the breast cancer. This study shows that MSH is notably best for avoiding postoperative recurrence of cancer when managed along with the thermo-chemotherapy [71].

7.3.5 Ionic Strength-Responsive Hydrogels

Hydrogels that experience conformational variations in response to ions like Na^+, Ca^{2+}, and K^+ are known as ionic strength-sensitive hydrogels. Commonly used materials for preparation of such hydrogels are polypeptides, deacetylated gellan gum, poly(acrylic acid), carboxymethyl dextran, poly(itaconic acid), alginate, carboxybetaine and sulfobetaine derivatives, and other ionizable and zwitterionic polymers. Few ionic strength-responsive hydrogels have been proposed thus far as cancer treatment platforms. For instance, doxorubicin-loaded hydroged composed using, poly (L-glutamic acid-co-L-lysine) hydrogels have shown sensitivity towards ionic strength, as evidenced by an improvement in the hydrogel's swelling ratio proportionate to an increase in the ionic strength of the surrounding environment. Na^+ and Cl^- shield the COO^- and $NH3^+$ polypeptide groups at high ionic strength, avoiding static interactions and encouraging swelling. These hydrogel systems indicated pH and enzyme-sensitive responses. Hydrogels responsive to ionic strength stimuli are used for delivering drugs, biosensors, and environmental sensing [72].

Xiang, Tao and colleagues synthesized zwitterionic hydrogel with ionic-strength responsiveness through free-radical copolymerization of sulfobetaine methacrylate (SBMA) with N-isopropylacrylamide, sodium styrene sulfonate (NaSS), and acrylic acid. The resulting PSBMA hydrogel demonstrated endothermic swelling characteristics, showing increased swelling ratios at higher ionic strengths under isothermal conditions [73].

7.4 MultiStimuli-Responsive Hydrogels

Multi-stimuli-responsive hydrogels can respond to a combination of different stimuli, providing a more complex and precisely controllable behavior. These hydrogels offer a high level of tunability and responsiveness by integrating multiple stimuli-responsive elements. The combination of stimuli can include temperature, pH, light, and more, allowing for highly sophisticated applications in drug delivery, smart textiles, and bioelectronics [74].

Stimuli-responsive hydrogels present an exciting and rapidly emerging field of study within the broader realm of smart delivery systems. The ability of these hydrogels to react to particular environmental cues, such as pH, temperature, ions, and light, has paved the way for pioneering advancements in several applications. By harnessing these specific properties, researchers are making an endeavor to create innovative ways for delivering the drug, tissue engineering, sensors, and beyond. As our understanding of stimuli-responsive hydrogels deepens and technology continues to progress, we can anticipate even more tailored and multifaceted applications of these extraordinary materials in the near future.

Ma, Guanglong et al, engineered a triple-responsive zwitterionic hydrogel composed of natural amino acids (glutamic acid and lysine), responding to enzymes, pH, and ionic strength. This smart hydrogel loaded with doxorubicin & diclofenac fully breaks down in the presence of trypsin - an enzyme associated with cancer and inflammation. The system also cleverly responds to pH changes to control drug release through electrical charge interactions, making it a promising candidate for targeted delivery in cancer [75].

Fathi et al. developed a dual-responsive (temperature/pH) hydrogel for breast cancer treatment by ionically crosslinking poly(NIPAAm-co-itaconic acid) with chitosan using glycerophosphate. These doxorubicin-loaded smart hydrogels have a remarkable therapeutic potential due to their stimuli-triggered drug release capabilities. According to this study, by adding a different kind of engineered smart copolymer to natural-based hydrogels as an injectable formulation for breast cancer therapy, it would be possible to create numerous de-novo hydrogels with improved sensitivity to environmental stimuli [76].

8 Patents and Clinical Trials

Over time, several hydrogel-based formulations for the diagnosis, treatment, and prevention of cancer have been developed; some of these formulations are currently undergoing clinical trials. Tables 11.1 and 11.2 enlist the various patents granted and marketed products approved and clinical studies in recent years, respectively.

Table 11.1 List patents on stimuli-responsive hydrogels

Patent No.	Title	Conclusion	References
US7786220B2	pH and temperature-sensitive hydrogel	The copolymer undergoes a temperature- and pH-sensitive sol-gel transition, ensuring stability for medical and drug delivery uses.	[77]
US20090117656 A1	Stimuli-responsive degradable gel	The stimuli-responsive hydrogel enables safe collection of a fully biological 3D cellular structure, making it ideal for regenerative medicine.	[78]
EP2130555 A1	Stimuli-responsive hydrogel	The hydrogel enables controlled drug release via modulation of polypeptide interactions, offering injectable and orally adjustable delivery.	[79]
US20100203141 A1	Stimuli-responsive low solubility hydrogel co-polymers of N-isopropyl acrylamide and synthesis method	The copolymer hydrogel is synthesized via polymerization of N-isopropylacrylamide and insoluble monomers, enabling stimuli-responsive properties	[80]
US20100098762 A1	Thermosensitive pluronic derivative hydrogel with high biodegradability and biocompatibility for tissue regeneration and preparation method thereof	The temperature-sensitive derivative of pluronic-based hydrogel enhances biocompatibility, biodegradability, and cell interaction, making it ideal for tissue regeneration	[80]
US7658947 B2	Thermogelling composition	The thermo-gelling methylcellulose-citric acid composition enables controlled delivery of active agents and soft tissue replacement	[81]
US8048450 B2	Aqueous dispersion of hydrogel nanoparticle with inverse thermogelling gelation	The IPN hydrogel nanoparticles enable uniform drug distribution in liquid form and controlled time-dependent release in solid form	[82]
WO2012159106 A2	pH-responsive self-healing hydrogels formed by boronate catechol complex	The biocompatible, pH-responsive, and self-healing catechol-borate hydrogels offer versatile applications in surgery and drug delivery	[83]
US20120020932 A1	Thermosensitive hydrogel composition and method	The thermosensitive hydrogel delivers therapeutic agents locally, forming a stable gel at body temperature for targeted treatment	[84]

(continued)

Table 11.1 (continued)

Patent No.	Title	Conclusion	References
US20120231072 A1	Thermoresponsive hydrogel composition	The thermo-responsive hydrogel enables controlled drug release by solidifying at physiological temperature	[85]
EP2455104 A1	Bio functionalized stimulus responsive dissolvable PEG hydrogel	The stimuli-responsive soluble PEG-hydrogel supports tissue regeneration and wound healing across various medical applications	[86]

Table 11.2 List of marketed products

Name	Manufacturer	Discription	Reference/ clinical trial no.
UGN-101®/ Jelmyto™	UroGen Pharma, Israel	Mitomycin C-loaded formulation for upper urothelial cancer therapy	[87]
Vantas®	Endo Pharmaceuticals, USA	Histrelin acetate formulation for prostate cancer therapy	[88]
ELIGARD®	Atrix Laboratories, USA	Leuprolide acetate for palliative therapy in advanced prostate cancer	[89]
SpaceOAR®	Boston Scientific Corporation, Marlborough, Massachusetts	Hydrogel-based organ spacer of PEG utilized to avoid injury to the rectum during radiation therapy in a patient with prostate cancer	[90]
Brachy Gel	Brachy Foam, LLC	PEG implant for cervical cancer	NCT04499521
PTX-022	Palvella Therapeutics, Inc.	Topical carbomer 940 hydrogel for *Pachyonychia congenita*, basal cell carcinomas in Gorlin syndrome patients	NCT05643872 (Ph 3)
LifePear	Grupo Espanol Multidisciplinario del Cancer Digestivo	Intraarterial injection of PEG microgel suspension for Metastatic colorectal cancer, liver metastasis	NCT04595266 (Ph II)
TraceIT hydrogel	University of Washington Cancer Consortium	Bladder cancer radiation therapy	NCT03125226

9 Future Perspective

Stimuli-responsive formulations like hydrogels have significant potential to revolutionize the delivery of anticancer agents by enabling precise, regulated, and sustained drug release within the tumor microenvironment. These smart hydrogels, designed to react to particular internal or external stimuli such as temperature, enzymes, pH, and light. Have the ability to enhance therapeutic efficacy significantly while reducing adverse effects & toxicity. Their ability to adapt to tumor-specific conditions allows for improved drug bioavailability and increased retention at the target site, leading to better treatment outcomes [91].

Despite their promising advantages, the clinical translation of these hydrogels has been hindered by several challenges. Issues such as limited biodegradability, inconsistent drug release kinetics, potential immunogenicity, and challenges in large-scale manufacturing have restricted their widespread adoption in cancer therapy. Additionally, ensuring the stability and reproducibility of these materials in physiological conditions remains a significant hurdle [92, 93].

Enhancing stimuli-responsive hydrogels requires overcoming current limitations through novel biomaterials and advanced fabrication techniques. At the same time, developing smart hydrogels with multi-stimuli responsiveness offers great potential for achieving more personalized and accurate delivery strategies of drugs. Microfluidics and 3D bioprinting enable the development of complex hydrogels that closely mimic the natural environment of cells, improving biocompatibility and drug delivery precision. A key future challenge is to develop drug delivery systems that respond to biomarkers in precise concentration ranges while integrating multiple functions such as detection, diagnosis, and therapy within a single nanoparticle-based system. Although stimuli-responsive hydrogels have gained considerable attention in academic research, their clinical translation remains complex. Nevertheless, overcoming these challenges could lead to groundbreaking advancements in cancer therapy. This paves the way for designing the systems that can effortlessly adjust to the changing microenvironment of the human body [9, 94].

10 Conclusion

Stimuli-responsive hydrogels are a remarkable innovation in science, playing a significant role in the delivery of anti-cancer entities and biomedical applications. To reduce the adverse effects of conventional chemotherapy, these smart delivery systems have been designed to enable tumor-specific drug release through exogenous and endogenous stimuli. However, significant challenges remain, including the development of multi-stimuli systems, improved tumor-targeting efficiency, biodegradation, and toxicity concerns. Ongoing research targets to address the limitations & advance combinational therapy for cancer treatment.

Acknowledgement The author would like to acknowledge Panjab University, Chandigarh, for providing the necessary facilities and support.

Author Contributions *Shivani Sharma*—Major contribution to drafting the original manuscript

Sunil Bharti—Significant involvement in drafting, revising, and editing the manuscript
Namarta Thakur—Critical review and content validation
Pemal Preet Kaur—Contribution to manuscript revision and refinement
Amita Sarwal—Supervision, investigation, and final review and approval of work

Declaration The author used Quill-Bot during the preparation of this chapter to avoid similarity. The author reviewed and edited the work as necessary and takes full responsibility for the final version.

References

1. Pravin S, Sudhir A. Integration of 3D printing with dosage forms: a new perspective for modern healthcare. Biomed Pharmacother. 2018;107:146–54.
2. Brown JS, Amend SR, Austin RH, Gatenby RA, Hammarlund EU, Pienta KJ. Updating the definition of cancer. Mol Cancer Res. 2023;21(11):1142–7.
3. Upadhyay A. Cancer: an unknown territory; rethinking before going ahead. Genes Dis. 2021;8(5):655–61.
4. Debela DT, Muzazu SG, Heraro KD, Ndalama MT, Mesele BW, Haile DC, et al. New approaches and procedures for cancer treatment: current perspectives. SAGE Open Med. 2021;9:20503121211034370.
5. Liu B, Zhou H, Tan L, Siu KTH, Guan XY. Exploring treatment options in cancer: tumor treatment strategies. Signal Transduct Target Ther. 2024;9(1)
6. Fayyaz A, Haqqi A, Khan R, Irfan M, Khan K, Reiner Ž, et al. Revolutionizing cancer treatment: the rise of personalized immunotherapies. Discov Oncol. 2024;15(1)
7. Zhao X, Bai J, Yang W. Stimuli-responsive nanocarriers for therapeutic applications in cancer. Cancer Biol Med. 2021;18(2):319–35.
8. Ho TC, Chang CC, Chan HP, Chung TW, Shu CW, Chuang KP, et al. Hydrogels: properties and applications in biomedicine. Molecules (Basel, Switzerland). 2022;27(9)
9. Abedi F, Ghandforoushan P, Adeli F, Yousefnezhad M, Mohammadi A, Moghaddam SV, et al. Development of stimuli-responsive nanogels as drug carriers and their biomedical application in 3D printing. Mater Today Chem. 2023;29:101372.
10. Sharma S, Tiwari S. Retracted: a review on biomacromolecular hydrogel classification and its applications. Int J Biol Macromol. 2020;162:737–47.
11. Dong R, Pang Y, Su Y, Zhu X. Supramolecular hydrogels: synthesis, properties and their biomedical applications. Biomater Sci. 2015;3(7):937–54.
12. Varaprasad K, Raghavendra GM, Jayaramudu T, Yallapu MM, Sadiku R. A mini review on hydrogels classification and recent developments in miscellaneous applications. Mater Sci Eng C. 2017;79:958–71.
13. Mantha S, Pillai S, Khayambashi P, Upadhyay A, Zhang Y, Tao O, et al. Smart hydrogels in tissue engineering and regenerative medicine. Materials. 2019;12
14. Sepantafar M, Maheronnaghsh R, Mohammadi H, Radmanesh F, Hasani-sadrabadi MM, Ebrahimi M, et al. Engineered hydrogels in cancer therapy and diagnosis. Trends Biotechnol. 2017;35(11):1074–87.
15. Lee SC, Kwon IK, Park K. Hydrogels for delivery of bioactive agents: a historical perspective. Adv Drug Deliv Rev. 2013;65(1):17–20.
16. Shukla A, Singh AP, Dubey T, Hemalatha S, Maiti P. Third generation Cyclodextrin graft with polyurethane embedded in hydrogel for a sustained drug release: complete shrinkage of melanoma. ACS Appl Biomater. 2019;2(4):1762–71.
17. Capanema NSV, Mansur AAP, Carvalho SM, Martins T, Gonçalves MS, Andrade RS, et al. Nanosilver-functionalized hybrid hydrogels of Carboxymethyl cellulose/Poly(Vinyl Alcohol) with antibacterial activity for prevention and therapy of infections of diabetic chronic wounds. Polymers. 2023;15(23)
18. Gonsalves A, Tambe P, Le D, Thakore D, Wadajkar AS, Yang J, et al. Synthesis and characterization of a novel pH-responsive drug-releasing nanocomposite hydrogel for skin cancer therapy and wound healing. J Mater Chem B. 2021;9(46):9533–46.
19. Amatya R, Hwang S, Park T, Chung YJ, Ryu S, Lee J, et al. BSA/Silver nanoparticle-loaded hydrogel film for local photothermal treatment of skin cancer. Pharm Res. 2021;38(5):873–83.

20. Minhas MU, Ahmad M, Anwar J, Khan S. Synthesis and characterization of biodegradable hydrogels for oral delivery of 5-fluorouracil targeted to colon: screening with preliminary in vivo studies. Adv Polym Technol. 2018;37(1):221–9.
21. Xu Y, Shen M, Sun Y, Gao P, Duan Y. Polymer nanocomposites based thermo-sensitive gel for paclitaxel and temozolomide co-delivery to glioblastoma cells. J Nanosci Nanotechnol. 2015;15(12):9777–87.
22. Ullah F, Othman MBH, Javed F, Ahmad Z, Akil HM. Classification, processing and application of hydrogels: a review. Mater Sci Eng C. 2015;57:414–33.
23. Mahinroosta M, JomehFarsangi Z, Allahverdi A, Shakoori Z. Hydrogels as intelligent materials: a brief review of synthesis, properties and applications. Mater Today Chem. 2018;8:42–55.
24. Ashok Sharma SS, Bashir S, Kasi R, Subramaniam RT. The significance of graphene based composite hydrogels as smart materials: a review on the fabrication, properties, and its applications. FlatChem. 2022;33:100352.
25. Mano JF. Stimuli-responsive polymeric systems for biomedical applications. Adv Eng Mater. 2008;10(6):515–27.
26. Gholamali I. Stimuli-responsive polysaccharide hydrogels for biomedical applications: a review. Regener Eng Transl Med. 2021;7(1):91–114.
27. Lin X, Zhao X, Xu C, Wang L, Xia Y. Progress in the mechanical enhancement of hydrogels: fabrication strategies and underlying mechanisms. J Polym Sci. 2022;60(17):2525–42.
28. Yuk H, Wu J, Zhao X. Hydrogel interfaces for merging humans and machines. Nat Rev Mater. 2022;7(12):935–52.
29. Li J, Mooney DJ. Designing hydrogels for controlled drug delivery. Nat Rev Mater. 2016;1(12):16071.
30. Thambi T, Phan VHG, Lee DS. Stimuli-sensitive injectable hydrogels based on polysaccharides and their biomedical applications. Macromol Rapid Commun. 2016;37(23):1881–96.
31. Dharmesh E, Stealey S, Salazar MA, Elbert D, Zustiak SP. Nanosilicate-hydrogel microspheres formed by aqueous two-phase separation for sustained release of small molecules. Front Biomater Sci. 2023;2(April):1–16.
32. Kilic Boz R, Aydin D, Kocak S, Golba B, Sanyal R, Sanyal A. Redox-responsive hydrogels for tunable and "on-demand" release of biomacromolecules. Bioconjug Chem. 2022;33(5):839–47.
33. Municoy S, Álvarez Echazú MI, Antezana PE, Galdopórpora JM, Olivetti C, Mebert AM, et al. Stimuli-responsive materials for tissue engineering and drug delivery. Int J Mol Sci. 2020;21(13):1–39.
34. Salem HF, Kharshoum RM, Gamal FA, Abo El-Ela FI, Abdellatif KRA. Treatment of breast cancer with engineered novel pH-sensitive triaryl-(Z)-olefin niosomes containing hydrogel: an in vitro and in vivo study. J Liposome Res. 2020;30(2):126–35.
35. Kaushik P, Priyadarshini E, Rawat K, Rajamani P, Bohidar HB. pH responsive doxorubucin loaded zein nanoparticle crosslinked pectin hydrogel as effective site-specific anticancer substrates. Int J Biol Macromol. 2020;152:1027–37.
36. Ning L, You C, Zhang Y, Li X, Wang F. Synthesis and biological evaluation of surface-modified nanocellulose hydrogel loaded with paclitaxel. Life Sci. 2020;241(December 2019):117137.
37. Lin CY, Battistoni CM, Liu JC. Redox-responsive hydrogels with decoupled initial stiffness and degradation. Biomacromolecules. 2021;22(12):5270–80.
38. Fukino T, Yamagishi H, Aida T. Redox-responsive molecular systems and materials. Adv Mater. 2017;29(25):1603888.
39. Nair DP, Podgorski M, Chatani S, Gong T, Xi W, Fenoli CR, et al. The thiol-Michael addition click reaction: a powerful and widely used tool in materials chemistry. Chem Mater. 2014;26(1):724–44.
40. Huang D, Zhu J, Xu MF, Chen J, Gao X, Zhao L, et al. One-pot preparation of pH-and redox-responsive polymeric microgel as an efficient carrier for improved breast cancer therapy. Colloids Surf A Physicochem Eng Asp. 2024;685:133320.
41. Robby AI, Yang JH, Jin EJ, Park SY. Tumor Microenvironment-Selective Sol–Gel Mineralization of ROS-Responsive Stretchable and Conductive Hydrogel. Adv Funct Mater. 2024;34(37):2402367.

42. He L, Fullenkamp DE, Rivera JG, Messersmith PB. PH responsive self-healing hydrogels formed by boronate-catechol complexation. Chem Commun. 2011;47(26):7497–9.
43. Sakai S, Kawakami K. Synthesis and characterization of both ionically and enzymatically cross-linkable alginate. Acta Biomater. 2007;3(4):495–501.
44. Sobczak M. Enzyme-responsive hydrogels as potential drug delivery systems—state of knowledge and future prospects. Int J Mol Sci. 2022;23(8):4421.
45. Su T, Tang Z, He H, Li W, Wang X, Liao C, et al. Glucose oxidase triggers gelation of N-hydroxyimide-heparin conjugates to form enzyme-responsive hydrogels for cell-specific drug delivery. Chem Sci. 2014;5(11):4204–9.
46. Zhao Z, Shen J, Zhang L, Wang L, Xu H, Han Y, et al. Injectable postoperative enzyme-responsive hydrogels for reversing temozolomide resistance and reducing local recurrence after glioma operation. Biomater Sci. 2020;8(19):5306–16.
47. Luo J, Cao S, Chen X, Liu S, Tan H, Wu W, et al. Super long-term glycemic control in diabetic rats by glucose-sensitive LbL films constructed of supramolecular insulin assembly. Biomaterials. 2012;33(33):8733–42.
48. Wang J, Wang Z, Yu J, Kahkoska AR, Buse JB, Gu Z. Glucose-responsive insulin and delivery systems: innovation and translation. Adv Mater. 2020;32(13):1902004.
49. Volpatti LR, Matranga MA, Cortinas AB, Delcassian D, Daniel KB, Langer R, et al. Glucose-responsive nanoparticles for rapid and extended self-regulated insulin delivery. ACS Nano. 2019;14(1):488–97.
50. Li X, Fu M, Wu J, Zhang C, Deng X, Dhinakar A, et al. pH-sensitive peptide hydrogel for glucose-responsive insulin delivery. Acta Biomater. 2017;51:294–303.
51. Khan A, Othman MBH, Razak KA, Akil HM. Synthesis and physicochemical investigation of chitosan-PMAA-based dual-responsive hydrogels. J Polym Res. 2013;20(10)
52. Farjadian F, Rezaeifard S, Naeimi M, Ghasemi S, Mohammadi-Samani S, Welland ME, et al. Temperature and pH-responsive nano-hydrogel drug delivery system based on lysine-modified poly (vinylcaprolactam). Int J Nanomedicine. 2019;14:6901–15.
53. Yue Z, Che Y, Jin Z, Wang S, Ma Q, Zhang Q, et al. A facile method to fabricate thermo- and pH-sensitive hydrogels with good mechanical performance based on poly(ethylene glycol) methyl ether methacrylate and acrylic acid as a potential drug carriers. J Biomater Sci Polym Ed. 2019;30(15):1375–98.
54. Pang X, Liang S, Wang T, Yu S, Yang R, Hou T, et al. Engineering thermo-ph dual responsive hydrogel for enhanced tumor accumulation, penetration, and chemo-protein combination therapy. Int J Nanomedicine. 2020;15:4739–52.
55. Garrett MC, O'Shea TM, Wollenberg AL, Bernstein AM, Hung D, Staarman B, et al. Injectable diblockcopolypeptide hydrogel provides platform to deliver effective concentrations of paclitaxel to an intracranial xenograft model of glioblastoma. PLoS One. 2020;15(7):e0219632.
56. Peng K, Tomatsu I, Kros A. Light controlled protein release from a supramolecular hydrogel. Chem Commun. 2010;46(23):4094–6.
57. Qiu Y, Park K. Environment-sensitive hydrogels for drug delivery. Adv Drug Deliv Rev. 2001;53(3):321–39.
58. Wang C, Zhang G, Liu G, Hu J, Liu S. Photo- and thermo-responsive multicompartment hydrogels for synergistic delivery of gemcitabine and doxorubicin. J Control Release. 2017;259:149–59.
59. Hou M, Yang R, Zhang L, Zhang L, Liu G, Xu Z, et al. Injectable and natural humic acid/agarose hybrid hydrogel for localized light-driven photothermal ablation and chemotherapy of cancer. ACS Biomater Sci Eng. 2018;4(12):4266–77.
60. Khan ZM, Wilts E, Vlaisavljevich E, Long TE, Verbridge SS. Electroresponsive hydrogels for therapeutic applications in the brain. Macromol Biosci. 2022;22(2):2100355.
61. Murdan S. Electro-responsive drug delivery from hydrogels. J Control Release. 2003;92(1):1–17.
62. Bansal M, Dravid A, Aqrawe Z, Montgomery J, Wu Z, Svirskis D. Conducting polymer hydrogels for electrically responsive drug delivery. J Control Release. 2020;328(August):192–209.

63. Ying Z, Wang Q, Xie J, Li B, Lin X, Hui S. Novel electrically-conductive electro-responsive hydrogels for smart actuators with a carbon-nanotube-enriched three-dimensional conductive network and a physical-phase-type three-dimensional interpenetrating network. J Mater Chem C. 2020;8(12):4192–205.
64. Servant A, Methven L, Williams RP, Kostarelos K. Electroresponsive polymer-carbon nanotube hydrogel hybrids for pulsatile drug delivery in vivo. Adv Healthc Mater. 2013;2(6):806–11.
65. Rahimi N, Swennen G, Verbruggen S, Scibiorek M, Molin DG, Post MJ. Short stimulation of electro-responsive PAA/fibrin hydrogel induces collagen production. Tissue Eng Part C Method. 2014;20(9):703–13.
66. Zheng Z, Guo Z, Zhong F, Wang B, Liu L, Ma W, et al. A dual crosslinked hydrogel-mediated integrated peptides and BMSC therapy for myocardial regeneration. J Control Release. 2022;347:127–42.
67. Frachini ECG, Petri DFS. Magneto-responsive hydrogels: Preparation, characterization, biotechnological and environmental applications. J Braz Chem Soc. 2019;30(10):2010–28.
68. Wang Y, Kohane DS. External triggering and triggered targeting strategies for drug delivery. Nat Rev Mater. 2017:2.
69. Adedoyin AA, Ekenseair AK. Biomedical applications of magneto-responsive scaffolds. Nano Res. 2018;11(10):5049–64.
70. Gao F, Xie W, Miao Y, Wang D, Guo Z, Ghosal A, et al. Magnetic hydrogel with optimally adaptive functions for breast cancer recurrence prevention. Adv Healthc Mater. 2019;8(14):1900203.
71. Wu H, Song L, Chen L, Zhang W, Chen Y, Zang F, et al. Injectable Magnetic Supramolecular Hydrogel with Magnetocaloric Liquid-Conformal Property Prevents the Post-operative Recurrence in a Breast Cancer Model Abstract Locoregional recurrence of breast cancer after tumor resection represents several clinical c. 2018;(86):1–26.
72. Andrade F, Roca-Melendres MM, Durán-Lara EF, Rafael D, Schwartz S. Review stimuli-responsive hydrogels for cancer treatment: the role of ph, light, ionic strength and magnetic field. Cancer. 2021;13(5):1–17.
73. Xiang T, Lu T, Zhao WF, Zhao CS. Ionic-strength responsive zwitterionic copolymer hydrogels with tunable swelling and adsorption behaviors. Langmuir. 2018;35(5):1146–55.
74. Pourjavadi A, Heydarpour R, Tehrani ZM. Multi-stimuli-responsive hydrogels and their medical applications. New J Chem. 2021;45(35):15705–17.
75. Ma G, Lin W, Yuan Z, Wu J, Qian H, Xu L, et al. Development of ionic strength/pH/enzyme triple-responsive zwitterionic hydrogel of the mixed l-glutamic acid and l-lysine polypeptide for site-specific drug delivery. J Mater Chem B. 2017;5(5):935–43.
76. Fathi M, Alami-Milani M, Geranmayeh MH, Barar J, Erfan-Niya H, Omidi Y. Dual thermo- and pH-sensitive injectable hydrogels of chitosan/(poly(N-isopropylacrylamide-co-itaconic acid)) for doxorubicin delivery in breast cancer. Int J Biol Macromol. 2019;128:957–64.
77. Shim S, Han Y, City SL, Sun J, Sang M, Dai H, et al. (12) United States Patent. 2010;2(12).
78. Application F, Data P. (12) Patent Application Publication (10) Pub. No.: US 2009/0117656A1. 2009;1(19).
79. Martin FASDP. European Patent Application. EP Patent 0879946A2. 1998;1(19):1–14.
80. Katsuhisa Kitano HA, Hamaguchi S. (12) Patent Application Publication (10) Pub. No.: US 2010/0292345 A1 Patent Application Publication. 2010;1(19):1–5.
81. Yang S, Examiner P, Silverman EE, Kung T. (12) United States Patent. 2010;2(12).
82. Delivery AD, Delivery AD, Langmuir BZ. (12) United States Patent. 2011;80(12):789–805.
83. Messer Smith et al. WO2012159106A2.pdf. 2012.
84. Yao et al. (12) Patent Application Publication (10) Pub. No.: US 2012/0286657 A1 Patent Application Publication. 2012;1(19):10–3.
85. Kang Miller et al. (12) Patent Application Publication (10) Pub. No.: US 2012/0286657 A1 Patent Application Publication. 2012;1(19):10–3.
86. Tomakidi P, et al. Designated extension states. Office. 2007;1(19):1–18.

87. Jacob JM, Woldu SL, Linehan J, Labbate C, Rose KM, Sexton WJ, et al. First analysis of the safety and efficacy of UGN-101 in the treatment of ureteral tumors. Urol Oncol Semin Orig Invest. 2024;42(1):20.e17–23.
88. Woolen S, Holzmeyer C, Nesbitt E, Siami PF. Long-term efficacy and tolerability of abdominal once-yearly histrelin acetate subcutaneous implants in patients with advanced prostate cancer. Prostate Cancer. 2014;2014
89. Abulateefeh SR. Long-acting injectable PLGA/PLA depots for leuprolide acetate: successful translation from bench to clinic. Drug Deliv Transl Res. 2023;13(2):520–30.
90. te Velde BL, Westhuyzen J, Awad N, Wood M, Shakespeare TP. Late toxicities of prostate cancer radiotherapy with and without hydrogel SpaceAOR insertion. J Med Imaging Radiat Oncol. 2019;63(6):836–41.
91. Prabaharan M, Mano JF. Stimuli-responsive hydrogels based on polysaccharides incorporated with thermo-responsive polymers as novel biomaterials. Macromol Biosci. 2006;6(12):991–1008.
92. Almawash S, Osman SK, Mustafa G, El Hamd MA. Current and future prospective of injectable hydrogels—design challenges and limitations. Pharmaceuticals. 2022;15(3)
93. Chao Y, Chen Q, Liu Z. Smart injectable hydrogels for cancer immunotherapy. Adv Funct Mater. 2020;30(2):1902785.
94. Mauri E, Giannitelli SM, Trombetta M, Rainer A. Synthesis of nanogels: current trends and future outlook. Gels. 2021;7(2):1–23.

Stimuli-Responsive Extracellular Vesicles/Exosomes

12

Ashish Dilip Sutar and Rahul Shukla

Abstract

Extracellular vesicles/exosomes (EVs), in particular, are stimuli-responsive nanosystems that have shown effect in cancer treatment due to their biocompatibility and capacity to transport bioactive materials. This chapter provides a thorough overview of the EVs'/exosomes' biogenesis, composition, isolation techniques, and stimuli-responsive characteristics. Different EV subtypes are described, such as oncosomes, apoptotic bodies, migrasomes, and ectosomes; exosomes are discussed in particular because of their importance in intercellular communication. The preparation and separation procedures for EVs, including density gradient ultracentrifugation, differential centrifugation, polymer precipitation, and microfluidics-based approaches, are elaborated upon, with a focus on the significance of isolation techniques in preserving the integrity and functionality of EVs. The chapter also describes the stimuli-responsive characteristics of EVs/exosomes, both exogenous (such as light, magnetic field, temperature, and ultrasound responsiveness) and endogenous (such as enzyme, pH, and redox responsiveness), emphasizing their potential for precise and regulated drug administration. It also looks at how stimuli-responsive EVs/exosomes are used in cancer therapy, including drug administration, immunotherapy, gene therapy, and diagnostic imaging, showing how important they are to the development of individualized and successful cancer treatment plans.

Keywords

Endogenous stimuli · Exogeneous stimuli · Extracellular vesicles · Exosomes · Cancer

A. D. Sutar · R. Shukla (✉)
Department of Pharmaceutics, National Institute of Pharmaceutical Education and Research (NIPER), Raebareli, Uttar Pradesh, India
e-mail: rahul.shukla@niperraebareli.edu.in

P. V. Devarajan et al. (eds.), *Advances in Stimuli-Responsive Nanosystems for Cancer Therapy*, https://doi.org/10.1007/978-981-95-2744-1_12

Abbreviations

CCs-SF/DOX	Silk fibroin and loaded with doxorubicin within calcium carbonates
CD147	Cluster of differentiation 147
CPQ	Carboxypeptidase Q
CTCs	Circulating tumor cells
DOX	Doxorubicin
DPPC	1,2-dipalmitoyl-sn-glycero-3-phosphocholine
DTX	Docetaxel
DVDMS	Sinoporphyrin sodium
EGFR	Epidermal growth factor receptor
EOGT	EGF domain-specific *O*-linked N-acetylglucosamine transferase
ESCRT	Endosomal sorting complex required for transport
EV	Extracellular vesicles
GSH	Glutathione
HDEA	3-(diethylamino)propylamine
HIPEC	Hyperthermic intraperitoneal chemotherapy
ILVs	Intraluminal vesicles
ISEV	International society for extracellular vesicles
LAMP1	Lysosomal-associated membrane protein 1
LIFU	Low-intensity focused ultrasound
MMPs	Matrix metalloproteinases
MNPs	Magnetic nanoparticles
MV	Microvesicles
MVBs	Multivesicular bodies
NIR	Near-infrared
PIGK	Phosphoglycerate kinase
PTX	Paclitaxel
ROS	Reactive oxygen species
SBC	Sodium bicarbonate
TME	Tumor microenvironment
VEGF	Vascular endothelial growth factor

1 Introduction

EVs are a wide array of nanoparticles discharged by cells, including microvesicles (MVs), exosomes, and apoptotic bodies [1]. These EVs differentiate each species with unique fabrication patterns, release profiles, sizes, compositions, and functionalities (Fig. 12.1) [2, 110]. These EVs, which are made up of nucleic acids, lipids, and proteins, are essential for intercellular communication in macro- and microenvironments [3]. EVs have varied protein profiles, which makes typical isolation methods more difficult to use, even if they lack discrete protein markers for differentiation [4]. The name "EV" is recommended by the International Society for Extracellular Vesicles (ISEV) when conclusive proof of biogenesis is lacking [2].

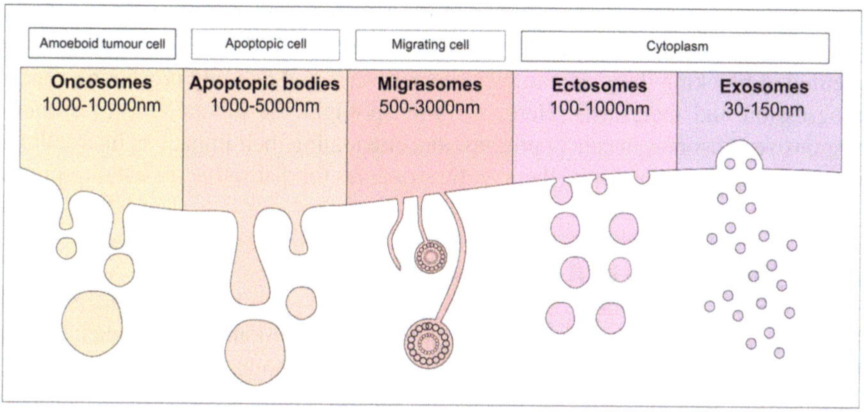

Fig. 12.1 Classification of extracellular vesicles by size and origin source

Cells most likely make EVs universally, which represent a variety of biological events and necessitate active sorting mechanisms [5]. Different from cytoplasmic material, its payload consists of DNA, RNA, proteins, lipids, and carbohydrates [1, 5]. It affects biological processes, including the genesis of cancer, and intercellular communication [6]. EVs use biomolecule transfer to aid in the development and progression of tumors. Furthermore, as targets, immunomodulators, and drug delivery vehicles, healthy and sick cell-derived EVs have therapeutic potential [7].

Intraluminal vesicles (ILVs) accumulate in multivesicular bodies (MVBs) as a result of the inward budding of the endosomal membrane, forming exosomes having a size from 30 to 150 nm [8]. Conversely, microvesicles, which are also called shedding microvesicles, are immediately discharged from the plasma membrane through outward budding and have a size range of 100 to 1000 nm [9]. The largest EVs, known as apoptotic bodies (1000–5000 nm) [10]. They are discharged from apoptotic cells and function as "corpse clearers," eliminating damaged cell detritus. Understanding the critical function of stimuli-responsive EVs, in particular exosomes, is essential to understanding cancer progression and metastasis [2]. Tumor growth, angiogenesis, and the creation of pre-metastatic niches are all aided by exosomes, which are loaded with tumor-specific proteins such as CD147 [11, 12]. Through oncogenic cargo, they modify the behavior of recipient cells, promoting invasion, migration, and the change from epithelium to mesenchymal tissue [12]. Their function in intercellular communication sets the stage for tumor cell colonization in distant organs, accelerating the spread of metastatic disease and changing the dynamics of the tumor microenvironment [13]. As a result, exosomes become essential players in the development of cancer metastasis and progression, influencing the intricate environment of tumor spread and disease progression [14]. Cancer cells and the tumor microenvironment, in response to different stimuli, release chemical messengers, setting up complex intercellular communication networks [14]. Packed with bioactive cargo, including proteins, lipids, and nucleic acids, these EVs act as channels for the transmission of pro-tumorigenic signals, promoting tumor growth, invasion, and the formation of distant metastases [15]. Taking advantage of the stimuli-responsive nature of EVs offers a sophisticated method of blocking these communication pathways and possibly preventing cancer progression and metastasis. Here, we will spotlight the pivotal role of stimuli-responsive exosomes in cancer progression, elucidating their impact on intercellular communication and potential therapeutic strategies for disrupting metastatic spread.

2 Biogenesis and Composition of EVs

EVs have overlapping diameters, comparable compositions within subclasses, a lack of particular markers, and difficulties with separation procedures that make classification difficult [16]. The process by which EVs are released from their parent cells, known as biogenesis, is essential to understanding how to classify them [15]. Differential vesicle biogenesis—which mainly involves the discharge of vesicles through the fusion of MVBs with the plasma membrane, budding from the plasma membrane—allows EVs to be classified [2, 17]. Microvesicles and exosomes are two different EV populations with different biogenesis processes [2, 18]. Production of exosomes takes place when MVBs connect with the plasma membrane, whereas microvesicles are created when the plasma membrane fissions and buds outside [19]. Gaining an understanding of these biogenesis pathways is essential to appreciating the various functions that EVs perform in molecular packing, cross-kingdom signaling, and intercellular communication.

2.1 Oncosomes

Emitted by malignant cells, large-sized oncosomes are a unique subtype of EVs that carry oncogenic cargo, including nucleic acids, proteins, and lipids [20]. These EVs are essential to the growth of tumors [21]. These oncosomes, having a diameter (1000–10,000 nm), are only released by amoeboid tumor cells, showing that these cells use a variety of migratory and metastatic strategies [2, 22]. Their biogenetically driven emergence is characterized by the dynamic shedding of non-apoptotic blebs [23]. Large oncosomes, which have a distinct molecular signature, contain RNAs that promote tumor metastasis, such as miR-125a and miR-1227, coupled with metalloproteinases, caveolin-1, and ADP ribosylation factor 6 [2]. They also include enzymes linked to metabolic pathways important for the advancement of cancer, containing a complicated yet specific cargo composition necessary for their harmful effect [2, 4].

2.2 Apoptotic Bodies

A unique class of EVs known as apoptotic bodies is released during the process of pre-decided cell death, or apoptosis [24]. Their origins, which reflect the dynamic nature of their release, are membrane blebs of apoptotic cells (1000–5000 nm) [25].

Apoptotic bodies, which are defined by nuclear chromatin condensation, caspase-mediated breaking of intracellular components, DNA fragmentation, and the creation of membrane blebs, are biogenetically the result of apoptosis [26]. Apoptotic bodies are packed with a variety of cargo, including whole organelles like mitochondria and DNA fragments, suggesting their possible function in tissue homeostasis and intercellular communication after cellular death [2, 26, 27].

2.3 Migrasomes

The oval-shaped, membrane-bound structures known as migratory cells (500–3000 nm) [2, 28]. They are formed at retraction fiber junctions or migration tips by migracytosis; they have vesicles similar to MVBs but do not expresslysosomal-associated membrane protein 1 (LAMP1) [29]. They enhance movement by integrin enrichment, actin polymerization, and microdomain clustering, particularly TSPAN4 (Tetraspanin4). NDST1 (glucosaminyl N-deacetylase/N-sulfotransferase), PIGK, CPQ, and EOGT are among the cargo; these proteins are essential for migration, tumor invasiveness, and cell-ECM dynamics [2, 29, 30].

2.4 Shedding Microvesicles/or Ectosomes

Ectosomes are a particular kind of EVs (100–1000 nm). They are also known as shedding microvesicles. When the plasma membrane evaginates and then scissions, they are discharged into the extracellular environment. Changes in the distribution of phospholipids within the plasma membrane and cytoskeleton contraction are involved in the production of ectosomes [31]. It is interesting to note that ectosome creation is facilitated by the same ESCRT mechanism that is essential for exosome generation. Matrix metalloproteinases (MMPs), Vascular endothelial growth factor (VEGF), and EGFR are among the various payloads carried by these vesicles [32]. This suggests that they are involved in important biological processes like angiogenesis, extracellular matrix remodeling, and proliferation [2].

2.5 Exosomes

Exosomes originate from late endosomes through the inward budding of the multivesicular body membrane, forming ILVs within the large multivesicular body membrane [33]. Proteins are incorporated into the invaginating membrane, and ILVs enclose cytosolic elements [33, 34]. Most ILVs are discharged into the extracellular environment as exosomes after fusing with the plasma membrane, although some may be sent to lysosomes for destruction [2, 33–35].

MVB formation and cargo sorting are facilitated by endosomal sorting complex required for transport (ESCRT) proteins, which include ESCRTs 0 to III [36, 37]. To cause vesicle budding and ILV production, ubiquitinated proteins are sequestered to the endosomal membrane by ESCRT [36–38]. Vps4 assists the ESCRT-III complex in mediating bud cleavage. Exosomes contain ubiquitinated proteins and ESCRT components. The ESCRT protein-associated protein Alix facilitates cargo selection and endosomal membrane budding [39, 40].

Alternatively, raft-based microdomains enriched in sphingomyelinases are used by an ESCRT-independent mechanism to sort exosomal cargo into MVBs. The production of ceramides stimulates the lateral phase separation that leads to domain-induced budding and ILV creation [33, 41]. Tetraspanins, such as CD81, in tetraspanin-enriched microdomains facilitate the differentiation of intracellular cargo and target receptors into exosomes. Depending on the kind of cell, these processes guarantee that a particular cargo is differentiated into exosomes through either ESCRT-independent or -dependent pathways [33, 42].

3 Preparation and Isolation of EVs

EVs are produced by a variety of procedures, including size-exclusion chromatography, immunocapture, ultrafiltration, polymer precipitation, differential centrifugation, density gradient ultracentrifugation, and membrane affinity chromatography [43, 44]. Usually, a mix of these techniques is applied to isolate particular EV subclasses. The characteristics of EV outer membranes are used for isolation by membrane affinity chromatography and polymer precipitation, yet these methods may not be successful in enriching EVs with different biological origins or sizes [19]. At the same time, methods such as size-exclusion chromatography, density gradient ultracentrifugation, and differential centrifugation separate EVs according to their physical characteristics, such as size and density [45, 46]. However, these methods may not be able to distinguish between EVs of different sizes, like lipoproteins, microvesicles, and exosomes, or distinguish EVs from particular cell types [47]. Although extremely selective for isolating distinct EV subgroups, immunocapture is too expensive for large-scale separation. The technical difficulties of separating particular subsets of EVs are still being studied by researchers in spite of these obstacles [48, 49]. Progress in EV separation techniques is essential to comprehending their various biological roles and possible medicinal applications. Furthermore, advancements in isolation methods may result in the creation of EV-based treatment interventions and diagnostic tools for a range of illnesses. Below are various techniques utilized for the isolation of EVs, described briefly.

3.1 Density Gradient Ultracentrifugation

Density gradient ultracentrifugation employs a density-based method using an iodixanol or sucrose gradient [2, 50]. High purity and selectivity are provided, especially when isolating particular EV subpopulations. Notwithstanding its benefits, it has limitations like limited yield, possible damage to exosomes, immobility, and susceptibility to the length of centrifugation [51, 52].

3.2 Differential Centrifugation

By progressively raising centrifuge speeds, differential centrifugation separates components according to variations in density, size, and shape [18]. Its ease of use and capacity to process huge volumes without the need for extra reagents make it beneficial. But because of the shear forces involved, it poses a risk to EVs, is labor-intensive, and not very portable [18, 53].

3.3 Polymer Precipitation

Hydrophilic polymers are used in polymer precipitation techniques as a kind of "fishing net" to interact with and trap EVs. They provide benefits, including automation, mobility, speed, cost, and integration with current workflows [54]. Nevertheless, they have drawbacks such as low purity because non-EV components precipitate with the polymer, which can make it more difficult to find protein biomarkers later on [54, 55].

3.4 Microfluidics-based Method

By utilizing a variety of microscale manipulation approaches, including size, density, viscoelasticity, and immunoaffinity, microfluidics-based methods are able to separate EVs [56]. They have the following advantages: easy operation, high recovery rates, quick processing, mobility, affordability, and automation possibilities. However, they also have disadvantages like lower purity, longer processing times, smaller sample volumes, and the need for more sophisticated and costly equipment [56, 57].

3.5 Immunocapture

Particular interaction between fixed antibodies and EV surface antigens. Benefits include mobility and easy operation. It does, however, have poor processing volume, high antibody input costs, and exosome elution processes [18, 58, 59].

3.6 Ultrafiltration

Separation based on size, through the use of a molecularly weighted filter membrane. It provides a quick process, inexpensive equipment, and the capacity to work with a small sample. Limitations include potential damage brought on by shear stress and membrane blockage that results in EV mass loss [18, 60].

3.7 Size-Exclusion Chromatography

Separation based on size, utilizing a column packed with nanoporous beads, is known as size-exclusion chromatography. Benefits include strong repeatability, high throughput, reasonably high purity, and intact EV structure. On the other hand, recovery time is minimal, and it can take a while [18, 61].

4 Stimuli-Responsive Properties of EVs/Exosomes

Specificity, controlled delivery, stimulus response, and feedback of the treatment are among the constraints that intelligent EVs are suggested to overcome [62, 63]. These EVs would have responsive modules for on-demand drug discharge in response to either external (such as light, heat, ultrasound, or magnetic field) or internal (such as pH, redox environment, enzymes, or hypoxia) stimuli [18, 63]. This engineering strategy, which is heavily dependent on material qualities, seeks to give EVs intelligence capabilities, greatly increasing medicinal applications [64]. Biological, chemical, and physical engineering approaches are used to modify EVs; these techniques are further divided into membrane modification, genetic engineering, and hybridization [18].

4.1 Endogenous Stimuli Responsive EVs/Exosomes

Endogenous stimuli-responsive EVs are able to target treatments by taking advantage of specific pathophysiological changes associated with diseases, including differences in pH, redox state, metabolism, and gene expression [65–68].

4.1.1 Redox Responsive

Oxidative stress serves as a pivotal factor in the pathogenesis of numerous diseases, acting as a significant pathological marker. The association between neoplastic transformation and heightened levels of oxidative stress is well-documented, with tumor tissues exhibiting a redox potential markedly elevated compared to normal tissue [69]. To counteract the detrimental effects of reactive oxygen species (ROS), tumor cells often upregulate the expression of glutathione (GSH), a crucial antioxidant present in higher concentrations within cellular compartments like the cytosol and nucleus [18, 70]. This disparity underscores the importance of GSH in cellular defense mechanisms. Research in drug delivery has focused on vehicles responsive to ROS or GSH levels (Fig. 12.2b), aiming for conditionally triggered drug release [71]. Various nanomaterials like black phosphorus quantum dots, metal organic frameworks, and nano-selenium show inherent responsiveness to GSH, facilitating reduction-mediated drug release. However, these materials face degradation challenges in the bloodstream, necessitating encapsulation within protective shell structures like EVs. Moreover, redox stress extends beyond tumors, encompassing various diseases such as inflammation, aging, and metabolic disorders, thereby broadening the potential applications of redox-responsive therapy [18, 72].

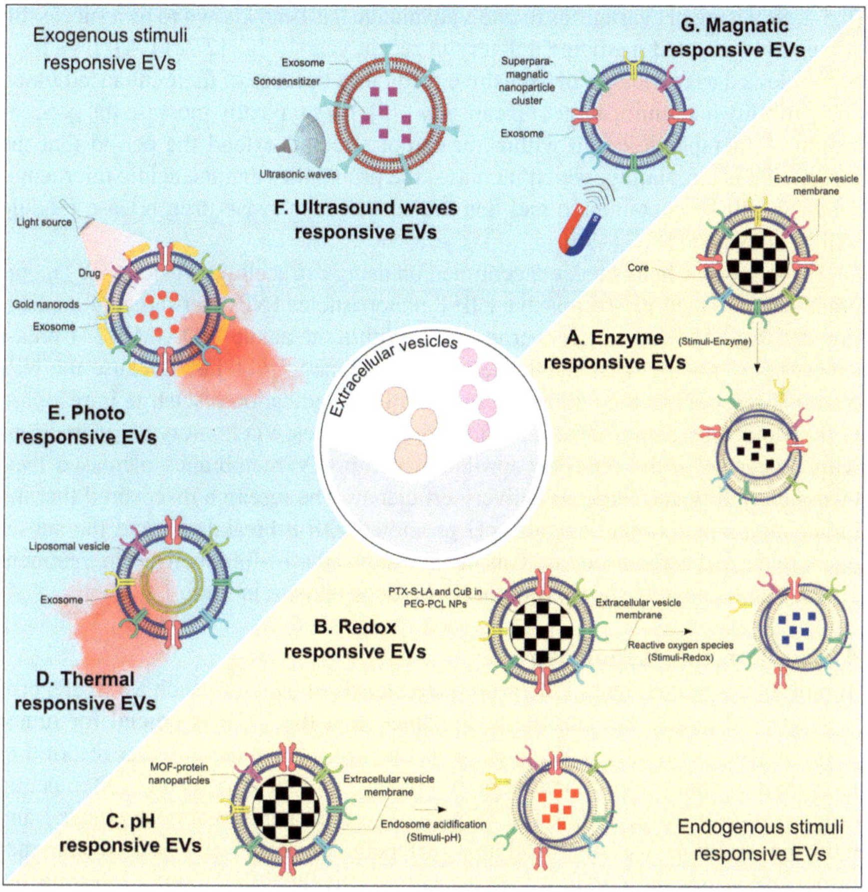

Fig. 12.2 Stimuli-responsive extracellular vesicles and their mechanism of drug release

In a study for the therapy of breast cancer, Wang et al. created an advanced drug delivery system that is triggered by ROS in paclitaxel–linoleic acid and cucurbitacin B thioether-linked compounds [73]. Through homotypic targeting, this approach, which was encapsulated into polymeric micelles and surrounded with exosome membrane, demonstrated strong efficacy against breast cancer cells. By regulating the focal adhesion kinase (FAK)/matrix metalloproteinase (MMP) pathway and removing circulating tumor cells (CTCs), the addition of EM improved tumor penetration, extended circulation, and reduced metastasis. This research proposes a viable approach to improve treatment outcomes for breast cancer patients.

4.1.2 pH responsive

The majority of cancerous cells experience the Warburg effect, sometimes referred to as aerobic glycolysis, a metabolic change that modifies glucose metabolism to promote fast growth. This change causes lactic acid to build up, which lowers the pH of the tumor microenvironment relative to normal physiological conditions

(6.5–6.8). Using pH variations to one's advantage has been known to be a successful strategy for targeted medicine delivery in recent years (Fig. 12.2c) [74]. EVs have been selected as carriers of pH-sensitive medicines because of their enhanced tissue compatibility and tumor-targeting capacity [75]. EVs typically increase the concentration of therapeutic cargo within the tumor site and extend the period that the medication is circulating, which increases the probability that the acidic microenvironment will be encountered and that the stimuli-responsive drug release module will be activated [18, 74–76].

The study by Cheng et al. concentrated on using extracellular vesicle (EV) membrane-camouflaged pH-responsive ZIF-8 nanoparticles (NPs) for targeted medication delivery in breast cancer therapy [77]. Within the acidic MDA-MB-231 breast cancer cells' microenvironment, they assembled cargo proteins. Because the NPs were pH-responsive, they allowed for controlled medication discharge in response to the tumor microenvironment's acidic circumstances. Via homotypic interactions with cancer cells, the NPs' encapsulation within EV membranes increased their biocompatibility and targeted delivery efficiency. The research discovered that the tumor microenvironment's acidic pH promoted ZIF-8 breakdown and the subsequent release of cargo proteins. This method showed a 14-fold increase in treatment efficacy and a notable suppression of tumor development in vivo.

4.1.3 Enzyme-responsive

Tumor microenvironment (TME): increased levels of enzymes such as MMPs provide specific targets for anti-tumor therapies, and the TME is crucial for tumor development, invasion, and immune evasion [78]. Nanomaterials that respond to enzyme-triggered stimuli present a promising approach to localize therapeutic drugs precisely to the tumor site (Fig. 12.2a), increasing treatment efficacy and reducing systemic toxicity. Building extracellular vesicle (EV)-based enzyme-responsive nanomaterials has received comparatively little attention, despite the shown efficacy of enzyme-responsive systems in tumor therapy [18, 79]. Additionally, novel hydrogels loaded with EV and possessing stimulus-responsive qualities are being developed. Advantages, including high specificity and concentration-dependent response intensity, are offered by these enzyme-responsive EV-based materials, which hold great promise for molecular diagnosis and tailored therapy approaches [18].

4.2 Exogenous Stimuli Responsive EVs/Exosomes

Differentiating between healthy and diseased tissues presents a problem for therapy targeting because of inter-lesion interaction and different endogenous activation. An alternative is provided by exogenous stimuli-responsive systems, which release drugs at precise target areas in response to external stimuli such as temperature, light, and ultrasound [80, 81]. These methods lack tissue selectivity, yet are successful in limiting off-target effects. Therapeutic medicines need to be highly targeted and histocompatible to address this. Such characteristics are best shown by EVs, which can target specific areas and are stable [82]. Tumor cell-derived EVs, for example, actively target tumor cells through a homing effect. Strategies to avoid immune clearance, such as overexpressing CD47 surface proteins, are examples of advancements. The incorporation of traits that respond to external stimuli into EVs has the potential to provide accurate diagnosis and treatment [78, 80–82].

4.2.1 Temperature Responsive

Maintaining the body's balance depends on temperature regulation, and external interventions can have an impact on physiological processes and the management of disease [83]. Interestingly, localized hyperthermia has become a powerful treatment for solid tumors by raising the temperature to a point where it causes cytotoxicity and tumor necrosis. Nonetheless, difficulties continue because of unequal heating, which reduces effectiveness against metastases and deep tumors [84]. A potential method to improve tumor targeting and medication release is to combine hyperthermia with thermoresponsive materials, such as those mediated by 1,2-dipalmitoyl-sn-glycero-3-phosphocholine (DPPC) [18, 85]. The potential of EVs is still not fully realized, despite the development of several materials such as liposomes and polymersomes. EVs can have their lipid bilayers changed using DPPC to resemble liposomes, although maintaining biological activity must be balanced with the EVs' functional needs as nanocarriers [86]. One kind of EVs that has advantages is called an exosome; these are EVs that have nucleic acid components that are stable despite heat stimulation. A thermosensitive exosome–liposome hybrid nanoparticle system is one of the most recent developments, showing enhanced drug delivery and effectiveness in the treatment of metastatic peritoneal cancer [87, 88, 111]. This novel method emphasizes how thermoresponsive EVs can help advance targeted therapy for difficult-to-treat cancers.

The development of a heat-responsive drug delivery system to treat metastatic peritoneal cancer was the focus of Qijun et al.'s creative study [89]. In this work, they developed engineered exosome thermosensitive liposome hybrid nanoparticles (gETL-NPs) that were loaded with docetaxel (DTX) or granulocyte-macrophage colony-stimulating factor (GM-CSF). This method released the medication in response to heat stimulation. gETL-NPs were shown by Qijun et al. to be an efficient inhibitor of tumor formation; the co-administration of hyperthermic intraperitoneal chemotherapy (HIPEC) drugs increased the efficacy of the treatment by about 39%. This work offers a promising way to enhance therapeutic outcomes and marks a major advancement in specific drug delivery for the treatment of metastatic peritoneal cancer.

4.2.2 Light Responsive

EVs that respond to light are a new development in targeted drug delivery and therapeutics. By utilizing the unique characteristics of light, especially near-infrared (NIR) wavelengths, these EVs provide targeted cargo release at precise locations with no tissue injury. Light energy triggers the release of drugs from EVs through processes such as photothermal effects, two-photon absorption, and up-conversion [18, 90]. Current developments emphasize the fusion of artificial and organic components to improve EV performance. Although issues like metabolic hazards and histocompatibility issues still exist, the addition of light-responsive mechanisms to EV-based drug delivery systems represents a viable path toward customized and efficient treatments [18, 90, 91].

Wang and colleagues have made significant contributions to the field of exosome-based therapeutic design through the creative integration of photo-responsive functions in their revolutionary study [92]. They have effectively created macrophage-derived exosomes with hybrid biological and synthetic features by utilizing a donor cell-assisted membrane modification method. Equipped with both folic acid and RGD ligand-mediated targeting, these engineered exosomes demonstrate exceptional accuracy in locating cancerous cells. Moreover, the incorporation of gold nanorods allows these exosomes to undergo photothermal conversion in response to light, which amplifies their potential as a therapeutic agent. Wang et al. have shown, notably, that the duration of therapy for preventing tumor relapse in HeLa tumor models may be extended in a programmable way. This is seen in the impressive increase to 30 days in comparison to the control group, which only managed to reach 15 days. This innovative discovery highlights the potential of photo-responsive capabilities in improving therapeutic outcomes in cancer therapy.

4.2.3 Magnetic Field-Responsive

A frontier in biomedical innovation is the integration of magnetic materials with EVs. Because of their biocompatibility and tunable reactions to external magnetic fields, magnetic nanoparticles (MNPs) provide a flexible platform for selective medication delivery and imaging [93]. Through the utilization of magnetic field-induced effects, MNPs improve the accuracy of EV-based targeting tactics, surmounting the drawbacks of traditional methods. Developments in animal models provide as an example of how this cooperation increases tissue-specific targeting and prolongs the half-life of therapeutic cargo. Furthermore, the potential for MRI-guided treatments highlights how revolutionary this multidisciplinary strategy can be, resulting in a new era of precision medicine [93–95].

Hongzhao et al. addressed important issues in exosome-based drug delivery with their innovative study, which created a dual-functional exosome-based superparamagnetic nanoparticle cluster for cancer therapy [96]. They greatly enhanced the superparamagnetic behavior and reactivity to external magnetic fields of each exosome by attaching several superparamagnetic nanoparticles via Tf – Tf receptor contact, allowing for accurate targeting of sick cells and effective separation from blood. Experiments conducted in vivo showed considerable growth reduction of tumors and improved cancer targeting. The presence of Tf resulted in a modest separation efficiency from serum, but replacing it with a different ligand suggests potential for improvement. The creation of exosome-based clusters, which gives delivery vehicles superparamagnetic and ferromagnetic properties, is the study's primary contribution. Through the integration of magnetism with exosomes, this discovery expands the medicinal applications and improves exosome purification for disease diagnostics, especially cancer.

4.2.4 Ultrasound Responsive

Engineering EVs to react to ultrasonic stimuli allows researchers to precisely control the release of drugs. This responsiveness usually refers to ways wherein ultrasonic waves induce the carrier medications to release by mechanical, thermal, or radiative forces [97]. This exact control minimizes off-target effects while maximizing treatment efficacy. The combination of EV-based drug delivery and ultrasound technology has the potential to completely transform medical care by providing focused, accurate, and non-invasive methods for treating a variety of illnesses [98–101].

In their research, Liu and their team introduced an exciting new approach to cancer treatment and diagnosis [102]. They developed a clever system that responds to ultrasound, using tiny particles called exosomes derived from cancer cells to carry a powerful substance called sinoporphyrin sodium (DVDMS). This innovative combo, dubbed EXO-DVDMS, not only proved stable and specific to tumors but also released its payload precisely when triggered by ultrasound. What's more, it showed an impressive ability to generate reactive oxygen species (ROS) for enhanced cancer therapy. But that's not all—EXO-DVDMS did not just attack tumors; it also provided imaging capabilities and even helped prevent cancer spread. By working together with ultrasound, this system effectively slowed tumor growth and triggered cancer cell death, hinting at a brighter future for personalized cancer treatment and diagnosis.

5 Applications in Cancer Therapy

Extracellular vesicle (EV)-based medication delivery systems that are responsive to stimuli show great promise for treating cancer [103]. Multimodal formulations are a more sophisticated tactic, even if unimodal techniques work well for focused medication administration and imaging. When several answers are combined into one formulation, we get more accuracy and complexity in the diagnosis and therapy [18, 104, 105]. Using stimuli-responsive EVs, several formulations have evolved at the forefront of cancer therapy. These combinations take advantage of various triggers to coordinate a multimodal strategy to fight cancer [7]. Temperature, pH, and enzymatic triggers are just a few examples of the components that work in harmony to enhance therapeutic efficiency. Within this paradigm, the formulations provide real-time monitoring and adaptive responses to the changing tumor microenvironment in addition to delivering medications with pinpoint accuracy [2]. Through the combination of therapeutic payloads and imaging modalities, these formulations offer unique insights into the course of disease while also delivering treatment. A new age in precision medicine is being brought in by the shift toward multimodal formulations, wherein medicines are customized with never-before-seen specificity to meet the demands of individual patients. These innovative techniques could transform cancer treatment as research progresses, bringing hope for better results and a higher standard of living. These are a few formulations created to treat cancer that take advantage of the potential of extracellular vesicles that respond to stimuli to transform therapeutic interventions and change the paradigm of cancer care [72, 89, 98].

In a study, Cao et al. formulated extracellular vesicle (EV)-based nano sonosensitizers that might be used for PA imaging-guided chemo-sonodynamic cancer treatment as well as dual stimuli-responsive drug release (Fig. 12.3a) [106]. These nanoplatforms were loaded with sodium bicarbonate (SBC) and paclitaxel (PTX) for pH-triggered drug release, as well as indocyanine green (ICG) for sonodynamic treatment and PA imaging. Improved cellular uptake of ICG by the EVs resulted in higher sonodynamic effects against breast cancer cells. Additionally, SBC-EV(ICG/PTX) demonstrated pH-responsive drug release that was enhanced by ultrasonic (US) irradiation and triggered by CO_2 production in an acidic environment. While SBC-EV(ICG/PTX) with US irradiation effectively inhibited tumor growth without systemic toxicity in tumor-bearing mice, high-resolution PA imaging demonstrated

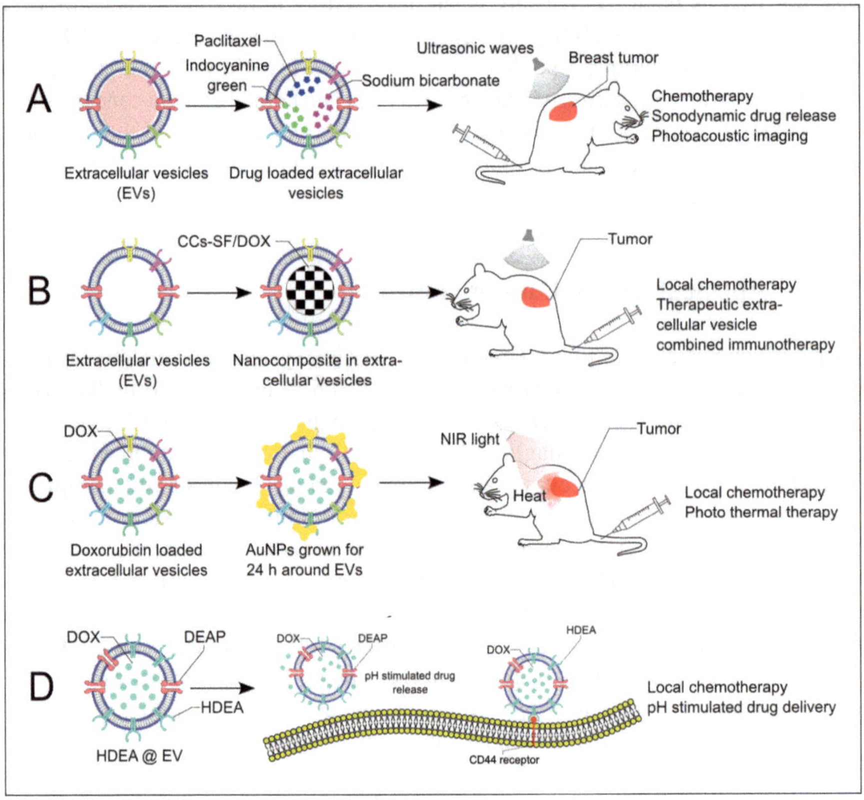

Fig. 12.3 Schematic illustration of: (**a**) A dual pH/US-responsive EVs co-encapsulating ICG, PTX, and SBC for sonotheranostic applications. (**b**) The low-intensity focused ultrasound-assisted nanocomposite for advanced triple cancer therapy, integrating local chemotherapy, therapeutic extracellular vesicles, and combined immunotherapy. (**c**) Self-grown gold nanopopcorn on extracellular vesicles for synergistic chemo-photothermal therapy. (**d**) pH-responsive hyaluronate-anchored extracellular vesicles for enhanced tumor-targeted drug delivery

preferential tumor accumulation of SBC-EV(ICG/PTX) in vivo. In addition to providing a unique method for cancer treatment, this study by Cao et al. highlights the promising potential of these dual stimuli-responsive EV-based nano sonosensitizers for safe and effective chemo-sonodynamic combination cancer therapy accompanying PA imaging.

Tan et al. devised a novel approach involving low-intensity focused ultrasound (LIFU) in conjunction with a nanocomposite for advanced triple cancer therapy, as depicted in Fig. 12.3b [107]. Their nanocomposite, based on silk fibroin and loaded with doxorubicin within calcium carbonates (CCs-SF/DOX), aimed to tackle the challenges associated with conventional nanoparticle-based drug delivery systems, such as low drug release and limited penetration. Through LIFU assistance, they achieved enhanced intracellular uptake of CCs-SF/DOX, facilitated either by acoustic perturbation-assisted delivery or endocytic uptake. The acidic microenvironment within cancer cells expedited the release of DOX, leading to efficient cytotoxicity. Moreover, residual CCs-SF/DOX or DOX from deceased cells were encapsulated within vesicles and expelled into the extracellular space, responding to the acidic conditions within the extracellular matrix (ECM) and thereby promoting deep drug penetration for EV therapy. Furthermore, CCs played a role in scavenging H+ ions, fostering M1-like macrophage polarization, consequently reversing the immunosuppressive tumor microenvironment (TME). Local release of chemotherapeutics further bolstered antitumor immune responses, thereby facilitating combined immunotherapy targeting PD1/PD-L1 checkpoints. Overall, this study underscores the potential of LIFU-activated nanocomposites for synergistic triple therapies, offering promising avenues for clinical antitumor applications pending further optimization in larger animal models.

In Zhang et al.'s work, EVs were utilized to build a novel technique for combinational chemo-photothermal therapy (Fig. 12.3c) [108]. In their investigation, doxorubicin (DOX), a chemotherapeutic drug, was encapsulated in EVs. Following that, the EVs were surrounded by self-growing gold nanoparticles, creating a distinctive nanostructure resembling popcorn. The cytotoxicity of DOX and the photothermal characteristics of the gold nanoparticles were preserved in this novel nanostructure. The gold nanopopcorn underwent near-infrared irradiation, which caused hyperthermia, tumor ablation, and the release of DOX, resulting in synergistic chemo-photothermal treatment. The nanoplatform demonstrated less side effects, better antitumor activity, regulated drug release, and an impressive tumor inhibitory rate of up to 98.6%. This work provides an easy-to-use and green way to create complex metal nanostructures, which is a major step forward in the field of nanomedicine.

Lee et al. developed pH-responsive hyaluronate-connected extracellular vesicles (HDEA@EVs) in their research to deliver drugs to tumors specifically [109]. These vesicles were created by grafting 3-(diethylamino)propylamine (HDEA) onto hyaluronic acid, then loading it with doxorubicin (DOX), a model anticancer medication. Through in vitro cell endocytosis, in vitro tumor cell toxicity, in vivo biodistribution, and in vivo tumor regression efficacy tests, the physicochemical properties of HDEA@EVs were assessed. The findings demonstrated that HDEA@ EVs actively coupled to CD44 receptors on HCT-116 tumor cells and effectively reacted to the acidic extracellular pH of tumors (pH 6.5). As a result, these EVs specifically prevented the growth of CD44+ tumor cells in vitro and the formation of CD44+ tumors in vivo. The promise of HDEA@EVs as a viable approach for targeted medication delivery in tumor therapy is highlighted by this study. This technology offers a unique technique for efficient anticancer treatment by using EVs designed with a particular ligand (HA) targeting tumor receptors and a pH-sensitive substance (DEAP) responsive to the acidic tumor microenvironment.

6 Future Perspective

Exosomes and EVs that respond to stimuli have great potential to advance clinical and scientific research in the biomedical fields. Many directions for further research become apparent as we learn more about their biogenesis, composition, and stimuli-responsive characteristics. First and foremost, it is crucial to better understand the molecular processes that underlie EVs' and exosomes' stimuli-responsive behavior. Even though a number of stimuli that can cause reactions in these vesicles have been identified with some success, a thorough knowledge of the signaling pathways involved is still missing. In addition to improving our basic understanding, research initiatives aimed at deciphering these complex dynamics will open the door to the sensible design of designed EVs with increased responsiveness. Furthermore, one intriguing direction for further study is the creation of novel methods for precisely modifying the stimulus responsiveness of EVs. The development of EVs with customizable reactivity to particular external or endogenous stimuli has the potential to transform tailored medication delivery techniques. Techniques like surface functionalization and genetic manipulation show promise in giving EVs the required stimuli-responsive characteristics, opening the door to customized therapeutic interventions for a range of illnesses, including cancer. Simultaneously, improvements in EV isolation methods will be essential to optimizing their medicinal potential. Scalability, yield, and purity issues hamper existing techniques like size-exclusion chromatography and density gradient ultracentrifugation. Future studies should concentrate on creating high-throughput, low-cost, and repeatable isolation methods that may effectively produce EV populations that are responsive to stimuli and appropriate for clinical use.

Furthermore, the creation of next-generation therapeutic platforms may benefit from the integration of stimuli-responsive EVs with cutting-edge technologies like microfluidics and nanotechnology. While nanotechnology presents prospects for developing EV-based drug delivery systems with improved targeting capabilities and regulated release profiles, microfluidic devices can provide fine control over EV isolation, purification, and functionalization procedures. Innovation in the field of stimuli-responsive EVs will be fueled by cooperative, multidisciplinary research projects that make use of this cutting-edge technology.

7 Conclusion

Research on exosomes and stimuli-responsive EVs is exciting and has significant implications for biological applications, especially in cancer therapy. These naturally occurring nanocarriers are excellent candidates for targeted medication delivery and precision medicine because they have built-in characteristics that enable them to react to a variety of endogenous and external stimuli. Many studies have been conducted on the formation and composition of EVs, exposing their varied biological origins and intricate chemical payload. The variety of EV subtypes, including exosomes, apoptotic bodies, migrasomes, and oncosomes, is further evidence of the population's heterogeneity. Each subtype of EV has unique traits and roles.

Numerous techniques have been devised, each with pros and cons, for the preparation and isolation of EVs. Ongoing work is aimed at improving these methods in order to produce uniform and pure EV populations that can be used in therapeutic settings. Many studies have been conducted on the stimuli-responsive characteristics of EVs, and significant progress has been made in our understanding of how responsive these cells are to external stimuli like temperature, light, magnetic fields, and ultrasound, as well as endogenous parameters like redox, pH, and enzymes. There is great potential for creating sophisticated drug delivery systems that can release therapeutic payloads in a targeted and regulated manner by utilizing these stimuli-responsive features.

Going forward, research efforts ought to concentrate on clarifying the fundamental molecular principles of EVs stimulus responsiveness, creating novel approaches for customizing EVs, improving procedures for separation and purification, and combining EVs with developing technologies. Through tackling these obstacles and leveraging chances for interdisciplinary cooperation, the field of stimuli-responsive EVs is well-positioned to achieve noteworthy advancements in achieving its therapeutic promise in cancer therapy and other areas.

Acknowledgments The authors acknowledge the Department of Pharmaceuticals, Ministry of Chemical and Fertilizers, Govt. of India for support. The NIPER-R communication number for the book chapter is NIPER-R/Communication/607.

Funding Statement None.

Conflict of Interest The authors declare no competing interests.

Data Availability Data will be made available on request.

References

1. Abels ER, Breakefield XO. Introduction to extracellular vesicles: biogenesis, RNA cargo selection, content, release, and uptake. Cell Mol Neurobiol. 2016;36(3):301–12.
2. Fonseka P, Marzan AL, Mathivanan S. Introduction to the community of extracellular vesicles. Subcell Biochem. 2021:3–18.

3. Di Bella MA. Overview and update on extracellular vesicles: considerations on exosomes and their application in modern medicine. Biology (Basel). 2022;11(6):804.
4. Théry C, Zitvogel L, Amigorena S. Exosomes: composition, biogenesis and function. Nat Rev Immunol. 2002;2(8):569–79.
5. Chulpanova DS, Kitaeva KV, James V, Rizvanov AA, Solovyeva VV. Therapeutic prospects of extracellular vesicles in cancer treatment. Front Immunol. 2018;3:9.
6. Palviainen M, Saari H, Kärkkäinen O, Pekkinen J, Auriola S, Yliperttula M, et al. Metabolic signature of extracellular vesicles depends on the cell culture conditions. J Extracell Vesicles. 2019;8(1)
7. Armstrong JPK, Holme MN, Stevens MM. Re-engineering extracellular vesicles as smart nanoscale therapeutics. ACS Nano. 2017;11(1):69–83.
8. Wang SE. Extracellular vesicles in cancer therapy. Semin Cancer Biol. 2022;86:296–309.
9. Keerthikumar S, Chisanga D, Ariyaratne D, Al Saffar H, Anand S, Zhao K, et al. ExoCarta: a web-based compendium of Exosomal cargo. J Mol Biol. 2016;428(4):688–92.
10. Kalra H, Drummen G, Mathivanan S. Focus on extracellular vesicles: introducing the next small big thing. Int J Mol Sci. 2016;17(2):170.
11. Uddin MJ, Mohite P, Munde S, Ade N, Oladosu TA, Chidrawar VR, et al. Extracellular vesicles: The future of therapeutics and drug delivery systems. Intelligent Pharmacy. 2024;
12. Walker S, Busatto S, Pham A, Tian M, Suh A, Carson K, et al. Extracellular vesicle-based drug delivery systems for cancer treatment. Theranostics. 2019;9(26):8001–17.
13. Sonoda H, Yokota-Ikeda N, Oshikawa S, Kanno Y, Yoshinaga K, Uchida K, et al. Decreased abundance of urinary exosomal aquaporin-1 in renal ischemia-reperfusion injury. Am. J. Physiol. Renal Physiol. 2009;297(4):F1006–16.
14. Campanella C, Caruso Bavisotto C, Logozzi M, Marino Gammazza A, Mizzoni D, Cappello F, et al. On the choice of the extracellular vesicles for therapeutic purposes. Int J Mol Sci. 2019;20(2)
15. Doyle LM, Wang MZ. Overview of extracellular vesicles, their origin, composition, purpose, and methods for exosome isolation and analysis. Cells. 2019;8(7)
16. Chuo STY, Chien JCY, Lai CPK. Imaging extracellular vesicles: current and emerging methods. J Biomed Sci. 2018;25(1):91.
17. Mathivanan S, Simpson RJ. ExoCarta: a compendium of exosomal proteins and RNA. Proteomics. 2009;9(21):4997–5000.
18. Fan Z, Jiang C, Wang Y, Wang K, Marsh J, Zhang D, et al. Engineered extracellular vesicles as intelligent nanosystems for next-generation nanomedicine. Nanoscale Horiz. 2022;7(7):682–714.
19. Zhang Y, Liu Y, Liu H, Tang WH. Exosomes: biogenesis, biologic function and clinical potential. Cell Biosci. 2019;9(1):19.
20. Di Vizio D, Kim J, Hager MH, Morello M, Yang W, Lafargue CJ, et al. Oncosome formation in prostate cancer: association with a region of frequent chromosomal deletion in metastatic disease. Cancer Res. 2009;69(13):5601–9.
21. Minciacchi VR, Freeman MR, Di Vizio D. Extracellular vesicles in cancer: exosomes, microvesicles and the emerging role of large oncosomes. Semin Cell Dev Biol. 2015;40:41–51.
22. Kim J, Morley S, Le M, Bedoret D, Umetsu DT, Di Vizio D, et al. Enhanced shedding of extracellular vesicles from amoeboid prostate cancer cells: potential effects on the tumor microenvironment. Cancer Biol Ther. 2014;15(4):409–18.
23. Minciacchi VR, You S, Spinelli C, Morley S, Zandian M, Aspuria PJ, et al. Large oncosomes contain distinct protein cargo and represent a separate functional class of tumor-derived extracellular vesicles. Oncotarget. 2015;6(13):11327–41.
24. Caruso S, Poon IKH. Apoptotic cell-derived extracellular vesicles: more than just debris. Front Immunol. 2018;9:1486.
25. Gavrilescu LC, Denkers EY. Apoptosis and the balance of homeostatic and pathologic responses to protozoan infection. Infect Immun. 2003;71(11):6109–15.
26. Kim HI, Park J, Zhu Y, Wang X, Han Y, Zhang D. Recent advances in extracellular vesicles for therapeutic cargo delivery. Exp Mol Med. 2024;56(4):836–49.

27. Cao Z, Li P, Li Y, Zhang M, Hao M, Li W, et al. Encapsulation of nano-bortezomib in apoptotic stem cell-derived vesicles for the treatment of multiple myeloma. Small. 2023;19(40)
28. Ma L, Li Y, Peng J, Wu D, Zhao X, Cui Y, et al. Discovery of the migrasome, an organelle mediating release of cytoplasmic contents during cell migration. Cell Res. 2015;25(1):24–38.
29. da Rocha-Azevedo B, Schmid SL. Migrasomes: a new organelle of migrating cells. Cell Res. 2015;25(1):1–2.
30. Svitkina T. The actin cytoskeleton and actin-based motility. Cold Spring Harb Perspect Biol. 2018;10:1.
31. Surman M, Stępień E, Hoja-Łukowicz D, Przybyło M. Deciphering the role of ectosomes in cancer development and progression: focus on the proteome. Clin Exp Metastasis. 2017;34(3–4):273–89.
32. Hankins HM, Baldridge RD, Xu P, Graham TR. Role of flippases, scramblases and transfer proteins in phosphatidylserine subcellular distribution. Traffic. 2015;16(1):35–47.
33. Sabatke B, Rossi IV, Sana A, Bonato LB, Ramirez MI. Extracellular vesicles biogenesis and uptake concepts: a comprehensive guide to studying host–pathogen communication. Mol Microbiol. 2023;27
34. Evans-Osses I, Reichembach LH, Ramirez MI. Exosomes or microvesicles? Two kinds of extracellular vesicles with different routes to modify protozoan-host cell interaction. Parasitol Res. 2015;114(10):3567–75.
35. Muhsin-Sharafaldine MR, Saunderson SC, Dunn AC, Faed JM, Kleffmann T, McLellan AD. Procoagulant and immunogenic properties of melanoma exosomes, microvesicles and apoptotic vesicles. Oncotarget. 2016;7(35):56279–94.
36. Liebana-Jordan M, Brotons B, Falcon-Perez JM, Gonzalez E. Extracellular vesicles in the fungi kingdom. Int J Mol Sci. 2021;22(13):7221.
37. Eliaz D, Kannan S, Shaked H, Arvatz G, Tkacz ID, Binder L, et al. Exosome secretion affects social motility in Trypanosoma brucei. PLoS Pathog. 2017;13(3):e1006245.
38. Moyano S, Musso J, Feliziani C, Zamponi N, Frontera LS, Ropolo AS, et al. Exosome biogenesis in the protozoa parasite Giardia lamblia: a model of reduced interorganellar crosstalk. Cells. 2019;8(12):1600.
39. Ferreira B, Lourenço Á, Sousa M do C. Protozoa-derived extracellular vesicles on intercellular communication with special emphasis on Giardia lamblia. Microorganisms. 2022;10(12):2422
40. Salas N, Coceres VM, Melo T dos S, Pereira-Neves A, Maguire VG, Rodriguez TM, et al. VPS32, a member of the ESCRT complex, modulates adherence to host cells in the parasite Trichomonas vaginalis by affecting biogenesis and cargo sorting of released extracellular vesicles. Cell Mol Life Sci 2022;79(1):11.
41. Jin Y, Ma L, Zhang W, Yang W, Feng Q, Wang H. Extracellular signals regulate the biogenesis of extracellular vesicles. Biol Res. 2022;55(1):35.
42. van Niel G, D'Angelo G, Raposo G. Shedding light on the cell biology of extracellular vesicles. Nat Rev Mol Cell Biol. 2018;19(4):213–28.
43. Sahu R, Kaushik S, Clement CC, Cannizzo ES, Scharf B, Follenzi A, et al. Microautophagy of cytosolic proteins by late endosomes. Dev Cell. 2011;20(1):131–9.
44. Record M. Intercellular communication by exosomes in placenta: a possible role in cell fusion? Placenta. 2014;35(5):297–302.
45. Fang X, Duan Y, Adkins GB, Pan S, Wang H, Liu Y, et al. Highly efficient exosome isolation and protein analysis by an integrated nanomaterial-based platform. Anal Chem. 2018;90(4):2787–95.
46. Yoo CE, Kim G, Kim M, Park D, Kang HJ, Lee M, et al. A direct extraction method for microRNAs from exosomes captured by immunoaffinity beads. Anal Biochem. 2012;431(2):96–8.
47. Zhu L, Sun HT, Wang S, Huang SL, Zheng Y, Wang CQ, et al. Isolation and characterization of exosomes for cancer research. J Hematol Oncol. 2020;13(1):152.
48. Oksvold MP, Neurauter A, Pedersen KW. Magnetic bead-based isolation of exosomes. Methods Mol Biol. 2015;1218:465–81.

49. Greening DW, Xu R, Ji H, Tauro BJ, Simpson RJ. A protocol for exosome isolation and characterization: evaluation of ultracentrifugation, density-gradient separation, and immunoaffinity capture methods. Methods Mol Biol. 2015;1295:179–209.
50. Haraszti RA, Miller R, Stoppato M, Sere YY, Coles A, Didiot MC, et al. Exosomes produced from 3D cultures of MSCs by tangential flow filtration show higher yield and improved activity. Mol Ther. 2018;26(12):2838–47.
51. Xu R, Greening DW, Zhu HJ, Takahashi N, Simpson RJ. Extracellular vesicle isolation and characterization: toward clinical application. J Clin Invest. 2016;126(4):1152–62.
52. Xu R, Greening DW, Rai A, Ji H, Simpson RJ. Highly-purified exosomes and shed microvesicles isolated from the human colon cancer cell line LIM1863 by sequential centrifugal ultrafiltration are biochemically and functionally distinct. Methods. 2015;87:11–25.
53. Li K, Wong DK, Hong KY, Raffai RL. Cushioned-density gradient ultracentrifugation (C-DGUC): a refined and high performance method for the isolation, characterization, and use of exosomes. Methods Mol Biol. 2018;1740:69–83.
54. Alvarez ML, Khosroheidari M, Kanchi Ravi R, DiStefano JK. Comparison of protein, microRNA, and mRNA yields using different methods of urinary exosome isolation for the discovery of kidney disease biomarkers. Kidney Int. 2012;82(9):1024–32.
55. Lee M, Liu T, Im W, Kim M. Exosomes from adipose-derived stem cells ameliorate phenotype of Huntington's disease in vitro model. Eur J Neurosci. 2016;44(4):2114–9.
56. Liu C, Guo J, Tian F, Yang N, Yan F, Ding Y, et al. Field-free isolation of exosomes from extracellular vesicles by microfluidic viscoelastic flows. ACS Nano. 2017;11(7):6968–76.
57. Wang J, Ma P, Kim DH, Liu BF, Demirci U. Towards microfluidic-based exosome isolation and detection for tumor therapy. Nano Today. 2021 Apr;37
58. Skoczylas Ł, Gawin M, Fochtman D, Widłak P, Whiteside TL, Pietrowska M. Immune capture and protein profiling of small extracellular vesicles from human plasma. Proteomics. 2023;15:e2300180.
59. Koksal AR, Ekmen N, Aydin Y, Nunez K, Sandow T, Delk M, et al. A single-step Immunocapture assay to quantify HCC exosomes using the highly sensitive fluorescence nanoparticle-tracking analysis. J Hepatocell Carcinoma. 2023;10:1935–54.
60. Wu TK, Liotta JL, Bowman DD. Comparison of extracellular vesicle isolation methods for the study of exosome cargo within Toxocara canis and Toxocara cati excretory secretory (TES) products. Exp Parasitol. 2024;261:108765.
61. An M, Wu J, Zhu J, Lubman DM. Comparison of an optimized ultracentrifugation method versus size-exclusion chromatography for isolation of exosomes from human serum. J Proteome Res. 2018;17(10):3599–605.
62. Baranyai T, Herczeg K, Onódi Z, Voszka I, Módos K, Marton N, et al. Isolation of exosomes from blood plasma: qualitative and quantitative comparison of ultracentrifugation and size exclusion chromatography methods. PLoS One. 2015;10(12):e0145686.
63. Truong Hoang Q, Nguyen Cao TG, Kang SJ, Lee M, Kang JH, Park HS, et al. Exosome membrane-sheathed and multi-stimuli-responsive MnO2 nanoparticles with self-oxygenation and energy depletion abilities potentiate the sonodynamic therapy of hypoxic tumors. Chem Eng J. 2023;472:144871.
64. Lehner R, Wang X, Marsch S, Hunziker P. Intelligent nanomaterials for medicine: carrier platforms and targeting strategies in the context of clinical application. Nanomedicine. 2013;9(6):742–57.
65. Liu C, Fan Z, He D, Chen H, Zhang S, Guo S, et al. Designer functional nanomedicine for myocardial repair by regulating the inflammatory microenvironment. Pharmaceutics. 2022;31;14(4)
66. Wang J, Tang W, Yang M, Yin Y, Li H, Hu F, et al. Inflammatory tumor microenvironment responsive neutrophil exosomes-based drug delivery system for targeted glioma therapy. Biomaterials. 2021;273:120784.
67. Fu Q, Feng H, Liu L, Li Z, Li J, Hu J, et al. Spatiotemporally controlled formation and rotation of magnetic nanochains in vivo for precise Mechanotherapy of tumors. Angew Chem Int Ed 2022 19;61(51).

68. Yang G, Xu L, Chao Y, Xu J, Sun X, Wu Y, et al. Hollow MnO2 as a tumor-microenvironment-responsive biodegradable nano-platform for combination therapy favoring antitumor immune responses. Nat Commun. 2017;8(1):902.
69. Li J, Sun C, Tao W, Cao Z, Qian H, Yang X, et al. Photoinduced PEG deshielding from ROS-sensitive linkage-bridged block copolymer-based nanocarriers for on-demand drug delivery. Biomaterials. 2018;170:147–55.
70. Wang S, Wang Z, Yu G, Zhou Z, Jacobson O, Liu Y, et al. Tumor-specific drug release and reactive oxygen species generation for cancer chemo/chemodynamic combination therapy. Adv Sci. 2019;6(5)
71. Luo C, Sun B, Wang C, Zhang X, Chen Y, Chen Q, et al. Self-facilitated ROS-responsive nanoassembly of heterotypic dimer for synergistic chemo-photodynamic therapy. J Control Release. 2019;302:79–89.
72. Ghassami E, Varshosaz J, Taymouri S. Redox sensitive polysaccharide based nanoparticles for improved cancer treatment: a comprehensive review. Curr Pharm Des. 2018;24(28):3303–19.
73. Wang K, Ye H, Zhang X, Wang X, Yang B, Luo C, et al. An exosome-like programmable-bioactivating paclitaxel prodrug nanoplatform for enhanced breast cancer metastasis inhibition. Biomaterials. 2020;257:120224.
74. Li X, Zheng BY, Ke MR, Zhang Y, Huang JD, Yoon J. A tumor-pH-responsive supramolecular photosensitizer for activatable photodynamic therapy with minimal *in vivo* skin phototoxicity. Theranostics. 2017;7(10):2746–56.
75. Pavlova NN, Thompson CB. The emerging hallmarks of cancer metabolism. Cell Metab. 2016;23(1):27–47.
76. Allen AE, Locasale JW. Glucose metabolism in cancer: The Saga of pyruvate kinase continues. Cancer Cell. 2018;33(3):337–9.
77. Cheng G, Li W, Ha L, Han X, Hao S, Wan Y, et al. Self-assembly of extracellular vesicle-like metal–organic framework nanoparticles for protection and intracellular delivery of biofunctional proteins. J Am Chem Soc. 2018;140(23):7282–91.
78. Wang M, Zhao J, Zhang L, Wei F, Lian Y, Wu Y, et al. Role of tumor microenvironment in tumorigenesis. J Cancer. 2017;8(5):761–73.
79. Wang L, Ding K, Zheng C, Xiao H, Liu X, Sun L, et al. Detachable nanoparticle-enhanced chemoimmunotherapy based on precise killing of tumor seeds and normalizing the growing soil strategy. Nano Lett. 2020;20(9):6272–80.
80. Kondapaneni LP, Bobde Y, Ghosh B. Exogenous stimuli–responsive polymeric micelles for drug delivery. In: Polymeric micelles for drug delivery. Elsevier; 2022. p. 409–28.
81. Sun Q, Wang Z, Liu B, He F, Gai S, Yang P, et al. Recent advances on endogenous/exogenous stimuli-triggered nanoplatforms for enhanced chemodynamic therapy. Coord Chem Rev. 2022;451:214267.
82. Wang L, Yu X, Zhou J, Su C. Extracellular vesicles for drug delivery in Cancer treatment. Biol Proced Online. 2023;25(1):28.
83. Tan CL, Knight ZA. Regulation of body temperature by the nervous system. Neuron. 2018;98(1):31–48.
84. Xiao YY, Gong XL, Kang Y, Jiang ZC, Zhang S, Li BJ. Light-, pH- and thermal-responsive hydrogels with the triple-shape memory effect. Chem Commun. 2016;52(70):10609–12.
85. Yuan Y, Raheja K, Milbrandt NB, Beilharz S, Tene S, Oshabaheebwa S, et al. Thermoresponsive polymers with LCST transition: synthesis, characterization, and their impact on biomedical frontiers. RSC Appl Polymer. 2023;1(2):158–89.
86. Zhang S, Cao C, Lv X, Dai H, Zhong Z, Liang C, et al. A H_2O_2 self-sufficient nanoplatform with domino effects for thermal-responsive enhanced chemodynamic therapy. Chem Sci. 2020;11(7):1926–34.
87. Ke Y, Zhou C, Zhou Y, Wang S, Chan SH, Long Y. Emerging thermal-responsive materials and integrated techniques targeting the energy-efficient smart window application. Adv Funct Mater. 2018;28(22)
88. Yang H, Leow WR, Chen X. Thermal-responsive polymers for enhancing safety of electrochemical storage devices. Adv Mater. 2018;30(13):e1704347.

89. Lv Q, Cheng L, Lu Y, Zhang X, Wang Y, Deng J, et al. Thermosensitive exosome–liposome hybrid nanoparticle-mediated chemoimmunotherapy for improved treatment of metastatic peritoneal cancer. Adv Sci. 2020;7(18)
90. Betzer O, Barnoy E, Sadan T, Elbaz I, Braverman C, Liu Z, et al. Advances in imaging strategies for in vivo tracking of exosomes. Wiley Interdiscip Rev Nanomed Nanobiotechnol. 2020;12(2):e1594.
91. Betzer O, Perets N, Angel A, Motiei M, Sadan T, Yadid G, et al. In vivo neuroimaging of exosomes using gold nanoparticles. ACS Nano. 2017;11(11):10883–93.
92. Wang J, Dong Y, Li Y, Li W, Cheng K, Qian Y, et al. Designer exosomes for active targeted chemo-photothermal synergistic tumor therapy. Adv Funct Mater. 2018;28(18)
93. Cardoso VF, Francesko A, Ribeiro C, Bañobre-López M, Martins P, Lanceros-Mendez S. Advances in magnetic nanoparticles for biomedical applications. Adv Healthc Mater. 2018;7(5)
94. Wan Y, Cheng G, Liu X, Hao SJ, Nisic M, Zhu CD, et al. Rapid magnetic isolation of extracellular vesicles via lipid-based nanoprobes. Nat. Biomed Eng. 2017:1.
95. Pooam M, Jourdan N, El Esawi M, Sherrard RM, Ahmad M. HEK293 cell response to static magnetic fields via the radical pair mechanism may explain therapeutic effects of pulsed electromagnetic fields. PLoS One. 2020;15(12):e0243038.
96. Qi H, Liu C, Long L, Ren Y, Zhang S, Chang X, et al. Blood exosomes endowed with magnetic and targeting properties for Cancer therapy. ACS Nano. 2016;10(3):3323–33.
97. Qian X, Zheng Y, Chen Y. Micro/nanoparticle-augmented sonodynamic therapy (SDT): breaking the depth shallow of photoactivation. Adv Mater. 2016;28(37):8097–129.
98. Canavese G, Ancona A, Racca L, Canta M, Dumontel B, Barbaresco F, et al. Nanoparticle-assisted ultrasound: a special focus on sonodynamic therapy against cancer. Chem Eng J. 2018;340:155–72.
99. Chen J, Luo H, Liu Y, Zhang W, Li H, Luo T, et al. Oxygen-self-produced nanoplatform for relieving hypoxia and breaking resistance to sonodynamic treatment of pancreatic cancer. ACS Nano. 2017;11(12):12849–62.
100. Qian X, Han X, Chen Y. Insights into the unique functionality of inorganic micro/nanoparticles for versatile ultrasound theranostics. Biomaterials. 2017;142:13–30.
101. Chen H, Zhou X, Gao Y, Zheng B, Tang F, Huang J. Recent progress in development of new sonosensitizers for sonodynamic cancer therapy. Drug Discov Today. 2014;19(4):502–9.
102. Liu Y, Bai L, Guo K, Jia Y, Zhang K, Liu Q, et al. Focused ultrasound-augmented targeting delivery of nanosonosensitizers from homogenous exosomes for enhanced sonodynamic cancer therapy. Theranostics. 2019;9(18):5261–81.
103. Ahirwar K, Kumar A, Srivastava N, Saraf SA, Shukla R. Harnessing the potential of nano-engineered siRNAs carriers for target responsive glioma therapy: recent progress and future opportunities. Int J Biol Macromol. 2024;266(Pt 1):131048.
104. Mule S, Khairnar P, Shukla R. Recent advances in nanocrystals heralding greater potential in brain delivery. Part Part Syst Charact. 2022;39(9)
105. Arora P, Shukla R. Microneedles along with conventional therapies: an in-depth observational review in alopecia areata treatment. J Drug Deliv Sci Technol. 2024;95:105627.
106. Nguyen Cao TG, Kang JH, Kim W, Lim J, Kang SJ, You JY, et al. Engineered extracellular vesicle-based sonotheranostics for dual stimuli-sensitive drug release and photoacoustic imaging-guided chemo-sonodynamic cancer therapy. Theranostics. 2022;12(3):1247–66.
107. Tan M, Chen Y, Guo Y, Yang C, Liu M, Guo D, et al. A low-intensity focused ultrasound-assisted nanocomposite for advanced triple cancer therapy: local chemotherapy, therapeutic extracellular vesicles and combined immunotherapy. Biomater Sci. 2020;8(23):6703–17.
108. Zhang D, Qin X, Wu T, Qiao Q, Song Q, Zhang Z. Extracellular vesicles based self-grown gold nanopopcorn for combinatorial chemo-photothermal therapy. Biomaterials. 2019;197:220–8.
109. Lee H, Park H, Noh GJ, Lee ES. pH-responsive hyaluronate-anchored extracellular vesicles to promote tumor-targeted drug delivery. Carbohydr Polym. 2018;202:323–33.

110. Bharti, Mangla Kanchan, Kohli Anuja Shashikant, Kamble Ashish Dilip, Sutar Rahul, Shukla Nanoerythrocytes in Cancer Therapy Advancing Targeted Chemotherapy and Theranostics Overview of Erythrocytes and Nanoerythrocytes CRC Press Boca Raton 1-29
111. Dyandevi, Mathure Ashish Dilip, Sutar Hemantkumar, Ranpise Atmaram, Pawar Rajendra, Awasthi (2023) Preparation and Optimization of Liposome Containing Thermosensitive In Situ Nasal Hydrogel System for Brain Delivery of Sumatriptan Succinate ASSAY and Drug Development Technologies 21(1) 3-16 10.1089/adt.2022.088

Dual and Multi-Stimuli Responsive Hydrogels for Cancer Therapy

13

Anil B. Jindal and Sanat Kumar Dash

Abstract

Cancer has emerged as a major global health threat, with both the incidence and mortality rates rising at an alarming rate. In 2022, the IARC reported approximately 20 million new cancer cases and 9.7 million cancer-related deaths worldwide. Despite significant advancements in cancer treatment, current therapies still face major challenges such as cancer cell resistance and severe side effects caused by high doses of chemotherapeutic agents. DSRHs and MSRHs represent a transformative approach in cancer therapy by responding to two or more stimuli at a time, thereby leveraging the unique TME for targeted drug delivery. These hydrogels respond to external and internal stimuli to enable controlled release of therapeutics, minimizing systemic toxicity while enhancing efficacy. Preclinical studies of these hydrogels have shown promising outcomes in cancer therapy, like sustained drug release, equal or slightly better therapeutic effect compared to clinically approved treatments, and synergistic antitumor activity. Manufacturing complexity, biocompatibility of synthetic polymers, and lack of regulatory guidelines are major barriers for clinical translation of these hydrogels. Despite these barriers, DSRHs and MSRHs show promise in addressing tumor heterogeneity, drug resistance, and off-target effects, emerged as next-generation platforms for cancer treatment.

A. B. Jindal (✉) · S. K. Dash
Department of Pharmacy, Birla Institute of Technology and Science Pilani (BITS Pilani), Pilani Campus, Pilani, Rajasthan, India
e-mail: anil.jindal@pilani.bits-pilani.ac.in

P. V. Devarajan et al. (eds.), *Advances in Stimuli-Responsive Nanosystems for Cancer Therapy*, https://doi.org/10.1007/978-981-95-2744-1_13

Keywords

Hydrogel · Cancer therapy · Tumor microenvironment · Dual-Stimuli-responsive hydrogel · Multi-Stimuli-responsive hydrogel · Controlled drug release · Targeted drug delivery

Abbreviations

AMF	alternating magnetic field
ATP	adenosine triphosphate
BAC	N,N′-bis (Acryloyl) cystamine
BTZ	bortezomib
CMC	carboxy methyl chitosan
CTD	chemodynamic therapy
CUR	curcumin
DCN	dual cross-linked network
DDSs	drug delivery systems
DMA	2-(dimethylaminoethyl) methacrylate
DNA	deoxyribonucleic acid
DOXO HCl	doxorubicin hydrochloride
DOXO	doxorubicin
DP	dopamine
DSRHs	dual-stimuli-responsive hydrogels
DTP	3,3′-dithiopropionic acid dihydrazide
DTT	1,4-dithiothreitol
EMF	electromagnetic force
Gela	gelatin
GO	graphene oxide
GSH	glutathione
HA	hyaluronic acid
HAPs	hypoxia-activated prodrugs
HIFs	hypoxia-inducible factors
HIFU	high-intensity focused ultrasound
HPMC	hydroxypropyl methylcellulose
IARC	International Agency for Research on Cancer
ICG	indocyanine green
LED	light-emitting diode
LIUS	low-intensity ultrasound
MA	maleic acid
MHT	magnetic hyperthermia
MMPs	matrix metalloproteinases
MNPs	magnetic nanoparticles

mRNA	messenger ribonucleic acid
MSRHs	multi-stimuli-responsive hydrogels
NIR	near-infrared region
NRF2	nuclear factor erythroid 2-related factor 2
OSA	oxidized alginate
PACT	photo-activated chemotherapy
PBA	phenylboronic acid
PDA NPs	polydopamine nanoparticles
PDA	polydopamine
PDPA	poly(diisopropylaminoethyl methacrylate)
PEI	polyethylenimine
PFH	perfluorohexane
PNAGA	poly(N-acryloyl glycinamide)
PNIPAM	poly(N-isopropylacrylamide)
Poly(MPC)	poly(2-methacryloyloxyethyl phosphorylcholine)
PS	photosensitizer
PTA	photothermal agents
PTD	photodynamic therapy
PTT	photothermal therapy
PVA	polyvinyl alcohol
ROS	reactive oxygen species
SF	silk fibroin
SRDDSs	stimuli-responsive drug delivery systems
SRHs	stimuli-responsive hydrogels
STD	sonodynamic therapy
TME	tumor-microenvironment
US	ultrasound
UV	ultraviolet
6-MP	6-Mercaptopurine

1 Introduction

The term "cancer" was coined by the Greek physician Hippocrates around 460–370 BCE. From then until the eighteenth century, understanding the cellular-level mechanism of cancer and its treatment remained unclear. In the nineteenth and twentieth centuries, significant improvement in cancer biology (in terms of cellular-level understanding and treatment) was observed. Figure 13.1 depicts significant milestones in cancer therapy between 1891 and 2017 [1–4].

Healthy cells adhere to a precise growth pattern or division and undergo natural death, which is crucial for maintaining physiological function. However, cancerous cells deviate from normal cell growth tendencies and undergo uncontrolled and abnormal proliferation within the body [5]. Almost all parts of the body can be affected by cancer. Generally, cancer cells tend to metastasize from one body part to another through the blood or lymph [6]. Based on origin, cancer can be categorized

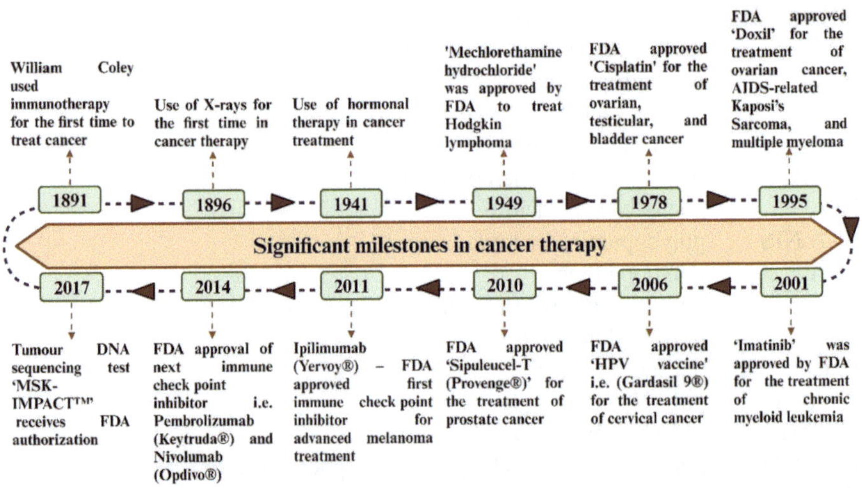

Fig. 13.1 Significant milestones in cancer therapy. (Created with BioRender.com)

into four different types, i.e., sarcoma (originates from connective tissue cells), carcinoma (originates from epithelial cells), lymphoma (originates from the immune system), and leukemia (originates from blood cells) [7]. Multiple factors are involved or interlinked with each other for inducing cancer, involving genetic, environmental, lifestyle, and other factors. Approximately 90–95% of cancer cases are triggered by environmental and lifestyle factors, while genetic factors account for only 5–10% [8].

Nowadays, cancer is a major global threat as the incidence of such cases and the mortality rate caused by such cases are increasing alarmingly day by day. According to the report issued in 2022 by the IARC, there were approximately 20 million cancer cases and 9.7 million deaths globally, with breast, lung, and colorectal cancers being among the major types of cancer. It is expected that the number of cancer cases around the world will rise by 77% between 2022 and 2050, and will exceed more than 35 million [9]. According to the American Cancer Society, 200.11 billion new cancer cases and 61.17 billion deaths are anticipated in the USA in 2024 [10]. Furthermore, the predicted number of cancer cases in India (including all states and union territories) in 2024 is 153.30 billion, with the highest and lowest cases in Uttar Pradesh and Lakshadweep, respectively [11].

Cancer was considered an incurable disease in the prehistoric era, with surgery being the only treatment option [12]. However, the complete removal of the tumor through surgery at that time was difficult, as acknowledged by ancient surgeons like Celsus and Galen, and often led to relapse. In the Middle Ages, cancer treatment was primarily palliative. Significant developments in cancer therapy were observed in the nineteenth and twentieth centuries with the introduction of anesthesia and antiseptics in cancer surgery, microscopic detection of tumors, radiotherapy, chemotherapy, hormone therapy, targeted therapy, and combination therapy [13]. Now, in the twenty-first century, cancer treatment is in a more advanced stage because of personalized treatment approaches due to advancements in molecular and genomic biology, the introduction of an immunotherapy approach, CAR-T cell, and CRISPR gene-mediated therapy [14]. Briefly, the current therapeutic options for cancer include surgery [15], chemotherapy [16], radiation therapy [17], immunotherapy [18], targeted therapy [19], hormone therapy [20], and stem cell or bone plate transplantation [21, 22].

Currently, nanoformulations like liposomes [23], polymeric nanoparticles [24], metallic nanoparticles [25], nucleic acid nanoparticles [26], spherical nucleic acids [27], extracellular vesicles [28], and nanocrystals [29] are widely used for cancer treatment at the preclinical level. However, the clinical success rate of such preclinically developed nanoformulations is very poor. Despite advancements in the cancer molecular biology field, several constraints persist with cancer therapy, including genetic diversity, resistance to treatment, treatment-related side effects, complex tumor microenvironment, failure of early detection and diagnosis, cost, and accessibility of available therapy [30, 31]. The biological behavior of each form of cancer differs from each other; therefore, treatment and management strategies for each type of cancer also differ from each other and stand as a hurdle. Looking at the current picture of cancer, it appears that diagnosis, treatment, and prevention of cancer hold a significant challenge for both healthcare professionals and affected patients. In addition to nanoformulations, hydrogel-based DDSs have also been extensively studied for cancer treatment. Hydrogel-based DDS offers several advantages, including high drug loading capacity [32], improved stability [33], prolonged retention capacity [34], enhanced tumor penetration [35], minimized burst release [36], and stimuli-responsive behavior [37]. SRHs can release the loaded drug at the TME in response to the external stimuli or internal physiological conditions present at the TME, thereby demonstrating enhanced antitumor activity. The present chapter describes the role of dual and multiple stimuli-responsive hydrogels in targeted drug delivery or the treatment of cancer.

2 Stimuli-Responsive Drug Delivery in Cancer Therapy

The resistance of cancer cells and high-dose-induced side effects of chemotherapeutics, are the major limitations of present cancer therapy [38]. SRDDS has shown promising results in the treatment of cancer in preclinical studies. SRDDSs are advanced drug delivery platforms engineered to release therapeutic agents at the desired site in response to external or internal stimuli [39]. External and internal stimuli are the triggering sources for the therapeutic activity of SRDDSs [40]. SRDDSs have significantly improved cancer therapy by addressing several limitations of conventional approaches, such as radiotherapy and chemotherapy, by providing sustained and controlled drug release, targeted delivery, enhancing therapeutic efficacy, improving drug stability, and reducing side effects [41, 42].

2.1 Stimuli Used in Cancer Therapy

Stimuli used in cancer therapy can be categorized into two types, i.e., external and internal stimuli (local stimuli) [43]. External stimuli include light, temperature, magnetic field, and ultrasound, while pH, redox potential, hypoxia, and enzymes are considered as internal stimuli [44]. Although both approaches involved in cancer therapy have their own merits and limitations, the selection of stimuli is influenced by several factors, like cancer types, patients, and their pathological condition. However, external stimuli-modulated cancer therapy is preferable to the internal type as the earlier one can be modulated manually in a controlled manner based on the patient's requirement [45].

2.1.1 External Stimuli

Light

Light or phototherapy is one of the most widely used techniques in cancer therapy, as it is a non-invasive approach [46]. The major light sources used in cancer therapy as an external stimulus include visible, UV, and infrared light [47]. However, UV-visible light has limited tissue penetration as it is absorbed by biological tissue [48]. Phototherapy also involves the use of lasers, fluorescent light, and LED as light sources. PTT [49], PTD [50], and PACT [51, 52] are major types of phototherapy involved in oncological treatment.

PTT is an innovative phototherapy approach involving the use of photothermal agents, which help in converting light to heat energy, thereby effectively targeting and destroying cancer cells [53]. PTA-like organic dyes (indocyanine green, heptamethine cyanine) [54], organic nanoparticles (semiconducting polymeric nanoparticle, porphysome) [55], gold nanomaterials (gold nanosphere, nanorod, nanocage, nanoprism, nanoshell) [56, 57], carbon nanomaterials (carbon nanotubes, graphene) [58], inorganic nanomaterials (quantum dots, metal oxide nanoparticles, silica nanoparticles) [59] are used in PTT therapy. Upon exposure to the NIR (800–2500 nm), PTA converts absorbed light to heat energy, raising tissue temperature and inducing local photocoagulation. Elevated tissue temperature results in cancer cell death through necrosis and apoptosis [60, 61]. The altered physiologic and biological nature of cancer cells, including enhanced membrane permeability, tumor vasculature, as well as changes in the intracellular TME induced by PTT, not only improves its therapeutic potency but also helps in enhancing the efficacy of other anticancer therapeutics [62, 63].

PDT is a clinically approved non-invasive or minimally invasive phototherapy that has gained attraction for malignancy treatment since 1970 [64]. PTD involves using a combination of light energy with a photosensitizing agent, which exerts a cytotoxic effect specifically on cancerous cells [65]. PTD consists of administering PS via the oral, topical, or systemic route, which then accumulates particularly on malignant cells. Following the accumulation of the PS, it is exposed to a specific wavelength of light for its excitation. Excitation of PS, in turn, produces highly reactive oxygen-free radicals known as ROS, which have cytotoxic effects on malignant cells and ultimately lead to their death through necrosis or apoptosis [66, 67]. Anticancer effects of PDT involve interconnected mechanisms, i.e., damage of tumor blood vessels, direct cytotoxic impact on malignant cells, and development of systemic immunity induced by the inflammatory reaction [68]. These interrelated mechanisms are influenced by several factors, like dose and type of PS, oxygen concentration at the tumor site, the time lapse between PS administration, and exposure duration of PS to the photon.

PACT is a novel approach in oncological treatment that combines the PDT principle with conventional chemotherapy. In PACT, the chemotherapeutics are initially caged with a photo-sensitive protective group, rendering them inactive or in prodrug form, which in turn restricts the interaction of chemotherapeutics with the cellular environment and circulates throughout the body. When the inactive drug reaches the tumor site, it is irradiated with light of a specific wavelength. This irradiation breaks the photo-labile chemical bond of the protecting group, thereby releasing the active chemotherapeutic at the tumor site, exerting a cytotoxic effect on malignant cells [69]. Metallic inorganic compounds are considered for oncology treatment using the PACT approach. PACT compounds are usually activated by three major modes, i.e., photo reduction in the reduced cellular environment (examples are Co^{3+} or Pt^{4+}), activation by photo substitution (examples are Re^{+}, Ru^{2+}, Ir^{3+} or Rh^{3+}), cleavage of the C–C bond via absorption of a photon by a nearby metal center [69, 70]. PACT is particularly promising in treating hypoxic tumors, which are less responsive to PDT [71].

Temperature

The use of temperature as a stimulus has gained consideration in cancer therapy. Hyperthermia [72] and cryotherapy [73] are the potential temperature-based techniques used in cancer treatment. Hyperthermia strategy for cancer treatment involves the rise of optimal physiological temperature to 40–45 ° C in combination with chemotherapy or radiotherapy. Stress induced in tumor cells because of the increase in body temperature results in perforation of the tumor cell membrane and enlargement of blood vessels, thereby enhancing the uptake of anti-cancer therapeutics via increased blood flow and permeability and making cancer cells more susceptible to the treatment [74, 75]. Hyperthermia affects cancer cells in several ways, such as altering cellular components' physical properties, affecting DNA repair pathways, and targeting nutrient-deprived and hypoxic cancer cells [76]. On the other hand, cryotherapy (cryoablation or cryosurgery) is a local antitumor treatment approach involving the use of extreme cold to destroy malignant cells [77]. Rapid freezing of cancer cells by argon or nitrogen cryogens generates ice crystals in cancer cells. The formation of ice crystals induces mechanical stress on the cell membrane. Ice crystals also block the microvasculature of cancer cells, resulting in cell necrosis because of the cut-off of nutrient and oxygen supply. Mitochondrial metabolic function is also affected by the rapid freeze-thaw cycle, leading to cell apoptosis [78]. The release of essential cellular components like tumor-specific antigens because of cell membrane damage is taken up by the macrophages, which further strengthens the immune system and helps in attacking the malignant cells [79].

Magnetic Field

Magnetic fields have also been used to alter or inhibit molecular processes within cancer cells. In 1957, MHT was used for the first time to treat cancer [80]. In MHT, MNPs accumulate at the tumor site. These MNPs, when exposed to an altered magnetic field, generate heat due to the Brownian and Neel relaxation processes. The generated heat destroys the malignant cells by denaturing the cytoplasmic and cell membrane proteins, ultimately leading to cell death [81]. Magnetic field-guided delivery of MNPs can be directly accumulated at the tumor site, enhancing the anticancer effect and reducing side effects [82]. Magnetic stimuli-mediated MNPs have a successful clinical record of treating prostate, pancreatic cancer, and glioblastoma [80, 83].

Ultrasound

The use of the US as an external stimulus in cancer therapy has turned into a promising strategy as it is non-toxic, non-invasive, and can penetrate deeper tissues [84]. Two types of US waves are used in cancer treatment, i.e., HIFU and LIUS [85]. In HIFU, ultrasonic energy is transferred to heat and mechanical energy, generating heat at the tumor site and leading to cell death because of necrosis [86]. In addition, the cavitation (non-thermal) effect caused by decreased acoustic frequency induces mechanical stress, thereby contributing to the destruction of tumor cells [87]. LIUS plays a vital role in SDT [88]. In SDT, sonosensitizers are exposed to LIUS and become activated. After activation, sonosensitizers generate ROS, which induces cancer cell apoptosis or necrosis [89]. The US also facilitates the delivery of chemotherapeutics at the targeted site by disrupting tumor vasculature or increasing cell membrane permeability [90].

2.1.2 Internal Stimuli

pH

Cancer cells depend on aerobic glycolysis (Warburg effect) rather than oxidative phosphorylation for energy production required for their cellular processes [91]. Relatively, aerobic glycolysis of cancer cells is inefficient compared to oxidative phosphorylation for ATP production to meet the energy requirement for cellular processes. As cancer cells proliferate rapidly, this, in turn, enhances glucose uptake and produces more lactic acid. This unique metabolic pattern of cancer cells makes TME acidic (pH between 6.0 and 6.8 compared to healthy cells [92–94]. The acidic microenvironment of the cancer cell provides the basis for the development of pH-responsive anticancer DDS. pH-responsive DDS, when exposed to an acidic environment, resulted in the release of anticancer cargo [95]. The drug release from the pH-responsive DDS is induced by protonation [96], acidic cleavage of susceptible bonds [97], and modification of surface charge [98, 99]. For instance, the protonation of certain functional groups like carboxyl, imidazolyl, and amino groups present within the matrix material of the nanocarrier could alter the hydrophobic-hydrophilic balance of the carrier, thereby altering structural changes followed by drug release when exposed to acidic pH [100]. Acidic cleavage of acid hydrolyzable bonds, like ketal, acetal, and hydrazone bonds of anticancer therapeutics in the acidic tumor environment, facilitates drug release at the site of action [101]. Surface charge often gets altered when the nanocarrier enters the acidic environment of cancer cells, thereby facilitating their cellular internalization [102].

Redox Potential

Redox potential is the second most used internal stimulus after pH for cancer treatment. Redox homeostasis plays a vital role in maintaining the physiological processes of both healthy and cancer cells [103]. ROS produced during mitochondrial respiration are removed by cellular antioxidants, thereby maintaining the cellular redox homeostasis in a dynamic equilibrium state [104]. Altered metabolic activity, elevated levels of ROS generation because of mitochondrial dysfunction, and accelerated peroxisomal activity are responsible for the imbalanced redox state of cancer cells [105]. Targeting the extra/intracellular redox state of cancer cells is a promising approach in cancer treatment as it triggers cellular apoptosis when it exceeds a certain threshold, thereby enhancing therapeutic efficacy [106].

By using pro-oxidant compounds (like selenium and melatonin), the oxidative stress of cancer cells can be further enhanced beyond their threshold tolerance level, either by enhancing the generation of ROS species or by disrupting their antioxidant defense mechanism, leading to cell damage followed by cell death [107]. In another strategy, when nanocarriers are exposed to the elevated ROS levels of cancer cells, the redox-susceptible linker within the nanocarrier is cleaved, releasing the drug at the tumor site and enhancing efficacy [108]. NRF2, a transcription factor, regulates antioxidant gene expression and maintains redox homeostasis. Modulation of NRF2 activity can alter the redox state of cancer cells. Cancer cells can undergo oxidative stress by inhibiting the NRF2 transcription factor, which will facilitate the effectiveness of anticancer therapy [103].

Hypoxia

Hypoxic TME is a typical feature of solid tumors [109]. Hypoxic conditions promote the proliferation and metastasis of cancer cells by influencing their metabolism, immune evasion, and angiogenesis [110]. Hypoxic cancer cells are therapeutically resistant to traditional radio and chemotherapy, which complicates the therapeutic strategy [111]. HAPs, or bio-reductive drug-based strategy, was found to be a promising approach for treating hypoxic tumor cells [112]. Under a typical oxygenic environment, HAPs remain inactive but are converted to their active cytotoxic form by reductase enzymes under a hypoxic tumor cell environment and target such cells [113]. Several classes of HAPs, including nitroaromatics, quinones, aliphatic, and hetero aromatic N-N-oxides, have developed. PR-104, TH-302 (evofosfamide), tirapazamine, SN30000, AQ4N (banoxantrone) are few representative examples of HAPs [112]. Hyperoxia therapy is another approach involving the administration of an excess amount of oxygen or carbogen to the cancer patient to overcome the negative effects of hypoxic tumor cells by facilitating ROS formation, reducing vasculature damage, reducing tumor cell growth and metastasis, as well as activating the immune response [114, 115]. HIFs regulate the transcription of several genes, which are responsible for the survival of tumor cells under a hypoxic environment. HIF inhibitor-induced targeting of HIFs is a promising approach for treating cancer. Such inhibitors inhibit mRNA expression, protein synthesis, protein stabilization and accumulation, HIF dimerization, DNA binding, and transcriptional activity, thereby promoting anticancer activity [116, 117].

Enzyme

Enzymes play a vital role in several biological processes because of their high specificity and catalytic activity. Compared to normal cells, elevated levels of enzymes are found in cancer cells. Nanocarriers are designed using matrix material containing enzyme-cleavable bonds or linkers that can be cleaved by specific enzymes (proteases by matrix metalloproteinase, ester by esterase, glycoside by glucosidases, hyaluronic acid by hyaluronidase) present in the TME, thereby providing targeted drug delivery at the cancer site [118]. Similarly, enzyme-cleavable prodrugs are designed for tumor treatment, which become cytotoxic after reaching the TME. For example, overexpressed phospholipases cleaved the phospholipase-drug conjugate, releasing the drug at the tumor site for anticancer effects [119]. Enzymes can indirectly alter tumor redox potential, pH, and expression of heat shock protein, thereby improving drug delivery to cancer cells [120].

3 Stimuli-Responsive Hydrogel in Cancer Therapy

Hydrogel consists of a three-dimensional polymeric arrangement of hydrophilic polymers that hold substantial amounts of water without losing their structural integrity [121]. It can absorb water up to a thousand times of their dry weight because of the hydrophilic group within the polymeric network [122]. It is often termed a biphasic material and contains a polymer matrix (solid) and water or biological fluid (liquid) [123]. Hydrogels can be prepared using natural polymers (like gelatin, chitosan, and hyaluronic acid) and synthetic polymers (like polyacrylate and PVA). Nowadays, the hydrogel system has emerged as a promising DDS because of its soft and flexible structure [124], biocompatibility [125], and controlled release kinetics [126]. Further, the exploration of SRHs in cancer therapy has gained significant attention as such DDS provides the platform for the targeted and controlled delivery of loaded anticancer drug at the tumor site, thereby minimizing systemic toxicity and enhancing therapeutic efficacy [127]. SRH is also a promising approach for the local delivery of anticancer therapeutics. It helps treat incompletely surgically removed or non-resectable tumors. SRHs undergo conformational changes upon exposure to external or internal stimuli, enabling drug delivery at the target cancer site, and curtail side effects [128]. Incorporating immunotherapeutics and chemotherapeutics within the hydrogel system offers a synergistic impact on affected cells [129]. Figure 13.2 depicts external and internal stimuli used in SRHs in cancer therapy. Dual and multi-stimuli responsive hydrogel systems are designed to release the drugs in TME in response to a combination of two or more stimuli by undergoing structural changes.

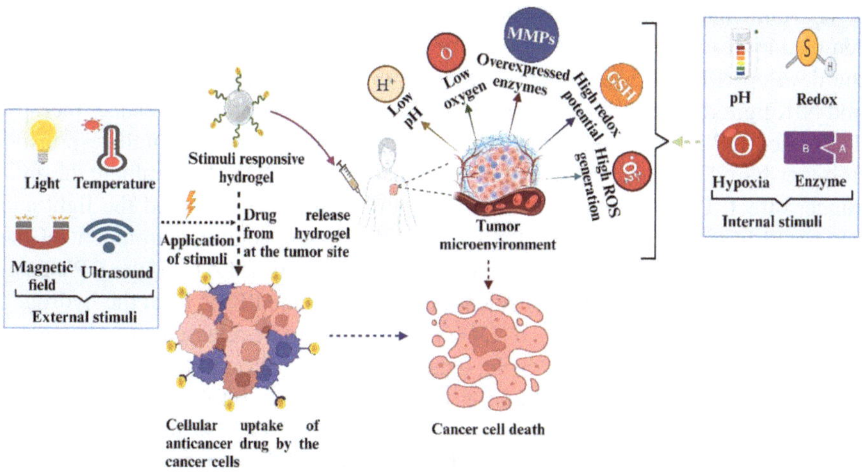

Fig. 13.2 External and internal stimuli used in SRHs in cancer therapy. (Created with BioRender.com)

3.1 Dual Stimuli-Responsive Hydrogels in Cancer Therapy

DSRHs are promising platforms in cancer therapy. These delivery systems are designed for targeted delivery of anticancer therapeutics at the cancer site in response to two distinct stimuli [130]. The dual stimuli-based hydrogel system provides advantages in cancer therapy, like offering controlled drug delivery at the TME, enhancing therapeutic efficacy, and minimizing side effects [131]. The dual system offers a multistage drug delivery platform where the sequential release of therapeutic agents occurs in response to specific stimuli across which they come in the TME. This treatment strategy, in turn, helps to overcome tumor biology complexity, particularly drug resistance and heterogeneity [132]. The dual stimuli-responsive system often provides a synergistic effect against cancer cells through a combined treatment approach [133]. Biological barriers usually hinder the delivery of therapeutics, mainly by inhibiting the cellular uptake of drugs and by other means. However, Dual stimuli-responsive DDS facilitates deeper tissue penetration of anticancer drugs, thereby improving therapeutic outcomes [134]. Dual-responsive DDS also offers better in vivo protection of drugs, thereby enabling longer systemic circulation duration [135].

Several preclinical studies based on DSRH systems have shown promising therapeutic outcomes in cancer treatment with reduced side effects. Bilalis et al. designed a dual (pH and enzyme) stimuli-based hydrogel loaded with gemcitabine to treat pancreatic cancer. The pH of the developed hydrogel was between 6.6 and 7.4. The hydrogel liquefied at the acidic microenvironment of the cancer cell, leading to drug release. A single dose of the developed hydrogel (loaded with 40% of the drug) exhibited equal or slightly better therapeutic effects than clinically approved treatment (two doses of peri-tumoral injection of 100% of free drug). Additionally, in vivo enzymatic biodegradation of the developed polypeptide hydrogel avoided its removal by resection after the delivery of the drug, which provided an advantage to the developed DDS [136]. In another study, GhavamiNejad et al. developed a pH and NIR light dual stimuli-based hydrogel for chemo-photo thermal cancer therapy. They used DP as a photothermal agent to develop BTZ-loaded stimuli-responsive polymer-based hydrogel alone and in combination with DOXO to evaluate the anti-cancer effect. Upon exposure of the hydrogel to light, DP absorbed the light and generated hyperthermia around the cancer cells. In the acidic environment of the cancer cells, the dissociation of the boronic acid group of BTZ from the DP's catechol group resulted in the higher release of surface-coated BTZ from the hydrogel. Under an acidic pH of 5.0, 81% of DOXO was found to be released after the end of the exposure of the fourth NIR cycle, which could be because of the increased solubility of the drug at a lower pH and also released prior to BTZ, which in turn was expected to provide a synergistic anticancer effect [137]. Zhou et al. fabricated a pH and magnetic field-responsive DOXO-loaded hydrogel for cancer therapy. At neutral pH, only 57.6% (with AMF) and 32.3% (without AMF) of DOXO was found to be released, while under acidic pH of 5.5, the cumulative percentage of drug release was found to be 78.8% (with AMF), and 41.7% (without AMF). Increased drug release in an acidic environment could be attributed to the enhanced DOXO and HPMC/Fe_3O_4 hydrogel solubility at lower pH. After 21 days of treatment, tumor recurrences were observed in only the thermal therapy receiving group, while such an incidence was not observed in the chemo-thermal receiving group. The superior therapeutic efficacy of the chemo-thermal treatment could be attributed to the enhanced release of DOXO during magnetic hyperthermia treatment, leading to a synergistic effect [138]. Dai et al. developed a pH/ROS dual-stimuli-based in-situ hydrogel system in integration with TiOx@CaO_2 (inorganic sonosensitizer) and aPD—L1 (immune checkpoint inhibitor) for US-mediated synergistic antitumor immunity. At pH 5.0, degradation of TiOx@CaO_2 generates ROS, i.e., O_2 and H_2O_2. Generated O_2 was helpful in reversing the tumor-induced immunosuppression by preventing tumor cells from undergoing a hypoxic environment. Additionally, the produced H_2O_2 contributed to enhancing the ROS-mediated antitumor effect. Released Ca^{2+} helped to enhance aPD—L1 mediated antitumor efficacy through the enhanced expression of tumor antigens. The in vitro cell line study revealed that TiOx@CaO_2 degradation-induced generation of ROS helped to balance the endogenous ROS deficiency, thereby providing a synergistic CDT/SDT effect. Degradation of the hydrogel at the TME facilitated the release of TiOx@CaO2 and aPD-L1 at the tumor site. US-mediated cellular uptake of TiOx@CaO_2 induces cell apoptosis via Ti3+-mediated CDT and TiO*x*-mediated SDT [139]. Mahdian et al. developed a curcumin CUR-loaded pH and temperature dual stimuli-responsive hydrogel for cancer therapy. An increase in temperature (30–50 °C) resulted in an increased release rate of CUR from the hydrogel, which could be due to the presence of thermosensitive PDA NPs within the nanocarrier. Similarly, after 48 hour of study at 50 °C, it was found that 50% and 70% of the drug were released from the hydrogel at pH 7.0 and 5.0, respectively. A higher percentage of drug release under acidic conditions could be attributable to the presence of chitosan nanocapsule within the nanocarrier [140]. Shen et al. developed DOXO. Hcl encapsulated redox/ultrasound dual stimuli-responsive nanogel for cancer treatment. US-responsive PFH was loaded into the center of the nanogel. The spherical structure of the hydrogel was maintained even after ultrasonic exposure to PFH, indicating its stability and homogeneity. In contrast, upon exposure to GSH, the PEI gel shell underwent degradation within 1 hour. A synergistic impact of redox/ultrasound dual stimuli was observed on the release of DOXO, and 90% of the drug was released within 10 min. Sequential stimuli-responsive controlled drug release was also observed, i.e., 15% (under the normal humoral condition in the first 4 hour), followed by 60% (after adding GSH in the next 4 hour), followed by the release of the remaining drug from the hydrophobic PNIPAM core (under ultrasonic response) [141]. Wang et al. designed an injectable hydrogel that responded to light/temperature dual stimuli for the local release of the loaded drug at the cancer site. Light/temperature dual stimuli-responsive chitosan hydrogel was developed by incorporating PNIPAM polymer (thermo-sensitive) and methacryloyl group (light-sensitive) of vinyl monomer to native chitosan polymer. Increased release of DOXO from the developed hydrogel was observed in response to a rise in temperature. Diffusion of the drug from the hydrogel matrix because of volume shrinkage of the hydrogel at higher temperatures could be attributable to higher drug release at higher temperatures. Similarly, NIR-irradiated hydrogel exhibited 78.5% drug release, which was 1.6 times higher than the non-irradiated hydrogel, signifying the importance of light stimuli in drug release [142]. Chen et al. developed a pH/redox-responsive hydrogel system for its application in tumor treatment using polysaccharide polymer, i.e., CMC, OSA, and DTP. The acylhydrazone and disulfide bond of the developed CMC-OSA-DTP hydrogel were stimuli-responsive and accountable for the sol-gel transformation. In the presence of the DTT reductant, the hydrogel turned to a colloidal sol state because of the reduction of the disulfide bond to the free thiol group, whereas, in the presence of the H_2O_2 oxidant, the sol again reformed to the hydrogel. Similarly, the acylhydrazone bond broke and reformed at lower and higher pH, respectively, indicating the pH responsiveness of the developed hydrogel system [143]. Li et al. developed a pH/glucose-responsive injectable hydrogel for doxorubicin delivery. The hydrogel system was developed by imine bond linkage resulting from the interaction between the aldehyde group of oxidized dextran and the amine group of chitosan. DOXO release at physiological pH was 52%, but it was enhanced to 70% at pH 6.5, which could be due to the susceptibility of the imine bond at lower pH. The charged PBA group interacted covalently with glucose through an esterase bond, exhibiting glucose responsiveness. As a result, adding 50 mM of glucose, even if at pH 7.4, enhanced the drug release to 61% [144]. Chen et al. have developed a pH/GSH-responsive PVA nanogel for enhanced cellular uptake and triggered release of DOXO for cancer treatment. The release rate of DOXO from the nanogel at pH 5.5 was 40% higher than at pH 7.4 under physiological conditions after 48 hours of study, which could be because of decreased electrostatic interaction between DOXO and the nanogel. However, under dual stimuli conditions, i.e., at pH 7.4 with 10 mM GSH and at pH 5.5 with 10 mM GSH, the DOXO release was found to be 58% in 48 hour and 95% in 10 hour, respectively. This higher drug release may be attributable to the rapid destabilization of the nanogel and the weaker electrostatic interaction of DOXO with the nanogel under dual stimuli conditions. The acid-labile terminal group permitted the nanogel for surface charge reversal from negative to positive under the extracellular tumor pH environment, thereby facilitating cellular internalization and uptake. Doxo-loaded charge reversal and redox-responsive hydrogel exhibited better cytotoxic effects than non-responsive hydrogel and free drug [145]. Table 13.1 briefly highlights a few examples of pre-clinically developed DSRHs for cancer treatment.

Table 13.1 Dual-stimuli responsive hydrogels for cancer treatment

Sl. No.	Stimuli used	Hydrogel composition	Experimental outcome	References
1	pH and enzyme	Gemcitabine and pentablock terpolypeptide i.e. PLys-b-(PHIS-co-PBLG)-PLys-b-(PHIS-co-PBLG)-b-PLys	A single dose of the developed hydrogel (loaded with 40% of the drug) exhibited equal or slightly better therapeutic effects than clinically approved treatment (two doses of peritumoral injection of 100% of free drug) and also underwent an in-vivo enzymatic biodegradation process	[136]
2	pH and NIR light	Dopamine, bortezomib, doxorubicin, poly(N-isopropyl acrylamide, and polyacrylamide	The developed nanocomposite hydrogel is a multi-modal platform offering controlled release of the loaded multidrug and was effective in killing cancer cells through photothermal and multidrug chemotherapy	[137]
3	pH and magnetic field	Doxorubicin, HPMC, PVA, and Fe_3O_4	The designed hydrogel exhibited higher efficacy and lowered the toxicity in response to the dual stimuli	[138]
4	pH and ROS	TSPBA, PVA, TiOx@CaO_2, and aPD-L1	Longer retention of the hydrogel at the tumor site offered synergistic antitumor immunity and prevented pulmonary metastasis	[139]
5	pH and temperature	Curcumin, gelatin, glucose, chitosan, PDA NPs	Under NIR irradiation and an acidic environment, 1.43-fold better release of curcumin from the developed hydrogel was observed compared to alkaline pH	[140]
6	Redox and ultrasound	Doxorubicin, perfluorohexane, PNIPAM, and PEI	The developed hydrogel system exhibited synergistic and sequential controlled drug release in response to the applied dual stimuli	[141]
7	Light and temperature	Doxorubicin, chitosan, PNIPAM, vinyl monomer	The developed hydrogel provided accelerated on-demand drug delivery in response to light and temperature dual stimuli	[142]

(continued)

Table 13.1 (continued)

Sl. No.	Stimuli used	Hydrogel composition	Experimental outcome	References
8	pH and redox	CMC, OSA, and DTP	The dynamic acylhydrazone and disulfide bond within the developed hydrogel system make it responsive to dual stimuli. Also, the excellent self-healing capacity exhibited because of the debonding and rebonding behavior of the dynamic bond makes it a promising system for future application in tumor therapy	[143]
9	pH and glucose	Doxorubicin, dextran, chitosan, phenyl boronic acid	The drug release rate from the developed hydrogel system was either enhanced by decreasing pH or increasing the glucose concentration around the cell. Also, the gel suppressed the growth of L929 fibroblast cells, indicating its anticancer potency	[144]
10	pH and GSH	Doxorubicin, PVA	The developed dual stimuli-responsive hydrogel offered a promising platform for targeted and controlled release of potent chemotherapeutics	[145]

3.2 Multi-Stimuli-Responsive Hydrogels in Cancer Therapy

Similar to DSRH, MSRH is also a promising platform for cancer treatment. MSRH differs from DSRH in that it responds to more than two stimuli simultaneously to exert its therapeutic effect [146]. Compared to DSRH, MSRH exhibited enhanced functionality because of its response to at least three or more stimuli, allowing it a highly acceptable DDS for treating cancer [147]. Response to multiple stimuli of MSRH enabled it to exhibit a dynamic response with varied physiological environments, thereby holding more precise control over drug release at the target site and minimizing off-target effects than DSRH [148]. MSRH mimics a closer biological environment, like the extracellular matrix, than DSRH, facilitating cellular adhesion, proliferation, and differentiation, which is particularly useful in cancer treatment [149].

Gang He et al. designed a multi-stimulus (pH, enzyme, and infrared light) responsive injectable hydrogel containing DOXO-loaded PDA and Gela nanoparticles for synergistic anticancer activity. Both PDA and Gela had a negative charge at pH 12. However, at pH 9, the Gela became positive and interacted with negatively charged PDA, resulting in sediment formation and exhibiting sol-gel behavior. The release study revealed that the DOXO release increased by 30.8% when the pH was decreased from 7.4 to 5.0. The increased drug release under an acidic environment could be attributed to the increased repulsion between DOXO and Gela nanoparticles and the increased solubility of DOXO. Enzymatic breakdown of proteins by MMPs in the TME is a key marker of tumor pathogenesis. More than 60% of DOXO was released from the Gela/PDA/DOXO hydrogel over 7 days upon the addition of protease enzyme, indicating enzyme-driven drug release because of the gradual degradation of the Gela nanoparticle. When the temperature increased from 37 °C to 50 °C, 30% higher DOXO release was observed, possibly due to Gela's thermosensitivity. Exposing the hydrogel to the NIR light at different pH values demonstrated an enhanced release of DOXO [150]. Bin Lei et al. developed a triple-stimuli (pH/redox/ultrasound) responsive hydrogel with a controlled drug release profile for cancer treatment. The developed hydrogel system released only 20% of DOXO at pH 7.4; however, at acidic conditions (i.e., pH 3.0 and pH 5.0), about 80–90% of the drug was released, which could be because of the presence of the pH-sensitive diblock (PDPA-b-PEI) copolymer. Furthermore, only 12% of the drug was released from the hydrogel in the media without GSH; however, upon the addition of 10 mM of GSH, it increased to 21%, which was assumed to be contributed by gel shell-loaded DOXO. Similarly, 47% and 61% higher release of DOXO was observed at pH 3.0 and pH 5.0, respectively, in combination with GSH than in acidic pH without GSH, which could be because of the decomposition of the disulfide bonds of PDPA by GSH, thereby facilitating drug release. After exposure to GSH/ultrasound dual stimuli, 80% of the drug was released within 20 min, and complete drug release was achieved quickly. However, in the absence of ultrasound, only 20% and 10% of the drug were released with and without GSH, respectively [151]. Zhao et al. developed a pH, sugar, and H_2O_2-responsive DCN hydrogel for cancer treatment. The developed DCN hydrogel consisted of a zwitterionic poly(MPC) as a backbone and imine and benzoxaboronic ester as dual linkages. The developed hydrogel, under acidic conditions, converted to a liquid state and reconstructed to gel form upon exposure to a basic environment. The developed hydrogel degraded 100% in sugar solution after 6 hour. Similarly, when DCN hydrogel was immersed in 10 mM of H_2O_2 in PBS solution, 100% of the hydrogel was dissociated because of benzoxaborolate complex cleavage in the oxidizing environment [152]. Gou et al. designed a multi-stimuli-responsive SF hydrogel for drug release in TME. Compared to physiological pH, 40% higher DOXO was released in an acidic buffer. The hydrogel exhibited concentration-dependent release behavior, and 90% and 65% of DOXO were released in response to the release media containing 1 mM H_2O_2 and 10 mM of GSH. Upon exposure to NIR light for 10 min, the hydrogel released 61% of the drug, whereas the hydrogel without exposure to NIR light released only 37% of the drug. Incubation of the hydrogel at 50 °C in the buffer of pH 7.4 resulted in increased drug release, indicating the hydrogel's temperature sensitivity. The photo-thermal effect of the incorporated cyanine-7 in the developed hydrogel was retained as NIR exposure increased the hydrogel temperature to 56 °C, which was far exceeding the tumor cells' threshold temperature for their ablation. Similarly, the fluorescent intensity of the cyanine-7 hydrogel was increased after NIR exposure, indicating the maintenance of the photodynamic properties of the fluorescent compound in the developed hydrogel. DOXO/cyanine—7- hydrogel exhibited the highest degree of tumor damage, and the treated group survived for more than 60 days [153]. Ping Ji et al. designed a pH, light, temperature, and redox-responsive PNIPAM/GO (PG) and PNIPAM/SGO (PSG) nanogel for a synergistic anti-cancer effect. Exposure of PG and PSG solutions to NIR light increased the temperature from 20 °C to 70 °C and 78 °C, respectively, because of the presence of light-absorptive GO. However, PSG exhibited higher photothermal conversion than PG and pure GO. NIR light-stimulated enhanced DOXO release (47%) was observed in PSG nanogel in a controlled manner compared to non-exposed PSG nanogel, which could be because of the decrease in particle size of the nanoparticles and the change of their nature from hydrophilic to hydrophobic with increasing temperature. Compared to physiological pH, under the acidic pH environment, i.e., at pH 5.0 and 6.5, an increased drug release was observed, which could be because of the protonation of the amine group present in the PNIPAM polymer and DOXO. PSG nanogel also exhibited redox-responsive drug release behavior, and with increasing GSH concentration, the amount of drug released from the nanogel increased. The MTT study revealed that PSG nanogel exhibited higher cytotoxic activity, and NIR treatment further enhanced the cytotoxic impact of PSG nanogel than PG nanogel, indicating a synergistic anticancer effect exhibited by the multi-stimuli hydrogel systems [154]. Hayati et al. designed a pH/temperature/magnetic field-responsive hydrogel for anticancer drug delivery. Under the acidic pH environment (pH 5.4), more than 90% of DOXO was released compared to ~60% at pH 7.4. The sensitivity of the hydrogel for pH could be because of electrostatic repulsion between the dimethyl amino groups of DMA. A 7% difference in drug release from the hydrogel at room temperature (89%) and physiological temperature (98%) was observed at pH 5.4, signifying the temperature-dependent drug release behavior of the hydrogel. However, in the presence of EMF, a decreased drug release percentage was observed than in the absence of EMF. The highest 77% of the drug was released from the hydrogel under acidic pH at 37 °C and without EMF [155]. Yuan et al. have developed a light/magnet/enzyme-responsive nanocomposite hydrogel for tumor treatment. The developed hydrogel (PNAGA-DMP-Fe_3O_4-GO) contains PNAGA (a thermo-sensitive polymer) and Fe_3O_4-GO nanocomposite (a magneto-photo thermal agent for the conversion of light and magnetic field to heat) as its formulation component. The study indicated that the magneto-thermal properties of the hydrogel could be adjusted by altering the intensity of the applied magnetic field or Fe_3O_4-SiO_2-GO content. After five turn-on and turn-off cycles of the magnetic field, the hydrogel maintained its magneto-thermal properties, which could be because of the existing strong interaction between PNAGA polymer and Fe_3O_4-SiO_2-GO (magneto photo-thermal agent), as well as the hindrance in leakage of Fe_3O_4-SiO_2-GO from the hydrogel induced by the stearic effect of the Fe_3O_4-SiO_2-GO. With increasing the temperature from 37 °C to 55 °C, the values of both loss and storage modulus decreased and became equal at 55 °C, indicating a stimuli-responsive gel–sol transition. DOXO release in the presence of esterase enzyme from the DOX-MSNs-PE (DMP) nano-carrier was more than 60% at pH 5.0 after 24 hour of incubation, whereas at pH 6.8 and pH 7.4, less than 15% of the drug was released. MTT assay revealed that after 10 days of incubation of LoVo cells with the developed hydrogel, only 5% of cells survived. An in vivo study revealed that 100% of mice survived after 35 days post-treatment of PNAGA-DMP-Fe_3O_4-GO hydrogel [156]. Ghorbani et al. developed a pH/temperature/redox-responsive PEGylated hybrid gold/nano gels (Au/NGs) to deliver DOXO and 6-MP. Compared to a normal tissue environment (pH 7.4, 37 °C, and 2 μM GSH), where only 25% and 12% of DOXO and MP were released, respectively, in a cancerous cell environment (pH 6, 41 °C and 10 mM GSH), 78% and 88% of DOXO and MP were released, respectively. Compared to 6-MP, at pH 6.0, higher DOXO release was observed. The higher DOXO release in acidic pH could be attributed to the dissociation of the ionic bond between DOXO and MA. In contrast, the AU-S bond of 6-MP was stable and resulted in lower drug release. At higher temperatures, increased release of both drugs was observed, which could be due to the collapse of the thermosensitive PNIPAM polymer. Similarly, higher drug release with high GSH concentration could be due to redox-responsive BAC monomer within nanogel formulation [157]. Yi-jon Jo et al. have designed DOXO-loaded pH/light/redox responsive HA-based hydrogel for cancer treatment. The hydrogel was formed by an inverse electron demand Diels–Alder reaction between the norbornene and tetrazine groups. Higher drug release at pH 5.0 could be attributed to the better solubility of DOXO at lower pH and higher swelling of HA because of the protonation of the carboxylic, hydroxyl, and amide groups. When the hydrogel was exposed to the NIR light, a rise in temperature and ROS generation because of the interaction between NIR and ICG resulted in an initial burst release. Similarly, the initial burst release after exposure to 10 mM DTT and 0.5% H_2O_2 is probably due to the rapid interaction of the diselenide part of the cross-linker with DTT and H_2O_2. However, the DOXO-ICG-loaded hydrogel exhibited slightly better anti-tumor efficacy ($p < 0.05$) than the DOXO-loaded hydrogel, indicating synergistic anti-tumor and photo-thermal activity [158]. Chi-Yao et al. have developed a pH/temperature/enzyme-responsive DOXO-loaded nanogel to treat cancer. The developed nanogel exhibited higher drug release in response to increasing temperature, lower pH, and high DNase I concentration. The morphological study of the hydrogel under low pH and high temperature revealed that the integrity of the DNA layer was maintained. From this study, they found that the weak DOXO and DNA interaction because of the competition of H^+ ions with DOXO for negatively charged DNA molecules and the diffusion of the drug at high temperatures were responsible for DOXO dissociation from the nanogel. After that, the nuclease treatment disintegrates the DNA layer, facilitating DOXO release [159]. Table 13.2 briefly highlights a few examples of preclinically developed MSRHs for cancer treatment.

To summarize, it can be said that both DSRHs and MSRHs have gained attention in cancer therapy and have shown promising preclinical outcomes. Their ability to respond to multiple stimuli has enabled them to release the drug in the TME, thereby providing targeted delivery, enhancing therapeutic efficacy, and reducing the side effects. Figure 13.3 highlights few preclinical outcomes of DSRHs and MSRHs in cancer therapy.

Table 13.2 Multi-stimuli-responsive hydrogels for cancer treatment

Sl. No.	Stimuli used	Hydrogel composition	Experimental outcome	References
1	pH, enzyme, and NIR light	Doxorubicin, gelatine, and polydopamine	The developed Gela/PDA/DOXO hydrogel exhibited synergistic antitumor activity through a combined chemo-photothermal approach	[150]
2	pH, redox, and ultrasound	Doxorubicin, PDPA-b-PEI copolymer	The developed hydrogel system exhibited slow and controlled drug release in response to pH and redox stimuli, whereas faster drug release in response to ultrasound	[151]
3	pH, sugar, and H_2O_2	4-formylphenyl methacrylate, 5-methacrylamido-1,2-benzoxaborole, 2-lactobionamidoethyl methacrylamide, 2-aminoethylmethacrylamide hydrochloride	The pH-dependent dissociation of the DCN hydrogel revealed that it could be used as an anticancer DDS	[152]
4	pH, ROS, GSH, temperature, and light	Doxorubicin, cyanine—7, silk fibroin	The developed hydrogel showed multi-stimuli responsive enhanced drug release profile, and also exhibited long-term retention behavior after intratumoral injection	[153]
5	pH, light, temperature, and redox	Doxorubicin, graphene oxide, PNIPAM polymer	The developed hybrid nanogel exhibited multi-stimuli responsive enhanced drug release behavior and synergistic anticancer activity	[154]
6	pH, temperature, and magnetic field	Doxorubicin, sodium alginate, 2-dimethylamino ethyl methacrylate, N-isopropylacrylamide	The designed hydrogel exhibited sustained drug release for 12.5 days	[155]
7	Light, magnetic field, and enzyme	Doxorubicin, PNAGA polymer, Graphene oxide, Fe_3O_4, mesoporous silica nanoparticles, and polyester	The developed hydrogel system provides tumor site-specific drug delivery by avoiding premature drug release, and helps to overcome the existing chemotherapy limitations	[156]

(continued)

Table 13.2 (continued)

Sl. No.	Stimuli used	Hydrogel composition	Experimental outcome	References
8	pH, temperature, and redox	Doxorubicin, 6-Mercaptopurine, PEG, Maleic acid, 2-hydroxyethyl methacrylate, N-isopropyl acrylamide, and N, N (dimethylamino) ethyl methacrylate, BAC	The developed DOXO and 6-MP loaded nanogel exhibited stimuli responsive increased drug release at the TME and enhanced cellular uptake	[157]
9	pH, light, and redox	Doxorubicin, indocyanine green, Hyaluronic acid, diselenide-di-tetrazine cross-linker, and 5-norbornene-2-methylamine	In vitro activity revealed that the DOXO-loaded hydrogel exhibited similar antitumor activity to the free drug, whereas the DOXO-ICG-loaded hydrogel exhibited slightly better synergistic antitumor activity	[158]
10	pH, temperature, and enzyme	Doxorubicin, Fe_3O_4, Cir-C DNA template, Bovine serum albumin, DNA polymerase	The developed DNA nanogel exhibited controlled drug release in response to multiple stimuli, good biocompatibility, enhanced cellular uptake, and improved therapeutic efficacy	[159]

4 Translational Challenges in Stimuli-Responsive Hydrogel

Although the preclinical studies of SRHs have shown potential results, the lab-to-clinic translation has not been seen. The complex manufacturing process of SRHS is one of the possible reasons for the lack of clinical translation. The design of SRHS involves complex chemical reactions and depends on multiple process parameters like cross-linker density, polymer composition and concentration, and reaction conditions. All these process parameters are interrelated for the successful development of SRHs. Minor variations in the process parameters resulted in inconsistent drug release profiles and affected the hydrogel's mechanical properties, leading to the variations in in vivo performances of SRHs [160]. The selection of material for the hydrogel design also plays a crucial role, affecting its biocompatibility and biodegradability. Even if the designing of hydrogels involves the use of natural polymers like alginate, chitosan, etc., there is still a need to keep a perfect balance between biocompatibility and ensuring its desired mechanical and biochemical properties [161, 162]. Hydrogel degrades in the biological environment and generates degradation products that should be non-toxic. However, most of the preclinically developed hydrogels are based on synthetic polymers. Biodegradation of synthetic polymer-based hydrogels could produce toxic degradation products, thereby limiting clinical translation [163]. The developed hydrogel system must

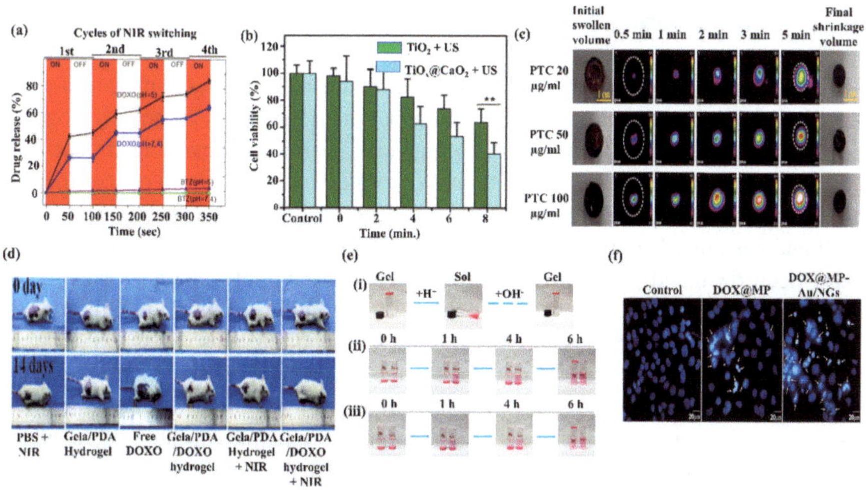

Fig. 13.3 SRHs for anticancer drug delivery (**a**) In vitro drug release profiles of DOXO and BTZ from composite hydrogels in PBS at pH 5.0 and 7.4, with or without NIR irradiation, at 37 °C. (Reproduced with permission from ref. GhavamiNejad et al. [137]). Copyright Scientific Reports; (**b**) Cell viability of 4T1 cells incubated with 100 mg/mL TiO_2 and TiOx@CaO_2 at different US irradiation times. (Reproduced with permission from ref. Dai et al. [139]. Copyright Materials Today); (**c**) Thermal imaging of thermo-responsive hybrid hydrogel with different PTC concentrations after exposure to NIR irradiation at 808 nm for different time points. (Reproduced with permission from ref. Wang et al. [142]. Copyright Elsevier); (**d**) Relative tumor volume in mice treated with different treatment regimens. (Reproduced with permission from ref. He et al. [150]. Copyright Royal Society of Chemistry); (**e**) (i) Reversible gel–sol–gel transition by changing pH, (ii) Sugar-responsiveness of the gel by adding fructose at different time intervals (left: in PBS, right: in PBS containing fructose), (iii) ROS-responsiveness of the gel by adding H_2O_2 at different time intervals (left: in PBS, right: in PBS containing H_2O_2). (Reproduced with permission from ref. Zhao et al. [152]. Copyright American Chemical Society); (**f**) DAPI staining of MCF-7 cell line treated with the mixture of drugs (DOX@MP) and drugs-loaded hybrid Au/NGs (DOX@MP-Au/NGs). (Reproduced with permission from ref. Ghorbani and Hamishehkar [157]. Copyright Elsevier)

meet the safety and biocompatibility aspects as per the standards of ISO 10993 for its regulatory approval [164]. However, most of the pre-clinically developed hydrogels failed to fulfill such criteria. Another significant challenge in the clinical translation of hydrogel is that no regulatory guidelines are available for manufacturing these DDS. Stimuli-mediated controlled drug release in response to external and internal stimuli is also a challenge as minor variation in such stimuli alters drug release behavior from the hydrogel, thereby affecting their therapeutic efficacy, which, in turn, complicates their translation from lab to industry [165]. Compared to the conventional dosage form, hydrogel development involves a complex process requiring high-cost investment. Balancing the manufacturing cost with respect to clinical benefit curtails its potential for market translation [166].

5 Conclusion

DSRHs and MSRHs have shown promising preclinical outcomes in cancer therapy. The hydrophilic group in the hydrophilic polymeric network of the hydrogel is responsible for its high water-holding capacity, thereby maintaining its soft and flexible structure. Hydrogel is an attractive drug delivery system in cancer therapy because of its safety, biocompatibility, enhanced drug loading capacity, and localized and sustained/controlled drug delivery. DSRHs and MSRHs in cancer therapy have been designed to release the drug by undergoing sol–gel transformation in TME in response to dual or multiple stimuli (external/internal). Although the preclinical outcomes of DSRHs and MSRHs are significant, such success has not been observed in the clinical stage. Complex manufacturing processes, biocompatibility, and toxicity issues arise because of the in vivo degradation of the synthetic polymer used in hydrogel manufacturing, altered therapeutic outcomes because of the minor variation in stimuli, failure to meet regulatory criteria required for the product approval, and investment to therapeutic benefit ratio are some of the constraints involved with translation of preclinically developed hydrogel to the clinical stage.

References

1. The History of Cancer Therapy, an Interactive Timeline [Internet]. [cited 2024 Aug 18]. Available from: https://static.scientificamerican.com/sciam/assets/media/multimedia/msk-timeline/index.html.
2. Do HH, Kim S. History of radiation therapy technology. Prog Med Phys. 2020;31(3):124–34.
3. Quintão NLM, Santin JR, Stoeberl LC, Corrêa TP, Melato J, Costa R. Pharmacological treatment of chemotherapy-induced neuropathic pain: PPARγ agonists as a promising tool. Front Neurosci. 2019;13:907.
4. Lehmann F, Wennerberg J. Evolution of nitrogen-based alkylating anticancer agents. Processes. 2021;9(2):1–10.
5. Brown JS, Amend SR, Austin RH, Gatenby RA, Hammarlund EU, Pienta KJ. Updating the definition of cancer. Mol Cancer Res. 2023;21(11):1142–7.
6. Fares J, Fares MY, Khachfe HH, Salhab HA, Fares Y. Molecular principles of metastasis: a hallmark of cancer revisited. Signal Transduct Target Ther. 2020;5(1):28.
7. Hoadley KA, Yau C, Hinoue T, Wolf DM, Lazar AJ, Drill E, et al. Cell-of-origin patterns dominate the molecular classification of 10,000 tumors from 33 types of cancer. Cell. 2018;173(2):291–304.e6.
8. Anand P, Kunnumakara AB, Sundaram C, Harikumar KB, Tharakan ST, Lai OS, et al. Cancer is a preventable disease that requires major lifestyle changes. Pharm Res. 2008;25(9):2097–116.
9. Global cancer burden growing, amidst mounting need for services [Internet]. [cited 2024 Aug 14]. Available from: https://www.who.int/news/item/01-02-2024-global-cancer-burden-growing%2D%2Damidst-mounting-need-for-services.
10. Siegel RL, Giaquinto AN, Jemal A. Cancer statistics, 2024. CA Cancer J Clin. 2024;74(1):12–49.
11. State/UT-wise Number of Estimated Incidence of Cancer Cases in India by different Both Sexes All sites from 2021 to 2024 [Internet]. 2024 [cited 2024 Aug 14]. Available from: https://www.data.gov.in/resource/stateut-wise-number-estimated-incidence-cancer-cases-india-different-both-sexes-all-sites.

12. Hajdu SI. A note from history: landmarks in history of cancer, part 1. Cancer. 2011;117(5):1097–102.
13. DeVita VT, Rosenberg SA. Two hundred years of cancer research. N Engl J Med. 2012;366(23):2207–14.
14. Zafar A, Khan MJ, Abu J, Naeem A. Revolutionizing cancer care strategies: immunotherapy, gene therapy, and molecular targeted therapy. Mol Biol Rep. 2024;51(1):219.
15. Mee T, Kirkby NF, Defourny NN, Kirkby KJ, Burnet NG. The use of radiotherapy, surgery and chemotherapy in the curative treatment of cancer: results from the FORTY (Favourable Outcomes from RadioTherapY) project. Br J Radiol. 2023;96(1152):20230334.
16. DeVita VT, Chu E. A history of cancer chemotherapy. Cancer Res. 2008;68(21):8643–53.
17. Baskar R, Lee KA, Yeo R, Yeoh KW. Cancer and radiation therapy: current advances and future directions. Int J Med Sci. 2012;9(3):193–9.
18. Yang L, Ning Q, Tang SS. Recent advances and next breakthrough in immunotherapy for cancer treatment. J Immunol Res. 2022;2022:1.
19. Lee YT, Tan YJ, Oon CE. Molecular targeted therapy: treating cancer with specificity. Eur J Pharmacol. 2018;834(July):188–96.
20. Deli T, Orosz M, Jakab A. Hormone replacement therapy in cancer survivors—review of the literature. Pathol Oncol Res. 2020;26(1):63–78.
21. Debela DT, Muzazu SGY, Heraro KD, Ndalama MT, Mesele BW, Haile DC, et al. New approaches and procedures for cancer treatment: current perspectives. SAGE Open Med. 2021;9:20503121211034366.
22. Cancer Treatment [Internet]. [cited 2024 Aug 16]. Available from: https://www.mayoclinic.org/tests-procedures/cancer-treatment/about/pac-20393344.
23. Hamad I, Harb AA, Bustanji Y. Liposome-based drug delivery systems in cancer research: an analysis of global landscape efforts and achievements. Pharmaceutics. 2024;16(3)
24. Xiao X, Teng F, Shi C, Chen J, Wu S, Wang B, et al. Polymeric nanoparticles—promising carriers for cancer therapy. Front Bioeng Biotechnol. 2022;10(October):1–20.
25. Roshani M, Rezaian-Isfahni A, Lotfalizadeh MH, Khassafi N, Abadi MHJN, Nejati M. Metal nanoparticles as a potential technique for the diagnosis and treatment of gastrointestinal cancer: a comprehensive review. Cancer Cell Int. 2023;23(1):1–23.
26. Lu W, Chen T, Xiao D, Qin X, Chen Y, Shi S. Application and prospects of nucleic acid nanomaterials in tumor therapy. RSC Adv. 2023;13(37):26288–301.
27. Tao Z, Zhang H, Wu S, Zhang J, Cheng Y, Lei L, et al. Spherical nucleic acids: emerging amplifiers for therapeutic nanoplatforms. Nanoscale. 2024;16(9):4392–406.
28. Chen J, Tan Q, Yang Z, Jin Y. Engineered extracellular vesicles: potentials in cancer combination therapy. J Nanobiotechnol. 2022;20(1):1–22.
29. Patel A, Patel K, Patel V, Rajput MS, Patel R, Rajput A. Nanocrystals: an emerging paradigm for cancer therapeutics. Future J Pharm Sci. 2024;10(1)
30. Elmore LW, Greer SF, Daniels EC, Saxe CC, Melner MH, Krawiec GM, et al. Blueprint for cancer research: critical gaps and opportunities. CA Cancer J Clin. 2021;71(2):107–39.
31. Chu X, Tian W, Ning J, Xiao G, Zhou Y, Wang Z, et al. Cancer stem cells : advances in knowledge and implications for cancer therapy. Springer US; 2024.
32. Zang C, Tian Y, Tang Y, Tang M, Yang D, Chen F, et al. Hydrogel-based platforms for site-specific doxorubicin release in cancer therapy. J Transl Med. 2024;22:879.
33. Akmal MH, Kalashgrani MY, Mousavi SM, Rahmanian V, Sharma N, Gholami A, et al. Recent advances in synergistic use of GQD-based hydrogels for bioimaging and drug delivery in cancer treatment. J Mater Chem B. 2024;12(21):5039–60.
34. Goyal PK, Khurana S, Mittal A. Hydrogel-bound cytotoxic drug delivery system for breast cancer. Health Sci Rev. 2023;9(June):100140.
35. Luo FQ, Xu W, Zhang JY, Liu R, Huang YC, Xiao C, et al. An injectable nanocomposite hydrogel improves tumor penetration and cancer treatment efficacy. Acta Biomater. 2022;147:235–44.
36. Cirillo G, Spizzirri UG, Curcio M, Nicoletta FP, Iemma F. Injectable hydrogels for cancer therapy over the last decade. Pharmaceutics. 2019;11(9)

37. Mateti T, Likhith K, Laha A, Thakur G. A critical analysis of the recent developments in multi-stimuli responsive smart hydrogels for cancer treatment. Curr Opin Biomed Eng. 2023;25:100424.
38. Senapati S, Mahanta AK, Kumar S, Maiti P. Controlled drug delivery vehicles for cancer treatment and their performance. Signal Transduct Target Ther. 2018;3(1):1–19.
39. Fatima M, Almalki WH, Khan T, Amirhossein Sahebkar PK. Harnessing the power of stimuli-responsive nanoparticles as an effective therapeutic drug delivery system. Adv Mater. 36(24):2312939.
40. Mohapatra A, Uthaman S, Park IK. External and internal stimuli-responsive metallic nanotherapeutics for enhanced anticancer therapy. Front Mol Biosci. 2021;7(January):1–32.
41. Sun T, Jiang C. Stimuli-responsive drug delivery systems triggered by intracellular or subcellular microenvironments. Adv Drug Deliv Rev. 2023;196:114773.
42. Rahim MA, Jan N, Khan S, Shah H, Madni A, Khan A, et al. Recent advancements in stimuli responsive drug delivery platforms for active and passive cancer targeting. Cancer. 2021;13(4):1–52.
43. Sharma P, Poonia A, Jangra M, Ankur. Application of stimuli-responsive polymers in cancer therapy. In: Handbook of oxidative stress in cancer: therapeutic aspects. Singapore: Springer; 2022. p. 909–22.
44. Rajendrakumar SK, Cherukula K, Park HJ, Uthaman S, Jeong YY, Il LB, et al. Dual-stimuli-responsive albumin-polyplex nanoassembly for spatially controlled gene release in metastatic breast cancer. J Control Release. 2018;276(November 2017):72–83.
45. Wang Y, Shim MS, Levinson NS, Hsing-Wen Sung YX. Stimuli-responsive materials for controlled release of theranostic agents. Adv Funct Mater. 2014;24(27):4206–20.
46. Shah BP, Pasquale N, De G, Tan T, Ma J, Lee KB. Core-shell nanoparticle-based peptide therapeutics and combined hyperthermia for enhanced cancer cell apoptosis. ACS Nano. 2014;8(9):9379–87.
47. Zhang X, Wang S, Cheng G, Yu P, Chang J. Light-responsive nanomaterials for cancer therapy. Engineering. 2022;13:18–30.
48. Lee GH, Moon H, Kim H, Lee GH, Kwon W, Yoo S, et al. Multifunctional materials for implantable and wearable photonic healthcare devices. Nat Rev Mater. 2020;5(2):149–65.
49. Varon E, Blumrosen G, Sinvani M, Haimov E, Polani S, Natan M, et al. An engineered nanocomplex with photodynamic and photothermal synergistic properties for cancer treatment. Int J Mol Sci. 2022;23(4)
50. Ozog DM, Rkein AM, Fabi SG, Gold MH, Goldman MP, Lowe NJ, et al. Photodynamic therapy: a clinical consensus guide. Dermatologic Surg. 2016;42(7):804–27.
51. Murugan B, Sagadevan S, Fatimah I, Oh WC, Motalib Hossain MA, Johan MR. Smart stimuli-responsive nanocarriers for the cancer therapy - nanomedicine. Nanotechnol Rev. 2021;10(1):933–53.
52. Overchuk M, Weersink RA, Wilson BC, Zheng G. Photodynamic and photothermal therapies: synergy opportunities for nanomedicine. ACS Nano. 2023;17(9):7979–8003.
53. Beniwal N, Verma A, Putta CL, Rengan AK. Recent trends in bio-nanomaterials and non-invasive combinatorial approaches of photothermal therapy against cancer. Nanotheranostics. 2024;8(2):219–38.
54. Ma X, Huang Y, Abedi S, Kim H, et al. Rational design and application of an indolium-derived heptamethine cyanine with record-long NIR-II emission. CCS Chem. 2021;4(6):1–38. https://doi.org/10.31635/ccschem.021.202101630
55. Zhang J, Chen J, Ren J, Guo W, Li X, Chen R, et al. Biocompatible semiconducting polymer nanoparticles as robust photoacoustic and photothermal agents revealing the effects of chemical structure on high photothermal conversion efficiency. Biomaterials. 2018;181:92–102.
56. Vines JB, Yoon JH, Ryu NE, Lim DJ, Park H. Gold nanoparticles for photothermal cancer therapy. Front Chem. 2019;7(APR):1–16.
57. Jain S, Hirst DG, O'Sullivan JM. Gold nanoparticles as novel agents for cancer therapy. Br J Radiol. 2012;85(1010):101–13.

58. Chen Y, Zhou F, Wang C, Hu L, Guo P. Nanostructures as photothermal agents in tumor treatment. Molecules. 2023;28(1)
59. Chen J, Ning C, Zhou Z, Yu P, Zhu Y, Tan G, Mao C. Nanomaterials as photothermal therapeutic agents. Prog Mater Sci. 2019;99. https://doi.org/10.1016/j.pmatsci.2018.07.005.
60. Tong L, Zhao Y, Huff TB, Hansen MN, Wei A, Cheng JX. Gold nanorods mediate tumor cell death by compromising membrane integrity. Adv Mater. 2007;19(20):3136–41.
61. Pattani VP, Shah J, Atalis A, Sharma A, Tunnell JW. Role of apoptosis and necrosis in cell death induced by nanoparticle-mediated photothermal therapy. J Nanopart Res. 2015;17(1)
62. Gormley AJ, Larson N, Banisadr A, Robinson R, Frazier N, Ray A, et al. Plasmonic photothermal therapy increases the tumor mass penetration of HPMA copolymers. J Control Release. 2013;166(2):130–8.
63. Fay BL, Melamed JR, Day ES. Nanoshell-mediated photothermal therapy can enhance chemotherapy in inflammatory breast cancer cells. Int J Nanomedicine. 2015;10:6931–41.
64. Kessel D. Photodynamic therapy: a brief history. J Clin Med. 2019;8(10)
65. Jiang W, Liang M, Lei Q, Li G, Wu S. The current status of photodynamic therapy in cancer treatment. Cancer. 2023;15(3)
66. Felsher DW. Cancer revoked: oncogenes as therapeutic targets. Nat Rev Cancer. 2003;3(5):375–80.
67. Mroz P, Yaroslavsky A, Kharkwal GB, Hamblin MR. Cell death pathways in photodynamic therapy of cancer. Cancer. 2011;3(2):2516–39.
68. Agostinis P, Berg K, Cengel KA, Foster TH, Girotti AW, Gollnick SO, Hahn SM, Hamblin MR, Juzeniene A. Photodynamic therapy of cancer: an update. CA Cancer J Clin. 2012;61(4):250–281. https://doi.org/10.3322/caac.20114
69. Bonnet S. Why develop photoactivated chemotherapy? Dalton Trans. 2018;47(31):10330–43.
70. Farrer NJ, Salassa L, Sadler PJ. Photoactivated chemotherapy (PACT): the potential of excited-state d-block metals in medicine. Dalton Trans. 2009;44(48):10690–701.
71. Chen Y, Bai L, Zhang P, Zhao H, Zhou Q. The development of ru(Ii)-based photoactivated chemotherapy agents. Molecules. 2021;26(18):1–20.
72. Dunne M, Regenold M, Allen C. Hyperthermia can alter tumor physiology and improve chemo- and radio-therapy efficacy. Adv Drug Deliv Rev. 2020;163–164:98–124.
73. Theodorescu D. Cancer cryotherapy. Rev Urol. 2004;6:9–19.
74. Cheng Y, Weng S, Yu L, Zhu N, Yang M, Yuan Y. The role of hyperthermia in the multidisciplinary treatment of malignant tumors. Integr Cancer Ther. 2019;18
75. Li Z, Deng J, Sun J, Ma Y. Hyperthermia targeting the tumor microenvironment facilitates immune checkpoint inhibitors. Front Immunol. 2020;11(November):1–12.
76. Crezee J, Franken NA, Oei AL. Hyperthermia-based anti-cancer treatments. Cancers. 2021;13:7–10.
77. Kwak K, Yu B, Lewandowski RJ, Kim DH. Recent progress in cryoablation cancer therapy and nanoparticles mediated cryoablation. Theranostics. 2022;12(5):2175.
78. Baust JG, Gage AA, Johansen TEB, Baust JM. Mechanisms of cryoablation: clinical consequences on malignant tumors. Cryobiology. 2014;68(1):1–11. https://doi.org/10.1016/j.cryobiol.2013.11.001
79. Baust JG, Snyder KK, Santucci KL, Robilotto AT, Buskirk R, Baust JM. Cryoablation: physical and molecular basis with putative immunological consequences. Int J Hyperthermia. 2019;36(sup. 1):10–16. https://doi.org/10.1080/02656736.2019.1647355
80. Liu X, Zhang Y, Wang Y, Zhu W, Li G, Ma X, et al. Comprehensive understanding of magnetic hyperthermia for improving antitumor therapeutic efficacy. Theranostics. 2020;10(8):3793–815.
81. Spoială A, Ilie CI, Motelica L, Ficai D, Semenescu A, Oprea OC, et al. Smart magnetic drug delivery systems for the treatment of cancer. Nanomaterials. 2023;13(5)
82. Mekaru H, Ichiyanagi Y, Tamanoi F. Magnetic nanoparticles and alternating magnetic field for cancer therapy. In: Cell-inspired materials and engineering. Fundamental Biomedical Technologies. Cham: Springer. p. 165–79.

83. Espinosa A, Kolosnjaj-Tabi J, Abou-Hassan A, Sangnier AP, Curcio A, Silva AKA, Di Corato R, Neveu S, Pellegrino T, Liz-Marzán LM, Wilhelm C. Magnetic (Hyper)Thermia or Photothermia? progressive comparison of iron oxide and gold nanoparticles heatingin water, in cells, and in vivo. Adv Funct Mater. 2018;28:1803660.
84. Tharkar P, Varanasi R, Wong WSF, Jin CT, Chrzanowski W. Nano-enhanced drug delivery and therapeutic ultrasound for cancer treatment and beyond. Front Bioeng Biotechnol. 2019;7(November):324.
85. Zhu Y, Liu Y, Khan K, Arkin G, Tareen AK, Xie Z, et al. Ultrasound combined with nanomaterials for cancer therapy. Mater Today Adv. 2023;17:100330.
86. Panzone J, Byler T, Bratslavsky G, Goldberg H. Applications of focused ultrasound in the treatment of genitourinary cancers. Cancer. 2022;14(6)
87. Awad NS, Paul V, Alsawaftah NM, Ter Haar G, Allen TM, Pitt WG, et al. Ultrasound-responsive nanocarriers in cancer treatment: a review. ACS Pharm Transl Sci. 2021;4(2):589–612.
88. Sun S, Wu M. Sonodynamic therapy: another "light" in tumor treatment by exogenous stimulus. Smart Mater Med. 2021;2(May):145–9.
89. Zhang Y, Zhao Y, Zhang Y, Liu Q, Zhang M, Tu K. The crosstalk between sonodynamic therapy and autophagy in cancer. Front Pharmacol. 2022;13(August):1–23.
90. Johansen PM, Hansen PY, Mohamed AA, Girshfeld SJ, Feldmann M, Lucke-Wold B. Focused ultrasound for treatment of peripheral brain tumors. Explor Drug Sci. 2023:107–25.
91. Lee H, Woo SM, Jang H, Kang M, Kim SY. Cancer depends on fatty acids for ATP production: a possible link between cancer and obesity. Semin Cancer Biol. 2022;86(P2):347–57.
92. Liao S, Wu G, Xie Z, Lei X, Yang X, Huang S, et al. pH regulators and their inhibitors in tumor microenvironment. Eur J Med Chem. 2024;267(January):116170.
93. Liberti MV, Locasale JW. The Warburg effect: how does it benefit cancer cells? Trends Biochem Sci. 2016;41(3):211.
94. Hao G, Xu ZP, Li L. Manipulating extracellular tumour pH: an effective target for cancer therapy. RSC Adv. 2018;8(39):22182–92.
95. Verkhovskii RA, Ivanov AN, Lengert EV, Tulyakova KA, Shilyagina NY, Ermakov AV. Current principles, challenges, and new metrics in pH-responsive drug delivery systems for systemic cancer therapy. Pharmaceutics. 2023;15(5)
96. Mu Y, Gong L, Peng T, Yao J, Lin Z. Advances in pH-responsive drug delivery systems. OpenNano. 2021;5(December 2021):100031.
97. Ding H, Tan P, Fu S, Tian X, Zhang H, Ma X, et al. Preparation and application of pH-responsive drug delivery systems. J Control Release. 2022;348(June):206–38.
98. Chu S, Shi X, Tian Y, Gao F. pH-responsive polymer nanomaterials for tumor therapy. Front Oncol. 2022;12(March):1–22.
99. Sun Y, Davis E. Nanoplatforms for targeted stimuli-responsive drug delivery: a review of platform materials and stimuli-responsive release and targeting mechanisms. Nanomaterials. 2021;11(3):1–103.
100. Alsawaftah NM, Awad NS, Pitt WG, Husseini GA. pH-responsive nanocarriers in cancer therapy. Polymers. 2022;14(5)
101. He Q, Chen J, Yan J, Cai S, Xiong H, Liu Y, et al. Tumor microenvironment responsive drug delivery systems. Asian J Pharm Sci. 2020;15(4):416–48.
102. Chang D, Ma Y, Xu X, Xie J, Ju S. Stimuli-responsive polymeric nanoplatforms for cancer therapy. Front Bioeng Biotechnol. 2021;9(June):1–26.
103. Li Y, Zhang X, Wang Z, Li B, Zhu H. Modulation of redox homeostasis: a strategy to overcome cancer drug resistance. Front Pharmacol. 2023;14(March):1–16.
104. Xing F, Hu Q, Qin Y, Xu J, Zhang B, Yu X, et al. The relationship of redox with hallmarks of cancer: the importance of homeostasis and context. Front Oncol. 2022;12(April):1–21.
105. Kuo CL, Ponneri Babuharisankar A, Lin YC, Lien HW, Lo YK, Chou HY, et al. Mitochondrial oxidative stress in the tumor microenvironment and cancer immunoescape: foe or friend? J Biomed Sci. 2022;29(1):1–25.
106. Wu X, Zhou Z, Li K, Liu S. Nanomaterials-induced redox imbalance: challenged and opportunities for nanomaterials in cancer therapy. Adv Sci. 2024;11(16)

107. Chaiswing L, Clair WHS, Clair DKS. Redox paradox: a novel approach to therapeutics-resistant cancer. Antioxid Redox Signal. 2018;29(13):1237–72.
108. Dai X, Wang D, Zhang J. Programmed cell death, redox imbalance, and cancer therapeutics. Apoptosis. 2021;26(7–8):385–414.
109. Zhuang Y, Liu K, He Q, Gu X, Jiang C, Wu J. Hypoxia signaling in cancer: implications for therapeutic interventions. MedComm. 2023;4(1):1–37.
110. Jing X, Yang F, Shao C, Wei K, Xie M, Shen H, et al. Role of hypoxia in cancer therapy by regulating the tumor microenvironment. Mol Cancer. 2019;18(1):1–15.
111. Kabakov AE, Yakimova AO. Hypoxia-induced cancer cell responses driving radioresistance of hypoxic tumors: approaches to targeting and radiosensitizing. Cancer. 2021;13(5):1–52.
112. Li Y, Zhao L, Feng LX. Targeting hypoxia: hypoxia-activated prodrugs in cancer therapy. Front Oncol. 2021;11(July):1–13.
113. Bernauer C, Man YKS, Chisholm JC, Lepicard EY, Robinson SP, Shipley JM. Hypoxia and its therapeutic possibilities in paediatric cancers. Br J Cancer. 2021;124(July 2020):539.
114. Graham K. Overcoming tumor hypoxia as a barrier to radiotherapy, chemotherapy and immunotherapy in cancer treatment. Int J Nanomedicine. 2018;13:6049–58.
115. Zhuang Y, He Q. Hypoxia signaling in cancer: implications for therapeutic interventions. MedComm. 2023;4(s):1–37.
116. Bui BP, Nguyen PL, Lee K, Cho J. Hypoxia-Inducible Factor-1: a novel therapeutic target for the management of cancer, drug resistance, and cancer-related pain. Cancers. 2022;14:1–26.
117. Qannita RA, Alalami AI, Harb AA, Aleidi SM, Taneera J, Abu-gharbieh E, et al. Targeting Hypoxia-Inducible Factor-1 (HIF-1) in cancer: emerging therapeutic strategies and pathway regulation. Pharmaceuticals (Basel). 2024;17(2):195.
118. Thomas RG, Surendran SP, Jeong YY. Tumor microenvironment-stimuli responsive nanoparticles for anticancer therapy. Front Mol Biosci. 2020;7(December):610533.
119. Zhao X, Bai J, Yang W. Stimuli-responsive nanocarriers for therapeutic applications in cancer nanocarriers. Cancer Biol Med. 2021;18:319–35.
120. Ding M, Zhang Y, Li J, Pu K. Bioenzyme - based nanomedicines for enhanced cancer therapy. Nano Convergence. 2022;9:7.
121. Bashir S, Hina M, Iqbal J, Rajpar AH, Mujtaba MA, Alghamdi NA, et al. Fundamental concepts of hydrogels: synthesis, properties, and their applications. Polymers. 2020;12(11):1–60.
122. Andrade F, Roca-Melendres MM, Durán-Lara EF, Rafael D, Schwartz S. Review stimuli-responsive hydrogels for cancer treatment: the role of ph, light, ionic strength and magnetic field. Cancer. 2021;13(5):1–17.
123. Zöller K, To D, Bernkop-Schnürch A. Biomedical applications of functional hydrogels: innovative developments, relevant clinical trials and advanced products. Biomaterials. 2025;312(February 2024):122718.
124. Zhang H, Zhai Y, Wang J, Zhai G. New progress and prospects: the application of nanogel in drug delivery. Mater Sci Eng C. 2016;60:560–8.
125. Hong Y, Lin Z, Yang Y, Jiang T, Shang J, Luo Z. Biocompatible conductive hydrogels: applications in the field of biomedicine. Int J Mol Sci. 2022;23(9)
126. Lin CC, Metters AT. Hydrogels in controlled release formulations: network design and mathematical modeling. Adv Drug Deliv Rev. 2006;58(12–13):1379–408.
127. Solanki R, Bhatia D. Stimulus-responsive hydrogels for targeted cancer therapy. Gels. 2024;10(7)
128. El-Husseiny HM, Mady EA, Hamabe L, Abugomaa A, Shimada K, Yoshida T, et al. Smart/stimuli-responsive hydrogels: cutting-edge platforms for tissue engineering and other biomedical applications. Mater Today Bio. 2022;13(September 2021):100186.
129. Xin H, Naficy S. Drug delivery based on stimuli-responsive injectable hydrogels for breast cancer therapy: a review. Gels. 2022;8(1):1–17.
130. Wei W, Lu P. Designing dual-responsive drug delivery systems: the role of phase change materials and metal–organic frameworks. Materials. 2024;17(13)
131. Jia R, Teng L, Gao L, Su T, Fu L, Qiu Z, et al. Advances in multiple stimuli-responsive drug-delivery systems for cancer therapy. Int J Nanomedicine. 2021;16:1525–51.

132. Li R, Peng F, Cai J, Yang D, Zhang P. Redox dual-stimuli responsive drug delivery systems for improving tumor-targeting ability and reducing adverse side effects. Asian J Pharm Sci. 2020;15(3):311–25.
133. Dai Z, Song XZ, Cao J, He Y, Wen W, Xu X, et al. Dual-stimuli-responsive TiOx/DOX nanodrug system for lung cancer synergistic therapy. RSC Adv. 2018;8(39):21975–84.
134. Li J, Han Y, Chen Q, Shi H, Ur Rehman S, Siddiq M, et al. Dual endogenous stimuli-responsive polyplex micelles as smart two-step delivery nanocarriers for deep tumor tissue penetration and combating drug resistance of cisplatin. J Mater Chem B. 2014;2(13):1813–24.
135. Kumar N, Singh S, Sharma P, Kumar B. Single-, Dual-, and Multi-stimuli-responsive nanogels for biomedical applications. Gels. 2024;10:1–40.
136. Bilalis P, Skoulas D, Karatzas A, Marakis J, Stamogiannos A, Tsimblouli C, et al. Self-healing pH- and enzyme stimuli-responsive hydrogels for targeted delivery of gemcitabine to treat pancreatic cancer. Biomacromolecules. 2018;19(9):3840–52.
137. GhavamiNejad A, Samarikhalaj M, Aguilar LE, Park CH, Kim CS. PH/NIR light-controlled multidrug release via a mussel-inspired nanocomposite hydrogel for chemo-photothermal cancer therapy. Sci Rep. 2016;6(June):1–12.
138. Zhou X, Wang L, Xu Y, Du W, Cai X, Wang F, et al. A pH and magnetic dual-response hydrogel for synergistic chemo-magnetic hyperthermia tumor therapy. RSC Adv. 2018;8(18):9812–21.
139. Dai Z, Zhang Q, Li X, Chen Q, Chen J, Wang M, et al. In situ forming pH/ROS-responsive niche-like hydrogel for ultrasound-mediated multiple therapy in synergy with potentiating anti-tumor immunity. Mater Today. 2023;65(May):62–77.
140. Mahdian M, Akbari Asrari S, Ahmadi M, Madrakian T, Rezvani Jalal N, Afkhami A, et al. Dual stimuli-responsive gelatin-based hydrogel for pH and temperature-sensitive delivery of curcumin anticancer drug. J Drug Deliv Sci Technol. 2023;84(December 2022):104537.
141. Shen N, Lei B, Wang Y, Xu S, Liu H. Redox/ultrasound dual stimuli-responsive nanogel for precisely controllable drug release. New J Chem. 2018;42(12):9472–81.
142. Wang L, Li B, Xu F, Xu Z, Wei D, Feng Y, et al. UV-crosslinkable and thermo-responsive chitosan hybrid hydrogel for NIR-triggered localized on-demand drug delivery. Carbohydr Polym. 2017;174:904–14.
143. Chen Y, Zhang R, Zheng B, Cai C, Chen Z, Li H, et al. A biocompatible, stimuli-responsive, and injectable hydrogel with triple dynamic bonds. Molecules. 2020;25(13):1–13.
144. Li J, Hu W, Zhang Y, Tan H, Yan X, Zhao L, et al. PH and glucose dually responsive injectable hydrogel prepared by in situ crosslinking of phenylboronic modified chitosan and oxidized dextran. J Polym Sci A Polym Chem. 2015;53(10):1235–44.
145. Chen W, Achazi K, Schade B, Haag R. Charge-conversional and reduction-sensitive poly(vinyl alcohol) nanogels for enhanced cell uptake and efficient intracellular doxorubicin release. J Control Release. 2015;205:15–24.
146. Garshasbi H, Soleymani S, Naghib SM, Mozafari M. Multi-stimuli-responsive hydrogels for therapeutic systems: an overview on emerging materials, devices, and drugs. Curr Pharm Des. 2024;30(26)
147. Pourjavadi A, Heydarpour R, Tehrani ZM. Multi-stimuli-responsive hydrogels and their medical applications. New J Chem. 2021;45(35):15705–17.
148. Knipe JM, Peppas NA. Multi-responsive hydrogels for drug delivery and tissue engineering applications. Regen Biomater. 2014;1(1):57–65.
149. Revete A, Aparicio A, Cisterna BA, Revete J, Luis L, Ibarra E, et al. Advancements in the use of hydrogels for regenerative medicine: properties and biomedical applications. Int J Biomater. 2022;2022:1.
150. He G, Chen S, Xu Y, Miao Z, Ma Y, Qian H, et al. Charge reversal induced colloidal hydrogel acts as a multi-stimuli responsive drug delivery platform for synergistic cancer therapy. Materials Horizons. 2019;6(4):711–6.
151. Lei B, Chen M, Wang Y, Zhang J, Xu S, Liu H. Double security drug delivery system DDS constructed by multi-responsive (pH/redox/US) microgel. Colloids Surf B: Biointerfaces. 2020;193(March):111022.

152. Zhao J, Diaz-Dussan D, Wu M, Peng YY, Wang J, Zeng H, et al. Dual-cross-linked network hydrogels with multiresponsive, self-healing, and shear strengthening properties. Biomacromolecules. 2021;22(2):800–10.
153. Gou S, Xie D, Ma Y, Huang Y, Dai F, Wang C, et al. Injectable, thixotropic, and multiresponsive silk fibroin hydrogel for localized and synergistic tumor therapy. ACS Biomater Sci Eng. 2020;6(2):1052–63.
154. Ji P, Zhang W, Ai S, Zhang Y, Liu J, Liu J, et al. Hybridization of graphene oxide into nanogels to acquire higher photothermal effects for therapeutic delivery. Nanotechnology. 2019;30(11):115701.
155. Hayati M, Rezanejade Bardajee G, Ramezani M, Hosseini SS, Mizani F. Temperature/pH/magnetic triple-sensitive nanogel–hydrogel nanocomposite for release of anticancer drug. Polym Int. 2020;69(2):156–64.
156. Yuan P, Yang T, Liu T, Yu X, Bai Y, Zhang Y, et al. Nanocomposite hydrogel with NIR/magnet/enzyme multiple responsiveness to accurately manipulate local drugs for on-demand tumor therapy. Biomaterials. 2020;262(September):120357.
157. Ghorbani M, Hamishehkar H. A novel multi stimuli-responsive PEGylated hybrid gold/nanogels for co-delivery of doxorubicin and 6-mercaptopurine. Mater Sci Eng C. 2018;92(July):599–611.
158. Jo YJ, Gulfam M, Jo SH, Gal YS, Oh CW, Park SH, et al. Multi-stimuli responsive hydrogels derived from hyaluronic acid for cancer therapy application. Carbohydr Polym. 2022;286(February):119303.
159. Yao C, Yuan Y, Yang D. Magnetic DNA nanogels for targeting delivery and multistimuli-triggered release of anticancer drugs. ACS Appl Bio Mater. 2018;1(6):2012–20.
160. Mooney JD, Jianyu L. Designing hydrogels for controlled drug delivery. Nat Rev Mater. 2016;1(12):1–38.
161. Zhao L, Zhou Y, Zhang J, Liang H, Chen X, Tan H. Natural polymer-based hydrogels: from polymer to biomedical applications. Pharmaceutics. 2023;15(10):2514.
162. Ahmadi F, Oveisi Z, Samani SM, Amoozgar Z. Chitosan based hydrogels: characteristics and pharmaceutical applications. Res Pharm Sci. 2015;10(February):1–16.
163. Raina N, Pahwa R, Bhattacharya J, Paul AK, Nissapatorn V, de Pereira ML, et al. Drug delivery strategies and biomedical significance of hydrogels: translational considerations. Pharmaceutics. 2022;14(3):1–31.
164. Kejlová K, Labský J, Jírová D, Bendová H. Hydrophilic polymers - biocompatibility testing in vitro. Toxicol In Vitro. 2005;19(7):957–62.
165. Lu P, Ruan D, Huang M, Tian M, Zhu K, Gan Z, et al. Harnessing the potential of hydrogels for advanced therapeutic applications: current achievements and future directions. Signal Transduct Target Ther. 2024;9(1):166.
166. Mandal A, Clegg JR, Anselmo AC, Mitragotri S. Hydrogels in the clinic. Bioeng Transl Med. 2020;5(2):1–12.

[illegible]

[illegible]

[illegible]

[illegible]

[illegible] 2016;6(1):[illegible]

52. Eberhardt M, Hamidi-chal [illegible] gatekeepers for co-delivery of [illegible] Sci Rep. 2018;8(1):[illegible]

53. [illegible] from hyaluronic acid [illegible] Carbohydr Polym. [illegible]

[illegible]

[illegible]

[illegible]

[illegible] Pharmaceutics. [illegible]

[illegible]

[illegible]

[illegible]

Preclinical Studies on Stimuli-Responsive Nanosystems for Cancer Therapy

14

Vasavi Pasupuleti, Shanmukh Kanchumurthy,
and Dharmendra Kumar Khatri

Abstract

Cancer is a complex and multifaceted, significant global challenge. The quest for a perfect and universally effective therapy for cancer remains elusive. In recent years, transformative advances have emerged in cancer therapy. Nanoparticles offer precise treatment and alternative chemotherapeutics. With the limitations of nanocarrier systems, stimuli-responsive nanoparticles garnered significant interest in cancer therapy, with target site specificity in addition to on-demand and handy spatiotemporal control over the drug release of anticancer agents in specific tumor sites. There are diverse stimuli speculated to promote anticipated activities, which were sorted as external and internal stimuli. The external stimuli include ultrasound, light, thermoresponsive, magnetic, and electric fields, whereas the internal stimuli bring in enzyme sensitivity, pH variations, and redox reactions. Internal stimuli response nanoparticles dynamically adapt to conditions, activated by variations in physiological pH, redox potential of cells, and enzyme-sensitive stimuli. Ultrasound-responsive nanoparticles permit non-invasive, spatially controlled release, particularly in photothermal therapy light responsive counterparts provide precise control of drug release and imaging. Nanoparticles undergo structural alterations with temperature to achieve thermo-response. Externally directed magnetic field-responsive nanoparticles, targeted

V. Pasupuleti · S. Kanchumurthy
Department of Pharmacology and Toxicology, National Institute of Pharmaceutical Education and Research (NIPER), Hyderabad, Telangana, India

D. K. Khatri (✉)
Department of Pharmacology and Toxicology, National Institute of Pharmaceutical Education and Research (NIPER), Hyderabad, Telangana, India

Molecular and Cellular Neuroscience Lab, Department of Pharmacology and Toxicology, NIMS Institute of Pharmacy, NIMS University Rajasthan, Jaipur, Rajasthan, India

P. V. Devarajan et al. (eds.), *Advances in Stimuli-Responsive Nanosystems for Cancer Therapy*, https://doi.org/10.1007/978-981-95-2744-1_14

delivery to intricate biological systems. Preclinical studies bridge the gap between innovative stimuli-responsive nanosystems and potential clinical applications, pioneering the future of cancer therapy. Testing the stimuli-responsive nanosystems in animal models (In-vivo) and controlled laboratory settings, like in cells (in vitro). This chapter will emphasize precisely the preclinical studies in the design and development of stimuli-responsive nanosystems as a promising chemo-therapeutic option in different cancer types, along with the future perspective and challenges. Stimuli-responsive nanoparticles showed promise in the treatment of breast cancer, colorectal, lung, and gastrointestinal delivering therapeutic drugs specifically with precision and causing the least collateral damage. Studies also demonstrated the effectiveness of the most aggressive cancer, like glioblastoma multiforme.

Keywords

Stimuli-responsive nanosystems · Cancer · Ultrasound · Thermal · Magnetic · Light · pH · Enzyme-responsive · Redox system

Abbreviations

AuNP	gold nanoparticles
AIPH	2,2`-azobis[2-(2-imidazolin-2-yl)propane]dihydrochloride
CET	cetuximab
ChNP	chitosan nanoparticles
CPP-CPT	cell-permeable peptide-camptothecin
CPT11	camptosar
CT	computed tomography
CTX	cyclophosphamide
DTCPP	2,3-dimethylmaleic anhydride
DOX	doxorubicin
EGFR	epidermal growth factor receptor
FDA	food and drug administration
GEM	gemcitabine
GNR	gold nanorods
HACPN	hyaluronic acid-chitosan-g-poly(N-isopropyl acrylamide)
HIFU	high-intensity focused ultrasound
LAT1	L-type amino acid transporter 1
LCST	lower critical solution temperature
LFUS	low-frequency ultrasound
LTSL	lysolipid-containing temperature-sensitive liposomes
MSN	mesoporous silica nanoparticles

MRI	magnetic resonance imaging
Nrf2	nuclear factor E2-related factor 2
NIR	near infrared radiation
NSCLC	non-small cell lung carcinoma
PBA	phenylboronic acid
PDT	photodynamic therapy
PEO-b-P(DEA-stat-MEMA	poly (ethyleneoxide)-block-poly(2-(dimethylamino) ethylmethacrylate)-stat-poly(methoxyethyl methacrylate)
PET	positron emission tomography
PFC	phase-shift perfluorocarbon
PNIPAAM	poly(N-isopropyl acrylamide)
Pox	poly(2-oxazoline)s
PTT	photothermal therapy
PTX	paclitaxel
ROS	reactive oxygen species
SET	sonodynamic therapy
SLP2	stomatin-like protein 2
SPION	superpara magnetic iron-oxide nanoparticles
TML	thermosensitive magnetic lipososmes
TMSP	tumor microenvironment-sensitive polype
TNF	tumor necrosis factor
TOP	topotecan
TRAIL	tumor necrosis factor-related apoptosis-inducing ligand
UCST	upper critical solution temperature
US	ultrasound

1 Introduction

Cancer, also termed a malignant tumor, is the leading cause of death worldwide. Chemotherapy is the standard therapeutic option for cancer, but is often accompanied by unintended adverse effects in addition to low tumor specificity and selectivity. Fortunately, the swift increase in nanotechnology has given a ray of hope for resolving this dilemma [1]. The nanoparticles, ranging between 1 and 100 nm in diameter, are ideally used for targeted or non-targeted drug delivery with enhanced efficiency. These nanocarriers are categorized into either inorganic or organic systems, accounting for the mesoporous silica nanoparticles, magnetic nanoparticles, quantum dots, carbon nanotubes, layered double hydroxide nanoparticles, and the polymeric micelles, dendrimers, liposomes, and protein nanoparticles, respectively [2]. The nanoparticles were ideally known for their diagnostic, biomedical imaging, and therapeutic activity by acting as drug-delivery vehicles [3].

Response to a particular stimulus is the characteristic feature of living organisms. This multifactorial response is incorporated in designing drug carriers to enhance the drug molecule's specificity, sensitivity, and selectivity. Entrenched with this concept, synthetic stimuli-responsive nanosystems were developed that unveiled an innovative approach for delivering drug molecules. These nanosystems respond to changes within the human body, specifically in the tumor microenvironment, by chemical organization of the biomaterials that compose them [4]. The characteristic tumor microenvironment features, like decreased tumor pH, increased glutathione concentration, reactive oxygen species (ROS), and overexpression of specific enzymes, act as guides for drug targeting [5]. Since its introduction into the scientific community, this nanotechnology platform has been greatly explored for targeted drug delivery in various preclinical cancer studies. This chapter briefly dwells on the different stimuli involved in activating therapeutic nano vehicles and the several preclinical studies involving stimuli-responsive nanosystems in various malignancies.

2 Principles of Stimuli-Responsive Nanosystems

The nano-carrier system, acknowledged for its promising drug delivery in combination with a stimuli-responsiveness, enhanced the target site specificity in addition to on-demand and handy spatiotemporal control over the drug release [6, 7]. There are diverse stimuli speculated to promote anticipated activities, which were sorted as external and internal stimuli centered around a physical driver and pathological deformities, respectively [7]. The external stimuli include ultrasound, light, thermo-responsive, magnetic, and electric fields, whereas the internal stimuli bring in enzyme sensitivity, p^H variations, and redox reactions. Table 14.1 below signifies the stimuli-specific nanoparticle response along with their action within the tumor microenvironment.

3 Preclinical Research and Its Significance

Before a drug is brought for human use, the toxicity assessment analysis is to be performed to ensure the drug's safety profile, which is where preclinical studies mark its point. The preclinical studies are merely postulated as the experimental analysis concerning the root cause of the disease and testing of the hypothetical treatment options through laboratory experimentation involving both in vitro and in vivo studies, as depicted in Fig. 14.1. These studies enable the researcher to center on a single molecule from millions of molecules, advancing to human studies [11, 12]. The Food and Drug Administration (FDA) specifies preclinical research as

Table 14.1 Stimuli-responsive nanosystems classification

Stimuli-responsive nanosystem	Stimuli	Description	Reference
Ultrasound-responsive nanosystems	Ultrasound (US)	Low-frequency US wave-associated radiation/cavitation phenomenon induced mechanical & thermal effects via gas bubble formation along with the thermal conversion of acoustic (>20 KHz) energy upon exposure to US vibration, resulting in increased membrane permeability and drug release	[8–10]
Thermo-responsive nanosystems	Thermal energy	These physiologically inactive nanocarrier systems exhibit sensitivity toward tumor cell temperatures (~40–42 °C), resulting in sharp modifications in the properties of nanocarriers and drug release	[8]
Magnetic-responsive nanosystems	Magnetic energy	The enhanced drug accumulation within the tumor is marked by extracorporeal focusing of the target position by an external magnetic field upon injection of a magnetic nanocarrier.	[8]
Light-responsive nanosystems	Light energy	Light-responsive stimuli provide access to the remote pathological regions in consort with its non-invasive nature enhancing the drug release upon illumination with near-infrared, ultraviolet, and visible light	[8]
pH-responsive nanosystems	pH variation	The site-specific drug release at the pathological sites within the organs and intracellular organelles by variation in physiological pH, particularly in cases of ailments like cancer, inflammation, etc.	[8]
Enzyme-responsive nanosystems	Enzymatic variations	These nanoparticles take advantage of the catalytic activity of enzymes to trigger specific responses or changes in the nanomaterials at the tumor microenvironment.	[8]
Redox-responsive nanosystems	Altered Redox potential	The increased reactive oxygen species generation, along with the significant rise in glutathione concentration observed in tumor cells, promoted cellular drug release by altered redox potential across the extracellular and intracellular tumor environment	[5, 8]

laboratory and animal testing of chemical entities following good laboratory practices to answer questions on dosing, safety, and toxicity [13]. The coming sections briefly describe various preclinical studies involving the various stimuli-responsive nanosystems in cancer therapy.

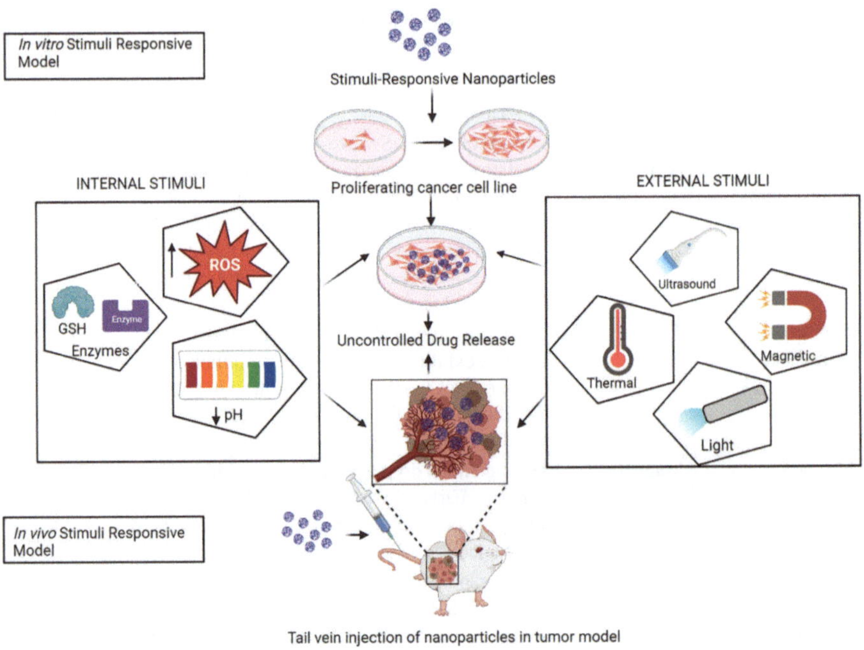

Fig. 14.1 Diagrammatic representation of preclinical stimuli-responsive nanosystem. In vitro and in vivo schematic of the introduction of modified stimuli-responsive nanoparticles to the tumor cell lines and tumor-established animal model, respectively, producing controlled drug release upon activation by internal and external stimuli

4 Preclinical Studies

4.1 External Stimuli-Responsive Nanosystem

4.1.1 Ultrasound (US) Responsive Nanocarrier System

The non-invasive characteristic of the US-responsive nanocarrier systems is one of the assuring triggers for their application in tumor therapeutic studies [10]. This US-responsive system provides spatiotemporal control over the drug release due to its low tissue scattering (<100um) and several centimeters of tissue penetrability [14]. The major mechanism of action includes the involvement of gas-filled nano−/micro- bubbles composed of surface modification ligands, shell, cargo, and core containing the drug. Upon stimulation by the US, these microparticles contract and expand, triggering cavitation due to the acoustic difference between the blood and the gas core [2].

US-controlled nanosystems like microbubbles (1–8μm spheres) are employed for controlled drug release by partially damaging the membrane integrity upon stimulation [15]. Jablonowski et al. optimized US microbubbles with different shell compositions targeting the tumor cells sensitive (MDA-MB-231) and resistant (MCF7) to Tumor necrosis factor (TNF)-related apoptosis-inducing ligand (TRAIL) by combining with DOX. Nanoshards, upon stimulation by the US, shattered into large numbers of fine particles with a greater surface area and predominant apoptosis [16]. Fan et al. employed the mechanics of US-activated microbubbles loaded with DOX conjugated to iron oxide nanoparticles. These microbubbles promoted targeted drug delivery by invading the blood-brain barrier through US and magnetic resonance imaging [17]. Lee et al. employed the US-responsive chlorin e6 expressing microbubbles for the precise delivery of the DOX enclosed in human serum albumin nanoparticles. These nanoparticles, upon US focusing, reported targeted release by cavitation together with extravasation of DOX nanoparticles and chlorin e6 liposome by sonoporation [18]. In vivo studies on oxygen-enclosing microbubbles designed by Eisenbrey et al. in mice with human breast adenocarcinoma tumors to alleviate the radio and chemoresistance anticipated from tumor hypoxia [19].

Polymeric nanoparticles exhibiting excellent US responsiveness were explored by Wu et al., who designed curcumin-loaded pluronic P123/F127 polymeric micelles where US triggers drug release associated with targeted action and longer circulation [15, 20]. Wei et al. designed US and pH-responsive polymersome of poly (ethyleneoxide)-block-poly(2-(dimethylamino)ethylmethacrylate)-stat-poly(methoxyethyl methacrylate) [PEO-b-P(DEA-stat-MEMA] where MEMA is the US-responsive segment. The in vitro report confirmed endosomal escape, targeted drug release, and reduced cytotoxicity, whereas the in vivo studies highlighted the increased release rate of the US-responsive polymersomes [21]. The pH and US-responsive polypeptide-DOX nanoconjugates were designed by Wang et al. to overcome DOX resistance. The in vivo DOX-resistant breast cancer studies demonstrate the targeted release in addition to increased cellular uptake and reduced deleterious effects upon stimulation by US [22].

US-responsive liposomes are aqueous cores enclosing lipid bilayers. These nanoparticles were framed to study their drug targeting and overcoming the therapeutic side effects of the drug DOX in both in vitro (MDA-MB-231 breast cancer cell line) and in vivo (MDA-MB-231 xenograft mouse) studies by Kim and the team. The final study outcome demonstrated 56% more tumor suppression compared to the liposomal formulation with US exposure [23]. The cancer therapy is marked by sonodynamic therapy (SDT) that employs a combination of sonosensitizers and low-intensity US that, upon activation within the tumor cell, release reactive oxygen species (ROS), ensuing cellular apoptosis [24]. Lin et al. fabricated the US-responsive liposomes for US imaging and synergic sonodynamic and gas cancer therapy. This team designed an AIPH (2,2′-azobis[2-(2-imidazolin-2-yl)propane]dihydrochloride) enclosed in a liposome that generates gas bubbles and ROS upon UV stimulation in the preclinical studies, in addition to being a potent imaging aid [25]. Kheirolomoom et al. developed lysolipid-containing temperature-sensitive liposomes (LTSLs) enclosing a pH-sensitive DOX and copper complex whose release is triggered upon exposure to ultrasound. This insonication suppressed the tumor growth over 28 days in the breast cancer tumor model, with no indication of systemic toxicity [26]. Elamir et al. fabricated trastuzumab-coated liposomes (immunoliposomes) with DOX and calcein core, which, upon exposure to low-frequency ultrasound (LFUS), triggered drug release and an increased uptake by HER+ cells (SKBR3) [27].

The US-responsive phase-shift perfluorocarbon (PFC) nanodroplets of <200 nm upon activation are supposed to exhibit long circulation time and prevent ingression into the healthy tissues. Baghbani et al. worked to overcome multidrug resistance in ovarian cancer by designing a US-responsive DOX/curcumin combination in alginate nanodroplets with the release of DOX and curcumin at 28 kHz and 1 MHz, respectively. This combination could efficiently enhance the cytotoxicity of adriamycin-resistant A2780 ovarian cancer cells [28].

The hydrogels are a 3-dimensional hydrophilic network expressing four major properties: controlled drug release, biocompatibility, biodegradability, and drug-loading ability [29]. US-responsive hydrogels demonstrate sonoporation upon stimulation by US [30]. Kim et al. designed free radicals-producing mechanophores enclosed in hydrogels, which, upon activation by the high-intensity focused US (HIFU) exhibited cytotoxicity in murine breast (E0771) and melanoma (B16F10) tumor cells [31]. Huebsch and the team performed in vitro (MCF7 and MDA-MB-231 breast cell lines) and in vivo cancer studies using alginate US-responsive hydrogels embedded with mitoxantrone that suppressed tumor growth [32].

The nanobubbles constitute submicronic molecules of 150–500 nm and exhibit enhanced drug distribution and aggregation compared to the microbubbles [33]. Yin et al. fabricated US-responsive nanobubbles embedded with anti-apoptotic gene targeting siRNA (BCL-2 siRNA) and paclitaxel (PTX) for hepatocellular carcinoma. This study reported targeted delivery and tumor suppression as a result of the synergistic action of two pharmacological agents at low doses of PTX [34]. This siRNA-embedded US-responsive chitosan-shelled nanobubble was employed to silence nuclear factor E2-related factor 2 (Nrf2) in the treatment of melanoma by Argenziano and the team. The study reported that Nrf2 silencing in M14 cells enhanced sensitivity to cisplatin treatment, aiding in bypassing melanoma chemoresistance [35]. Xie et al. designed cell-permeable peptide-camptothecin (CPP-CPT) into a nanobubble that, upon activation by US, resulted in enhanced tumor targeting and safety in both in vitro (HeLa cell line) and in vivo (xenografted HeLa cells tumor) [36].

4.1.2 Thermo-Responsive Nanocarrier System

Slight variations in the temperature trigger the thermo-responsive nanosystems. These nanosystems comprise thermosensitive polymers, polymeric micelles, nanotubes, microbeads, hydrogels, and liposomes as summarized in Table 14.2 [86].

The thermo-responsive polymers exhibit a phase transition in response to temperature change. This polymeric group includes a family of poly(2-oxazoline)s (POxs), cellulosic polymers of natural origin, and poly(N-isopropyl acrylamide)s (PNIPAAMs) expressing either lower critical solution temperature (LCST) or an upper critical solution temperature (UCST) [87]. Wadajkar et al. developed poly (N-isopropylacrylamide-acrylamide-allylamine)-coated iron-oxide nanoparticles coupled with prostate cancer-specific R11 peptides having an LCST at 40 °C. The in vitro (LNCaP and PC3 pancreatic cancer cell lines) study reports a dose-dependent therapeutic activity, whereas the in vivo biodistribution reports its tumor-specific localization, providing promising theranostics for prostate cancer [37]. Thermoresponsive polymers like n-isopropyl acrylamide were designed by

Table 14.2 Summary of preclinical studies in stimuli-responsive nanosystems in cancer therapy

Formulation	Payload	In vitro/in vivo	Result	Reference
Ultrasound-responsive nanosystems				
Microbubbles	DOX	MDA-MB-231, MCF7 cell lines	Induced apoptosis	[16]
Microbubbles with iron oxide nanoparticles	DOX	C6 cell line, in vivo C6 glioma model	Targeted drug release	[17]
Chlorin e6 expressing microbubbles with Human serum albumin nanoparticles	DOX	MIA-paca-2 cells, in vivo MIA-paca-2 cells subcutaneous tumor model	Targeted drug release via cavitation and sonoporation	[18]
Microbubbles	Oxygen	Mice with human breast adenocarcinoma	Alleviate hypoxia-induced chemo and radioresistance in breast cancer	[19]
P123/F127 polymeric micelles	Curcumin	MDA-MB-231, 4T1 cell lines, 4T1 xenograft mice	Targeted drug release and longer circulation time	[20]
PEO-b-P(DEA-stat-MEMA	–	L02, HeLa cells, HeLa tumor-bearing nude mice	Endosomal escape, targeted drug release, reduced cytotoxicity, increased release rate	[21]
Polypeptide	DOX	MCF-7/ADR and MCF-7 cells, in vivo MCF-7/ADR tumor cells implanted mice	Overcome DOX resistance in breast cancer	[22]
Liposome	DOX	MDA-MB-231& MDA-MB-231 mouse xenograft	Overcome the therapeutic side effects of DOX	[23]
Liposome	AIPH	MCF-7 cells, MCF-7 tumor xenograft nude mice	Study sonodynamic and gas cancer therapy	[25]
Temperature sensitive-liposome	DOX& copper complex	NDL cells, NDL-tumor-bearing mice	Tumor growth suppression in breast cancer	[26]

(continued)

Table 14.2 (continued)

Formulation	Payload	In vitro/in vivo	Result	Reference
Trastuzumab-coated liposomes	DOX & calcein	SKBR3 cell line	Drug release and increased cellular uptake	[27]
Perfluorocarbon nanodroplets	DOX/Curcumin	A2780 Cell line	Overcome multidrug resistance	[28]
Hydrogel	Free radicals producing mechanophores	E0771, B16F10	Cytotoxicity in breast and melanoma	[31]
Hydrogel	Mitoxantrone	MCF7, MDA-MB-231 and in vivo	Tumor suppression	[32]
Nanobubbles	BCL-2 siRNA, PTX	HepG2 cell, in vivo *HepG2* tumor-induced nude mice	Tumor suppression and targeted delivery	[34]
Nanobubbles	Nrf2 siRNA, Cisplatin	M14	Bypassing melanoma chemoresistance	[35]
Nanobubbles	CPP-CPT	HeLa cell line, HeLa xenograft	Tumor targeting	[36]
Thermo-Responsive Nanosystems				
PNIPAAM-coated iron nanoparticles	Prostate cancer-specific R11 peptides	LNCap, PC3 and in vivo xenograft	Dose-dependent activity, diagnostics	[37]
n-isopropyl acrylamide	DOX	Morris hepatoma 7777 cell line in vitro, in vivo induced xenograft	Hyperthermia-induced drug release and targeted drug release	[38]
Indocyanine green liposomes	Polyinosinic: polycytidylic acid containing TLRs	CT-26, B16 cell line induced in vitro, in vivo BALB/c and C57BL/6 mice tumor models, respectively	Immuno-photothermal therapy	[39]
Liposomes	GEM	MiaPaCa-2 cell line in vitro, in vivo induced tumor model	Pancreatic cancer therapy	[40]

Liposomes	Cisplatin	MDA-MB-231, A549,SKOV-3 and lung cancer animal model	Increased cytotoxicity and cellular uptake	[41]
Liposomes	LAT1	HeLa	Cellular uptake above LCST	[42]
HACPN hydrogel	Graphene oxide conjugated DOX	MCF-7, Estrogen-responsive MCF-7 xenograft tumor model	Controlled drug release in breast cancer	[43]
PCNA hydrogel	Gold nanorods	3T3 cells, 4T1 cells-induced breast cancer models	Targeted drug release and reduced tumor recurrence	[44]
PDLLA-PEG-PDLLA, PLEL hydrogel	CTX	CT26	Anti-tumor immune responses	[45]
CPN hydrogel	CPT-11, CET, shRNA	EGFR overexpressing U87, in vivo studies	Glioblastoma therapy	[46]
Magnetic-responsive nanosystems				
Fe_3O_4 nanoparticles	DOX	Human melanoma cells induced in vivo tumor model	Hyperthermia-induced tumor suppression in multiple melanoma	[47]
Fe_5C_2@Fe3O_4 nanoparticles	–	4T1 293T, L02 and L929 cells, 4T1induced tumor model	Tumor suppression and improved MRI imaging	[48]
Chitosan-coated SPIONs	Bortezomib	SiHa, HeLa	Upregulation of pro-apoptotic, anti-apoptotic, p53 genes	[49]
Polydopamine coated SPION	–	HeLa, HepG2	Cell apoptosis upon NIR irradiation	[50]
SPION	Epirubicin	WM266, HaCaT	Dose-dependent tumor suppression	[51]

(continued)

Table 14.2 (continued)

Formulation	Payload	In vitro/in vivo	Result	Reference
Liposomes	DOX	MCF-7	Antiproliferative action	[52]
Liposome coated with Fe_3O_4	CET	U87	To recognize primary glioblastoma	[53]
Fe3O4/P(NIPAM-AA-MAPEG) magnetic nanogel	DOX with Herceptin antibody	Breast cancer cell lines	Targeted drug release and tumor diagnosis	[54]
Alginate-coated SPION nano gel	DOX	HepG2 and LO2 cells	Tumor imaging	[55]
Iron oxide nanogels	DOX, MTX	MCF7, MDA-MB-231	Cytotoxic activity	[56]
Light-responsive nanosystems				
Visible light-sensitive chitosan nanoparticles	5-FU	HepG2	Cytotoxic activity	[57]
UV light-sensitive liposome-mediated AuNPs	Vincristine	HeLa, cervical carcinoma-bearing BALB/c mice	Antitumor activity	[58]
pH-Responsive Nanosystems				
PDPL polymeric micelle	SPIONs and PTX	Bel-7402	Lysosomal-induced apoptosis and MRI sensitivity	[59]
Poly(ethyleneglycol)-b-polycarbonate-b-oligo([R]-3-hydroxybutyrate) Copolymer	DOX and 8-hydroxyquinoline glucose and galactose	HCT-116 and MCF-7 cell	Study cytotoxic activity	[60]
DOPE composed Liposome	DOX	MCF7, MDA-MB-231, MDA-MB-435, and MDA-MB-435S mouse model	pH-controlled drug release, targeted delivery, increased cellular uptake, and cytotoxicity	[61]
Liposome	Afatinib	H-1975, HCC-827	Drug release kinetics and Anti-tumor action	[62]
DTCPP-conjugated MSNs	DOX	$\alpha_v\beta_3$-positive HeLa cell line	Cytotoxic activity	[63]

MSNs	TOP	MC3T3-E1, LNCaP cell lines	Enhanced tumor cell affinity and cytotoxicity	[64]
Mesoporous silica-chitosan-gold aptamer nanosystem	Curcumin	MUC-1-positive tumor cells than MUC-1-negative cells	Targeted drug release	[65]
MSNs	RB & HA	4T1	Enhanced pH-sensitivity and tumor targeting	[66]
MSNs	DOX	SKBR3, MCF7	Tumor targeting	[67]
Enzyme-Responsive Nanosystems				
Polymeric micelles	Cabazitaxel	In vitro, and prostate cancer mice	Tumor suppression	[68]
Co-polymer	DOX	4T1 cell line, 4T1 induced tumor model	Cytotoxic effects	[69]
Dendrimer	DOX	SKOV-3 xenograft models	Drug pharmacokinetics study	[70]
Dendrimer	GEM	4T1 cell line, 4T1 murine breast cancer model	Drug pharmacokinetics study	[71]
Phospholipase C-responsive phosphoramidite dendrimers	DOX	In vitro MCF-7 and adriamycin resistant cell line MCF-7/ADR and in vivo Nude mice bearing xenografts of MCF-7/ADR breast cancer	Cytotoxicity studies	[72]
Polysaccharide-based hydrogel	Imatinib	Colon cancer cell lines and in vivo models	Enhanced drug availability studies	[73]
Lipid nanocarrier	TMSP	HT-1080, A549, KB	Improved efficacy, specificity, and tumor-suppressive action, along with longer circulation time	[74]

(continued)

Table 14.2 (continued)

Formulation	Payload	In vitro/in vivo	Result	Reference
β-cyclodextrin (β-CD) MMP-2 responsive liposome	Pirfenidone and GEM	Panc-1, COS7, and Pancreatic stellate cells, Panc-1 xenograft models	Antitumor activity	[75]
Phospholipase A2- enzyme-sensitive nanosystems	Colchicinoid	PANC-1, BxPC-3, COLO-357, and HaCaT cell line	Improved antiproliferative activity	[76]
Redox-responsive nanosystems				
Micelles	CPT	L929 cell line	Increased solubility, reduced cytotoxicity, and enhanced specificity	[77]
Arginine/glucose oxidase shell composed micelle	DOX	WSU HN6, HaCaT cell lines & WSU HN6 xenograft mice	Overcome multidrug resistance	[78]
Heparosan-deoxycholic acid nanoscale micelle	DOX	FaDu, COS-7 cell lines	Drug sensitivity studies	[79]
PEG and poly(ε)PEG and poly(ε-caprolactone) micelle	DOX, Fe_3O_4	HeLa, HepG2, MCF-7, A549 & H22 tumor-bearing mice	Anti-proliferative activity	[80]
Nanosystem	DEX & IND	MCF-7/ADR cells & tumor-bearing mice	Multi-drug resistance	[81]
Albumin (ABD035)-modified micelle	PTX	MBA-MB-231 cells & mice model	Drug accumulation, retention, and uptake studies	[82]
Liposomal nanohybrid cerasome	DOX	MCF7, SMMC-7721 cells	Intracellular drug delivery and antitumor activity studies	[83]
DDA liposome	DOX	C26 cell line and mice model	Cytotoxicity studies	[84]
Phosphatidylcholine-based liposomes	Irinotecan	MCF7, A549 cell line & 4T-1mice model	Anti-tumor activity	[85]

Purushotham et al. to coat DOX-encapsulated composite magnetic nanoparticles for multi-model treatment of hepatocellular carcinoma. This study reports the hyperthermia-induced drug release and target site drug release in in vivo studies [38].

Thermo-responsive liposomes composed of Indocyanine green embedded with polyinosinic: polycytidylic acid containing toll-like receptors (TLRs) were developed by Xu et al. for cancer therapy, reporting the anti-metastatic action of this immuno-photothermal therapy [39]. Affram et al. employed thermoresponsive liposomes for the delivery of gemcitabine (GEM) for the treatment of pancreatic cancer. This team has also designed gadolinium-conjugated thermoresponsive liposomes to increase the magnetic resonance imaging (MRI) contrast of the tumor. This study reports a 65% drug release at 40–42 °C within 10 minutes and a significant accumulation in tumor tissue [40]. Shah et al. have developed thermoresponsive liposomes encapsulating Cisplatin constituted by varied hydrogenated soy phosphatidylcholine and dipalmitoyl phosphatidylcholine ratios and fixed amounts of Cisplatin, distearyl phosphoethanolamine, and polyethylene glycol$_{2000}$. The in vitro cytotoxicity studies in MDA-MB-231(breast cancer), A549 cells (lung carcinoma), and SKOV-3 (ovarian cancer) cells reported increased cytotoxicity at 39 °C along with enhanced internalization by the tumor cells. The animal studies in albino mice induced with lung cancer demonstrated enhanced treatment efficacy [41]. These thermoresponsive liposomes were further employed in understanding the cellular uptake of the L-type amino acid transporter 1 (LAT1) by Maekawa-Matsuura and the team. The team developed egg phosphatidylcholine and dioleoyl phosphatidylethanolamine composed liposome conjugated with L-tyrosine-conjugated poly(N-isopropylacrylamide-co-N,N-dimethylacrylamide) a LAT1 targeting thermoresponsive polymer. The in vitro HeLa cancer model reports the cellular uptake of these liposomes at temperatures above LCST [42].

Thermoresponsive hydrogels express phase transition from liquid at >25 °C to gel at <37 °C, i.e., they exhibit solubility below LCST, and above this, they express increased insolubility, forming the gel. These thermoresponsive hydrogels provide spatial and temporal control over the drug release, increasing the therapeutic index of the desired drug molecules [88]. To enhance the cellular uptake of the chemotherapeutics, taking advantage of the pH-sensitive drug release of graphene oxide (GO), Fong et al. designed folic acid conjugated GO with DOX (GOFA-DOX), followed by hyaluronic acid-chitosan-g-poly(N-isopropyl acrylamide) [HACPN] hydrogel encapsulation. This study reported a targeted and controlled drug release in the preclinical breast cancer studies [43]. To shut out post-operative breast cancer recurrence, biodegradable methoxylpoly(ethylene glycol)-poly(ε-caprolactone)-acryloyl chloride (PECA)/glycidyl methacrylate chitooligosaccharide (COS-GMA)/N-isopropyl acrylamide (NIPAm)/acrylamide (AAm) (PCNA) hydrogels encapsulating gold nanorods (GNR) with DOX (DOX-PCNA-GNR) complex were developed by Ying Qu et al. reported the targeted drug release upon stimulation with near-infrared radiation (NIR) and reduced tumor relapse [44]. Yang and the team demonstrated the design of a cancer vaccine using cyclophosphamide (CTX) enclosed thermoresponsive poly (D, l-lactide)-poly (ethylene glycol)-poly (D, l-lactide) (PDLLA-PEG-PDLLA, PLEL) hydrogel system by injecting CTX-hydrogel into mice with colorectal carcinoma (CT26 cell) in combination with an injection of hydrogel enclosing cytosine-phosphate-guanine oligonucleotide with tumor lysates promoting anti-tumor immune responses. This study reported decreased CTX-cytotoxicity and decreased tumor growth [45]. Lu et al. studied the targeted drug delivery of thermoresponsive hydrogel for glioblastoma therapy by designing a polymer solution composed of irinotecan (CPT-11) bound to cetuximab (CET)-conjugated GO (GO-CET/CPT11), stomatin-like protein 2 (SLP2) short hairpin RNA (shRNA), and chitosan-g-poly(N-isopropylacrylamide) (CPN) that is studied in both in vitro (epidermal growth factor receptor [EGFR] overexpressing U87 cell line) and in vivo studies [46].

4.1.3 Magnetic-Responsive Nanocarrier System

Magnetic-responsive nanosystems promote tumor targeting by the intrinsic response to the magnetic field in addition to drug release as a result of altering the current magnetic field (ACMF)-triggered hyperthermia. The magnetic responsive nanocarriers include magnetic nanoparticles, polymeric micelles, superparamagnetic iron-oxide nanoparticles (SPIONs), magnetic nanogels, and liposomes, as summarized in Table 14.2 [89].

Magnetic nanoparticles involved in drug delivery studies in multiple myeloma were performed by formulating Fe_3O_4 with DOX, with a rise in temperature above 44 °C, concerning ACMF-induced hyperthermia, resulting in tumor suppression [47]. Yu et al. took advantage of the H_2O_2 excess associated acidic tumor microenvironment for the generation of ROS using iron carbide (Fe5C2@Fe3O4) nanoparticles, reporting enhanced tumor suppression in both in vitro and in vivo studies, along with improved T_2-weighted magnetic resonance imaging (MRI) [48]. Unsoy et al. fabricated chitosan-coated SPIONs enclosing Bortezomib that, upon treatment in SiHa and HeLa cervical carcinoma, reported an upregulation in pro-apoptotic genes (PUMA, NOXA) and anti-apoptotic genes (BCL-2, SURVIVIN, cIAP-2) along with an increase in p53 tumor suppressor gene [49].

Photothermal therapy (PTT) has recently been employed in cancer therapy, where light energy is converted into heat energy [90]. The SPION coated with poly(dopamine) is directed as a theranostic agent with a combination of PTT and MRI imaging properties by Wu and the team. This in vitro study (HeLa, HepG2) reported that the outer poly(dopamine) coating induced apoptosis upon NIR irradiation, whereas the inner core SPION produced MRI relaxivity along with excessive cellular uptake upon stimulation by an external magnetic field [50]. The SPION-conjugated epirubicin transdermal formulation was fabricated by Rao et al., who reported an enhanced cellular uptake of this combination in melanoma (WM266) and human keratinocyte (HaCaT) cell lines, along with dose-dependent tumor suppression and deep tissue penetration driven by external magnetic stimuli [51].

Magnetic-responsive liposomes were therapeutically applied for tumor targeting and understanding drug efficacy studies. Lin et al. complexed cell-penetrating peptides conjugated DOX with thermosensitive magnetic liposomes that exhibit enhanced anti-proliferative activity in the MCF-7 cell line, along with anti-tumor activity in the MCF-7 xenograft model [52]. A conjugated hybrid nanosystem of magnetic and thermal dual activity was designed by Lu et al. that is composed of thermosensitive magnetic liposomes (TML) encapsulating Camptosar (CPT11) along with citric acid-coated Fe_3O_4 within the inner core and outer surface conjugated Cetuximab (CET) [TML-CPT11-CET] to recognize EGFR in primary glioblastoma. The preclinical studies report enhanced uptake of the TML-CPT11-CET by the U87 cells and enhanced drug biodistribution with temperature rise [53].

Nanogels signify a promising way of targeting and controlling the release of anticancer drugs, offering stability and a small particle size. Nanogels of superparamagnetic compounds respond to externally applied magnetic fields, promoting drug transport and MRI imaging [91]. Zhang et al. designed a magnetic nanogel possessing theranostic action in the treatment and diagnosis of breast cancer. They formulated iron oxide/poly(N-isopropyl acrylamide-propyl acrylic acid-methacrylate polyethyleneglycol [Fe3O4/P(NIPAM-AA-MAPEG)] magnetic nano gels (MNLs) comprising DOX conjugated to Herceptin antibody, which aided in targeted drug release and tumor diagnosis via MRI imaging [54]. Peng et al. fabricated an ideal theranostic approach composed of an alginate-coated SPION nano gel-DOX combination where the alginate coating acts in response to acidic p^H and high glutathione concentration of tumor microenvironment, safeguarding the SPION properties for MRI imaging [55]. Salehi et al. designed 2 stimuli-responsive iron oxide nano gels [poly(N-isopropylacrylamide-dimethylaminoethyl methacrylate quaternary ammonium alkyl halide-methacrylic acid-hydroxyethyl methacrylate) P(NIPAAm-DMAEMAQ-MAA-HEMA) and poly(N-isopropylacrylamide-dimethylaminoethyl methacrylate quaternary ammonium alkyl halide-methacrylic acid) P(NIPAAm-DMAEMAQ-MAA)] copolymers embedded with DOX and Methotrexate (MTX). This nanogel exhibits pH sensitivity along with temperature and supramagnetic properties, providing maximum cytotoxicity to MCF7 and MDA-MB-231 cell lines [56].

4.1.4 Light-Responsive Nanosystems

Light energy is considered to be one of the external stimuli for controlled drug delivery. This includes the use of visible or ultraviolet-sensitive nanocarriers but is marked with few pitfalls like minimum tissue penetration, and phototoxicity that provoked the use of NIR displaying improved tissue penetration, controlled drug release, and lower phototoxicity. The light-responsive therapy where the quantum energy is converted to thermal energy upon NIR irradiation, is termed photothermal therapy (PTT). In contrast, the release of ROX upon NIR irradiation upon delivery of photodynamic agents from nanocarriers is termed photodynamic therapy (PDT). A diverse range of nanocarriers, including gold nanoparticles, transition metal dichalcogen++ides, carbon nanomaterials, and polymeric nanomaterials, were employed as drug carriers in light-responsive nanosystems as described in Table 14.2 [92].

Gold nanoparticles are nowadays considered to be an attractive application for cancer therapy as they are characterized by a marked peak resonance upon excitation at specific light wavelengths [93]. A combination of Chemo and PTT was put forth by Salem and the team, where they prepared chitosan nanoparticles (ChNPs) enclosing 5-fluorouracil (5-FU) loaded by gold nanoparticles (AuNPs), exhibiting synergistic cytotoxic activity on HepG2 (hepatocellular carcinoma) in vitro assays upon excitation by visible light [57]. Liposomes mediating the delivery of AuNPs encapsulating a vinca alkaloid named vincristine were fabricated by Liu et al. to improve the site specificity and drug release kinetics. This study reported elevated intracellular uptake along with vincristine antitumor activity upon UV light illumination in both in vitro HeLa cells (cervical carcinoma) and in cervical carcinoma-bearing BALB/c nude mice, compared to vincristine sulfate [58].

4.2 Internal Stimuli-Responsive Nanosystems

4.2.1 pH-Responsive Nanosystems

The pH-responsive nanosystems were acknowledged for their characteristic disease imaging and therapeutic approaches [94]. The pH-responsive nanocarriers include polymeric micelles, liposomes, and mesoporous silica nanoparticles, as summarized in Table 14.2 [95].

Polymeric micelles are marked with a hydrophilic outer shell that encapsulates hydrophobic drugs, protecting them from the harsh external environment within a 10–200 nm size [96, 97]. Xiao et al. fabricated a biodegradable poly(amino acid)-PEG block copolymer mPEG-PAsp(DIP)-co-PLLeu (PDPL) assembled into a micelle loaded with SPIONs and PTX that were justified as an approach delivering SPION and PTX in the lysosomal compartment, inducing apoptosis. At the same time, SPION enhances the MRI sensitivity when treated in Bel-7402 hepatocarcinoma cell lines [59]. Dominski et al. developed a 25 nm triblock p^H-responsive copolymer structured with poly(ethyleneglycol)-b-polycarbonate-b-oligo([R]-3-hydroxybutyrate), which is embedded with DOX and 8-hydroxyquinoline glucose and galactose. The in vitro cytotoxicity studies in HCT-116 and MCF-7 cell lines exhibited significant cytotoxic results [60].

The pH-responsive liposomes are biocompatible, nontoxic, and biodegradable nanoparticles enabling in replicating of physiological membranes, thereby promoting cellular uptake [97]. The concept of decreased tumor cell p^H encouraged Zhai et al. to design a p^H-responsive liposome composed of Dvar7 (DACEEQNOWARYLEWLFPTETLLLEL) an acid-sensitive peptide with 1,2-diol eoyl-sn-glycerol-3-phosphoethanolamine (DOPE) enclosing DOX. The in vitro breast cancer studies in MCF-7, MDA-MB-231, and MDA-MB-435, and in vivo MDA-MB-435S mouse model confirmed the pH-responsive controlled drug release, targeted drug delivery, and increased cellular uptake, along with enhanced cytotoxicity and tumor suppression [61]. Almurshedi et al. performed Non-small-cell lung carcinoma (NSCLC) targeting by designing cationic and pH-sensitive liposomes encapsulating Afatinib. Human lung cancer cells (H-1975 and HCC-827) were investigated to understand the drug release characteristics, which were reported as sustained drug release at p^H 7.5 and fast drug release at p^H 5.5, in addition to eminent antitumor activity [62].

pH-responsive mesoporous silica nanoparticles (MSN) were incorporated for cancer therapy for their superior characteristics related to enhanced drug solubility and drug retention at the tumor site. Cheng and the team, taking into consideration these noteworthy aspects, designed DOX encapsulating mesoporous nanoparticles with tumor-targeted fusion peptide conjugated with 2,3-dimethylmaleic anhydride (DTCPP) and a therapeutic peptide conjugated with 2,3-dimethylmaleic anhydride (DTPP) (DOX@MSN-ss-DTPP&DTCPP). This study reports a positive association of DOX@MSN-ss-DTPP&DTCPP with cytotoxicity in $\alpha_v\beta_3$-positive HeLa cell lines [63]. Martínez-Carmona et al. employed p^H-sensitive MSN for the targeted delivery of topotecan (TOP) to overcome drug inactivation at physiological p^H, promoting the drug release at acidic cellular organelles, stating that the in vitro studies report enhanced tumor cell affinity and cytotoxicity [64]. Esmaeili et al. developed a nano theranostic system using mesoporous silica-chitosan-gold (MCM@CS@Au) aptamer nanosystem encapsulating curcumin. This system acts as an on/off biosensor that is the DNA aptamer upon binding to the MUC-1 receptor of tumor cells, leading to fluorescence and targeted drug release [65]. These MSNs were employed in chemo-photodynamic combination therapy by Chen et al., where hollow MSNs (HMSNs) encapsulating DOX in combination with Rose bengal (RB) (photosensitizer) and hyaluronic acid (HA) [RB-DOX@HMSNs-N = C-HA] were developed. This combination exhibited good drug-loading capability in addition to high p^H sensitivity and target selectivity when examined in 4T1 cells [66]. The aptamer-conjugated p^H-sensitive MSNs were employed by Shen et al. for targeting HER2-positive breast cancer. The team reported positive in vitro results depicting targeted release of DOX and enhanced cytotoxicity in SKBR3 cells overexpressing HER2 compared to the MCF7 cell line [67].

4.2.2 Enzyme-Responsive Nanosystems

Enzyme-responsive nanosystems exhibit altered physiochemical reactions aided by enzymes, the constituents of the biotechnology toolbox [98, 99]. Enzyme classes like oxidoreductases, proteases, and phospholipases were involved in designing enzyme-responsive nanosystems. This class counts in several nanoscaled drug delivery systems, including polymers, liposomes, organic molecules, and organic/inorganic hybrid molecules, as summarized in Table 14.2 [99].

The enzyme-responsive polymeric nanosystems constitute micelles, inorganic polymeric hybrids, hydrogels, and dendrimers [99]. Of these polymeric nanosystems enzyme-sensitive polymeric micelles enclosing cabazitaxel were fabricated by Barve and the team as a therapeutic agent for prostate cancer. This formulation, constituted with PEG and an enzyme-responsive peptide, is cleaved by MMP-2 overexpressed in prostate cancer, upon which cabazitaxel is released. The in vivo study reports tumor suppression in prostate cancer mice [68]. Further, a p^H- and enzyme-responsive copolymer conjugated to DOX was fabricated by Wei et al. as a therapeutic delivery system in breast cancer. The preclinical in vitro 4T1 cell line cytotoxicity studies report improved DOX accumulation in tumor cells, whereas the in vivo 4T1 tumor models report fewer toxic effects [69]. In another study, an enzyme-responsive mPEGylated peptide dendrimer-tetrapeptide sequence-DOX (dendrimer-GFLG-DOX) nanocarrier was designed to improve the drug delivery for ovarian carcinoma. The in vivo study results demonstrated enhanced drug maintenance and accumulation and improved imaging in SKOV-3 xenograft models [70]. With more emphasis on the targeted and optimum drug delivery, Zhang et al. developed a peptide dendrimer-GEM prodrug conjugate. The in vitro study reported improved release kinetics of GEM specifically in cathepsin B-expressing breast cancer cells [71]. Multidrug resistance has been considered to be an important drawback for impaired clinical outcomes, which was targeted to overcome by Zhang et al. by designing phospholipase C-responsive phosphoramidate dendrimers. The preclinical cytotoxicity studies reported improved DOX circulation time and drug release, resulting in overcoming drug resistance [72]. Chemotherapeutic drugs like Imatinib, though efficient anticancer agents, exhibit poor solubility; therefore, admixture with mesylate salt that results in high dose requirements, limiting its therapeutic outcome. To overcome this, Wang and the team designed an intestine enzyme-responsive polysaccharide-based hydrogel for oral delivery of imatinib against colon cancer. The in vitro studies reported equivalent lethality in both Imatinib hydrogel and Imatinib mesylate salt, whereas the in vivo studies reported a six-fold higher tumor suppression with hydrogel [73].

Concerning low drug penetration and identification of chemotherapeutics for malignant tumors, Gao et al. designed a folic acid and tumor microenvironment-sensitive polypeptide (TMSP) modified lipid nanocarrier targeting overexpressing folate receptor and MMP-2. The HT-1080, A549, and KB cell line studies reported improved efficacy, specificity, and tumor-suppressive action along with a lengthy circulation time [74]. Further research by Ji et al. on the β-cyclodextrin (β-CD) MMP-2 responsive liposome constituted with pirfenidone and GEM as a chemotherapeutic agent against pancreatic cancer. In vitro studies on pancreatic cancer cells (Panc-1), COS7, and in vivo pancreatic stellate cells, Panc-1 xenograft models reported increased drug sustenance and penetration with no observed side effects [75]. An elaborative physicochemical and enzymological study performed by Shchegravina et al. involving phospholipase A2- enzyme-sensitive nanosystems reported enhanced anti-proliferative activity of colchicinoid in PANC-1, BxPC-3, COLO-357 pancreatic cell lines and HaCaT (nontransformed immortalized keratinocytes) [76].

4.2.3 Redox-Responsive Nanosystems

Redox-responsive nanosystems primarily work on compounds expressing disulfide bonds that are reduced to sulfhydryl (-SH) groups by the glutathione (GSH) present at high concentrations within the tumor microenvironment. This nanosystem includes micelles, nanoparticles, liposomes, and nanogels as summarized in Table 14.2 [100].

Redox and ROS-responsive CPT embedded nanosystem triggering drug release at high GSH and H_2O_2 respectively, designed by Chiang and the team to protect CPT activity. The CPT-loaded micelles exhibited increased solubility, reduced cytotoxicity along with enhanced specificity in the L929 cell line [77]. To overcome the multidrug resistance, Jin et al. fabricated a tumor-selective multifunctional micelle composed of an arginine/glucose oxidase shell, DOX core, and redox-responsive linker disulfide bond against oral and pharyngeal cancer. The in vitro study reports on WSU HN6 oral and pancreatic cells and the HaCaT human epidermal keratinocyte cell line depicted controlled drug release and enhanced cytotoxic effect. The in vivo WSU HN6 xenograft mice exhibited superior cytotoxicity and tumor specificity [78]. Sun et al. designed heparosan-deoxycholic acid nanoscale micelle embedded with DOX against laryngopharyngeal cancer. This team reported that the cell line studies reported tumor suppression in association with ten-fold selectivity for FaDu tumor cells compared to normal COS-7 cells [79]. To achieve efficient delivery of chemotherapeutic agents, Tang and the team brought about a magnetic star-shaped micelle made of PEG and poly(ε)PEG and poly(ε-caprolactone) with disulfide linkers. This nanosystem was embedded with DOX and Fe_3O_4 along with phenylboronic acid (PBA) ligand encapsulation. This study reports dual and active tumor targeting and excellent anti-proliferative activity of in vitro salic acid-positive HeLa, HepG2, and salic acid-negative MCF-7, A549 cell studies, along with tumor suppressive activity in H22 tumor-bearing mice [80]. To counter the multidrug resistance (MDR) of tumor cells, Zhou et al. came up with a redox-responsive dextran (DEX) and indomethacin (IND) based nanosystem to reduce MDR-associated proteins (MRPs) chemotherapeutic efflux. The preclinical studies reported significantly high standard stability, cellular bioavailability, targeting, cytotoxicity in the MCF-7/ADR cells, and tumor suppression in MCF-7/ADR-bearing mice [81]. Zhang et al. developed a tumor cell-targeting albumin (ABD035)-modified micelle embedded with disulfide-containing spacer conjugated PTX delineating the enhanced drug retention, accumulation, and uptake in both in vitro (MBA-MB-231 cells) and in vivo (MBA-MB-231cells induced mice) triple-negative breast cancer [82].

Liposomes, though, protect the hydrophobic drug molecules from degradation; they exhibit poor stability and leakage, limiting their usage. To overcome this, Zhou et al. synthesized a redox-responsive liposomal nanohybrid cerasome (RRC) composed of cleavable disulfide bonds and DOX. RRC/DOX expressed cytotoxicity in MCF-7 breast cancer cell lines and SMMC-7721 human hepatoma cells, promoting intracellular drug delivery and antitumor activity [83]. Mirhandi and the team employed redox-responsive 10,10′-diselanediylbis decanoic acid (DDA) liposome as a DOX carrier that reported increased cellular uptake and cytotoxicity in C26 (murine colon cancer cells) cell lines, as well as in C26 tumor-bearing mice [84]. Wang and the team worked on the redox-responsive disulfide phosphatidylcholine-based liposomes for improved irinotecan delivery to the tumor site. The in vitro studies reported irinotecan-induced cytotoxicity in MCF-7 and A549 cells and antitumor activity in 4T-1 mice at 5 mg/kg irinotecan [85].

5 Future Perspectives and Challenges for Preclinical Studies on Stimuli-Responsive Nanosystems

As described in the earlier chapters, stimuli-responsive nanosystems bring about a revolutionalized frontier in the field of nanotechnology associated with diagnosis and controlled drug delivery, and therapy. The advancing research pushes the boundaries of nanotechnology, unfolding the potential of these stimuli-responsive nanosystems, their future perspectives, and challenges emerging in preclinical studies. The introduction of versatile multi-stimuli-responsive nanosystems at cellular, tissue, extracorporeal, and subcellular levels that overcame the physiological and pathological barriers enhanced the drug delivery and real-time drug pharmacokinetic monitoring. In addition to the above-stated advantages, this technology offers a promising platform for the reversal of multidrug resistance and co-delivery of therapeutic agents [101]. The advancing imaging technologies ideally play a prominent role in preclinical studies. Researchers employ cutting-edge imaging technologies like Magnetic resonance imaging (MRI), Photoacoustic (PA) imaging, Positron emission tomography (PET), and Computed tomography (CT) to precisely track and visualize multifunctional stimuli-responsive nanoparticles in vivo, enabling the effective monitoring of therapeutic management [101].

6 Challenges

Ensuring the long-term stability of stimuli-responsive nanosystems, both in terms of physical integrity and responsiveness, poses a challenge. Several inorganic nanoparticles are employed for the synthesis and development of stimuli-responsive nanosystems, but exhibit a relatively low margin of safety, providing an outstanding challenge for drug delivery [102]. The physicochemical properties of the nanomaterials, like the size, charge, composition, and biocompatibility, affect cellular adhesion, interaction, and therapeutic or toxic effects, paving the way for the engineering of stimuli-responsive nanosystems for therapeutic applications [103]. In addition to earlier challenges, tissue penetration, controlled drug release, loading capacity, biodegradability, biocompatibility, and bioavailability also need to be addressed [104].

Another challenge in drug delivery is the nanotoxicity arising from undesired toxic effects upon unintended interactions with biological entities. Lack of clinical translation is one of the prominent setbacks in stimuli-responsive nanotechnology as a result of the poor understanding of the nano-bio interfacial interactions. Speaking about the commercial large-scale manufacturing of nanoparticles is quite challenging as a result of marked variations in the nanoformulation physiochemical characteristics with every batch, in addition to the complex manufacturing process of these nanotherapeutics. Therefore well-designed manufacturing process is essential. In addition to the above-mentioned offsets, the regulatory approval of these nanotherapeutics is very challenging as there are no predefined guidelines set by the Food and Drug Administration (FDA) [103]. To address these challenges associated with stimuli-responsive nanoparticles, understanding and developing the strategies associated with oncotherapeutics, pharmacological aspects, tumor cell physiology, and microenvironment is essential.

7 Conclusion

In light of the above information, the preclinical studies on stimuli-responsive nanosystems in cancer therapy provided an opportunistic and promising strategy for revolutionizing the landscape of oncotherapeutics. The innovative design of the stimuli-responsive nanocarriers, which respond to the tumor microenvironment-specific stimuli, paves the way for a wide range of applications in diagnostics and therapeutics. As discussed in the earlier sections, the stimuli-responsive nanosystem is modulated in response to external and internal stimuli like ultrasound, thermal, magnetic, electroresponsive, light, and redox potential, enzymatic modifications, and pH, respectively. This target-specific action demonstrates exceptionally remarkable tumor cell sensitivity, promoting tumor-specific release of therapeutic payloads.

The evidence gathered from the preclinical studies highlighted the ability of the stimuli-responsive nanosystems to overcome the setbacks with conventional oncotherapeutics, such as off-target effects and inadequate drug delivery to the target tissue. Moreover, the preclinical studies have demonstrated the potential of stimuli-responsive nanosystems to enhance the therapeutic index by decreasing the therapeutic adverse effects. This targeted approach model improved the treatment outcomes, improving the patient's quality of life.

While the preclinical results are quite promising, there is also a need for acknowledging further research to address challenges associated with drug release kinetics by different nanoformulations like polymers, hydrogels, nanotubes, and liposomes, along with scalability, long-term stability, and clinical translation. Continued exploration of the translational potential of these nanosystems will be crucial for validating their efficacy and safety in clinical settings.

In summary, the preclinical findings on stimuli-responsive nanosystems in cancer therapy underscore their potential as a groundbreaking strategy for improving the precision and effectiveness of cancer treatments. As we move forward, bridging the gap between preclinical research and clinical applications will be vital to realizing the full therapeutic benefits of these innovative nanotechnologies in the fight against cancer.

References

1. Qin X, Li Y. Strategies to design and synthesize polymer-based stimuli-responsive drug-delivery nanosystems. ChemBioChem [Internet]. 2020 May 4 [cited 2023 Dec 6];21(9):1236–53. Available from: https://onlinelibrary.wiley.com/doi/full/10.1002/cbic.201900550.
2. Ayana G, Ryu J, Choe SW. Ultrasound-responsive nanocarriers for breast cancer chemotherapy. Micromachines. 2022 Sep 11 [cited 2023 Oct 30];13(9):1508. Available from: https://www.mdpi.com/2072-666X/13/9/1508/htm.
3. Pham SH, Choi Y, Choi J. Stimuli-responsive nanomaterials for application in antitumor therapy and drug delivery. Pharmaceutics [Internet]. 2020 Jul 1 [cited 2023 Dec 6];12(7):1–19. Available from: /pmc/articles/PMC7408499/.
4. Naziris N, Pippa N, Pispas S, Demetzos C. Stimuli-responsive drug delivery nanosystems: from bench to clinic. Curr Nanomed. 2016;6(999):1.

5. Zhang J, Lin Y, Lin Z, Wei Q, Qian J, Ruan R, et al. Stimuli-responsive nanoparticles for controlled drug delivery in synergistic cancer immunotherapy. Adv Sci [Internet]. 2022 Feb 1 [cited 2023 Dec 8];9(5):2103444. Available from: https://onlinelibrary.wiley.com/doi/full/10.1002/advs.202103444.
6. Lopes CM, Barata P, Oliveira R. Stimuli-responsive nanosystems for drug-targeted delivery. In: Drug targeting and stimuli sensitive drug delivery systems. William Andrew; 2018. p. 155–209.
7. Chen Y, Chen N, Feng X. The role of internal and external stimuli in the rational design of skin-specific drug delivery systems. Int J Pharm. 2021;592:120081.
8. Mura S, Nicolas J, Couvreur P. Stimuli-responsive nanocarriers for drug delivery. Nat Mater. 2013 Oct 23 [cited 2023 Oct 19];12(11):991–1003. Available from: https://www.nature.com/articles/nmat3776.
9. Zhao X, Bai J, Yang W. Stimuli-responsive nanocarriers for therapeutic applications in cancer. Cancer Biol Med [Internet]. 2021 May 5 [cited 2023 Oct 30];18(2):319. Available from: /pmc/articles/PMC8185873/.
10. Entzian K, Aigner A. Drug delivery by ultrasound-responsive nanocarriers for cancer treatment. Pharmaceutics. 2021 Jul 26 [cited 2023 Oct 27];13(8):1135. Available from: https://www.mdpi.com/1999-4923/13/8/1135/htm.
11. Sun D, Gao W, Hu H, Zhou S. Why 90% of clinical drug development fails and how to improve it? Acta Pharm Sin B [Internet]. 2022 Jul 1 [cited 2023 Oct 20];12(7):3049–62. Available from: https://www.als.net/news/what-is-preclinical-research/.
12. Steinmetz KL, Spack EG. The basics of preclinical drug development for neurodegenerative disease indications. BMC Neurol [Internet]. 2009 Jun 12 [cited 2023 Oct 20];9(SUPPL. 1):1–13. Available from: https://bmcneurol.biomedcentral.com/articles/10.1186/1471-2377-9-S1-S2.
13. Step 2: Preclinical Research | FDA [Internet]. [cited 2023 Dec 8]. Available from: https://www.fda.gov/patients/drug-development-process/step-2-preclinical-research.
14. Cui X, Han X, Yu L, Zhang B, Chen Y. Intrinsic chemistry and design principle of ultrasound-responsive nanomedicine. Nano Today. 2019;28:100773.
15. Tu L, Liao Z, Luo Z, Wu YL, Herrmann A, Huo S. Ultrasound-controlled drug release and drug activation for cancer therapy. Exploration [Internet]. 2021 Dec 1 [cited 2023 Oct 31];1(3):20210023. Available from: https://onlinelibrary.wiley.com/doi/full/10.1002/EXP.20210023.
16. Jablonowski LJ, Conover D, Teraphongphom NT, Wheatley MA. Manipulating multifaceted microbubble shell composition to target both TRAIL-sensitive and resistant cells. J Biomed Mater Res A [Internet]. 2018 Jul 1 [cited 2023 Oct 30];106(7):1903–15. Available from: https://onlinelibrary.wiley.com/doi/full/10.1002/jbm.a.36389.
17. Fan CH, Ting CY, Lin HJ, Wang CH, Liu HL, Yen TC, et al. SPIO-conjugated, doxorubicin-loaded microbubbles for concurrent MRI and focused-ultrasound enhanced brain-tumor drug delivery. Biomaterials. 2013;34(14):3706–15.
18. Lee H, Han J, Shin H, Han H, Na K, Kim H. Combination of chemotherapy and photodynamic therapy for cancer treatment with sonoporation effects. J Control Release. 2018;283:190–9.
19. Eisenbrey JR, Shraim R, Bin LJ, Li J, Stanczak M, Oeffinger B, et al. Sensitization of hypoxic tumors to radiation therapy using ultrasound-sensitive oxygen microbubbles. Int J Radiat Oncol Biol Phys. 2018;101(1):88–96.
20. Wu P, Jia Y, Qu F, Sun Y, Wang P, Zhang K, et al. Ultrasound-responsive polymeric micelles for sonoporation-assisted site-specific therapeutic action. ACS Appl Mater Interfaces [Internet]. 2017 Aug 9 [cited 2023 Oct 31];9(31):25706–16. Available from: https://pubs.acs.org/doi/full/10.1021/acsami.7b05469.
21. Wei P, Sun M, Yang B, Xiao J, Du J. Ultrasound-responsive polymersomes capable of endosomal escape for efficient cancer therapy. J Control Release. 2020;322:81–94.
22. Wang Z, He Q, Zhao W, Luo J, Gao W. Tumor-homing, pH- and ultrasound-responsive polypeptide-doxorubicin nanoconjugates overcome doxorubicin resistance in cancer therapy. J Control Release. 2017;264:66–75.

23. Kim YS, Ko MJ, Moon H, Sim W, Cho AS, Gil G, et al. Ultrasound-responsive liposomes for targeted drug delivery combined with focused ultrasound. Pharmaceutics [Internet]. 2022 Jul 1 [cited 2023 Oct 31];14(7):1314. Available from: https://www.mdpi.com/1999-4923/14/7/1314/htm.
24. Awad NS, Paul V, Alsawaftah NM, Ter Haar G, Allen TM, Pitt WG, et al. Ultrasound-responsive nanocarriers in cancer treatment: a review. ACS Pharmacol Transl Sci [Internet]. 2021 Apr 9 [cited 2023 Oct 31];4(2):589–612. Available from: https://pubs.acs.org/doi/full/10.1021/acsptsci.0c00212.
25. Lin X, Qiu Y, Song L, Chen S, Chen X, Huang G, et al. Ultrasound activation of liposomes for enhanced ultrasound imaging and synergistic gas and sonodynamic cancer therapy. Nanoscale Horiz [Internet]. 2019 Apr 23 [cited 2023 Oct 31];4(3):747–56. Available from: https://pubs.rsc.org/en/content/articlehtml/2019/nh/c8nh00340h.
26. Kheirolomoom A, Lai CY, Tam SM, Mahakian LM, Ingham ES, Watson KD, et al. Complete regression of local cancer using temperature-sensitive liposomes combined with ultrasound-mediated hyperthermia. J Control Release. 2013;172(1):266–73.
27. Elamir A, Ajith S, Sawaftah NA, Abuwatfa W, Mukhopadhyay D, Paul V, et al. Ultrasound-triggered herceptin liposomes for breast cancer therapy. Sci Rep 2021 Apr 6 [cited 2023 Oct 31];11(1):1–13. Available from: https://www.nature.com/articles/s41598-021-86860-5.
28. Baghbani F, Moztarzadeh F. Bypassing multidrug resistant ovarian cancer using ultrasound responsive doxorubicin/curcumin co-deliver alginate nanodroplets. Colloids Surf B Biointerfaces. 2017;153:132–40.
29. Sun Z, Song C, Wang C, Hu Y, Wu J. Hydrogel-based controlled drug delivery for cancer treatment: a review. Mol Pharm [Internet]. 2020 Feb 3 [cited 2023 Nov 1];17(2):373–91. Available from: https://pubs.acs.org/doi/full/10.1021/acs.molpharmaceut.9b01020.
30. Yeingst TJ, Arrizabalaga JH, Hayes DJ. Ultrasound-induced drug release from stimuli-responsive hydrogels. Gels. 2022 Sep 1 [cited 2023 Nov 1];8(9):554. Available from: https://www.mdpi.com/2310-2861/8/9/554/htm.
31. Kim G, Wu Q, Chu JL, Smith EJ, Oelze ML, Moore JS, et al. Ultrasound controlled mechanophore activation in hydrogels for cancer therapy. Proc Natl Acad Sci U S A [Internet]. 2022 Jan 25 [cited 2023 Nov 1];119(4):e2109791119. Available from: https://www.pnas.org/doi/abs/10.1073/pnas.2109791119.
32. Huebsch N, Kearney CJ, Zhao X, Kim J, Cezar CA, Suo Z, et al. Ultrasound-triggered disruption and self-healing of reversibly cross-linked hydrogels for drug delivery and enhanced chemotherapy. Proc Natl Acad Sci U S A [Internet]. 2014 Jul 8 [cited 2023 Nov 1];111(27):9762–7. Available from: https://www.pnas.org/doi/abs/10.1073/pnas.1405469111.
33. Su C, Ren XJ, Nie F, Li T, Lv W, Li H, et al. Current advances in ultrasound-combined nanobubbles for cancer-targeted therapy: a review of the current status and future perspectives. RSC Adv [Internet]. 2021 Apr 6 [cited 2023 Nov 1];11(21):12915–28. Available from: https://pubs.rsc.org/en/content/articlehtml/2021/ra/d0ra08727k.
34. Yin T, Wang P, Li J, Wang Y, Zheng B, Zheng R, et al. Tumor-penetrating codelivery of siRNA and paclitaxel with ultrasound-responsive nanobubbles hetero-assembled from polymeric micelles and liposomes. Biomaterials. 2014;35(22):5932–43.
35. Argenziano M, Bessone F, Dianzani C, Cucci MA, Grattarola M, Pizzimenti S, et al. Ultrasound-Responsive Nrf2-Targeting siRNA-loaded nanobubbles for enhancing the treatment of melanoma. Pharmaceutics. 2022 Jan 31 [cited 2023 Nov 1];14(2):341. Available from: https://www.mdpi.com/1999-4923/14/2/341/htm.
36. Xie X, Lin W, Liu H, Deng J, Chen Y, Liu H, et al. Ultrasound-responsive nanobubbles contained with peptide–camptothecin conjugates for targeted drug delivery. Drug Deliv [Internet]. 2016 Oct 12 [cited 2023 Nov 1];23(8):2756–64. Available from: https://www.tandfonline.com/doi/abs/10.3109/10717544.2015.1077289.
37. Wadajkar AS, Menon JU, Tsai YS, Gore C, Dobin T, Gandee L, et al. Prostate cancer-specific thermo-responsive polymer-coated iron oxide nanoparticles. Biomaterials. 2013;34(14):3618–25.

38. Purushotham S, Chang PEJ, Rumpel H, Kee IHC, Ng RTH, Chow PKH, et al. Thermoresponsive core–shell magnetic nanoparticles for combined modalities of cancer therapy. Nanotechnology [Internet]. 2009 Jul 7 [cited 2023 Nov 2];20(30):305101. Available from: https://iopscience.iop.org/article/10.1088/0957-4484/20/30/305101.
39. Xu L, Zhang W, Bin PH, Kwak M, Oh J, Lee PCW, et al. Indocyanine green and poly I:C containing thermo-responsive liposomes used in immune-photothermal therapy prevent cancer growth and metastasis. J Immunother Cancer [Internet]. 2019 Aug 14 [cited 2023 Nov 2];7(1):1–14. Available from: https://jitc.biomedcentral.com/articles/10.1186/s40425-019-0702-1.
40. Affram K, Udofot O, Singh M, Krishnan S, Reams R, Rosenberg J, et al. Smart thermosensitive liposomes for effective solid tumor therapy and in vivo imaging. PLoS One [Internet]. 2017 Sep 1 [cited 2023 Nov 2];12(9):e0185116. Available from: https://journals.plos.org/plosone/article?id=10.1371/journal.pone.0185116.
41. Shah H, Madni A, Filipczak N, Jan N, Khan MM, Khan S, et al. Cisplatin-loaded thermoresponsive liposomes for enhanced anticancer efficacy. J Drug Deliv Sci Technol. 2023;84:104509.
42. Maekawa-Matsuura M, Fujieda K, Maekawa Y, Nishimura T, Nagase K, Kanazawa H. LAT1-targeting thermoresponsive liposomes for effective cellular uptake by cancer cells. ACS Omega [Internet]. 2019 Apr 8 [cited 2023 Nov 2];4(4):6443–51. Available from: https://pubs.acs.org/doi/full/10.1021/acsomega.9b00216.
43. Fong YT, Chen CH, Chen JP. Intratumoral delivery of doxorubicin on folate-conjugated graphene oxide by in-situ forming thermo-sensitive hydrogel for breast cancer therapy. Nanomaterials. 2017 Nov 14 [cited 2023 Nov 2];7(11):388. Available from: https://www.mdpi.com/2079-4991/7/11/388/htm.
44. Qu Y, Chu BY, Peng JR, Liao JF, Qi TT, Shi K, et al. A biodegradable thermo-responsive hybrid hydrogel: therapeutic applications in preventing the post-operative recurrence of breast cancer. NPG Asia Mater. 2015 Aug 14 [cited 2023 Nov 2];7(8):e207. Available from: https://www.nature.com/articles/am201583.
45. Yang F, Shi K, Hao Y, Jia Y, Liu Q, Chen Y, et al. Cyclophosphamide loaded thermo-responsive hydrogel system synergize with a hydrogel cancer vaccine to amplify cancer immunotherapy in a prime-boost manner. Bioact Mater. 2021;6(10):3036–48.
46. Lu YJ, Lan YH, Chuang CC, Lu WT, Chan LY, Hsu PW, et al. Injectable thermo-sensitive chitosan hydrogel containing CPT-11-Loaded EGFR-Targeted Graphene Oxide and SLP2 shRNA for localized drug/gene delivery in glioblastoma therapy. Int J Mol Sci. 2020 Sep 26 [cited 2023 Nov 2];21(19):7111. Available from: https://www.mdpi.com/1422-0067/21/19/7111/htm.
47. Hayashi K, Nakamura M, Miki H, Ozaki S, Abe M, Matsumoto T, et al. Magnetically responsive smart nanoparticles for cancer treatment with a combination of magnetic hyperthermia and remote-control drug release. Theranostics [Internet]. 2014 [cited 2023 Nov 3];4(8):834. Available from: /pmc/articles/PMC4063981/.
48. Yu J, Zhao F, Gao W, Yang X, Ju Y, Zhao L, et al. Magnetic reactive oxygen species nanoreactor for switchable magnetic resonance imaging guided cancer therapy based on pH-sensitive Fe5C2@Fe3O4 nanoparticles. ACS Nano [Internet]. 2019 Sep 24 [cited 2023 Nov 3];13(9):10002–14. Available from: https://pubs.acs.org/doi/full/10.1021/acsnano.9b01740.
49. Unsoy G, Yalcin S, Khodadust R, Mutlu P, Onguru O, Gunduz U. Chitosan magnetic nanoparticles for pH responsive Bortezomib release in cancer therapy. Biomed Pharmacother. 2014;68(5):641–8.
50. Wu M, Zhang D, Zeng Y, Wu L, Liu X, Liu J. Nanocluster of superparamagnetic iron oxide nanoparticles coated with poly (dopamine) for magnetic field-targeting, highly sensitive MRI and photothermal cancer therapy. Nanotechnology [Internet]. 2015 Feb 27 [cited 2023 Nov 3];26(11):115102. Available from: https://iopscience.iop.org/article/10.1088/0957-4484/26/11/115102.
51. Rao YF, Chen W, Liang XG, Huang YZ, Miao J, Liu L, et al. Epirubicin-loaded superparamagnetic iron-oxide nanoparticles for transdermal delivery: cancer therapy by circumventing

the skin barrier. Small [Internet]. 2015 Jan 1 [cited 2023 Nov 3];11(2):239–47. Available from: https://onlinelibrary.wiley.com/doi/full/10.1002/smll.201400775.
52. Lin W, Xie X, Yang Y, Fu X, Liu H, Yang Y, et al. Thermosensitive magnetic liposomes with doxorubicin cell-penetrating peptides conjugate for enhanced and targeted cancer therapy. Drug Deliv [Internet]. 2016 Nov 21 [cited 2023 Nov 3];23(9):3436–43. Available from: https://www.tandfonline.com/doi/abs/10.1080/10717544.2016.1189983.
53. Lu YJ, Chuang EY, Cheng YH, Anilkumar TS, Chen HA, Chen JP. Thermosensitive magnetic liposomes for alternating magnetic field-inducible drug delivery in dual targeted brain tumor chemotherapy. Chem Eng J. 2019;373:720–33.
54. Zhang X, Wei P, Wang Z, Zhao Y, Xiao W, Bian Y, et al. Herceptin-Conjugated DOX-Fe3O4/P(NIPAM-AA-MAPEG) nanogel system for HER2-Targeted breast cancer treatment and magnetic resonance imaging. ACS Appl Mater Interfaces [Internet]. 2022 Apr 13 [cited 2023 Nov 3];14(14):15956–69. Available from: https://pubs.acs.org/doi/full/10.1021/acsami.1c24770.
55. Peng N, Ding X, Wang Z, Cheng Y, Gong Z, Xu X, et al. Novel dual responsive alginate-based magnetic nanogels for onco-theranostics. Carbohydr Polym. 2019;204:32–41.
56. Salehi R, Rasouli S, Hamishehkar H. Smart thermo/pH responsive magnetic nanogels for the simultaneous delivery of doxorubicin and methotrexate. Int J Pharm. 2015;487(1–2):274–84.
57. Salem DS, Hegazy SF, Obayya SSA. Nanogold-loaded chitosan nanocomposites for pH/light-responsive drug release and synergistic chemo-photothermal cancer therapy. Colloid Interface Sci Commun. 2021;41:100361.
58. Liu Y, He M, Niu M, Zhao Y, Zhu Y, Li Z, et al. Delivery of vincristine sulfate-conjugated gold nanoparticles using liposomes: a light-responsive nanocarrier with enhanced antitumor efficiency. Int J Nanomedicine [Internet]. 2015 Apr 22 [cited 2023 Nov 10];10:3081–95. Available from: https://www.tandfonline.com/action/journalInformation?journalCode=dijn20.
59. Xiao H, Li X, Zheng C, Liu Q, Sun C, Huang J, et al. Intracellular pH-responsive polymeric micelle for simultaneous chemotherapy and MR imaging of hepatocellular carcinoma. J Nanoparticle Res [Internet]. 2020 May 1 [cited 2023 Nov 7];22(5):1–15. Available from: https://link.springer.com/article/10.1007/s11051-020-04821-x.
60. Domiński A, Krawczyk M, Konieczny T, Kasprów M, Foryś A, Pastuch-Gawołek G, et al. Biodegradable pH-responsive micelles loaded with 8-hydroxyquinoline glycoconjugates for Warburg effect based tumor targeting. Eur J Pharm Biopharm. 2020;154:317–29.
61. Zhai L, Luo C, Gao H, Du S, Shi J, Wang F. A dual PH-responsive dox-encapsulated liposome combined with glucose administration enhanced therapeutic efficacy of chemotherapy for cancer. Int J Nanomed [Internet]. 2021 [cited 2023 Nov 7];16:3185–99. Available from: https://www.tandfonline.com/action/journalInformation?journalCode=dijn20.
62. Almurshedi AS, Radwan M, Omar S, Alaiya AA, Badran MM, Elsaghire H, et al. A novel pH-sensitive liposome to trigger delivery of afatinib to cancer cells: impact on lung cancer therapy. J Mol Liq. 2018;259:154–66.
63. Cheng YJ, Qin SY, Ma YH, Chen XS, Zhang AQ, Zhang XZ. Super-pH-sensitive mesoporous silica nanoparticle-based drug delivery system for effective combination cancer therapy. ACS Biomater Sci Eng [Internet]. 2019 Apr 8 [cited 2023 Nov 7];5(4):1878–86. Available from: https://pubs.acs.org/doi/full/10.1021/acsbiomaterials.9b00099.
64. Martínez-Carmona M, Lozano D, Colilla M, Vallet-Regí M. Selective topotecan delivery to cancer cells by targeted pH-sensitive mesoporous silica nanoparticles. RSC Adv [Internet]. 2016 May 23 [cited 2023 Nov 7];6(56):50923–50932. Available from: https://pubs.rsc.org/en/content/articlehtml/2016/ra/c6ra07763c.
65. Esmaeili Y, Khavani M, Bigham A, Sanati A, Bidram E, Shariati L, et al. Mesoporous silica@chitosan@gold nanoparticles as "on/off" optical biosensor and pH-sensitive theranostic platform against cancer. Int J Biol Macromol. 2022;202:241–55.
66. Chen K, Chang C, Liu Z, Zhou Y, Xu Q, Li C, et al. Hyaluronic acid targeted and pH-responsive nanocarriers based on hollow mesoporous silica nanoparticles for chemo-photodynamic combination therapy. Colloids Surf B Biointerfaces. 2020;194:111166.

67. Zhang J, Shen Y, Li M, Liu T, Liu J, Xie Y, et al. A dual-functional her2 aptamer-conjugated, ph-activated mesoporous silica nanocarrier-based drug delivery system provides in vitro synergistic cytotoxicity in her2-positive breast cancer cells. Int J Nanomed [Internet]. 2019 [cited 2023 Nov 7];14:4029–44. Available from: https://www.tandfonline.com/action/journalInformation?journalCode=dijn20.
68. Barve A, Jain A, Liu H, Zhao Z, Cheng K. Enzyme-responsive polymeric micelles of cabazitaxel for prostate cancer targeted therapy. Acta Biomater. 2020;113:501–11.
69. Wei X, Luo Q, Sun L, Li X, Zhu H, Guan P, et al. Enzyme- and pH-Sensitive branched polymer-doxorubicin conjugate-based nanoscale drug delivery system for cancer therapy. ACS Appl Mater Interfaces [Internet]. 2016 May 11 [cited 2023 Nov 8];8(18):11765–78. Available from: https://pubs.acs.org/doi/full/10.1021/acsami.6b02006.
70. Zhang C, Pan D, Luo K, Li N, Guo C, Zheng X, et al. Dendrimer–doxorubicin conjugate as enzyme-sensitive and polymeric nanoscale drug delivery vehicle for ovarian cancer therapy. Polym Chem [Internet]. 2014 Aug 12 [cited 2023 Nov.8];5(18):5227–35. Available from: https://pubs.rsc.org/en/content/articlehtml/2014/py/c4py00601a.
71. Zhang C, Pan D, Li J, Hu J, Bains A, Guys N, et al. Enzyme-responsive peptide dendrimer-gemcitabine conjugate as a controlled-release drug delivery vehicle with enhanced antitumor efficacy. Acta Biomater. 2017;55:153–62.
72. Zhang Z, Zhou Y, Zhou Z, Piao Y, Kalva N, Liu X, et al. Synthesis of enzyme-responsive phosphoramidate dendrimers for cancer drug delivery. Polym Chem [Internet]. 2018 Jan 23 [cited 2023 Nov 9];9(4):438–49. Available from: https://pubs.rsc.org/en/content/articlehtml/2018/py/c7py01492a.
73. Wang CY, Sun M, Fan Z, Du JZ. Intestine enzyme-responsive polysaccharide-based hydrogel to open epithelial tight junctions for oral delivery of imatinib against colon cancer. Chin J Polym Sci (English Edition) [Internet]. 2022 Oct 1 [cited 2023 Nov 9];40(10):1154–64. Available from: https://link.springer.com/article/10.1007/s10118-022-2726-0.
74. Gao W, Meng T, Shi N, Zhuang H, Yang Z, Qi X. Targeting and microenvironment-responsive lipid nanocarrier for the enhancement of tumor cell recognition and therapeutic efficiency. Adv Healthc Mater [Internet]. 2015 Apr 1 [cited 2023 Nov 8];4(5):748–59. Available from: https://onlinelibrary.wiley.com/doi/full/10.1002/adhm.201400675.
75. Ji T, Li S, Zhang Y, Lang J, Ding Y, Zhao X, et al. An MMP-2 responsive liposome integrating antifibrosis and chemotherapeutic drugs for enhanced drug perfusion and efficacy in pancreatic cancer. ACS Appl Mater Interfaces [Internet]. 2016 Feb 10 [cited 2023 Nov 8];8(5):3438–45. Available from: https://pubs.acs.org/doi/full/10.1021/acsami.5b11619.
76. Shchegravina ES, Tretiakova DS, Alekseeva AS, Galimzyanov TR, Utkin YN, Ermakov YA, et al. Phospholipidic colchicinoids as promising prodrugs incorporated into enzyme-responsive liposomes: chemical, biophysical, and enzymological aspects. Bioconjug Chem [Internet]. 2019 Apr 17 [cited 2023 Nov 8];30(4):1098–113. Available from: https://pubs.acs.org/doi/full/10.1021/acs.bioconjchem.9b00051.
77. Chiang YT, Yen YW, Lo CL. Reactive oxygen species and glutathione dual redox-responsive micelles for selective cytotoxicity of cancer. Biomaterials. 2015;61:150–61.
78. Jin R, Liu Z, Liu T, Yuan P, Bai Y, Chen X. Redox-responsive micelles integrating catalytic nanomedicine and selective chemotherapy for effective tumor treatment. Chin Chem Lett. 2021;32(10):3076–82.
79. Sun C, Li X, Du X, Wang T. Redox-responsive micelles for triggered drug delivery and effective laryngopharyngeal cancer therapy. Int J Biol Macromol. 2018;112:65–73.
80. Tang Z, Zhang L, Wang Y, Li D, Zhong Z, Zhou S. Redox-responsive star-shaped magnetic micelles with active-targeted and magnetic-guided functions for cancer therapy. Acta Biomater. 2016;42:232–46.
81. Zhou Y, Wang S, Ying X, Wang Y, Geng P, Deng A, et al. Doxorubicin-loaded redox-responsive micelles based on dextran and indomethacin for resistant breast cancer. Int J Nanomedicine. 2017;12:6153–68.
82. Zhang Y, Guo Z, Cao Z, Zhou W, Zhang Y, Chen Q, et al. Endogenous albumin-mediated delivery of redox-responsive paclitaxel-loaded micelles for targeted cancer therapy. Biomaterials. 2018;183:243–57.

83. Zhou G, Li L, Xing J, Jalde S, Li Y, Cai J, et al. Redox responsive liposomal nanohybrid cerasomes for intracellular drug delivery. Colloids Surf B Biointerfaces. 2016;148:518–25.
84. Mirhadi E, Mashreghi M, Askarizadeh A, Mehrabian A, Alavizadeh SH, Arabi L, et al. Redox-sensitive doxorubicin liposome: a formulation approach for targeted tumor therapy. Sci Rep. 2022 Jul 4 [cited 2023 Nov 10];12(1):1–17. Available from: https://www.nature.com/articles/s41598-022-15239-x.
85. Wang T, He W, Du Y, Wang J, Li X. Redox-sensitive irinotecan liposomes with active ultra-high loading and enhanced intracellular drug release. Colloids Surf B Biointerfaces. 2021;206:111967.
86. Karimi M, Sahandi Zangabad P, Ghasemi A, Amiri M, Bahrami M, Malekzad H, et al. Temperature-responsive smart nanocarriers for delivery of therapeutic agents: applications and recent advances. ACS Appl Mater Interfaces [Internet]. 2016 Aug 8 [cited 2023 Nov 1];8(33):21107. Available from: /pmc/articles/PMC5003094/.
87. Kim YJ, Matsunaga YT. Thermo-responsive polymers and their application as smart biomaterials. J Mater Chem B [Internet]. 2017 Jun 14 [cited 2023 Nov 1];5(23):4307–21. Available from: https://pubs.rsc.org/en/content/articlehtml/2017/tb/c7tb00157f.
88. Rafael D, Melendres MMR, Andrade F, Montero S, Martinez-Trucharte F, Vilar-Hernandez M, et al. Thermo-responsive hydrogels for cancer local therapy: challenges and state-of-art. Int J Pharm. 2021;606:120954.
89. Mi P. Stimuli-responsive nanocarriers for drug delivery, tumor imaging, therapy and theranostics. Theranostics [Internet]. 2020 [cited 2023 Nov 3];10(10):4557. Available from: /pmc/articles/PMC7150471/.
90. Zhao L, Zhang X, Wang X, Guan X, Zhang W, Ma J. Recent advances in selective photothermal therapy of tumor. J Nanobiotechnology [Internet]. 2021 Dec 1 [cited 2023 Nov 3];19(1):1–15. Available from: https://jnanobiotechnology.biomedcentral.com/articles/10.1186/s12951-021-01080-3.
91. Rezaei SJT, Norouzi K, Hesami A, Malekzadeh AM, Ramazani A, Amani V, et al. Au(III) complexes loaded pH-responsive magnetic nanogels for cancer therapy. Appl Organomet Chem [Internet]. 2018 Apr 1 [cited 2023 Nov 3];32(4):e4303. Available from: https://onlinelibrary.wiley.com/doi/full/10.1002/aoc.4303.
92. Saneja A, Kumar R, Arora D, Kumar S, Panda AK, Jaglan S. Recent advances in near-infrared light-responsive nanocarriers for cancer therapy. Drug Discov Today. 2018;23(5):1115–25.
93. Kim HS, Lee DY. Near-infrared-responsive cancer photothermal and photodynamic therapy using gold nanoparticles. Polymers. 2018 Aug 30 [cited 2023 Nov 10];10(9):961. Available from: https://www.mdpi.com/2073-4360/10/9/961/htm.
94. Mu Y, Gong L, Peng T, Yao J, Lin Z. Advances in pH-responsive drug delivery systems. OpenNano. 2021;5:100031.
95. Torchilin V. Smart pharmaceutical nanocarriers. Smart Pharm Nanocarriers [Internet]. 2015 Dec 31 [cited 2023 Nov 4];1–564. Available from: https://books.google.com/books/about/Smart_Pharmaceutical_Nanocarriers.html?id=U1O2DAAAQBAJ.
96. Ahmad Z, Shah A, Siddiq M, Kraatz HB. Polymeric micelles as drug delivery vehicles. RSC Adv [Internet]. 2014 Apr 7 [cited 2023 Nov 7];4(33):17028–38. Available from: https://pubs.rsc.org/en/content/articlehtml/2014/ra/c3ra47370h.
97. Alsawaftah NM, Awad NS, Pitt WG, Husseini GA. pH-responsive nanocarriers in cancer therapy. Polymers. 2022 Feb 26 [cited 2023 Nov 7];14(5):936. Available from: https://www.mdpi.com/2073-4360/14/5/936/htm.
98. Shahriari M, Zahiri M, Abnous K, Taghdisi SM, Ramezani M, Alibolandi M. Enzyme responsive drug delivery systems in cancer treatment. J Control Release. 2019;308:172–89.
99. Fathi M, Safary A, Barar J. Therapeutic impacts of enzyme-responsive smart nanobiosystems. Bioimpacts [Internet]. 2020 [cited 2023 Nov 8];10(1):1. Available from: /pmc/articles/PMC6977590/.
100. Raza A, Hayat U, Rasheed T, Bilal M, Iqbal HMN. Redox-responsive nano-carriers as tumor-targeted drug delivery systems. Eur J Med Chem. 2018;157:705–15.

101. Jia R, Teng L, Gao L, Su T, Fu L, Qiu Z, et al. Advances in multiple stimuli-responsive drug-delivery systems for cancer therapy. Int J Nanomedicine [Internet]. 2021 [cited 2023 Dec 17];16:1525. Available from: /pmc/articles/PMC7920594/.
102. Paris JL, Baeza A, Vallet-Regí M. Overcoming the stability, toxicity, and biodegradation challenges of tumor stimuli-responsive inorganic nanoparticles for delivery of cancer therapeutics. Expert Opin Drug Deliv [Internet]. 2019 Oct 3 [cited 2023 Dec 19];16(10):1095–1112. Available from: https://www.tandfonline.com/doi/abs/10.1080/17425247.2019.1662786.
103. Navya PN, Kaphle A, Srinivas SP, Bhargava SK, Rotello VM, Daima HK. Current trends and challenges in cancer management and therapy using designer nanomaterials. Nano Converg [Internet]. 2019 [cited 2023 Dec 19];6(1). Available from: https://pubmed.ncbi.nlm.nih.gov/31304563/.
104. Alsehli M. Polymeric nanocarriers as stimuli-responsive systems for targeted tumor (cancer) therapy: recent advances in drug delivery. Saudi Pharm J. 2020;28(3):255–65.

MIX
Papier aus verantwortungsvollen Quellen
Paper from responsible sources
FSC® C105338

If you have any concerns about our products,
you can contact us on
ProductSafety@springernature.com

In case Publisher is established outside the EU,
the EU authorized representative is:
Springer Nature Customer Service Center GmbH
Europaplatz 3, 69115 Heidelberg, Germany

Printed by Libri Plureos GmbH
in Hamburg, Germany